2/6/90

Practical Handbook of
Agricultural Science

Edited by

A. A. Hanson, Ph.D.

Director
W-L Research, Inc.
Bakersfield, California

CRC Press, Inc.
Boca Raton, Florida

Cover design: the tractor silhouette
is superimposed upon an aerial view
of farms in Lancaster County, Pennsylvania,
an area that has been cultivated intensively
for more than 200 years.
(From Land, Yearbook of Agriculture,
U.S. Department of Agriculture,
Washington, D.C., 1976.)

Practical handbook of Agricultural science / edited by A. A. Hanson.
    p. cm.
  Bibliography: p.
  Includes index.
  ISBN 0-8493-3706-2
  1. Agriculture—Handbooks, manuals, etc. 2. Agriculture—United
States—Handbooks, manuals, etc. I. Hanson, A. A. (Angus
Alexander), 1922-
S501.2.P73 1990
630—dc20

89-10007
CIP

This book represents information obtained from authentic and highly regarded sources. Reprinted material is quoted with permission, and sources are indicated. A wide variety of references are listed. Every reasonable effort has been made to give reliable data and information, but the author and the publisher cannot assume responsibility for the validity of all materials or for the consequences of their use.

Direct all inquiries to CRC Press, Inc., 2000 Corporate Blvd., N.W., Boca Raton, Florida, 33431.

International Standard Book Number 0-8493-3706-2

Library of Congress Card Number 89-10007
Printed in the United States

PREFACE

This handbook was prepared as a "quick reference" to a variety of topics pertaining to soils and to the production and use of plants and animals. Emphasis has been devoted to providing information of broad significance in responding to questions on soils, plant adaptation, seeds, and major field and horticultural crops, as well as on selected factors involved in the production and utilization of animal products. Also, a substantial amount of illustrative data are included on the composition and nutritive value of plants for both animal feeding programs and human consumption.

Most of the tabular information has been assembled from the CRC Handbook Series in Agriculture and from other appropriate CRC handbooks. Although more recent data have been added for some subject matter areas, original published tables have been retained where it was felt that they were of merit in responding to general questions. Rather detailed background information has been included for certain topics (e.g., seed storage, seed germination, and rates of seeding) where an overly abridged presentation would be restricted in value. In addition, similar or somewhat similar information may appear in more than one table, either to underscore one or more aspects of the same data or to furnish readers with information in a more readily accessible format. The final product is designed to assist users in locating critical reference data where there is an immediate need for such information, expecially in the absence of more comprehensive source materials.

The book was developed for the most part with the nonspecialist in mind, and, for that reason, most tables are either self-explanatory or require limited experience for adequate interpretation. On the assumption that the handbook will serve primarily as a ready source of facts and figures, tabular materials are presented without an accompanying explanatory text. Particular attention is given to those situations where prospective users must locate, with minimal effort, either a specific "general value" or "term" in the course of their endeavors. This explains the attention given to the inclusion of both common and scientific plant names, various conversion tables, tabular material used in the interpretation of certain statistical tests, and a glossary, albeit abridged, of common technical terms encountered in the improvement and management of soils, plants, and animals.

It would be surprising if some users did not question the treatment given to various topics covered in the handbook. Differences in emphasis can be attributed in varying degrees to limitations imposed by space, to the availability of what were judged to be acceptable summary tables, and to the selection process exercised by the editor. Information on pesticides, their application and relative effectiveness, has been excluded because current recommendations should be obtained from state agricultural extension services. All applications of pesticides, including herbicides, should be made in strict accordance with instructions provided on labels. Recognizing that the handbook provides more of an overview of selected topics than an encyclopedic treatment of "practical agriculture", it is my hope that the resulting compilation can serve as a valuable resource in the broad field of agricultural science.

A. A. Hanson

THE EDITOR

A. A. Hanson, Ph.D., is a Director of W-L Research, Inc., Bakersfield, California. Dr. Hanson received his B.S.A. degree from the University of British Columbia, Vancouver in 1944, the M.Sc. degree from McGill University, Montreal, Quebec in 1946, and the Ph.D. degree from the Pennsylvania State University, University Park, in 1951. He served as Lecturer and Assistant Professor at Macdonald College of McGill University from 1946 to 1949, and as Agent (Geneticist) and Agronomist at the U.S. Regional Pasture Research Laboratory, State College, Pennsylvania from 1949 to 1952. In 1953 he transferred to the U.S. Department of Agriculture's Research Center, Beltsville, Maryland, where he served as Research Leader for Grass and Turf Investigations, from 1953 to 1965; as Chief, Forage and Range Research Branch, from 1965 to 1972; and as Director, Beltsville Agricultural Research Center, from 1972 to 1979. In 1980 he was appointed Research Director and Vice-President, W-L Research, Inc. He served in this position through 1985, when he was named Vice-President for Research. He continues his association with the company as a member of the board of directors.

Dr. Hanson is a member of the American Association for the Advancement of Science, the American Society of Agronomy, the Crop Science Society of America, Sigma Xi, the Cosmos Club, and the Council for Agricultural Science and Technology. He served as President, Crop Science Society of America, 1966 to 1967, Editor of *Crop Science,* 1962 to 1964; Editor, *Journal of Environmental Quality,* 1971 to 1973; and Senior Editor of two monographs published by the American Society of Agronomy, namely, "Turfgrass Science and Technology" and "Alfalfa and Alfalfa Improvement". He is a Fellow of the American Society of Agronomy, the Crop Science Society of America, and the American Association for the Advancement of Science. He has received the Superior Service Award, Distinguished Service Award, and Merit Certificate for Outstanding Performance from the U.S. Department of Agriculture; the Merit Certificate, Medallion Award, and Distinguished Grasslander Awards from the American Forage and Grassland Council; Alfalfa Industry Award, Certified Alfalfa Seed Council; Honorary Membership in the Association of Official Seed Certifying Agencies; has held visiting lectureships sponsored by the Biology Committee for the Atlantic Provinces, and the Northwestern College of Agriculture, Shaanxi, Peoples Republic of China; and had guest status at the Golden Jubilee Meeting of the V.I. Lenin All-Union Academy of Agricultural Sciences, Moscow, U.S.S.R.

Dr. Hanson is the author of more than 130 technical and review papers, coauthor of one textbook, and has acted as Series Editor of the CRC Handbook Series in Agriculture.

ACKNOWLEDGMENTS

The majority of the tables and figures constituting this volume have been previously published in CRC handbooks. Those which were previously published in a non-CRC source are published here with the appropriate credit to that source.

I am grateful to the Literary Executor of the late Sir Ronald A. Fisher, F.R.S., to Dr. Frank Yates, F.R.S., and the Longman Group Ltd., London for permission to reprint Table 419 (Table VII) from their book, *Statistical Tables for Biological, Agricultural and Medical Research* (6th Edition 1974).

That the interested reader may refer to the CRC handbook source for additional references and related information, there follows a list of the handbooks utilized, each with a list of specific chapters drawn upon and the tables and/or figures excerpted from each.

Finkel, H. J., Ed., *CRC Handbook of Irrigation Technology,* Vol. 1, CRC Press, Boca Raton, Fla., 1982.

Benami, A., Sprinkler irrigation, 193—245; Tables 126 and 127.

Finkel, H. J., Gravity irrigation, 339—358; Tables 124 and 125.

Nir, D. and Finkel, H. J., Water requirements of crops and irrigation rates, 61—78; Tables 6, 7, 116—118, and 122.

Nir, Z., Hydraulics of open channels, 93—143; Table 123.

Ravina, I., Soil-water relationships, 11—47; Table 16.

Ravina, I., Soil salinity and water quality, 79—91; Tables 119—121.

Finkel, J. J., Ed., *CRC Handbook of Irrigation Technology* Vol. 2, CRC Press, Boca Raton, Fla., 1982.

Finkel, H. J., Drainage of irrigated fields, 13—34; Tables 132 and 133.

Finkel, H. J., Irrigation of cereal crops, 159—190; Tables 106—108.

Finkel, H. J., Irrigation of alfalfa, 191—198; Table 105.

Nir, D., Economics and costing of irrigation, 61—76; Table 128.

Finney, E. E., Jr., *CRC Handbook of Transportation and Marketing in Agriculture,* Vol. 1, *Food Commodities,* CRC Press, Boca Raton, Fla., 1981.

Ashby, B. H., Transportation practices for fresh produce, 355—361; Table 188.

Brant, A. W., Quality indicators for shell eggs, 89—94; Table 411.

Hardenburg, R. E., Storage recommendations, shelf life, and respiration rates for horticultural crops, 261—285; Table 189.

Kastner, C. L., Carcasses, primals, and subprimals, 239—258; Tables 398 and 399.

Nichols, M. L., Jr., Regulations governing poultry grading operations, 155—162; Tables 408 and 409.

Sink, J. D. and Kotula, A. W., Exsanguination and evisceration of meat animals, 209—217; Tables 395.

Finney, E. E., Jr., *CRC Handbook of Transportation and Marketing in Agriculture,* Vol. 2, *Field Crops,* CRC Press, Boca Raton, Fla., 1981.

Brook, R. C. and Foster, G. H., Drying, cleaning, and conditioning, 63—110; Tables 258—269, and 273—275.

Davis, R., Stored-grain insects and their control, 111—123; Table 280.

Pomeranz, Y., Composition of cereal grains, 33—45; Tables 208—210, and 212—217.

Welch, G. B., Boyd, A. H., and Davila, S., Cleaning and conditiong seeds for marketing, 127—157, Tables 276—279; Figures 16 and 17.

Woodstock, L. W., Seed testing: purity, germination, vigor, and variety, 159—205; Table 87.

Kilmer, V. J., Ed., *CRC Handbook of Soils and Climate in Agriculture,* CRC Press, Boca Raton, Fla., 1982.

Christenson, D. R., Lime, lime materials, and other soil amendments, 331—348; Tables 24, 26—32, 44, 45, 64, 70—72; Figure 7.

Munson, R. D., Soil fertility, fertilizers, and plant nutrition, 269—292; Tables 73, 76, 81—83; Figure 11.

Shaw, R. H., Climate of the United States, 1—101; Tables 1, 2, 4, 5, and 115.

Simonson, R. W., Soil classification, 103—129; Tables 8—16; Figures 3 and 6.

Stolzy, L. H. and Jury, W. A., Soil physics, 131—158; Figures 4 and 5.

Terman, G. L., Fertilizer sources and composition, 295—329; Tables 84—87.

Thomas, G. W., Soil chemistry, 159—175; Tables 18—23, 25, 33, 36—39, 78—80.

Tiedje, J. M. and Dazzo, F. B., Soil microbiology, 177—209; Table 58.

Volk, B. G. and Loeppert, R. H., Soil organic matter, 211—268; Table 46—50, 53—57; Figure 9.

Pimentel, D., Ed., *CRC Handbook of Energy Utilization in Agriculture,* CRC Press, Boca Raton, Fla., 1980.

Batty, J. C. and Keller, J., Energy Requirements for irrigation, 35—44; Table 129.

Peart, R. M., Brook, R., and Okos, M. R., Energy requirements for various methods of crop drying, 49—54; Tables 270—272.

Pimentel, D., Ed., *CRC Handbook of Pest Management in Agriculture,* Vol., 1, CRC Press, Boca Raton, Fla., 1981.

Chandler, J. M., Estimated losses of crops to weeds, 95—109; Table 113.

James, W. C., Estimated losses of crops from plant pathogens, 79—94; Tables 159—163.

Kommedahl, T., The environmental control of plant pathogens using eradication, 297—315; Tables 164 and 165.

Palm, E. W., Estimated crop losses without the use of fungicides and nematicides and without nonchemical controls, 139—158; Table 158.

Schwartz, P. H. and Klassen, W., Estimate of losses by insects and mites to agricultural crops, 15—77; Table 154.

Steelman, C. D., Environmental control of arthropod pests of livestock, 509—526; Table 178.

Teetes, G. L., The environmental control of insects using planting times and plant spacing, 209—221; Tables 155 and 156.

Pimentel, D., Ed., *CRC Handbook of Pest Management in Agriculture,* Vol. 2, CRC Press, Boca Raton, Fla., 1981.

Roberts, D. A., Using regulatory programs, biological methods, cultural practices, and resistant varieties to control diseases of plants, 69—78; Table 147.

Minotti, P. L. and Sweet, R. D., Role of crop competition in limiting losses from weeds, 351—367; Tables 174 and 175.

Pimentel, D., Ed., *CRC Handbook of Pest Management in Agriculture,* Vol. 3, CRC Press, Boca Raton, Fla., 1981.

Hooker, A. L., Pest management systems for corn diseases, 275—283; Table 167.

Leath, K. T., Pest management systems for alfalfa diseases, 293—302, Table 166.

Pedigo, L. P., Higgins, R. A., Hammond, R. B., and Bechinski, E. J., Soybean pest management, 417—538; Tables 168, 169, 171—173.

Peters, D. C. and Starks, K. J., Pest management systems for sorghum insects, 549—562; Table 157.

Rechcigl, M., Jr., Ed., *CRC Handbook of Agricultural Productivity,* Vol. 1, *Plant Productivity,* CRC Press, Boca Raton, Fla., 1982.

    Arkley, R. J., Transpiration and productivity, 209—211; Table 114.

    Carter, D. L., Salinity and plant productivity, 117—133; Tables 40 and 43.

    Couto, W., Soil pH and plant productivity, 71—83; Tables 34 and 35.

    Eagles, C. F. and Wilson, D., Photosynthetic efficiency and plant productivity, 213—247; Tables 136—138.

    Hall, A. E., Humidity and plant productivity, 23—40; Tables 101 and 102.

    Hardy, R. W. F., Nitrogen fixation and crop productivity, 103—116; Tables 59—61; Figure 10.

    Hellmers, H. and Warrington, I., Temperature and plant productivity, 11—21; Table 99.

    Kramer, P. J., Water and plant productivity or yield, 41—47; Tables 103 and 104; Figure 12.

    Olson, R. A., Soil fertility and plant productivity, 85—101; Tables 74 and 77.

    Sumner, D. R., Crop rotation and plant productivity, 273—313; Tables 145, 146, and 148—150.

    Taylor, H. M. and Terrell, E. E., Rooting pattern and plant productivity, 185—200; Table 142.

    Triplett, G. B., Jr., Tillage and crop productivity, 251—262; Tables 143 and 144.

    Willey, R. W., Plant population and crop yield, 201—207; Tables 139 and 140.

    Witt, M. and Barfield, B. J., Environmental stress and plant productivity, 347—374; Tables 100, 110—112.

Rechcigl, M., Jr., Ed., *CRC Handbook of Agricultural Productivity,* Vol. 2, *Animal Productivity,* CRC Press, Boca Raton, Fla., 1982.

    Bondi, A., Nutrition and animal productivity, 195—212; Tables 344, 345, 412, and 413; Figure 18.

    Byerly, T. C., Agricultural productivity: potential and constraints, 265—304; Tables 109, 342, 343, and 346.

    Christison, G. I. and Williams, C. M., Effects of cold on animal production, 69—76; Tables 352, 358, 363, and 382; Figure 19.

    Johnson, H. D. and Hahn, G. L., Climate and animal productivity, 3—53; Tables 333—336, 349, 351, 354—357, 364—367, 371—375, 381, 383, and 384.

    Smith, A. H., Gravity and animal productivity, 119—134; Tables 347 and 348.

    Stenning, B. C., Housing and animal productivity, 135—159; Tables 353 and 390.

Wolff, I. A., Ed., *CRC Handbook of Processing and Utilization in Agriculture,* Vol. 1, *Animal Products,* CRC Press, Boca Raton, Fla., 1982.

    Brink, M. F. and Lofgren, P. A., Composition and nutritional value of milk and dairy products, 291—313; Tables 405—407.

    Chandan, R. C. and Shahani, K. M., Cultured milk products, 365—377; Table 403.

    Davis, G. W., Slaughter of meat producing animal, 3—27; Tables 392—394, 396, and 397.

    Walstra, P. and Geurts, T. J., Milk fat and butter, 411—430; Table 404.

Wolff, I. A., Ed., *CRC Handbook of Processing and Utilization in Agriculture,* Vol. 2 (1), *Plant Products,* CRC Press, Boca Raton, Fla., 1982.

    Anderson, R. A. and Watson, S. A., The corn milling industry, 31—63; Tables 255 and 256.

    Burdick, D., Spencer, R. R., and Kohler, G. O., Processing and utilization of forage crops, 361—484; Tables 281, 285—297, 299—306, 308, 310—325.

Burger, W. C., Barley, 187—222; Tables 211, 218, and 219.

Chen, G. C. P., Cane Sugar, 487—577, Tables 134, 135, and 414.

Fisher, C. H. and Chen, J. C. P., Processing and utilization of sugar beets, 579—610, Tables 326, 328, and 329.

Lorenz, K., Triticale processing and utilization: comparison with other cereal grains, 277—327, Tables 224, 231.

Miller, D. L. and Eisenhauer, R. A., Agricultural by-products and residues, 691—709; Tables 52, 52, 63, and 307.

Oelke, E. A. and Strait, J., Wild-rice production and processing, 329—357; Tables 236—238.

Rooney, L. W., Earp, C. F., and Khan, M. N., Sorghum and millets, 123—137; Tables 239—245.

Shellenberger, J. A., Processing and utilization: wheat, 91—121; Tables 220—223.

Smith, B. A., Sweet sorghum, 611—621; Table 327.

Watson, S. A., Corn: amazing maize. General properties, 3—29; Tables 247—254, 309.

Youngs, V. L., Processing and utilization: oats, 223—242; Tables 232—235.

Wolff, I. A., Ed., *CRC Handbook of Processing and Utilization in Agriculture,* Vol. 2 (2), *Plant Products,* CRC Press, Boca Raton, Fla., 1983.

Cherry, J. P. and Berardi, L. C., Cottonseed, 187—256; Tables 205 and 206.

Chortyk, O. T., Tobacco: from seed to smoke, 577—605; Tables 330 and 331.

Dorrell, D. G., Sunflowers, 145—156; Tables 203, 204 and 207.

Dull, G. G. and Fuller, G., Production, marketing, and transport of processed horticultural crops, 301—395; Table 190.

Liptrap, D. O., Johns, J. T., Lane, G. and Johnson, T. H., 83—90; Tables 196—200.

Pryde, E. H., The soybean: its importance and use distribution, 3—14; Table 195.

TABLE OF CONTENTS

Climatic factors 1
Soil 29
  Classification 29
  Structure and Composition 35
  Acidity and Crop Response 43
  Salinity and Crop Response 53
  Organic Matter, Crop Residues, and N_2 Fixation 67
  Animal Manures and Sewage Sludge 81
  Plant Nutrient Uptake 87
  Commercial Fertilizers 98
Plants 109
  Names and Distribution 109
  Seeds and Seed Certification 124
  Seeding Rates 171
  Environment 176
  Water Stress and Irrigation 178
  Productivity 198
  Roots 219
Tillage and Rotations 223
Pest and Other Losses 231
  Losses in Plants 231
  Losses in Animals 266
Horticultural Crops 273
  Storage 273
  Selected Nutrients 280
Oilseeds — Composition and Feeding 285
Cereal Grains 295
  Composition and Feeding 295
    General 295
    Wheat 306
    Triticale 309
    Oats and Wild-Rice 314
    Sorghum 318
    Corn 323
  Physical Properties and Drying 331
Seed Conditioning 341
Forage 353
  Hay Grades 353
  Harvesting, Composition, and Feeding 355
  Dehydrated Forage 365
Leaf Protein 369
Roughages 371
Silage Crops — Composition and Feeding 373
Sugar Crops and Tobacco — Composition 385
Animals 389
  General 389
    Reproduction 389
    Feeding 393
    Environment 402

    Production and Feeding 409
      Cattle 409
        Beef Cattle 409
        Dairy Cattle 411
      Swine 418
      Sheep 422
      Poultry 423
      Horses 427
Environmental Control of Livestock 431
Meat, Milk, and Eggs — Grades and Composition 433
Comparative Productivity 453
Conversion Factors 455
Common Statistical Terms 461
Statistical Tables 463
Glossary 473
Index 487

CLIMATIC FACTORS

Table 1
CLIMATE GROUPS

Group	Description of climate
A	Tropical — frost free; may be rainy all year (r) or winter dry (w)
B	Dry — potential evaporation exceeds rainfall; may be steppe or desert
C	Subtropical — 8 to 12 months 50°F (10°C) or above; (f) has no dry season, (s) has dry summer
D	Temperate — 4 to 7 months 50°F (10°C) or above; (c) is continental, (o) is oceanic, (a) has hot summer, (b) has cool summer
E	Boreal — 1 to 3 months 50°F (10°C) or above
F	Polar — all months below 50°F; (t) is tundra

REFERENCE
Trewartha, G. T., *An Introduction to Climate,* 4th ed. McGraw-Hill, New York, 1968.

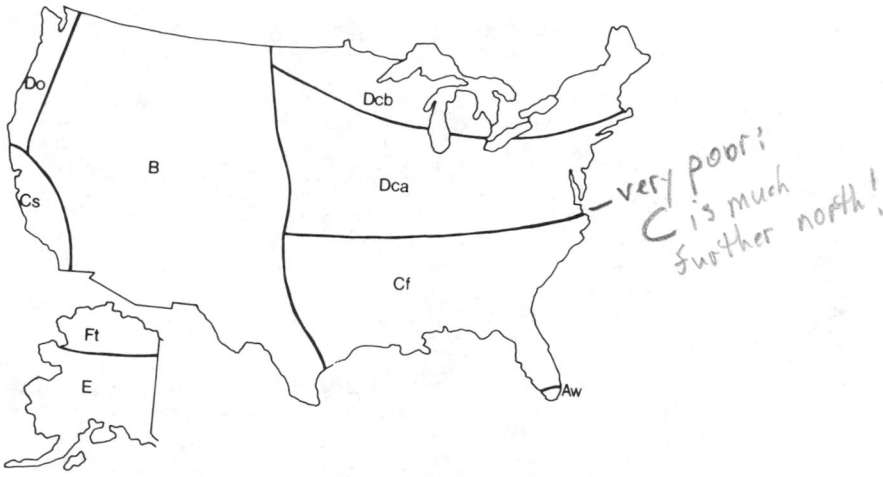

FIGURE 1. Climatic zones of the United States according to the Trewartha climate system. The Hawaiian Islands (not shown) would generally have A climates on the windward sides with some B climates on the leeward sides. C types may be found in the mountains.

Table 2
AVERAGE INCHES OF PRECIPITATION IN THE U.S. AND CERTAIN PACIFIC ISLANDS
(NORMALS 1941-1970)

Location	Years	Jan.	Feb.	Mar.	April	May	June	July	Aug.	Sept.	Oct.	Nov.	Dec.	Annual
Alabama														
Birmingham	30	4.84	5.28	6.17	4.62	3.62	4.00	5.22	4.31	3.64	2.58	3.72	5.23	53.23
Huntsville	30	5.13	5.16	5.79	4.79	3.86	3.97	4.88	3.46	3.29	2.57	3.88	5.38	52.15
Mobile	30	4.71	4.76	7.07	5.59	4.52	6.09	8.86	6.93	6.59	2.55	3.39	5.92	66.98
Montgomery	30	4.02	4.30	6.02	4.45	3.47	4.03	5.09	3.47	4.41	2.24	3.43	4.93	49.86
Alaska														
Anchorage	30	0.84	0.84	0.56	0.56	0.59	1.07	2.07	2.32	2.37	1.43	1.02	1.07	14.74
Annette	30	10.42	9.27	9.15	8.77	6.20	5.05	5.42	7.15	10.01	17.85	12.87	12.14	114.30
Barrow	30	0.23	0.20	0.19	0.21	0.17	0.35	0.88	1.04	0.58	0.55	0.30	0.19	4.89
Barter Island	30	0.55	0.33	0.26	0.23	0.31	0.53	1.12	1.28	0.89	0.81	0.45	0.29	7.05
Bethel	30	0.54	0.74	0.79	0.43	0.83	1.24	1.98	3.97	2.42	1.32	0.96	0.62	15.84
Bettles	30	0.72	0.77	0.82	0.63	0.62	1.22	1.79	2.77	1.78	1.23	0.81	0.82	14.18
Big Delta	30	0.36	0.27	0.33	0.31	0.94	2.20	2.49	1.92	1.23	0.56	0.41	0.42	11.44
Cold Bay	30	2.42	2.59	1.93	1.54	2.19	1.84	2.22	3.89	3.95	4.31	3.90	2.45	33.23
Fairbanks	30	0.60	0.53	0.48	0.33	0.65	1.42	1.90	2.19	1.08	0.73	0.66	0.65	11.22
Gulkana	30	0.58	0.47	0.34	0.22	0.63	1.34	1.84	1.58	1.72	0.88	0.75	0.76	11.11
Homer	30	1.70	1.54	1.22	1.09	0.91	1.06	1.70	2.56	2.85	3.38	2.76	2.29	23.08
Juneau	30	3.94	3.44	3.57	2.99	3.31	2.93	4.69	5.00	6.90	7.85	5.53	4.52	54.67
King Salmon	30	0.94	0.99	1.16	0.90	1.13	1.44	2.18	3.46	3.07	2.00	1.43	1.05	19.75
Kodiak	30	5.01	4.89	3.85	3.81	4.35	4.12	3.54	4.30	6.11	6.29	5.41	5.03	56.71
Kotzebue	30	0.29	0.30	0.33	0.33	0.40	0.52	1.55	2.26	1.43	0.61	0.41	0.33	8.78
McGrath	30	0.85	0.90	0.86	0.66	0.80	1.70	2.28	3.28	2.14	1.22	1.03	1.02	16.74
Nome	30	0.90	0.84	0.79	0.73	0.70	0.95	2.42	3.57	2.40	1.42	0.98	0.74	16.44
St. Paul Island	30	1.84	1.37	1.26	1.09	1.28	1.25	2.23	3.44	3.02	3.15	2.58	2.03	24.54
Talkeetna	30	1.63	1.79	1.54	1.12	1.46	2.17	3.48	4.89	4.52	2.54	1.79	1.71	28.54
Unalakleet	30	0.50	0.53	0.60	0.49	0.71	0.92	2.32	3.89	2.20	1.03	0.56	0.41	14.16
Valdez	30	5.06	5.30	4.33	3.06	3.20	2.70	4.31	5.80	7.74	6.75	5.67	5.39	59.31
Yakutat	30	10.36	9.28	9.57	7.65	8.02	5.68	8.46	10.81	15.45	19.52	14.80	12.86	132.46
Arizona														
Flagstaff	30	1.89	1.47	1.83	1.33	0.56	0.57	2.48	2.65	1.68	1.37	1.33	2.15	19.31
Phoenix	30	0.71	0.60	0.76	0.32	0.14	0.12	0.75	1.22	0.69	0.46	0.46	0.82	7.05

Station	Yrs													Annual
Tucson	30	0.77	0.70	0.64	0.35	0.14	0.20	2.38	2.34	1.37	0.66	0.56	0.94	11.05
Winslow	30	0.42	0.38	0.44	0.36	0.28	0.28	1.18	1.50	0.85	0.69	0.36	0.59	7.33
Yuma	30	0.38	0.27	0.21	0.11	0.03	0.00	0.18	0.44	0.22	0.27	0.22	0.34	2.67
Arkansas														
Fort Smith	30	2.38	3.20	3.64	4.74	5.48	3.93	3.24	2.91	3.31	3.47	3.08	2.89	42.27
Little Rock	30	4.24	4.42	4.93	5.25	5.30	3.50	3.38	3.01	3.55	2.99	3.86	4.09	48.52
California														
Bakersfield	30	0.96	1.03	0.83	0.85	0.19	0.06	0.02	0.01	0.08	0.26	0.69	0.74	5.72
Bishop	30	1.20	1.06	0.43	0.41	0.27	0.09	0.17	0.10	0.10	0.26	0.58	1.05	5.72
Blue Canyon	30	13.66	9.42	8.55	5.47	3.14	0.99	0.10	0.27	0.52	4.14	9.04	12.28	67.58
Eureka	30	7.42	5.15	4.83	2.95	2.11	0.66	0.14	0.27	0.65	3.23	5.77	6.58	39.76
Fresno	30	1.84	1.72	1.62	1.24	0.32	0.06	0.00	0.02	0.07	0.42	1.22	1.71	10.24
Long Beach	30	2.26	2.16	1.50	0.89	0.07	0.04	0.00	0.02	0.09	0.19	1.38	1.65	10.25
Los Angeles														
International Airport	30	2.52	2.32	1.71	1.10	0.08	0.03	0.01	0.02	0.07	0.22	1.76	1.75	11.59
City	30	3.00	2.77	2.19	1.27	0.13	0.03	0.00	0.04	0.17	0.27	2.02	2.16	14.05
Mount Shasta	30	6.65	5.61	4.03	3.05	1.87	1.08	0.32	0.31	0.69	2.54	5.09	6.25	37.49
Oakland	30	4.03	2.83	2.32	1.58	0.55	0.14	0.01	0.03	0.18	1.08	2.37	3.57	18.69
Red Bluff	30	4.48	3.17	2.51	1.79	0.98	0.47	0.04	0.18	0.31	1.17	3.05	3.91	22.06
Sacramento	30	3.73	2.68	2.17	1.54	0.51	0.10	0.01	0.05	0.19	0.99	2.13	3.12	17.22
Sandberg	30	2.15	2.45	1.33	1.12	0.25	0.03	0.03	0.06	0.21	0.34	2.05	1.95	11.97
San Diego	30	1.88	1.48	1.55	0.81	0.15	0.05	0.01	0.07	0.13	0.34	1.25	1.73	9.45
San Francisco														
International Airport	30	4.37	3.04	2.54	1.59	0.41	0.13	0.01	0.03	0.16	0.98	2.29	3.98	19.53
City	30	4.51	2.97	2.77	1.63	0.54	0.17	0.01	0.05	0.17	1.06	2.60	4.18	20.66
Santa Maria	30	2.25	2.40	1.98	1.31	0.19	0.04	0.03	0.02	0.10	0.52	1.36	2.05	12.25
Stockton	30	2.91	2.11	1.96	1.37	0.42	0.07	0.01	0.03	0.17	0.72	1.72	2.68	14.17
Colorado														
Alamosa	30	0.25	0.26	0.35	0.63	0.62	0.52	1.17	1.15	0.71	0.69	0.24	0.35	6.94
Colorado Springs	30	0.31	0.34	0.77	1.45	2.12	2.31	3.10	2.58	1.11	0.92	0.45	0.27	15.73
Denver	30	0.61	0.67	1.21	1.93	2.64	1.93	1.78	1.29	1.13	1.13	0.76	0.43	15.51
Grand Junction	30	0.64	0.61	0.75	0.79	0.63	0.55	0.46	1.05	0.84	0.93	0.61	0.55	8.41
Pueblo	30	0.32	0.32	0.68	1.29	1.65	1.36	1.87	1.96	0.79	0.96	0.42	0.29	11.91
Connecticut														
Bridgeport	30	2.71	2.71	3.49	3.39	3.57	2.56	3.44	3.80	2.88	2.79	3.83	3.44	38.61
Hartford	30	3.28	3.17	3.82	3.75	3.50	3.53	3.41	3.94	3.55	3.03	4.33	4.06	43.37

Table 2 (continued)
AVERAGE INCHES OF PRECIPITATION IN THE U.S. AND CERTAIN PACIFIC ISLANDS (NORMALS 1941-1970)

Location	Years	Jan.	Feb.	Mar.	April	May	June	July	Aug.	Sept.	Oct.	Nov.	Dec.	Annual
Delaware														
Wilmington	30	2.85	2.75	3.74	3.20	3.35	3.24	4.31	3.98	3.42	2.60	3.49	3.32	40.25
Washington, D.C.														
Dulles Airport	30	2.84	2.61	3.48	2.96	3.68	3.61	4.12	4.25	3.29	2.74	3.06	3.47	40.11
National Airport	30	2.62	2.45	3.33	2.86	3.68	3.48	4.12	4.67	3.08	2.66	2.90	3.04	38.89
Florida														
Apalachicola	30	3.07	3.78	4.70	3.61	2.78	5.30	8.02	8.07	9.00	2.88	2.68	3.32	57.21
Daytona Beach	30	2.05	2.92	3.37	2.39	2.65	6.60	6.69	6.84	7.10	5.52	2.13	1.96	50.22
Fort Myers	30	1.64	2.03	3.06	2.03	3.99	8.89	8.90	7.72	8.71	4.37	1.31	1.30	53.95
Jacksonville	30	2.78	3.58	3.56	3.07	3.22	6.27	7.35	7.89	7.83	4.54	1.79	2.59	54.47
Key West	30	1.67	1.85	1.56	2.17	2.51	4.55	4.11	4.47	7.34	5.57	2.67	1.52	39.99
Lakeland	30	2.32	2.52	4.02	2.57	3.44	6.70	8.09	7.18	6.06	2.84	1.60	2.09	49.43
Miami	30	2.15	1.95	2.07	3.60	6.12	9.00	6.91	6.72	8.74	8.18	2.72	1.64	59.80
Orlando	30	2.28	2.95	3.46	2.72	2.94	7.11	8.29	6.73	7.20	4.07	1.56	1.90	51.21
Pensacola	30	4.37	4.69	6.31	4.99	4.25	6.30	7.33	6.67	8.15	3.13	3.37	4.66	64.22
Tallahassee	30	3.74	4.77	5.93	4.07	4.04	6.62	8.92	6.89	6.64	2.93	2.81	4.22	61.58
Tampa	30	2.33	2.86	3.89	2.10	2.41	6.49	8.43	8.00	6.35	2.54	1.79	2.19	49.38
West Palm Beach	30	2.60	2.60	3.32	3.51	5.17	8.14	6.52	6.91	9.85	8.75	2.48	2.21	62.06
Georgia														
Athens	30	4.76	4.55	5.70	4.37	4.02	4.27	5.23	3.56	3.70	2.73	3.38	4.33	50.60
Atlanta	30	4.34	4.41	5.84	4.61	3.71	3.67	4.90	3.54	3.15	2.50	3.43	4.24	48.34
Augusta	30	3.44	3.75	4.67	3.37	3.38	3.66	5.09	4.21	3.26	2.17	2.21	3.42	42.63
Columbus	30	4.34	4.40	6.03	4.50	4.05	4.12	5.75	4.22	3.67	1.97	2.96	4.93	50.94
Macon	30	3.67	4.38	5.15	3.55	3.57	3.86	4.52	3.65	3.10	2.31	2.39	4.31	44.46
Rome	30	4.99	5.19	6.17	4.77	3.93	3.74	4.74	3.49	3.92	2.90	3.76	4.99	52.59
Savannah	30	2.92	2.86	4.41	2.93	4.20	5.89	7.87	6.47	5.57	2.81	1.94	3.28	51.15
Hawaii														
Hilo	30	9.07	12.90	13.69	12.88	10.07	6.61	9.54	10.88	7.44	10.96	13.77	15.76	133.57
Honolulu	30	4.40	2.46	3.18	1.36	0.96	0.32	0.60	0.76	0.67	1.51	2.99	3.69	22.90
Kahului	30	3.59	2.64	2.87	1.25	0.62	0.21	0.39	0.34	0.27	1.08	2.33	2.84	18.43
Lihue	30	6.24	4.28	4.67	3.25	2.43	1.57	1.87	2.21	1.85	3.84	5.63	6.34	44.18

Location														
Idaho														
Boise	30	1.47	1.16	1.01	1.14	1.32	1.06	0.15	0.30	0.41	0.80	1.32	1.36	11.50
Lewiston	30	1.27	0.85	0.96	1.13	1.58	1.83	0.53	0.60	0.86	1.08	1.25	1.27	13.21
Pocatello	30	1.05	0.80	0.94	1.06	1.29	1.28	0.36	0.62	0.61	0.75	1.05	0.99	10.80
Illinois														
Cairo	30	3.99	3.77	4.74	4.37	5.18	4.39	3.24	3.50	3.47	2.68	3.85	3.94	47.12
Chicago														
O'Hare Airport	30	1.70	1.30	2.52	3.38	3.41	4.15	3.46	2.73	3.01	2.32	2.10	1.64	31.72
Midway Airport	30	1.85	1.59	2.73	3.75	3.41	3.95	4.09	3.14	3.00	2.62	2.20	2.11	34.44
Moline	30	1.66	1.29	2.57	3.82	3.92	4.42	4.58	3.37	3.84	2.69	1.87	1.77	35.80
Peoria	30	1.82	1.50	2.80	4.36	3.87	3.91	3.76	3.07	3.55	2.51	2.02	1.89	35.06
Rockford	30	1.79	1.29	2.65	3.85	3.86	4.42	4.27	3.66	4.00	2.85	2.37	1.71	36.72
Springfield	30	1.75	1.77	2.70	4.14	3.54	4.16	3.81	2.74	3.27	3.08	2.14	1.92	35.02
Indiana														
Evansville	30	3.40	3.27	4.69	4.06	4.38	3.57	3.77	2.95	2.80	2.52	3.17	3.30	41.88
Fort Wayne	30	2.49	2.05	2.90	3.56	3.85	3.91	3.89	2.90	2.56	2.84	2.65	2.20	35.80
Indianapolis	30	2.86	2.36	3.75	3.87	4.08	4.16	3.67	2.80	2.87	2.51	3.10	2.71	38.74
South Bend	30	2.37	1.94	2.75	4.01	3.17	3.72	3.67	3.26	3.07	3.06	2.71	2.47	36.20
Iowa														
Burlington	30	1.58	1.25	2.65	3.79	3.58	4.71	3.76	3.36	3.74	3.02	1.60	1.59	34.63
Des Moines	30	1.14	1.05	2.31	2.94	4.21	4.90	3.28	3.30	3.07	2.14	1.42	1.09	30.85
Dubuque	30	1.73	1.26	2.97	4.15	4.67	5.29	4.31	4.02	4.64	2.83	2.48	1.92	40.27
Sioux City	30	0.65	0.94	1.45	2.19	3.54	4.59	3.30	2.95	2.84	1.63	0.91	0.75	25.74
Waterloo	30	1.00	0.92	2.25	3.43	4.18	5.01	4.75	3.50	3.63	2.38	1.50	1.19	33.74
Kansas														
Concordia	30	0.66	0.93	1.55	2.26	4.23	5.00	3.31	3.15	3.13	1.82	0.78	0.78	27.60
Dodge City	30	0.50	0.63	1.13	1.71	3.13	3.34	3.08	2.64	1.67	1.65	0.59	0.51	20.58
Goodland	30	0.37	0.39	0.88	1.42	2.54	2.92	2.71	2.14	1.35	0.99	0.53	0.41	16.65
Topeka	30	0.97	0.98	2.17	3.62	4.01	5.80	4.21	4.18	3.28	2.65	1.26	1.53	34.66
Wichita	30	0.85	0.98	1.78	2.95	3.60	4.49	4.35	3.10	3.69	2.50	1.17	1.12	30.58
Kentucky														
Covington-Cincinnatti, Ohio Airport	30	3.34	3.04	4.09	3.64	3.74	3.81	4.12	2.62	2.55	2.15	3.08	2.86	39.04
Lexington	30	3.95	3.42	4.80	3.87	4.16	4.31	4.83	3.40	2.65	2.12	3.36	3.62	44.49
Louisville	30	3.53	3.47	5.05	4.10	4.20	4.05	3.76	2.99	2.94	2.35	3.33	3.34	43.11
Louisiana														
Baton Rouge	30	4.40	4.76	5.14	5.10	4.39	3.77	6.51	4.67	3.79	2.65	3.84	5.03	54.05
Lake Charles	30	4.04	4.47	3.84	4.33	5.06	5.04	6.55	4.75	4.13	3.48	4.08	5.70	55.47

Table 2 (continued)

AVERAGE INCHES OF PRECIPITATION IN THE U.S. AND CERTAIN PACIFIC ISLANDS (NORMALS 1941-1970)

Location	Years	Jan.	Feb.	Mar.	April	May	June	July	Aug.	Sept.	Oct.	Nov.	Dec.	Annual
New Orleans	30	4.53	4.82	5.49	4.15	4.20	4.74	6.72	5.27	5.58	2.26	3.88	5.13	56.77
Shreveport	30	4.04	3.71	4.10	5.19	5.04	3.34	2.89	2.68	3.07	2.90	3.57	4.19	44.72
Maine														
Caribou	30	2.04	2.11	2.20	2.42	2.96	3.41	3.98	3.78	3.49	3.31	3.50	2.62	35.82
Portland	30	3.38	3.52	3.60	3.34	3.33	3.10	2.61	2.60	3.09	3.31	4.86	4.06	40.80
Maryland														
Baltimore	30	2.91	2.81	3.69	3.07	3.61	3.77	4.07	4.21	3.12	2.81	3.13	3.26	40.46
Massachusetts														
Blue Hill Observatory	30	4.12	3.97	4.51	3.64	3.62	3.15	2.95	3.83	3.65	3.62	5.06	4.70	46.82
Boston	30	3.69	3.54	4.01	3.49	3.47	3.19	2.74	3.46	3.16	3.02	4.51	4.24	42.52
Worcester	30	3.35	3.18	3.85	3.83	3.97	3.60	3.62	4.19	3.52	3.54	4.66	3.93	45.24
Michigan														
Alpena	30	1.66	1.35	1.87	2.43	2.70	2.87	2.50	2.69	3.25	1.97	2.45	1.85	27.59
Detroit														
City Airport	30	1.93	1.80	2.33	3.08	3.43	3.04	2.99	3.04	2.30	2.52	2.31	2.19	30.98
Metro Airport	30	1.91	1.75	2.47	3.22	3.31	3.42	3.10	3.28	2.16	2.48	2.32	2.27	31.89
Flint	30	1.70	1.58	2.11	2.86	3.12	3.43	2.99	3.25	2.52	2.24	2.29	1.68	29.77
Grand Rapids	30	1.94	1.50	2.50	3.39	3.19	3.42	3.09	2.54	3.30	2.56	2.80	2.16	32.39
Houghton Lake	30	1.46	1.20	1.72	2.37	2.78	3.34	3.07	2.35	3.17	2.58	2.53	1.82	28.39
Lansing	30	1.91	1.62	2.36	2.90	3.32	3.47	2.82	2.79	2.63	2.31	2.26	2.00	30.39
Marquette	30	1.54	1.52	1.88	2.60	2.93	3.44	3.07	3.01	3.45	2.44	2.99	1.97	30.84
Muskegon	30	2.25	1.76	2.39	3.16	2.71	2.64	2.51	2.57	3.36	2.70	3.04	2.44	31.53
Sault Ste. Marie	30	1.92	1.48	1.74	2.22	3.01	3.31	2.60	3.10	3.85	2.85	3.26	2.36	31.70
Minnesota														
Duluth	30	1.16	0.85	1.76	2.55	3.41	4.44	3.73	3.79	3.06	2.30	1.73	1.40	30.18
International Falls	30	0.85	0.71	1.10	1.67	2.75	3.91	3.98	3.39	3.32	1.69	1.30	0.98	25.65
Minneapolis — St. Paul	30	0.73	0.84	1.68	2.04	3.37	3.94	3.69	3.05	2.73	1.78	1.20	0.89	25.94
Rochester	30	0.65	0.65	1.70	2.36	3.46	4.63	3.74	3.59	3.08	1.82	1.02	0.77	27.47
Saint Cloud	30	0.76	0.78	1.33	2.30	3.60	4.64	3.23	3.89	2.65	1.69	1.12	0.85	26.84
Mississippi														
Jackson	30	4.53	4.62	5.63	4.65	4.38	3.40	4.27	3.59	2.99	2.22	3.87	5.04	49.19

Location														
Meridian	30	4.33	4.86	6.21	5.10	3.84	3.68	5.12	3.89	3.29	2.18	3.51	5.57	51.58
Missouri														
Columbia	30	1.57	1.72	2.58	3.83	4.68	4.59	3.89	3.19	4.39	3.38	1.79	1.78	37.39
Kansas City	30	1.25	1.25	2.55	3.50	4.28	5.55	4.37	3.81	4.21	3.24	1.47	1.52	37.00
Saint Joseph	30	1.09	1.03	2.40	3.21	4.65	6.50	3.82	4.06	3.83	2.47	1.24	1.35	35.65
St. Louis	30	1.85	2.06	3.03	3.92	3.86	4.42	3.69	2.87	2.89	2.79	2.47	2.04	35.89
Springfield	30	1.67	2.22	2.99	4.27	4.93	4.72	3.62	2.94	4.11	3.44	2.34	2.45	39.70
Montana														
Billings	30	0.70	0.64	1.01	1.56	2.08	2.61	0.87	1.00	1.39	0.88	0.73	0.68	14.15
Glasgow	30	0.39	0.32	0.37	0.71	1.31	2.72	1.43	1.51	0.85	0.56	0.39	0.31	10.87
Great Falls	30	0.88	0.75	0.97	1.18	2.37	3.11	1.27	1.09	1.17	0.68	0.81	0.71	14.99
Havre	30	0.52	0.40	0.49	1.02	1.48	2.55	1.38	1.05	1.11	0.67	0.46	0.42	11.55
Helena	30	0.55	0.38	0.69	0.93	1.76	2.38	0.96	0.98	0.97	0.59	0.61	0.58	11.38
Kalispell	30	1.52	0.97	0.86	1.01	1.80	2.56	1.03	1.30	1.13	1.21	1.39	1.46	16.24
Miles City	30	0.49	0.51	0.65	1.26	2.06	3.32	1.55	1.20	1.19	0.71	0.51	0.48	13.93
Missoula	30	1.17	0.74	0.71	1.01	1.68	2.12	0.92	0.92	1.06	0.92	0.96	1.13	13.34
Nebraska														
Grand Island	30	0.52	0.76	1.18	2.47	3.78	4.40	3.00	2.54	2.51	1.08	0.61	0.56	23.41
Lincoln	30	0.73	1.14	1.54	2.59	3.87	5.16	3.24	3.63	3.25	1.69	0.98	0.79	28.61
Norfolk	30	0.62	0.78	1.37	2.15	3.69	4.88	3.18	2.66	2.41	1.33	0.62	0.63	24.32
North Platte	30	0.45	0.52	0.99	1.93	3.26	3.77	2.98	2.07	2.01	0.99	0.52	0.41	19.90
Omaha	30	0.76	0.98	1.59	2.97	4.11	4.94	3.71	3.97	3.27	1.93	1.11	0.84	30.18
Omaha (North)	30	0.69	0.96	1.62	2.82	3.99	4.93	3.71	4.01	3.16	1.76	1.01	0.77	29.43
Scottsbluff	30	0.32	0.33	0.76	1.52	2.81	3.36	1.76	0.95	1.15	0.82	0.39	0.40	14.57
Valentine	30	0.31	0.53	0.76	1.77	2.80	3.60	2.50	2.38	1.48	0.92	0.45	0.30	17.80
Nevada														
Elko	30	1.16	0.77	0.83	0.82	1.03	1.01	0.41	0.61	0.34	0.66	1.01	1.13	9.78
Ely	30	0.64	0.60	0.85	1.00	0.93	0.93	0.61	0.56	0.61	0.60	0.66	0.71	8.70
Las Vegas	30	0.45	0.30	0.33	0.27	0.10	0.09	0.44	0.49	0.27	0.22	0.43	0.37	3.76
Reno	30	1.21	0.86	0.70	0.47	0.66	0.40	0.26	0.22	0.23	0.42	0.68	1.09	7.20
Winnemucca		0.97	0.81	0.71	0.73	0.91	1.01	0.23	0.26	0.28	0.65	0.97	0.94	8.47
New Hampshire														
Concord	30	2.67	2.45	2.77	2.92	3.02	3.35	3.14	2.89	3.06	2.68	3.96	3.26	36.17
Mt. Washington	30	5.12	6.51	5.60	5.46	5.84	6.50	6.77	7.19	6.36	6.12	7.67	7.03	76.17
New Jersey														
Atlantic City	30	3.56	3.37	4.31	3.37	3.54	3.38	4.36	4.90	2.99	3.46	4.21	4.01	45.46
Newark	30	2.91	2.95	3.93	3.44	3.60	2.99	4.03	4.27	3.44	2.82	3.61	3.46	41.45
Trenton	30	2.76	2.70	3.81	3.15	3.40	3.21	4.74	4.17	3.17	2.53	3.25	3.28	40.17

Table 2 (continued)
AVERAGE INCHES OF PRECIPITATION IN THE U.S. AND CERTAIN PACIFIC ISLANDS
(NORMALS 1941-1970)

Location	Years	Jan.	Feb.	Mar.	April	May	June	July	Aug.	Sept.	Oct.	Nov.	Dec.	Annual
New Mexico														
Albuquerque	30	0.30	0.39	0.47	0.48	0.53	0.50	1.39	1.34	0.77	0.79	0.29	0.52	7.77
Clayton	30	0.28	0.36	0.67	1.27	2.64	1.75	2.76	2.67	1.77	1.11	0.33	0.30	15.91
Roswell	30	0.40	0.37	0.47	0.49	1.00	1.24	1.71	1.48	1.47	1.22	0.29	0.47	10.61
New York														
Albany	30	2.20	2.11	2.58	2.70	3.26	3.00	3.12	2.87	3.12	2.63	2.84	2.93	33.36
Binghamton	30	2.32	2.25	2.87	3.18	3.83	3.59	3.83	3.61	3.02	3.00	3.10	2.75	37.35
Buffalo	30	2.90	2.55	2.85	3.15	2.97	2.23	2.93	3.53	3.25	3.01	3.74	3.00	36.11
Central Park	30	2.71	2.92	3.73	3.30	3.47	2.96	3.68	4.01	3.27	2.85	3.76	3.53	40.19
JFK Airport	30	2.69	3.05	3.77	3.59	3.54	2.98	4.04	4.30	3.31	2.76	3.90	3.60	41.53
La Guardia Airport	30	2.88	3.10	3.98	3.56	3.36	2.89	3.85	4.48	3.15	2.96	3.77	3.63	41.61
Rochester	30	2.25	2.42	2.57	2.74	2.80	2.54	2.89	2.97	2.35	2.62	2.83	2.35	31.33
Syracuse	30	2.68	2.79	3.03	3.08	3.02	3.09	3.08	3.50	2.71	3.09	3.25	3.09	36.41
North Carolina														
Asheville	30	3.39	3.60	4.66	3.53	3.31	3.97	4.87	4.50	3.57	3.25	2.94	3.59	45.18
Cape Hatteras	30	4.26	4.15	3.84	3.07	3.28	4.83	5.90	6.75	5.76	4.79	4.45	4.54	55.62
Charlotte	30	3.51	3.83	4.52	3.40	2.90	3.70	4.57	3.96	3.46	2.69	2.74	3.44	42.72
Greensboro	30	3.22	3.37	3.72	3.15	3.04	3.91	4.39	4.30	3.55	2.94	2.62	3.15	41.36
Raleigh	30	3.22	3.32	3.44	3.07	3.32	3.67	5.08	4.93	3.78	2.81	2.82	3.08	42.54
Wilmington	30	3.18	3.39	4.05	2.94	3.97	5.58	8.34	6.82	5.64	3.31	3.01	3.36	53.59
North Dakota														
Bismarck	30	0.51	0.44	0.73	1.44	2.17	3.58	2.20	1.96	1.32	0.80	0.56	0.45	16.16
Fargo	30	0.50	0.44	0.83	2.08	2.29	3.20	3.16	2.85	1.84	1.09	0.72	0.62	19.62
Williston	30	0.59	0.53	0.63	1.24	1.62	3.25	2.04	1.56	1.21	0.64	0.53	0.49	14.33
Ohio														
Akron	30	2.69	2.16	3.15	3.32	3.87	3.50	3.80	2.77	2.60	2.37	2.50	2.40	35.13
Cincinnati	30	3.40	2.95	4.13	3.85	3.96	3.92	3.96	3.02	2.69	2.20	3.08	2.87	40.03
Cleveland	30	2.56	2.18	3.05	3.49	3.49	3.28	3.45	3.00	2.80	2.57	2.76	2.36	34.99
Columbus	30	2.87	2.32	3.44	3.71	4.10	4.13	4.21	2.86	2.41	1.89	2.68	2.39	37.01
Dayton	30	2.76	2.24	3.21	3.34	3.76	3.88	3.54	2.55	2.28	1.94	2.57	2.29	34.36
Mansfield	30	2.35	1.99	2.83	3.33	3.78	3.49	3.69	2.71	2.85	1.97	2.52	2.16	33.67

Location	Yrs	Jan	Feb	Mar	Apr	May	Jun	Jul	Aug	Sep	Oct	Nov	Dec	Annual
Toledo	30	2.08	1.75	2.52	2.95	3.33	3.38	3.23	3.07	2.40	2.24	2.32	2.24	31.51
Youngstown	30	2.94	2.42	3.24	3.67	3.90	3.59	3.90	3.23	2.64	2.94	2.97	2.55	37.99
Oklahoma														
Oklahoma City	30	1.11	1.32	2.05	3.47	5.20	4.22	2.66	2.56	3.55	2.57	1.40	1.26	31.37
Tulsa	30	1.43	1.72	2.52	4.17	5.11	4.69	3.51	2.95	4.07	3.22	1.87	1.64	36.90
Oregon														
Astoria	30	9.73	7.82	6.62	4.61	2.72	2.45	0.96	1.46	2.83	6.80	9.78	10.57	66.34
Burns	30	1.76	1.18	0.92	0.70	1.03	0.97	0.32	0.44	0.46	0.89	1.43	1.73	11.83
Eugene	30	7.54	4.67	4.43	2.31	2.06	1.28	0.26	0.58	1.26	4.00	6.53	7.64	42.56
Medford	30	3.54	2.15	1.64	1.00	1.44	0.89	0.25	0.33	0.56	2.05	3.10	3.69	20.64
Pendleton	30	1.60	1.07	1.00	1.01	1.24	1.01	0.26	0.34	0.64	1.11	1.50	1.53	12.31
Portland	30	5.88	4.06	3.64	2.22	2.09	1.59	0.47	0.82	1.60	3.59	5.61	6.04	37.61
Salem	30	6.90	4.79	4.33	2.29	2.09	1.39	0.35	0.57	1.46	3.98	6.08	6.85	41.08
Sexton Summit	30	6.54	4.05	3.43	2.06	2.22	1.26	0.33	0.44	1.03	3.63	5.61	6.19	36.79
Pacific														
Guam	30	5.16	4.26	2.94	4.03	4.49	5.19	9.59	12.16	14.08	14.40	8.51	5.85	90.66
Johnston Island	30	2.59	1.95	2.85	2.53	2.22	1.03	1.09	2.58	2.05	3.48	2.30	3.39	28.06
Koror Island	30	10.79	7.43	7.79	8.69	14.53	13.75	16.19	15.40	12.99	12.59	10.67	12.87	143.69
Kwajalein Island	30	3.74	2.35	6.21	5.91	9.71	9.93	9.70	10.20	10.87	11.97	11.46	9.23	101.28
Majuro, Marshall Is.	30	8.96	6.98	9.25	10.31	12.15	12.17	13.64	11.78	14.39	16.00	15.38	11.88	142.89
Pago Pago, American Samoa	30	12.27	14.16	12.07	12.26	10.69	9.41	6.94	7.40	6.08	12.10	10.31	14.85	128.54
Ponape Island	30	11.75	11.44	15.04	19.58	19.43	15.66	17.82	16.70	16.76	16.42	17.17	15.95	193.72
Truk, Caroline Islands	30	8.59	6.40	7.98	13.17	15.74	12.59	15.41	13.44	13.07	13.39	11.99	13.92	145.69
Wake Island	30	1.08	1.21	1.60	2.22	2.25	2.35	4.45	6.35	5.69	5.45	3.02	1.73	37.40
Yap Island	30	8.44	5.41	5.58	5.98	10.03	10.90	14.64	15.14	13.09	12.42	10.03	9.88	121.54
Pennsylvania														
Allentown	30	3.02	2.78	3.61	3.79	3.78	3.47	4.36	4.18	3.59	2.73	3.59	3.59	42.49
Erie	30	2.47	2.12	2.75	3.55	3.63	3.50	3.52	3.35	3.56	3.24	3.70	2.81	38.20
Harrisburg	30	2.57	2.42	3.22	2.98	3.76	3.11	3.70	3.22	2.66	2.57	3.19	3.07	36.47
Philadelphia	30	2.81	2.62	3.69	3.29	3.35	3.70	4.09	4.11	3.03	2.53	3.39	3.32	39.93
Pittsburgh														
International Airport	30	2.79	2.35	3.60	3.40	3.63	3.48	3.84	3.15	2.52	2.52	2.47	2.48	36.23
City	30	2.61	2.29	3.58	3.44	3.59	3.74	3.78	3.18	2.53	2.47	2.49	2.52	36.22
Avoca	30	2.04	1.96	2.50	3.06	3.50	3.40	4.09	3.21	2.82	2.71	3.01	2.51	34.81
Williamsport	30	2.52	2.58	3.53	3.42	3.99	3.25	4.19	3.44	3.03	3.20	3.74	3.10	39.99
Puerto Rico														
San Juan	30	3.73	2.50	2.04	3.40	6.54	5.64	6.41	6.98	6.07	5.64	5.49	4.71	59.15

Table 2 (continued)
AVERAGE INCHES OF PRECIPITATION IN THE U.S. AND CERTAIN PACIFIC ISLANDS
(NORMALS 1941-1970)

Location	Years	Jan.	Feb.	Mar.	April	May	June	July	Aug.	Sept.	Oct.	Nov.	Dec.	Annual
Rhode Island														
Block Island	30	3.41	3.32	3.88	3.51	3.25	2.20	2.74	3.86	3.00	2.88	4.35	4.11	40.51
Providence	30	3.52	3.45	3.99	3.72	3.49	2.65	2.85	3.90	3.26	3.27	4.52	4.13	42.75
South Carolina														
Charleston	30	2.90	3.27	4.75	2.95	3.81	6.30	8.21	6.44	5.17	3.05	2.13	3.14	52.12
Columbia	30	3.44	3.67	4.67	3.51	3.35	3.82	5.65	5.63	4.32	2.58	2.34	3.38	46.36
Greenville — Spartanburg	30	4.07	4.43	5.33	4.31	2.95	4.05	4.18	4.06	3.79	3.18	3.10	4.09	47.54
South Dakota														
Aberdeen	30	0.53	0.63	0.94	1.96	2.57	3.63	2.74	2.10	1.71	1.22	0.62	0.45	19.10
Huron	30	0.43	0.75	1.07	1.96	2.75	3.76	2.23	1.98	1.78	1.53	0.67	0.53	19.44
Rapid City	30	0.47	0.57	0.99	2.09	2.81	3.67	2.10	1.47	1.22	0.86	0.48	0.39	17.12
Sioux Falls	30	0.57	1.04	1.40	2.30	3.37	4.32	2.94	2.84	2.85	1.50	0.85	0.74	24.72
Tennessee														
Bristol — Johnson City	30	3.62	3.71	3.91	3.30	3.29	3.52	4.98	3.72	2.88	2.25	2.78	3.51	41.47
Chattanooga	30	5.38	5.19	5.63	4.42	3.43	3.68	5.14	3.22	3.69	2.95	3.94	5.25	51.92
Knoxville	30	4.67	4.71	4.86	3.61	3.28	3.63	4.70	3.24	2.78	2.67	3.56	4.47	46.18
Memphis	30	4.93	4.73	5.10	5.42	4.39	3.46	3.53	3.33	3.01	2.58	3.92	4.70	49.10
Nashville	30	4.75	4.43	5.00	4.11	4.10	3.38	3.83	3.24	3.09	2.16	3.46	4.45	46.00
Oak Ridge	30	5.25	5.24	5.45	4.21	3.52	3.94	5.67	3.85	3.34	2.72	4.05	5.36	52.60
Texas														
Abilene	30	1.02	0.97	0.98	2.47	3.86	2.82	2.34	2.05	2.26	2.60	1.20	1.02	23.59
Amarillo	30	0.54	0.56	0.77	1.23	2.83	3.45	2.95	2.93	1.93	1.83	0.53	0.73	20.28
Austin	30	1.88	3.09	1.89	3.49	3.97	3.13	1.88	2.20	3.68	3.02	2.04	2.22	32.49
Brownsville	30	1.35	1.48	0.69	1.28	2.51	2.80	1.19	2.66	5.23	3.32	1.34	1.24	25.09
Corpus Christi	30	1.58	1.95	1.10	2.15	3.17	2.67	1.88	3.20	4.90	2.77	1.63	1.53	28.53
Dallas — Fort Worth	30	1.80	2.36	2.54	4.30	4.47	3.05	1.84	2.26	3.15	2.68	2.03	1.82	32.30
Del Rio	30	0.60	1.02	0.72	1.57	2.41	2.03	1.02	1.22	3.05	2.07	0.65	0.52	16.88
El Paso	30	0.39	0.42	0.39	0.24	0.32	0.60	1.53	1.12	1.16	0.78	0.32	0.50	7.77
Galveston	30	3.02	2.67	2.60	2.63	3.16	4.05	4.41	4.40	5.60	2.83	3.16	3.67	42.20
Houston	30	3.57	3.54	2.68	3.54	5.10	4.52	4.12	4.35	4.65	4.05	4.03	4.04	48.19
Lubbock	30	0.55	0.50	0.89	1.08	3.17	2.78	2.23	1.87	2.19	2.05	0.49	0.61	18.41

Location		Jan	Feb	Mar	Apr	May	Jun	Jul	Aug	Sep	Oct	Nov	Dec	Annual
Midland — Odessa	30	0.59	0.56	0.59	0.85	2.16	1.49	1.82	1.52	1.54	1.38	0.49	0.52	13.51
Port Arthur	30	4.06	4.24	3.05	4.19	4.94	4.81	5.89	5.69	5.34	3.71	4.26	4.89	55.07
San Angelo	30	0.81	0.79	0.87	1.66	2.70	1.88	1.23	1.44	2.74	1.86	0.85	0.70	17.53
San Antonio	30	1.66	2.06	1.54	2.54	3.07	2.79	1.69	2.41	3.71	2.84	1.77	1.46	27.54
Victoria	30	1.76	2.20	1.89	2.65	3.96	3.31	2.79	3.15	4.61	3.63	2.31	1.95	34.29
Waco	30	1.87	2.38	2.36	4.02	4.60	2.73	1.47	1.81	3.19	2.55	2.27	2.01	31.26
Wichita Falls	30	1.07	1.16	1.62	3.16	4.58	3.39	2.16	.77	3.00	2.68	1.35	1.28	27.22
Utah														
Milford	30	0.61	0.70	1.04	0.90	0.61	0.56	0.51	0.68	0.61	0.78	0.67	0.73	8.40
Salt Lake City	30	1.27	1.19	1.63	2.12	1.49	1.30	0.70	0.93	0.68	1.16	1.31	1.39	15.17
Vermont														
Burlington	30	1.74	1.68	1.93	2.62	3.01	3.46	3.54	3.72	3.05	2.74	2.86	2.19	32.54
Virginia														
Lynchburg	30	2.77	2.79	3.46	2.73	3.22	3.43	4.05	4.05	3.30	2.60	2.66	3.21	38.27
Norfolk	30	3.35	3.31	3.42	2.71	3.34	3.62	5.70	5.92	4.20	3.06	2.94	3.11	44.68
Richmond	30	2.86	3.01	3.38	2.77	3.42	3.52	5.63	5.06	3.58	2.94	3.20	3.22	42.59
Roanoke	30	2.74	3.09	3.33	2.80	3.47	3.51	3.74	4.15	3.42	3.19	2.48	3.11	39.03
Washington														
Olympia	30	7.93	5.97	4.81	3.14	1.88	1.57	0.70	1.17	2.12	5.28	7.98	8.19	50.74
Quillayute	30	14.60	11.95	10.79	8.15	4.71	3.50	2.36	2.75	5.16	11.58	13.84	15.60	104.99
Seattle														
Urban Site	30	5.17	3.93	3.24	2.41	1.71	1.57	0.87	0.88	1.75	3.43	5.34	5.35	35.65
International Airport	30	5.79	4.19	3.61	2.46	1.70	1.53	0.71	1.08	1.99	3.91	5.88	5.94	38.79
Spokane	30	2.47	1.68	1.53	1.12	1.46	1.36	0.40	0.58	0.83	1.42	2.20	2.37	17.42
Stampede Pass	30	12.89	10.29	8.93	6.63	4.17	3.96	1.54	2.35	4.64	8.89	12.46	14.31	91.06
Walla Walla	30	2.07	1.40	1.37	1.43	1.58	1.18	0.33	0.45	0.85	1.49	1.89	1.97	16.01
Yakima	30	1.33	0.78	0.58	0.51	0.55	0.73	0.16	0.25	0.31	0.58	1.07	1.15	8.00
West Virginia														
Beckley	30	3.46	3.33	4.23	3.31	3.78	4.22	4.38	3.77	3.40	2.49	2.92	3.33	42.62
Charleston	30	3.39	3.11	4.03	3.33	3.48	3.31	5.04	3.68	2.94	2.45	2.81	3.18	40.75
Elkins	30	3.29	2.92	3.93	3.62	3.88	4.76	4.94	4.02	3.18	2.71	2.72	3.25	43.22
Huntington	30	3.15	2.90	4.07	3.26	3.82	3.37	4.19	3.34	2.86	2.09	2.86	2.97	38.88
Parkersburg	30	3.08	2.77	3.75	3.45	3.56	4.01	4.28	3.34	2.80	2.11	2.52	2.77	38.44
Wisconsin														
Green Bay	30	1.09	1.01	1.68	2.69	3.10	3.41	3.09	2.62	3.24	1.93	1.88	1.27	27.01
La Crosse	30	0.96	0.87	2.02	2.63	3.70	4.44	3.52	3.02	3.38	2.05	1.45	1.04	29.08
Madison	30	1.25	0.95	1.93	2.66	3.41	4.33	3.81	3.05	3.36	2.16	1.87	1.47	30.25
Milwaukee	30	1.63	1.13	2.24	2.76	2.88	3.58	3.41	2.68	3.02	1.98	2.01	1.75	29.07

Table 2 (continued)
AVERAGE INCHES OF PRECIPITATION IN THE U.S. AND CERTAIN PACIFIC ISLANDS
(NORMALS 1941-1970)

Location	Years	Jan.	Feb.	Mar.	April	May	June	July	Aug.	Sept.	Oct.	Nov.	Dec.	Annual
Wyoming														
Casper	30	0.50	0.50	0.91	1.45	1.94	1.44	0.95	0.57	0.87	0.92	0.68	0.49	11.22
Cheyenne	30	0.46	0.46	1.05	1.57	2.52	2.41	1.82	1.45	1.03	0.95	0.58	0.35	14.65
Lander	30	0.48	0.66	1.18	2.36	2.59	1.93	0.61	0.42	1.05	1.24	0.87	0.45	13.84
Sheridan	30	0.69	0.77	1.21	2.12	2.45	2.99	1.07	0.95	1.28	1.02	0.92	0.69	16.16

Source: U.S. Department of Commerce, Comparative Climatic Data through 1977, Environmental Data Service, National Oceanic and Atmospheric Administration, Asheville, N.C., 1978.

Table 3

COMPARISON OF AVERAGE INCHES OF PRECIPITATION IN THE U.S. FOR 1941—1970 AND 1951—1980

Location	Annual (in.) 1941—1970[a]	1951—1980[b]	Location	Annual (in.) 1941—1970[a]	1951—1980[b]
Alabama			Augusta	42.63	43.07
Birmingham C.O.	53.23	52.16	Columbus	50.94	51.09
Huntsville	52.15	54.72	Macon	44.46	44.86
Mobile	66.98	64.64	Savannah	51.15	49.70
Montgomery	49.86	49.16	Hawaii		
Alaska			Hilo	133.57	128.15
Anchorage	14.74	15.20	Honolulu	22.90	23.47
Barrow	4.89	4.75	Kahului	18.43	19.85
Fairbanks	11.22	10.37	Idaho		
Juneau	54.67	53.15	Boise	11.50	11.71
Kodiak	56.71	74.24	Lewiston	13.21	12.78
Arizona			Pocatello	10.80	10.86
Flagstaff	19.31	20.86	Illinois		
Phoenix	7.05	7.11	Cairo	47.12	47.65
Tucson	11.05	11.14	Chicago Midway AP	34.44	33.34
Winslow	7.33	7.64	Moline	35.80	37.17
Yuma	2.67	2.65	Peoria	35.06	34.89
Arkansas			Rockford	36.72	36.78
Fort Smith	42.27	39.91	Springfield	35.02	33.78
Little Rock	48.52	49.20	Indiana		
California			Evansville	41.88	41.55
Bakersfield	5.72	5.72	Fort Wayne	35.80	34.40
Eureka	39.76	38.51	Indianapolis	38.74	39.12
Fresno	10.24	10.52	South Bend	36.20	38.16
Los Angeles AP	11.59	12.08	Iowa		
Sacramento	17.22	17.10	Des Moines	30.85	30.83
San Diego	9.45	9.32	Dubuque	40.27	38.59
San Francisco AP	19.53	19.71	Sioux City	25.74	25.37
Stockton	14.17	13.77	Waterloo	33.74	33.10
Colorado			Kansas		
Denver	15.51	15.31	Concordia	27.60	27.11
Grand Junction	8.41	8.00	Dodge City	20.58	20.66
Pueblo	11.91	10.86	Goodland	16.65	16.31
Connecticut			Wichita	30.58	28.61
Bridgeport	38.61	41.56	Kentucky		
Hartford	43.37	44.39	Lexington	44.49	45.68
Delaware			Louisville	43.11	43.56
Wilmington	40.25	41.38	Louisiana		
District of Columbia			Baton Rouge	54.05	55.77
National Airport	38.89	39.00	Lake Charles	55.47	53.03
Florida			New Orleans	56.77	59.74
Daytona Beach	50.22	48.46	Shreveport	44.72	43.84
Fort Myers	53.95	53.64	Maine		
Jacksonville	54.47	52.76	Caribou	35.82	36.59
Miami	59.80	57.55	Portland	40.80	43.52
Orlando	51.21	47.82	Maryland		
Pensacola	64.22	61.16	Baltimore	40.46	41.84
Tallahassee	61.58	64.59	Massachusetts		
Tampa	49.38	46.73	Blue Hill	46.82	49.14
West Palm Beach	62.06	59.72	Boston	42.52	43.81
Georgia			Worcester	45.24	47.60
Athens	50.60	50.15	Michigan		
Atlanta	48.34	48.61	Detroit, CAP	30.98	30.97

Table 3 (continued)
COMPARISON OF AVERAGE INCHES OF PRECIPITATION IN THE U.S. FOR
1941—1970 AND 1951—1980

Location	Annual (in.) 1941—1970[a]	Annual (in.) 1951—1980[b]	Location	Annual (in.) 1941—1970[a]	Annual (in.) 1951—1980[b]
Flint	29.77	29.21	New York (JFK AP)	41.53	41.76
Grand Rapids	32.39	34.35	Rochester	31.33	31.27
Lansing	30.39	29.58	Syracuse	36.41	39.11
Marquette	30.84	37.13	North Carolina		
Muskegon	31.53	31.50	Asheville	45.18	47.71
Minnesota			Charlotte	42.72	43.16
Duluth	30.18	29.68	Greensboro	41.36	42.47
International Falls	25.65	24.35	Raleigh	42.54	41.76
Minneapolis	25.94	26.36	Wilmington	53.59	53.35
Rochester	27.47	28.25	North Dakota		
Saint Cloud	26.84	27.72	Bismark	16.16	15.36
Mississippi			Fargo	19.62	19.59
Jackson	49.19	52.82	Williston	14.33	13.85
Meridian	51.58	53.30	Ohio		
Missouri			Akron	35.13	35.90
Columbia	37.39	36.07	Cleveland	34.99	35.40
Kansas City	37.00	35.16	Columbus	37.01	36.97
St. Louis	35.89	33.91	Dayton	34.36	34.71
Springfield	39.70	39.47	Mansfield	33.67	34.87
Montana			Toledo	31.51	31.78
Billings	14.15	15.09	Youngstown	37.99	37.33
Great Falls	14.99	15.24	Oklahoma		
Havre	11.55	11.17	Oklahoma City	31.37	30.89
Helena	11.38	11.37	Tulsa	36.90	38.77
Miles City	13.93	14.11	Oregon		
Missoula	13.34	13.29	Astoria	66.34	69.60
Nebraska			Burns	11.83	10.13
Grand Island	23.41	23.31	Eugene	42.56	46.04
Lincoln	28.61	26.92	Medford	20.64	19.84
North Platte	19.90	19.47	Pendleton	12.31	12.20
Omaha (North)	29.43	29.94	Portland	37.61	37.39
Scottsbluff	14.57	14.59	Salem	41.08	40.35
Valentine	17.80	17.11	Pennsylvania		
Nevada			Allentown	42.49	44.31
Elko	9.78	9.30	Erie	38.20	39.39
Ely	8.70	9.02	Harrisburg	36.47	39.09
Las Vegas	3.76	4.19	Philadelphia	39.93	41.42
Reno	7.20	7.49	Pittsburg, City	36.22	36.30
Winnemucca	8.47	7.87	Williamsport	39.99	41.28
New Hampshire			Puerto Rico		
Concord	36.17	36.53	San Juan	59.15	53.99
Mt. Washington	76.17	89.92	Rhode Island		
New Jersey			Block Is.	40.51	41.91
Atlantic City	45.46	40.06	Providence	42.75	45.32
Newark	41.45	42.34	South Carolina		
New Mexico			Charleston AP	52.12	51.59
Albuquerque	7.77	8.12	Columbia	46.36	49.12
Clayton	15.91	14.12	Greenville	47.54	50.53
Roswell	10.61	9.70	South Dakota		
New York			Aberdeen	19.10	17.79
Albany	33.36	35.74	Huron	19.44	18.66
Binghamton	37.35	36.78	Rapid City	17.12	16.27
Buffalo	36.11	37.52	Sioux Falls	24.72	24.12

Table 3 (continued)
COMPARISON OF AVERAGE INCHES OF PRECIPITATION IN THE U.S. FOR 1941—1970 AND 1951—1980

Location	Annual (in.)		Location	Annual (in.)	
	1941—1970[a]	1951—1980[b]		1941—1970[a]	1951—1980[b]
Tennessee			Norfolk	44.68	45.22
Chattanooga	51.92	52.60	Richmond	42.59	44.07
Knoxville	46.18	47.29	Roanoke	39.03	39.15
Memphis	49.10	51.57	Washington		
Nashville	46.00	48.49	Olympia	50.74	50.96
Texas			Seattle C.O.	35.65	38.84
Abilene	23.59	23.26	Spokane	17.42	16.71
Amarillo	20.28	19.10	Walla Walla	16.01	15.96
Austin	32.49	31.50	Yakima	8.00	7.98
Brownsville	25.09	25.44	West Virginia		
Corpus Christi	28.53	30.18	Beckley	42.62	42.13
Dallas	32.30	29.46	Charleston	40.75	42.43
El Paso	7.77	7.82	West Virginia		
Galveston	42.20	40.24	Elkins	43.22	42.85
Lubbock	18.41	17.76	Huntington	38.88	40.74
Midland	13.51	13.70	Wisconsin		
San Angelo	17.53	18.19	Green Bay	27.01	28.00
San Antonio	27.54	29.13	La Crosse	29.08	30.25
Victoria	34.29	36.90	Madison	30.25	30.84
Waco	31.26	30.95	Milwaukee	29.07	30.94
Wichita Falls	27.22	26.73	Wyoming		
Utah			Casper	11.22	11.43
Milford	8.40	8.59	Cheyenne	14.65	13.31
Salt Lake City	15.17	15.31	Lander	13.84	13.16
Vermont			Sheridan	16.16	14.93
Burlington	32.54	33.69			
Virginia					
Lynchburg	38.27	39.91			

[a] See Table 2.
[b] From Comparative Climatic Data for the United States Through 1987, National Climatic Data Center, National Oceanic and Atmospheric Administration, Vol. 1, Normals, Asheville, N.C.

Table 4
MEAN DATE OF LAST 32°F TEMPERATURE IN SPRING, FIRST 32°F IN AUTUMN, AND MEAN LENGTH OF FREEZE-FREE PERIOD (DAYS)

State and station	Mean date last 32°F in spring	Mean date first 32°F in fall	Mean freeze-free period (no. days)
Alabama			
Birmingham	Mar. 19	Nov. 14	241
Mobile Uª	Feb. 17	Dec. 12	298
Montgomery U	Feb. 27	Dec. 3	279
Alaska			
Anchorage	May 18	Sept. 13	118
Barrow	June 27	July 5	8
Cordova	May 10	Oct. 2	145
Fairbanks	May 24	Aug. 29	97
Juneau	Apr. 27	Oct. 19	176
Nome	June 12	Aug. 24	73
Arizona			
Flagstaff	June 8	Oct. 2	116
Phoenix	Jan. 27	Dec. 11	317
Tucson	Mar. 6	Nov. 23	261
Winslow	Apr. 28	Oct. 21	176
Yuma U	Jan. 11	Dec. 27	350
Arkansas			
Fort Smith	Mar. 23	Nov. 9	231
Little Rock	Mar. 16	Nov. 15	244
California			
Bakersfield	Feb. 14	Nov. 28	287
Eureka U	Jan. 24	Dec. 25	335
Fresno	Feb. 3	Dec. 3	303
Los Angeles U	*ᵇ	*	*
Red Bluff	Feb. 25	Nov. 29	277
Sacramento	Jan. 24	Dec. 11	321
San Diego	*	*	*

State and station	Mean date last 32°F in spring	Mean date first 32°F in fall	Mean freeze-free period (no. days)
Indiana			
Evansville	Apr. 2	Nov. 4	216
Fort Wayne	Apr. 24	Oct. 20	179
Indianapolis U	Apr. 17	Oct. 27	193
South Bend	May 3	Oct. 16	165
Des Moines U	Apr. 20	Oct. 19	183
Dubuque U	Apr. 19	Oct. 19	184
Koekuk	Apr. 12	Oct. 26	197
Sioux City	Apr. 28	Oct. 12	167
Kansas			
Concordia U	Apr. 16	Oct. 24	191
Dodge City	Apr. 22	Oct. 24	184
Goodland	May 5	Oct. 9	157
Topeka U	Apr. 9	Oct. 26	200
Wichita	Apr. 5	Nov. 1	210
Kentucky			
Lexington	Apr. 13	Oct. 28	198
Louisville U	Apr. 1	Nov. 7	220
Louisiana			
Lake Charles	Feb. 18	Dec. 6	291
New Orleans	Feb. 13	Dec. 12	302
Shreveport	Mar. 1	Nov. 27	272
Maine			
Greenville	May 27	Sept. 20	116
Portland	Apr. 29	Oct. 15	169
Maryland			
Annapolis	Mar. 4	Nov. 15	225
Baltimore U	Mar. 28	Nov. 17	234

Location			
San Francisco U	*	*	*
Colorado			
Denver U	May 2	Oct. 14	165
Palisades	Apr. 22	Oct. 17	178
Pueblo	Apr. 28	Oct. 12	167
Connecticut			
Hartford	Apr. 22	Oct. 19	180
New Haven	Apr. 15	Oct. 27	195
District of Columbia			
Washington U	Apr. 10	Oct. 28	200
Florida			
Apalachicola U	Feb. 2	Dec. 21	322
Fort Myers	*	*	*
Jacksonville U	Feb. 6	Dec. 16	313
Key West	*	*	*
Lakeland	Jan. 10	Dec. 25	349
Miami	*	*	*
Orlando	Jan. 31	Dec. 17	319
Pensacola U	Feb. 18	Dec. 15	300
Tallahassee	Feb. 26	Dec. 3	280
Tampa	Jan. 10	Dec. 26	349
Georgia			
Atlanta U	Mar. 20	Nov. 19	244
Augusta	Mar. 7	Nov. 22	260
Macon	Mar. 12	Nov. 19	252
Savannah	Feb. 21	Dec. 9	291
Idaho			
Boise	Apr. 29	Oct. 16	171
Pocatello	May 8	Sept. 30	145
Salmon	June 4	Sept. 6	94
Illinois			
Cairo U	Mar. 23	Nov. 11	233
Chicago U	Apr. 19	Oct. 28	192
Freeport	May 8	Oct. 4	149
Peoria	Apr. 22	Oct. 16	177
Springfield U	Apr. 8	Oct. 30	205
Frederick	Mar. 24	Oct. 17	176
Massachusetts			
Boston	Apr. 16	Oct. 25	192
Nantucket	Apr. 12	Nov. 16	219
Michigan			
Alpena U	May 6	Oct. 9	156
Detroit	Apr. 25	Oct. 23	181
Escanaba U	May 14	Oct. 6	145
Grand Rapids U	Apr. 25	Oct. 27	185
Marquette U	May 14	Oct. 17	156
S. Ste. Marie	May 18	Oct. 3	138
Minnesota			
Albert Lee	May 3	Oct. 6	156
Big Falls R.S	June 4	Sept. 7	95
Brainerd	May 16	Sept. 24	131
Duluth	May 22	Sept. 24	125
Minneapolis	Apr. 30	Oct. 13	166
St. Cloud	May 9	Sept. 29	144
Mississippi			
Jackson	Mar. 10	Nov. 13	248
Meridian	Mar. 13	Nov. 14	246
Vicksburg U	Mar. 8	Nov. 15	252
Missouri			
Columbia	Apr. 9	Oct. 24	198
Kansas City	Apr. 5	Oct. 31	210
St. Louis U	Apr. 2	Nov. 8	220
Springfield	Apr. 10	Oct. 31	203
Montana			
Billings	May 15	Sept. 24	132
Glasgow U	May 19	Sept. 20	124
Great Falls	May 14	Sept. 26	135
Havre U	May 9	Sept. 23	138
Helena	May 12	Sept. 23	134
Kalispell	May 12	Sept. 23	135
Miles City	May 5	Oct. 3	150
Superior	June 5	Aug. 30	85

Table 4 (continued)
MEAN DATE OF LAST 32°F TEMPERATURE IN SPRING, FIRST 32°F IN AUTUMN, AND MEAN LENGTH OF FREEZE-FREE PERIOD (DAYS)

State and station	Mean date last 32°F in spring	Mean date first 32°F in fall	Mean freeze-free period (no. days)
Nebraska			
Grand Island	Apr. 29	Oct. 6	160
Lincoln	Apr. 20	Oct. 17	180
Norfolk	May 4	Oct. 3	152
North Platte	Apr. 30	Oct. 7	160
Omaha	Apr. 14	Oct. 20	189
Nevada			
Valentine Lakes	May 7	Sept. 30	146
Elko	June 6	Sept. 3	89
Las Vegas	Mar. 13	Nov. 13	245
Reno	May 14	Oct. 2	141
Winnemucca	May 18	Sept. 21	125
New Hampshire			
Concord	May 11	Sept. 30	142
New Jersey			
Cape May	Apr. 4	Nov. 15	225
Trenton U	Apr. 8	Nov. 5	211
New Mexico			
Albuquerque	Apr. 16	Oct. 29	196
Roswell	Apr. 9	Nov. 2	208
New York			
Albany	Apr. 27	Oct. 13	169
Binghamton U	May 4	Oct. 6	154
Buffalo	Apr. 30	Oct. 25	179
New York U	Apr. 7	Nov. 12	219
Rochester	Apr. 28	Oct. 21	176
Syracuse	Apr. 30	Oct. 15	168
N. Carolina			
Asheville U	Apr. 12	Oct. 24	195

State and station	Mean date last 32°F in spring	Mean date first 32°F in fall	Mean freeze-free period (no. days)
S. Dakota			
Huron U	May 4	Sept. 30	149
Rapid City U	May 7	Oct. 4	150
Sioux Falls U	May 5	Oct. 3	152
Tennessee			
Chattanooga U	Mar. 26	Nov. 10	229
Knoxville U	Mar. 31	Nov. 6	220
Memphis U	Mar. 20	Nov. 12	237
Nashville U	Mar. 28	Nov. 7	224
Texas			
Albany	Mar. 30	Nov. 9	224
Balmorhea	Apr. 1	Nov. 12	226
Beeville	Feb. 21	Dec. 6	288
College Station	Mar. 1	Dec. 1	275
Corsicana	Mar. 13	Nov. 27	259
Dalhart Exp. Sta	Apr. 23	Oct. 18	178
Dallas	Mar. 18	Nov. 22	249
Del Rio	Feb. 12	Dec. 9	300
Encinal	Feb. 15	Dec. 12	301
Houston	Feb. 5	Dec. 11	309
Lampasas	Apr. 1	Nov. 10	223
Matagorda	Feb. 12	Dec. 17	308
Midland	Apr. 3	Nov. 6	218
Mission	Jan. 30	Dec. 21	325
Mount Pleasant	Mar. 23	Nov. 12	233
Nacogdoches	Mar. 15	Nov. 13	243
Plainview	Apr. 10	Nov. 6	211
Presidio	Mar. 20	Nov. 13	238
Quanah	Mar. 31	Nov. 7	221

Location	Last spring freeze	First fall freeze	Days
Charlotte U	Mar. 21	Nov. 15	239
Greenville	Mar. 28	Nov. 5	222
Hatteras	Feb. 25	Dec. 18	296
Raleigh U	Mar. 24	Nov. 16	237
Wilmington U	Mar. 8	Nov. 24	262
N. Dakota			
Bismarck	May 11	Sept. 24	136
Devils Lake U	May 18	Sept. 22	127
Fargo	May 13	Sept. 27	137
Williston U	May 14	Sept. 23	132
Ohio			
Akron—Canton	Apr. 29	Oct. 20	173
Cincinnati (Abbe)	Apr. 15	Oct. 25	192
Cleveland	Apr. 21	Nov. 2	195
Columbus U	Apr. 17	Oct. 30	196
Dayton	Apr. 20	Oct. 21	184
Toledo	Apr. 24	Oct. 25	184
Oklahoma			
Okla. City U	Mar. 28	Nov. 7	223
Tulsa	Mar. 31	Nov. 2	216
Oregon			
Astoria	Mar. 18	Nov. 24	251
Bend	June 17	Aug. 17	62
Medford	Apr. 25	Oct. 20	178
Pendleton	Apr. 27	Oct. 8	163
Portland U	Feb. 25	Dec. 1	279
Salem	Apr. 14	Oct. 27	197
Pennsylvania			
Allentown	Apr. 20	Oct. 16	180
Harrisburg	Apr. 10	Oct. 28	201
Philadelphia U	Mar. 30	Nov. 17	232
Pittsburgh	Apr. 20	Oct. 23	187
Scranton U	Apr. 24	Oct. 14	174
Rhode Island			
Providence U	Apr. 13	Oct. 27	197
S. Carolina			
Charleston U	Feb. 19	Dec. 10	294
Columbia U	Mar. 14	Nov. 21	252
Greenville	Mar. 23	Nov. 17	239
San Angelo	Mar. 25	Nov. 15	235
Ysleta	Apr. 6	Oct. 30	207
Utah			
Blanding	May 18	Oct. 14	148
Salt Lake City	Apr. 12	Nov. 1	202
Vermont			
Burlington	May 8	Oct. 3	148
Virginia			
Lynchburg	Apr. 6	Oct. 27	205
Norfolk U	Mar. 18	Nov. 27	254
Richmond U	Apr. 2	Nov. 8	220
Roanoke	Apr. 20	Oct. 24	187
Washington			
Bumping Lake	June 17	Aug. 16	60
Seattle U	Feb. 23	Dec. 1	281
Spokane	Apr. 20	Oct. 12	175
Tatoosh Island	Jan. 25	Dec. 20	329
Walla Walla U	Mar. 28	Nov. 1	218
Yakima	Apr. 19	Oct. 15	179
W. Virginia			
Charleston	Apr. 18	Oct. 28	193
Parkersburg	Apr. 16	Oct. 21	189
Wisconsin			
Green Bay	May 6	Oct. 13	161
La Crosse U	May 1	Oct. 8	161
Madison U	Apr. 26	Oct. 19	177
Milwaukee U	Apr. 20	Oct. 25	188
Wyoming			
Casper	May 18	Sept. 25	130
Cheyenne	May 20	Sept. 27	130
Lander	May 15	Sept. 20	128
Sheridan	May 21	Sept. 21	123

a   U indicates urban.

b   * indicates occurrence in less than 1 year in 10. No freeze of record in Key West, Fla.

Charts and tabulation were derived from the Freeze Data tabulation in Climatography of the United States No. 60 - Climates of the States.

Source: U.S. Department of Commerce, Climatic Atlas of the United States, Environmental Science Services Administration, Environmental Data Service, Washington, D.C., 1968.

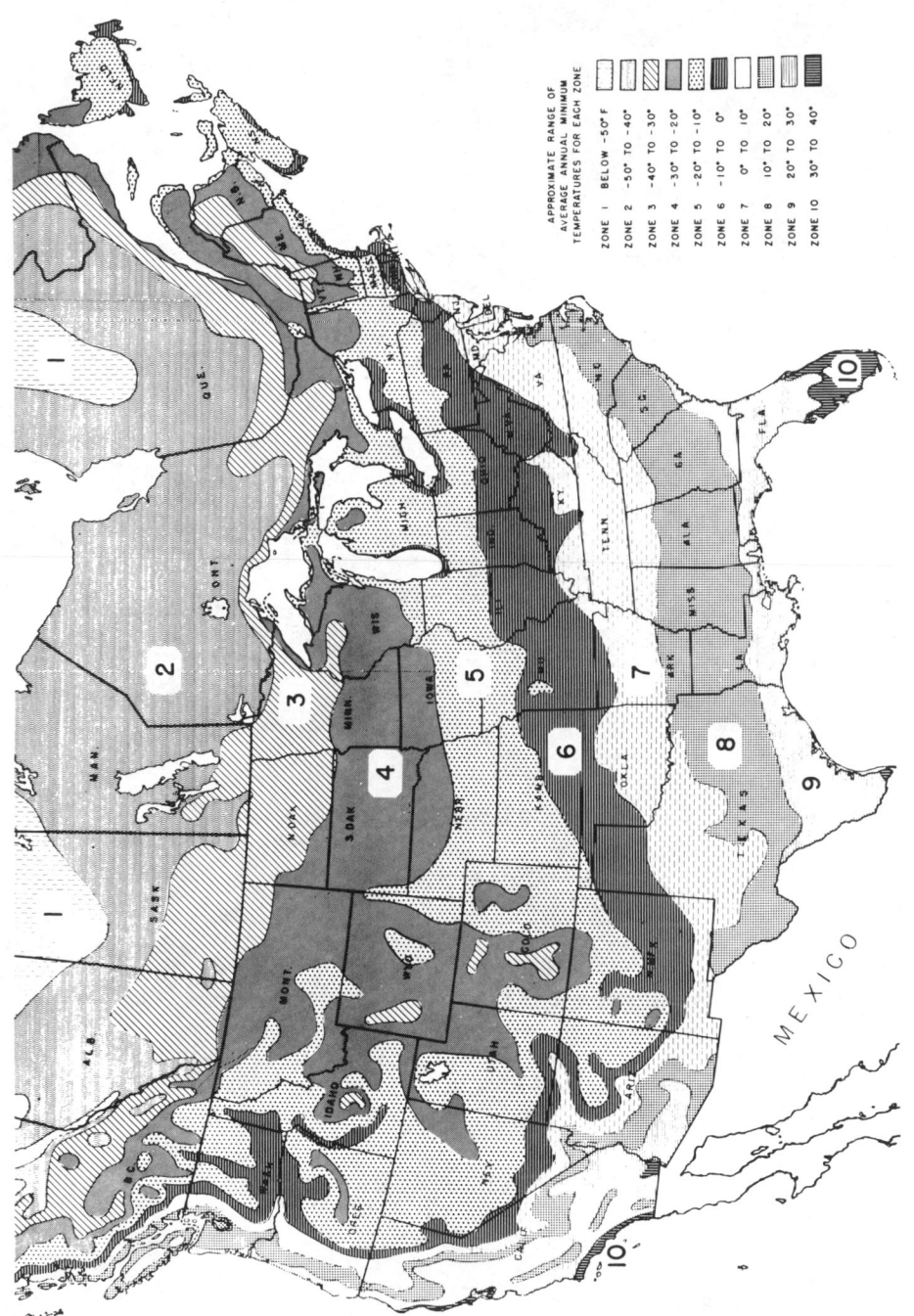

FIGURE 2.   Zones of plant hardiness. (Plant Hardiness Zone Map, *U.S. Dep. Agric. Res. Serv. Agric. Misc. Publ.*, No. 814, reprinted 1972.)

Table 5

MEAN GROWING DEGREE DAYS (ADJUSTED 50°F BASE°): ACCUMULATED WEEKLY MARCH 1 TO INDICATED DATES

Station	Mar. 8	Mar. 15	Mar. 22	Mar. 29	Apr. 5	Apr. 12	Apr. 19	Apr. 26	May 3	May 10
West										
Phoenix, Ariz.	76	160	252	352	458	579	710	839	974	1123
Fresno, Calif.	50	102	164	232	309	391	481	570	663	767
Denver, Colo.	13	30	52	83	117	152	199	255	311	378
Pocatello, Idaho	4	7	15	30	51	77	113	148	188	247
Glasgow, Mont.	1	2	7	18	35	54	85	117	159	209
Pendleton, Ore.	8	20	41	66	99	142	187	231	281	345
North Central										
Des Moines, Iowa	5	14	23	41	67	94	138	194	259	340
Minneapolis, Minn.	0	1	4	11	25	40	71	109	162	224
North Platte, Nebr.	10	24	39	67	100	133	180	236	296	366
Omaha, Nebr.	8	20	33	58	91	127	178	244	316	405
Fargo, N. Dak.	0	0	2	8	17	30	54	85	134	186
Huron, S. Dak.	2	5	12	26	46	70	108	154	210	272
Green Bay, Wis.	0	1	3	7	17	28	55	85	125	176
Madison, Wis.	1	3	8	17	33	53	89	132	186	255
South Central										
Fort Smith, Ark.	41	87	131	191	269	348	437	551	664	794
Dodge City, Kans.	20	46	76	121	172	222	286	365	442	537
Topeka, Kans.	18	42	63	101	148	194	254	337	418	519
Wichita, Kans.	22	50	77	121	177	231	297	388	474	582
Shreveport, La.	61	132	203	282	384	488	593	728	866	1020
Columbia, Mo.	18	39	58	93	137	184	245	330	417	519
St. Louis, Mo.	19	41	60	97	142	190	254	345	436	541
Oklahoma City, Okla.	35	76	119	174	246	318	400	505	605	728
Amarillo, Tex.	33	78	123	180	244	308	385	474	560	667
Corpus Christi, Tex.	95	210	324	445	587	728	878	1049	1229	1411
Fort Worth, Tex.	57	122	191	271	369	470	578	710	838	988
Midland, Tex.	57	128	200	282	376	473	578	700	822	964
Midwest										
Chicago, Ill.	5	10	18	35	36	81	124	177	238	313
Moline, Ill.	5	10	18	36	61	90	135	192	261	344

Table 5 (continued)

MEAN GROWING DEGREE DAYS (ADJUSTED 50°F BASE°): ACCUMULATED WEEKLY MARCH 1 TO INDICATED DATES

Station	Mar. 8	Mar. 15	Mar. 22	Mar. 29	Apr. 5	Apr. 12	Apr. 19	Apr. 26	May 3	May 10
Peoria, Ill.	8	16	25	46	74	105	152	214	284	367
Indianapolis, Ind.	12	26	39	64	95	128	175	250	324	411
Louisville, Ky.	22	48	71	108	156	207	272	370	466	569
Flint, Mich.	2	5	9	19	35	53	87	127	175	239
Grand Rapids, Mich.	1	5	10	20	37	55	89	131	185	252
Columbus, Ohio	12	26	38	64	95	129	176	246	320	404
Northeast										
Portland, Maine	0	1	1	4	12	24	41	65	94	135
Albany, N.Y.	1	4	6	16	33	51	84	130	179	244
Binghamton, N.Y.	1	3	5	14	28	41	66	105	145	198
Syracuse, N.Y.	2	5	8	19	36	51	83	126	174	236
Harrisburg, Pa.	8	17	28	52	83	116	163	228	295	377
Philadelphia, Pa.	8	20	31	57	91	125	175	238	302	384
Pittsburgh, Pa.	10	21	32	55	82	111	154	217	287	364
Williamsport, Pa.	4	10	15	33	58	84	126	182	243	318
Richmond, Va.	29	61	89	137	194	254	327	425	517	620
Southeast										
Macon, Ga.	58	127	195	278	375	477	580	715	858	999
Raleigh, N.C.	39	82	118	174	240	311	389	497	604	718
Memphis, Tenn.	39	85	127	183	258	337	422	539	659	793
Nashville, Tenn.	35	74	109	158	222	288	363	474	585	705

Station	May 17	May 24	May 31	June 7	June 14	June 21	June 28	July 5	July 12	July 19
West										
Phoenix, Ariz.	1274	1442	1614	1794	1976	2172	2378	2593	2817	3045
Fresno, Calif.	877	1008	1139	1276	1411	1569	1732	1895	2066	2243
Denver, Colo.	447	524	613	709	819	937	1068	1212	1360	1510
Pocatello, Idaho	305	378	456	540	630	734	846	965	1101	1244
Glasgow, Mont.	274	347	429	517	613	716	821	934	1073	1217
Pendleton, Ore.	418	504	592	698	795	913	1033	1159	1303	1460
North Central										
Des Moines, Iowa	428	525	629	753	897	1046	1201	1369	1535	1705

Minneapolis, Minn.	295	374	458	565	691	822	954	1103	1252	1409
North Platte, Nebr.	440	523	617	722	846	977	1115	1271	1424	1580
Omaha, Nebr.	502	607	722	854	1007	1165	1327	1504	1677	1855
Fargo, N. Dak.	250	325	404	503	612	724	839	968	1108	1253
Huron, S. Dak.	345	425	514	621	743	870	1001	1150	1305	1462
Green Bay, Wis.	234	302	374	466	573	683	797	923	1050	1185
Madison, Wis.	325	406	491	597	719	845	977	1119	1258	1406
South Central										
Fort Smith, Ark.	929	1081	1240	1406	1589	1774	1964	2159	2356	2553
Dodge City, Kans.	634	741	865	997	1154	1318	1487	1667	1847	2026
Topeka, Kans.	623	741	867	1006	1168	1333	1503	1685	1862	2042
Wichita, Kans.	691	816	952	1100	1271	1447	1628	1819	2010	2196
Shreveport, La.	1178	1348	1525	1706	1902	2100	2300	2504	2712	2916
Columbia, Mo.	622	741	864	1003	1168	1330	1501	1681	1857	2039
St. Louis, Mo.	647	770	899	1047	1220	1389	1566	1752	1932	2119
Oklahoma City, Okla.	849	988	1137	1294	1475	1658	1844	2039	2233	2428
Amarillo, Tex.	770	889	1019	1156	1311	1478	1649	1830	2010	2187
Corpus Christi, Tex.	1600	1794	1998	2200	2412	2626	2840	3055	3270	3487
Fort Worth, Tex.	1140	1308	1486	1670	1871	2072	2278	2488	2702	2916
Midland, Tex.	1110	1263	1431	1603	1788	1978	2171	2364	2558	2754
Midwest										
Chicago, Ill.	391	481	579	701	843	984	1142	1310	1470	1639
Moline, Ill.	429	526	629	755	898	1045	1200	1365	1524	1691
Peoria, Ill.	454	554	660	788	938	1085	1245	1415	1576	1746
Indianapolis, Ind.	499	602	708	834	986	1132	1293	1462	1623	1792
Louisville, Ky.	677	801	932	1073	1239	1399	1572	1752	1926	2110
Flint, Mich.	302	375	455	555	676	795	924	1064	1197	1338
Grand Rapids, Mich.	321	398	482	588	717	841	976	1121	1259	1408
Columbus, Ohio	490	593	697	818	964	1104	1260	1424	1581	1749
Northeast										
Portland, Maine	181	237	304	386	470	563	668	787	903	1032
Albany, N.Y.	313	393	478	586	702	820	963	1110	1248	1402
Binghamton, N.Y.	253	320	387	476	580	683	805	938	1058	1197
Syracuse, N.Y.	300	376	457	559	677	794	930	1075	1212	1360
Harrisburg, Pa.	463	563	667	792	936	1075	1239	1410	1573	1748
Philadelphia, Pa.	475	576	681	811	955	1094	1262	1436	1604	1785
Pittsburgh, Pa.	442	533	625	736	869	994	1140	1295	1438	1595
Williamsport, Pa.	397	485	576	686	814	937	1083	1237	1380	1537
Richmond, Va.	733	858	984	1120	1279	1430	1608	1786	1962	2147

Table 5 (continued)
MEAN GROWING DEGREE DAYS (ADJUSTED 50°F BASE):[a] ACCUMULATED WEEKLY MARCH 1 TO INDICATED DATES

Station	May 17	May 24	May 31	June 7	June 14	June 21	June 28	July 5	July 12	July 19
Southeast										
Macon, Ga.	1149	1314	1484	1656	1842	2029	2225	2421	2616	2814
Raleigh, N.C.	839	976	1113	1255	1420	1579	1757	1936	2113	2298
Memphis, Tenn.	935	1094	1257	1427	1618	1804	1999	2198	2398	2599
Nashville, Tenn.	831	974	1120	1276	1454	1626	1810	1998	2134	2376
(July 26)	*July 26*	*Aug. 2*	*Aug. 9*	*Aug. 16*	*Aug. 23*	*Aug. 30*	*Sept. 6*	*Sept. 13*	*Sept. 20*	*Sept. 27*
West										
Phoenix, Ariz.	3274	3499	3721	3943	4161	4373	4582	4792	4986	5168
Fresno, Calif.	2422	2603	2773	2946	3115	3274	3436	3592	3734	3878
Denver, Colo.	1663	1818	1971	2118	2256	2394	2512	2630	2730	2815
Pocatello, Idaho	1389	1535	1677	1815	1943	2062	2172	2283	2372	2452
Glasgow, Mont.	1366	1511	1658	1794	1925	2037	2135	2229	2300	2366
Pendleton, Ore.	1611	1768	1919	2071	2214	2337	2465	2584	2685	2785
North Central										
Des Moines, Iowa	1880	2057	2228	2384	2535	2688	2822	2942	3044	3133
Minneapolis, Minn.	1572	1732	1889	2030	2167	2303	2418	2519	2600	2666
North Platte, Nebr.	1740	1906	2068	2216	2357	2503	2625	2740	2840	2923
Omaha, Nebr.	2039	2222	2401	2566	2724	2889	3031	3159	3268	3362
Fargo, N. Dak.	1405	1551	1696	1828	1957	2082	2186	2278	2353	2413
Huron, S. Dak.	1623	1783	1942	2088	2229	2373	2491	2598	2687	2761
Green Bay, Wis.	1324	1463	1597	1714	1828	1948	2057	2151	2228	2291
Madison, Wis.	1557	1708	1853	1986	2117	2249	2368	2471	2560	2633
South Central										
Fort Smith, Ark.	2754	2958	3157	3349	3540	3726	3905	4070	4229	4378
Dodge City, Kans.	2211	2398	2587	2761	2929	3103	3257	3399	3531	3641
Topeka, Kans.	2231	2420	2608	2779	2949	3120	3274	3414	3542	3655
Wichita, Kans.	2392	2590	2788	2971	3148	3328	3493	3645	3784	3906
Shreveport, La.	3124	3333	3542	3747	3951	4150	4345	4529	4709	4881
Columbia, Mo.	2226	2414	2598	2769	2937	3107	3263	3404	3535	3651
St. Louis, Mo.	2313	2507	2696	2872	3046	3220	3381	3527	3661	3784
Oklahoma City, Okla.	2629	2831	3032	3226	3415	3603	3779	3941	4095	4241
Amarillo, Tex.	2368	2552	2736	2911	3080	3250	3407	3553	3690	3813

	Oct. 4	Oct. 11	Oct. 18	Oct. 25	Nov. 1	Nov. 8	Nov. 15	Nov. 22	Nov. 29	Dec. 6
Corpus Christi, Tex.	3704	3922	4138	4356	4573	4784	4998	5206	5411	5614
Fort Worth, Tex.	3132	3350	3567	3781	3990	4196	4397	4584	4766	4940
Midland, Tex.	2950	3149	3347	3541	3734	3920	4101	4270	4436	4593
Midwest										
Chicago, Ill.	1814	1989	2158	2316	2473	2631	2778	2905	3023	3123
Moline, Ill.	1863	2034	2198	2350	2499	2652	2787	2904	3014	3108
Peoria, Ill.	1922	2098	2266	2421	2574	2730	2871	2996	3110	3209
Indianapolis, Ind.	1967	2142	2308	2464	2618	2770	2915	3041	3161	3269
Louisville, Ky.	2301	2489	2673	2845	3018	3136	3353	3497	3630	3751
Flint, Mich.	1483	1628	1765	1893	2016	2145	2263	2364	2452	2527
Grand Rapids, Mich.	1560	1712	1858	1993	2125	2260	2387	2494	2590	2668
Columbus, Ohio	1917	2085	2247	2399	2553	2703	2849	2973	3087	3193
Northeast										
Portland, Maine	1164	1293	1411	1532	1646	1755	1856	1945	2018	2087
Albany, N.Y.	1557	1710	1851	1990	2123	2252	2378	2481	2568	2648
Binghamton, N.Y.	1334	1473	1601	1724	1844	1960	2077	2169	2245	2313
Syracuse, N.Y.	1511	1662	1805	1943	2072	2204	2333	2438	2525	2606
Harrisburg, Pa.	1925	2105	2270	2433	2593	2747	2902	3031	3148	3247
Philadelphia, Pa.	1969	2154	2323	2493	2660	2820	2985	3120	3244	3349
Pittsburgh, Pa.	1756	1917	2068	2214	2356	2497	2635	2751	2856	2946
Williamsport, Pa.	1697	1858	2006	2152	2294	2432	2569	2683	2784	2869
Richmond, Va.	2335	2521	2701	2881	3056	3224	3394	3539	3675	3794
Southeast										
Macon, Ga.	3014	3214	3415	3612	3808	3997	4185	4363	4532	4692
Raleigh, N.C.	2489	2678	2863	3046	3227	3399	3572	3725	3868	4995
Memphis, Tenn.	2804	3011	3214	3411	3604	3792	3974	4143	4303	4452
Nashville, Tenn.	2573	2770	2962	3145	3330	3508	3683	3842	3990	4125
West										
Phoenix, Ariz.	5347	5511	5664	5802	5930	6040	6141	6218	6297	6372
Fresno, Calif.	4016	4137	4250	4352	4446	4526	4596	4643	4683	4716
Denver, Colo.	2901	2981	3052	3111	3161	3193	3228	3252	3273	3292
Pocatello, Idaho	2525	2608	2663	2707	2739	2757	2771	2776	2780	2781
Glasgow, Mont.	2427	2479	2529	2563	2591	2606	2615	2620	2623	2624
Pendleton, Ore.	2863	2926	2981	3026	3057	3074	3087	3097	3109	3115

Table 9 (continued)

MEAN GROWING DEGREE DAYS (ADJUSTED 50°F BASE):ᵃ ACCUMULATED WEEKLY MARCH 1 TO INDICATED DATES

	Oct. 4	Oct. 11	Oct. 18	Oct. 25	Nov. 1	Nov. 8	Nov. 15	Nov. 22	Nov. 29	Dec. 6
North Central										
Des Moines, Iowa	3218	3289	3371	3427	3467	3489	3512	3527	3534	3539
Minneapolis, Minn.	2725	2777	2837	2875	2901	2912	2921	2927	2928	2929
North Platte, Nebr.	3004	3081	3156	3213	3261	3288	3316	3334	3349	3360
Omaha, Nebr.	3450	3527	3614	3674	3721	3748	3775	3792	3802	3810
Fargo, N. Dak.	2468	2515	2565	2595	2616	2627	2632	2635	2635	2635
Huron, S. Dak.	2830	2890	2955	2998	3030	3048	3063	3072	3076	3078
Green Bay, Wis.	2346	2389	2442	2474	2495	2506	2513	2520	2521	2522
Madison, Wis.	2703	2759	2823	2863	2890	2907	2920	2929	2931	2934
South Central										
Fort Smith, Ark.	4508	4628	4727	4843	4925	4981	5045	5097	5138	5169
Dodge City, Kans.	3749	3847	3941	4010	4069	4105	4144	4172	4195	4213
Topeka, Kans.	3756	3849	3947	4020	4077	4111	4150	4178	4197	4211
Wichita, Kans.	4019	4121	4224	4300	4359	4396	4437	4467	4488	4504
Shreveport, La.	5034	5173	5309	5422	5521	5595	5674	5751	5814	5864
Columbia, Mo.	3756	3847	3944	4014	4069	4103	4144	4173	4192	4207
St. Louis, Mo.	3892	3984	4081	4151	4205	4239	4278	4308	4326	4340
Oklahoma City, Okla.	4367	4482	4597	4685	4759	4808	4865	4908	4942	4971
Amarillo, Tex.	3926	4029	4127	4205	4273	4317	4372	4414	4451	4484
Corpus Christi, Tex	5798	5978	6154	6312	6452	6573	6691	6811	6917	7012
Fort Worth, Tex.	5094	5234	5376	5493	5592	5665	5746	5817	5877	5926
Midland, Tex.	4733	4864	4990	5093	5182	5246	5319	5381	5437	5483
Midwest										
Chicago, Ill.	3211	3283	3363	3415	3451	3473	3494	3512	3519	3524
Moline, Ill.	3193	3264	3345	3400	3439	3461	3484	3501	3508	3515
Peoria, Ill.	3295	3368	3448	3503	3542	3564	3589	3607	3615	3621
Indianapolis, Ind.	3360	3437	3519	3578	3618	3644	3672	3695	3706	3714
Louisville, Ky.	3859	3947	4041	4113	4166	4201	4241	4276	4296	4310
Flint, Mich.	2595	2652	2710	2752	2778	2796	2808	2820	2825	2828
Grand Rapids, Mich.	2740	2799	2862	2903	2931	2946	2961	2974	2879	2983
Columbus, Ohio	3282	3357	3436	3495	3533	3561	3585	3608	3620	3627

Northeast									
Portland, Maine	2145	2192	2236	2271	2293	2308	2315	2324	2328	2330
Albany, N.Y.	2717	2776	2833	2877	2905	2923	2934	2945	2950	2953
Binghamton, N.Y.	2372	2417	2468	2503	2524	2539	2548	2558	2562	2564
Syracuse, N.Y.	2677	2733	2790	2835	2863	2882	2896	2910	2917	2921
Harrisburg, Pa.	3336	3412	3485	3540	3578	3609	3631	3650	3659	3667
Philadelphia, Pa.	3444	3530	3606	3669	3712	3749	3775	3800	3815	3825
Pittsburgh, Pa.	3027	3092	3160	3211	3243	3267	3288	3310	3320	3327
Williamsport, Pa.	2946	3011	3076	3126	3158	3182	3197	3209	3215	3219
Richmond, Va.	3901	3997	4087	4160	4219	4272	4318	4363	4393	4415
Southeast									
Macon, Ga.	4839	4972	5095	5203	5292	5370	5441	5517	5574	5617
Raleigh, N.C.	4114	4217	4314	4395	4460	4518	4568	4623	4658	4683
Memphis, Tenn.	4584	4701	4816	4907	4981	5035	5093	5148	5187	5217
Nashville, Tenn.	4248	4354	4458	4540	4604	4650	4700	4746	4776	4800

ᵃ The mean growing degree day (GDD) values are based on the adjusted 50°F method and were calculated from records of daily maximum and minimum temperature for the period 1949 to 1968.

Table 6
DAYLENGTHS FOR EACH MONTH OF THE YEAR (N, hr)

Latitude North South	Jan. Jul.	Feb. Aug.	Mar. Sept.	Apr. Oct.	May Nov.	Jun. Dec.	Jul. Jan.	Aug. Feb.	Sept. Mar.	Oct. Apr.	Nov. May	Dec. Jun.
0	12.12	12.12	12.12	12.12	12.12	12.12	12.12	12.10	12.11	12.12	12.12	12.12
5	11.87	11.96	12.08	12.22	12.35	12.41	12.38	12.28	12.16	12.02	11.90	11.83
10	11.61	11.81	12.06	12.35	12.57	12.70	12.64	12.45	12.17	11.91	11.67	11.55
15	11.34	11.66	12.04	12.47	12.82	13.00	12.92	12.62	12.22	11.81	11.44	11.25
20	11.07	11.50	12.01	12.60	13.07	13.32	13.22	12.81	12.26	11.70	11.20	10.94
25	10.78	11.33	11.97	12.74	13.34	13.66	13.53	13.02	12.31	11.58	10.94	10.62
30	10.45	11.14	11.97	12.88	13.65	14.05	13.88	13.23	12.35	11.47	10.67	10.26
35	10.09	10.95	11.95	13.06	13.98	14.47	14.27	13.47	12.42	11.33	10.36	9.86
40	9.68	10.71	11.91	13.25	14.36	14.96	14.71	13.76	12.48	11.18	10.00	9.39
45	9.19	10.45	11.87	13.48	14.82	15.55	15.25	14.09	12.55	11.01	9.60	8.85
50	8.61	10.13	11.84	13.78	15.38	16.29	15.91	14.48	12.66	10.80	9.07	8.17
55	7.83	9.73	11.79	14.10	16.14	17.28	16.78	14.99	12.76	10.55	8.45	7.28
60	6.79	9.21	11.74	14.62	17.10	18.70	18.01	15.67	12.92	10.22	7.60	6.04

Table 7
MEAN DAILY PERCENTAGE (p) OF ANNUAL DAYTIME HOURS FOR DIFFERENT LATITUDES

Latitude North South[a]	Jan. July	Feb. Aug.	Mar. Sept.	Apr. Oct.	May Nov.	June Dec.	July Jan.	Aug. Feb.	Sept. Mar.	Oct. Apr.	Nov. May	Dec. June
60°	0.15	0.20	0.26	0.32	0.38	0.41	0.40	0.34	0.28	0.22	0.17	0.13
58	0.16	0.21	0.26	0.32	0.37	0.40	0.39	0.34	0.28	0.23	0.18	0.15
56	0.17	0.21	0.26	0.32	0.36	0.39	0.38	0.33	0.28	0.23	0.18	0.16
54	0.18	0.22	0.26	0.31	0.36	0.38	0.37	0.33	0.28	0.23	0.19	0.17
52	0.19	0.22	0.27	0.31	0.35	0.37	0.36	0.33	0.28	0.24	0.20	0.17
50	0.19	0.23	0.27	0.31	0.34	0.36	0.35	0.32	0.28	0.24	0.20	0.18
48	0.20	0.23	0.27	0.31	0.34	0.36	0.35	0.32	0.28	0.24	0.21	0.19
46	0.20	0.23	0.27	0.30	0.34	0.35	0.34	0.32	0.28	0.24	0.21	0.20
44	0.21	0.24	0.27	0.30	0.33	0.35	0.34	0.31	0.28	0.25	0.22	0.20
42	0.21	0.24	0.27	0.30	0.33	0.34	0.33	0.31	0.28	0.25	0.22	0.21
40	0.22	0.24	0.27	0.30	0.32	0.34	0.33	0.31	0.28	0.25	0.22	0.21
35	0.23	0.25	0.27	0.29	0.31	0.32	0.32	0.30	0.28	0.25	0.23	0.22
30	0.24	0.25	0.27	0.29	0.31	0.32	0.31	0.30	0.28	0.26	0.24	0.23
25	0.24	0.26	0.27	0.29	0.30	0.31	0.31	0.29	0.28	0.26	0.25	0.24
20	0.25	0.26	0.27	0.28	0.29	0.30	0.30	0.29	0.28	0.26	0.25	0.25
15	0.26	0.26	0.27	0.28	0.29	0.29	0.29	0.28	0.28	0.27	0.26	0.25
10	0.26	0.27	0.27	0.28	0.28	0.29	0.29	0.28	0.28	0.27	0.26	0.26
5	0.27	0.27	0.27	0.28	0.28	0.28	0.28	0.28	0.28	0.27	0.27	0.27
0	0.27	0.27	0.27	0.27	0.27	0.27	0.27	0.27	0.27	0.27	0.27	0.27

[a]   Southern latitudes: apply 6-month difference as shown.

From Doorenbos, J. and Pruitt, W. O., Crop Water Requirements, Irrigation and Drainage Paper No. 24, Food and Agriculture Organization, Rome, 1975. With permission of the Food and Agriculture Organization of the United Nations.

SOIL

SOIL CLASSIFICATION

Table 8
CATEGORIES OF SOILS IN THE SCHEME OF MARBUT WITH BRIEF DESCRIPTIONS AND THE NAMES

Category	Descriptions	Names
VI	Solum composition groups	Orders
V	Inorganic colloid composition groups	Suborders
IV	Broad environment groups	Great soil groups
III	Local environment groups	Families
II	Soil type groups	Soil series
I	Soil units	Soil types

Source: Marbut, C. F., Soils of the United States, in Atlas of American Agriculture, Part 3, Baker, O. E., Ed., U.S. Department of Agriculture, Washington, D.C., 1935, 12.

Table 9
ORDERS, SUBORDERS, AND GREAT SOIL GROUPS OF THE SYSTEM OUTLINED IN 1938

Orders	Suborders	Great soil groups
Zonal soils	Soils of the cold zone	Tundra soils
	Light-colored soils of arid regions	Desert soils
		Red desert soils
		Sierozems
		Brown soils
		Reddish brown soils
	Dark-colored soils of semi-arid, subhumid, and humid grasslands	Chestnut soils
		Reddish chestnut soils
		Chernozems
		Prairie soils
		Reddish prairie soils
	Soils of the forest-grass-land transition	Degraded chernozems
		Noncalcic brown soils
	Light-colored podzolized soils of the timbered regions	Podzols
		Brown podzolic soils
		Gray-brown podzolic soils
	Lateritic soils of forested warm-temperate and tropical regions	Yellow podzolic soils
		Red podzolic soils
		Yellowish brown lateritic soils
		Reddish brown lateritic soils
		Laterite soils
Intrazonal soils	Halomorphic (saline and alkali) soils of imperfectly drained arid regions and littoral deposits	Solonchak soils
		Solonetz soils
		Solodj soils
	Hydromorphic soils of marshes, swamps, seep areas, and flats	Wiesenböden (meadow soils)
		Alpine meadow soils
		Bog soils
		Planosols
		Ground-water podzols
		Ground-water laterite soils
	Calcimorphic soils	Brown forest soils
		Rendzinas

Table 9 (continued)
ORDERS, SUBORDERS, AND GREAT SOIL GROUPS OF THE SYSTEM OUTLINED IN 1938[10]

Orders	Suborders	Great soil groups
Azonal soils	(No suborders)	Lithosols
		Alluvial soils
		Dry sands

Source: Baldwin, M., Kellogg, C. E., and Thorp, J., Soil classification, in Soils and Men, U.S. Department of Agriculture, Washington, D.C., 1938, 979.

Table 10
ORDERS, SUBORDERS, AND GREAT SOIL GROUPS OF THE SYSTEM IN USE IN 1960[11]

Orders	Suborders	Great soil groups
Zonal soils	Soils of cold zones	Tundra soils
		Subarctic brown forest soils
	Light-colored soils of arid regions	Desert soils
		Red desert soils
		Sierozems
		Brown soils
		Reddish brown soils
	Dark-colored soils of semi-arid, subhumid, and humid grasslands	Chestnut soils
		Reddish chestnut soils
		Chernozems
		Brunizems
		Reddish prairie soils
	Soils of the forest-grassland transition	Noncalcic brown soils
	Light-colored podzolized soils of forested regions	Podzols
		Brown podzolic soils
		Gray wooded soils
		Sols bruns acides
		Gray-brown podzolic soils
	Lateritic soils of forested warm-temperate and tropical regions	Red-yellow podzolic soils
		Reddish brown lateritic soils
		Yellowish brown lateritic soils
		Groups of Latosols
Intrazonal soils	Halomorphic soils	Solonchak soils
		Solonetz soils
		Solodj soils
	Hydromorphic soils	Humic gley soils
		Alpine meadow soils
		Bog soils
		Low-humic gley soils
		Planosols
		Ground-water podzols
		Ground-water laterite soils
	Calcimorphic soils	Brown forest soils
		Rendzinas
		Grumusols
		Calcisols
	Dark soils on volcanic ash	Ando soils
Azonal soils	(No suborders)	Lithosols
		Regosols
		Alluvial soils

Source: Simonson, R. W. and Steele, J. G., Soil (great soil groups), in *Encyclopedia of Science and Technology,* Vol. 12, 2nd ed., McGraw-Hill, New York, 1966, 433.

Table 11
ORDERS, SUBORDERS, AND SPECIMEN GREAT GROUPS OF THE CURRENT AMERICAN SYSTEM

Orders	Suborders	Specimen great groups
Alfisols	Aqualfs	Albaqualfs
	Boralfs	Cryoboralfs
	Udalfs	Hapludalfs
	Ustalfs	Natrustalfs
	Xeralfs	Durixeralfs
Aridisols	Argids	Haplargids
	Orthids	Camborthids
Entisols	Aquents	Haplaquents
	Arents	(None defined)
	Fluvents	Udifluvents
	Orthents	Troporthents
	Psamments	Quartzipsamments
Histosols	Fibrists	Sphagnofibrists
	Folists	Borofolists
	Hemists	Medihemists
	Saprists	Cryosaprists
Inceptisols	Andepts	Hydrandepts
	Aquepts	Fragiaquepts
	Ochrepts	Eutrochrepts
	Plaggepts	(None defined)
	Tropepts	Humitropepts
	Umbrepts	Haplumbrepts
Mollisols	Albolls	Natralbolls
	Aquolls	Haplaquolls
	Borolls	Calciborolls
	Rendolls	(None defined)
	Udolls	Argiudolls
	Ustolls	Vermustolls
	Xerolls	Palexerolls
Oxisols	Aquox	Gibbsaquox
	Humox	Sombrihumox
	Orthox	Acrorthox
	Torrox	(None defined)
	Ustox	Haplustox
Spodosols	Aquods	Cryaquods
	Ferrods	(None defined)
	Humods	Haplohumods
	Orthods	Troporthods
Ultisols	Aquults	Ochraquults
	Humults	Tropohumults
	Udults	Paleudults
	Ustults	Haplustults
	Xerults	Palexerults
Vertisols	Torrerts	(None defined)
	Uderts	Pelluderts
	Usterts	Chromusterts
	Xererts	Chromoxererts

Source: Soil Survey Staff, Soil Taxonomy: a Basic System of Soil Classification for Making and Interpreting Soil Surveys, *U.S. Dep. Agric. Agric Handb.*, No. 436, 1975.

Table 12
CRITERIA FOR SUBDIVISION OF SOIL ORDERS IN THE CURRENT CLASSIFICATION SYSTEM

Soil orders	Number of suborders	Bases for subdivisions
Alfisols	5	Moisture and temperature regimes
Aridisols	2	Argillic horizon
Entisols	5	Moisture regime, particle sizes, stratification
Histosols	4	Moisture regime, nature of organic materials
Inceptisols	6	Moisture and temperature regimes, mineralogy, epipedon
Mollisols	7	Moisture and temperature regimes, mineralogy, albic horizon
Oxisols	5	Moisture regime, amount of organic matter
Spodosols	4	Moisture regime, nature of spodic horizon
Ultisols	5	Moisture regime, amount of organic matter
Vertisols	4	Moisture regime

Source: Soil Survey Staff, Soil Taxonomy: a Basic System of Soil Classification for Making and Interpreting Soil Surveys, *U.S. Dep. Agric. Agric Handb.,* No. 436, 1975.

Table 13
MAJOR BASES FOR SUBDIVISION OF SUBORDERS INTO GREAT GROUPS IN THE CURRENT AMERICAN SYSTEM

Orders	Great groups set apart on	
	Diagnostic horizons	Other features
Alfisols	18	19
Aridisols	9	2
Entisols		27
Histosols		20
Inceptisols	14	20
Mollisols	16	17
Oxisols	9	5
Spodosols	10	8
Ultisols	10	14
Vertisols		7

Source: Soil Survey Staff, Soil Taxonomy: a Basic System of Soil Classification for Making and Interpreting Soil Surveys, *U.S. Dep. Agric. Agric Handb.,* No. 436, 1975.

Table 14
**APPROXIMATE PERCENTAGES OF THE
LAND SURFACE OF THE EARTH IN NINE
BROAD SOIL REGIONS**

Dominant soil order in region	Percentage of land surface
Inceptisols	3.44
Spodosols and Histosols	9.89
Alfisols and Inceptisols	4.25
Ultisols	3.83
Mollisols	12.38
Vertisols	2.81
Aridisols	25.34
Oxisols, Inceptisols, and Ultisols	20.52
Soils of mountains — Entisols and Inceptisols	17.44

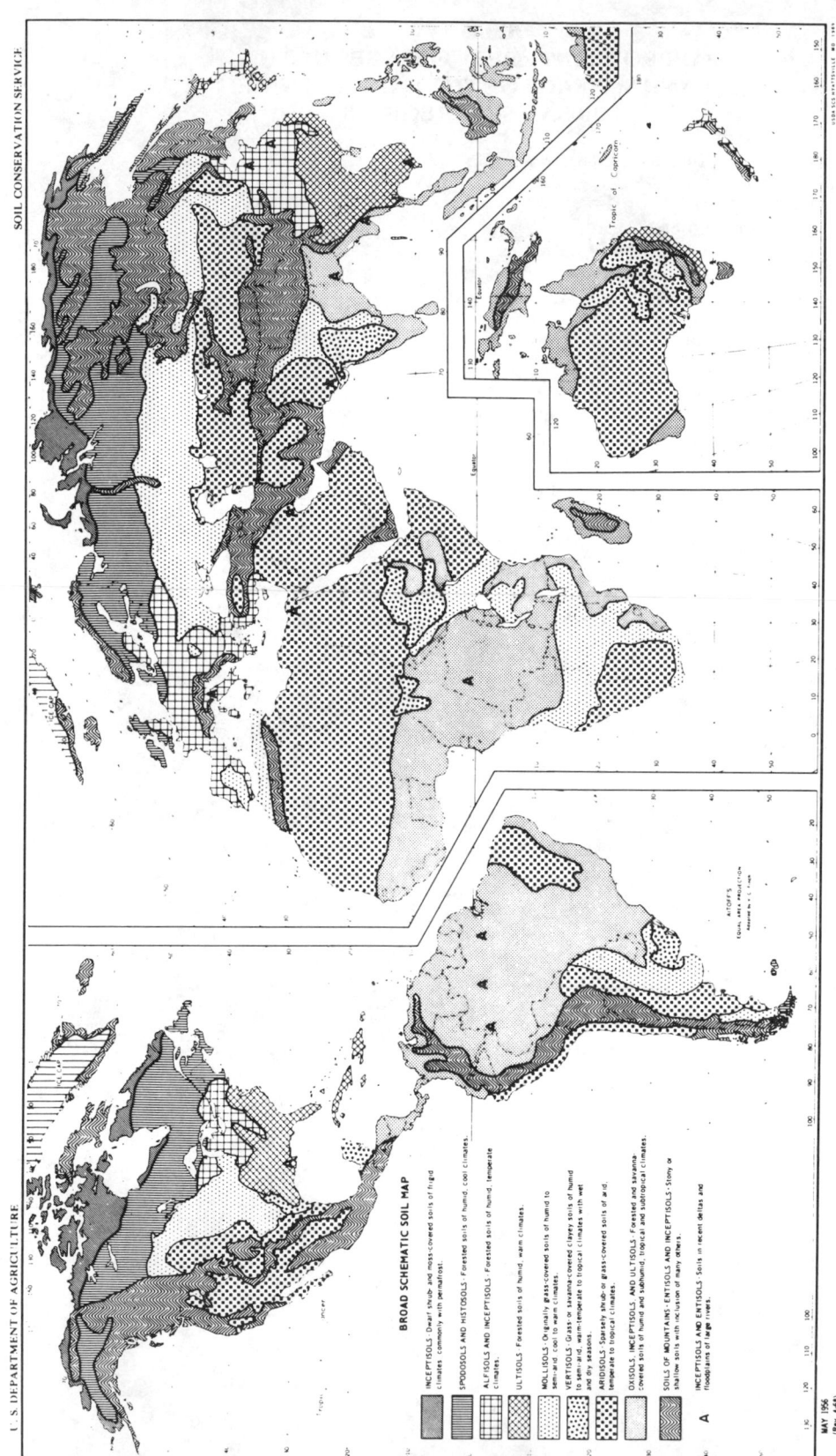

FIGURE 3.   Map of the world showing nine major soil regions. Proportions of the land surface of the earth in the broad regions are given in Table 14.

SOIL STRUCTURE AND COMPOSITION

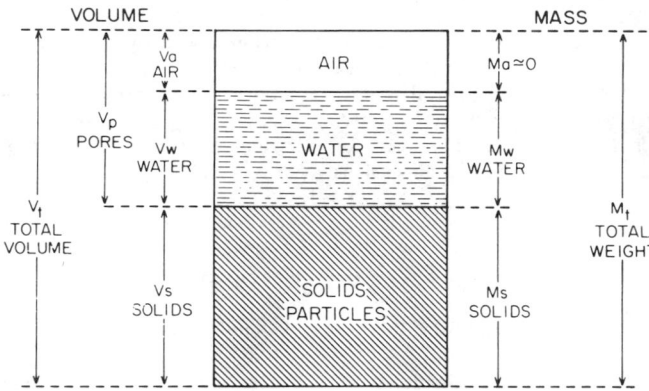

FIGURE 4. Schematic drawing of the soil as a three-phase physical model.

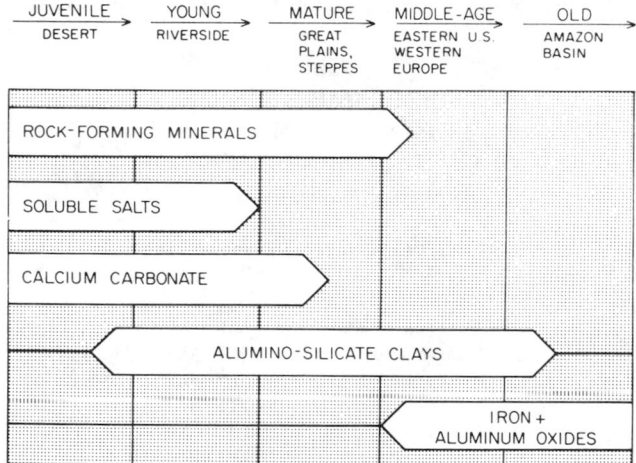

FIGURE 5. Some compositional features of soil. (Redrawn from Coleman, N. T., The good and not-so-good earth, 19th Annual Research Lecture, Riverside, Calif., 1970.)

Table 15
PRINCIPAL TYPES (SHAPES) AND SIZE CLASSES OF SOIL STRUCTURE

Platelike, with the vertical dimension much smaller than the other two; faces mostly horizontal; size ranges for vertical dimensions	Prismlike, with the vertical dimension much greater than the other two; vertical faces distinct; size ranges for horizontal dimensions	Blocklike, with all 3 dimensions the same; a polyhedron arranged around a point and having plane surfaces	Spheroidal, arranged around a point and without plane surfaces
Very fine platy < 1 mm	Very fine prismatic < 10 mm	Very fine blocky < 5 mm	Very fine granular < 1 mm
Fine platy 1—2 mm	Fine prismatic 10—20 mm	Fine blocky 5—10 mm	Fine granular 1—2 mm
Medium platy 2—5 mm	Medium prismatic 20—50 mm	Medium blocky 10—20 mm	Medium granular 2—5 mm
Coarse platy 5—10 mm	Coarse prismatic 50—100 mm	Coarse blocky 20—50 mm	Coarse granular 5—10 mm
Very coarse platy > 10 mm	Very coarse prismatic > 100 mm	Very coarse blocky > 50 mm	Very coarse granular > 10 mm

Source: Soil Survey Staff, Soil Survey Manual, *U.S. Dep. Agric. Agric. Handb.*, No. 18, 1951.

Table 16
SIZE LIMITS OF SOIL SEPARATES IN SCHEME OF U.S. DEPARTMENT OF AGRICULTURE

Name of separate	Diameter range (mm)
Very coarse sand	2.0—1.0
Coarse sand	1.0—0.5
Medium sand	0.5—0.25
Fine sand	0.25—0.10
Very fine sand	0.10—0.05
Silt	0.05—0.002
Clay	Below 0.002

Source: Soil Survey Staff, Soil Survey Manual, *U.S. Dep. Agric. Agric. Handb.*, No. 18, 1951.

Table 17
COMPARISON OF TEXTURAL CLASSIFICATIONS OF SOIL PARTICLES

Dia. mm.	USDA	International	European	Dia. mm.
2.0	Very coarse sand			2.0
1.0			Coarse sand	
	Coarse sand	Coarse sand		0.6
0.5				
	Medium sand		Medium sand	
0.25				0.2
	Fine sand			
0.1			Fine sand	
	Very fine sand	Fine sand		
0.05				0.06
			Coarse silt	
				0.02
	Silt		Medium silt	
		Silt		0.006
			Fine silt	
0.002				0.002
			Coarse clay	
				0.0006
	Clay	Clay	Medium clay	
				0.0002
			Fine clay	

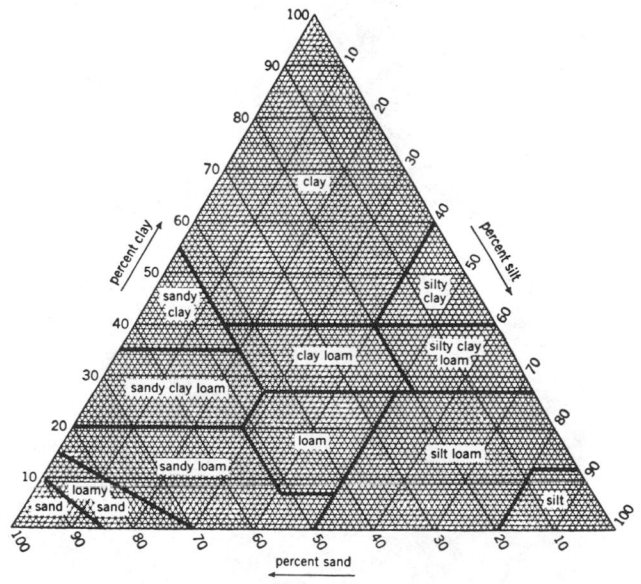

FIGURE 6.   Triangular graph showing the proportions of clay (below 0.002 mm), silt (0.05 to 0.002 mm), and sand (2.0 to 0.05 mm) in the basic classes of soil texture.

Table 18
PERCENTAGE COMPOSITION OF SOILS FROM THE NORTHEASTERN, SOUTHEASTERN, SOUTHWESTERN AND MIDWESTERN U.S. AND PUERTO RICO

State / Soil series name[a]	N.H. Gloucester sl		Pa. Penn sl		Pa. Hagerstown l		N.Y. Volusia sl		W. Va. Teas sl		Va. Iredell cl	N.C. White Store sl	N.C. Cecil sl		Puerto Rico Nipe cl	
Depth sampled (cm)	0—51	51—228	0—51	51—152	0—51	51—152	0—51	51—228	0—5	18—33	23—61	41—64	0—51	51—228	0—28	122-157
Constituent (%)																
SiO_2	65.68	73.80	74.33	71.76	70.99	66.49	75.12	74.64	69.2	68.3	50.30	51.10	88.57	55.69	23.7	10.0
Al_2O_3	14.15	13.24	11.00	14.36	11.39	14.80	10.49	12.26	9.6	15.5	11.50	17.35	5.74	24.42	25.3	25.6
Fe_2O_3	5.67	4.37	4.65	5.82	4.23	5.99	4.13	5.01	4.3	5.8	17.61	10.47	1.55	8.83	29.7	44.1
K_2O	2.16	2.22	1.57	1.50	2.71	3.58	1.40	1.99	1.2	2.3	0.21	1.42	0.82	1.06	0.1	<0.1
CaO	1.36	1.19	1.13	1.73	0.93	0.35	0.49	0.37	0.3	<0.1	1.17	0.35	0.39	0.40	0.1	<0.1
MgO	0.83	0.39	0.69	1.06	1.08	1.93	0.48	0.90	—	—	—	—	0.21	0.29	—	—
Na_2O	1.39	1.75	1.53	1.54	0.82	0.66	0.90	0.99	—	—	—	—	0.16	0.14	—	—
P_2O_5	0.15	0.11	0.16	0.10	0.19	0.16	0.18	0.15	—	—	—	—	0.08	0.07	—	—
SO_3	0.17	0.03	0.15	0.10	0.39	0.14	0.09	0.10	—	—	—	—	0.04	0.09	—	—
Mean annual temp (°C)	8	8	12	12	12	12	8	8	11	11	13	18	12	12	24	24
Mean annual precipitation (cm)	76	76	101	101	101	101	86	86	129	129	112	132	130	130	192	192

State / Soil series name[a]	Ala. Decatur cl		Miss. Shubuta sl		Texas Beaumont cl	Texas Houston Black cl	Texas Nacogdoches sc	Ind. Alford sl		Mo. Marshall sl		Minn. Clarion sl		Wis. Carrington l
Depth sampled (cm)	0—25	25—97	0-13	18-36	0-23	0-41	18-74	0-15	33-56	0-97	97-228	0-6	23-38	0-71
Constituent (%)														
SiO_2	72.35	74.81	86.2	69.3	65.50	48.00	59.80	80.8	73.4	73.61	71.43	71.3	74.5	73.50
Al_2O_3	8.89	12.80	4.8	15.3	11.37	6.68	12.55	8.0	11.7	9.67	13.44	9.5	10.9	9.10
Fe_2O_3	4.44	5.28	1.8	7.4	5.99	3.20	21.39	2.5	5.2	3.54	4.28	2.6	3.5	4.30
K_2O	0.67	0.75	1.2	1.5	0.37	0.37	0.46	2.0	2.2	2.28	1.40	1.6	1.7	2.03
CaO	0.63	0.40	0.3	<0.1	1.10	1.75	0.30	0.5	0.5	1.08	1.40	1.5	1.0	0.94
MgO	0.39	0.33	—	—	—	—	—	—	—	0.77	1.28	—	—	0.71
Na_2O	0.24	0.16	—	—	—	—	—	—	—	1.03	0.63	—	—	1.67
P_2O_5	0.18	0.15	—	—	—	—	—	—	—	0.22	0.16	—	—	0.24
SO_3	0.13	0.19	—	—	—	—	—	—	—	0.17	0.14	—	—	0.13
Mean annual temp (°C)	16	16	17	17	17	18	17	13	13	12	12	8	8	8
Mean annual precipitation (cm)	129	129	140	140	127	97	115	106	106	102	102	85	85	89

[a] Robinson, W. O. and Holmes, R. S., The chemical composition of soil colloids, Bull. 1311, U. S. Department of Agriculture, Washington, D.C., 1924.

[b] Soil Survey Staff, Soil Classification, a Comprehensive System, 7th Approximation, U.S. Department of Agriculture, Washington, D.C., 1960.

[c] Barnhisel, R. I., Ed., Analyses of clay, silt and sand fractions of selected soils from southeastern United States, *Southern Coop. Ser. Bull.*, 1978, 219.

[d] sl - silt loam; l - loam; cl - clay loam; sc - sandy clay.

Table 19
COMPOSITION OF CLAYS IN SOILS FROM VARYING CLIMATES, ARRANGED WITH INCREASING SiO₂/R₂O₃ RATIO

Soil type	Location	Mean ann. rainfall (cm)	Ann. mean temp. (°C)	SiO_2	Al_2O_3	Fe_2O_3 (%)	CaO	MgO	K_2O	$\dfrac{SiO_2}{R_2O_3}$
Cecil (subsoil)	Ga.	124	17	31.84	38.28	10.04	0.35	0.23	0.46	1.20
Cecil (surface)	Ga.	124	17	33.95	36.06	11.02	0.31	0.40	0.56	1.34
Vega Baja (surface)	P.R.	175	25	36.26	32.85	12.44	0.44	0.18	0.36	1.34
Susquehanna (surface)	Miss.	147	18	36.97	36.91	10.44	0.20	0.42	2.12	1.35
Chester (surface)	Va.	112	13	34.82	27.88	15.93	0.17	2.86	0.85	1.54
Norfolk (surface)	N.C.	127	11	38.25	31.21	11.25	0.54	0.53	0.26	1.66
Orangeburg (subsoil)	Miss.	134	17	40.35	33.27	10.08	0.45	0.67	0.81	1.71
Manor (surface)	Md.	101	12	38.86	31.11	10.33	0.64	1.23	1.37	1.74
Chester (surface)	Md.	109	12	39.00	29.98	11.27	0.93	0.22	1.24	1.77
Chester (subsoil)	Md.	109	12	40.03	30.58	11.34	0.36	0.78	1.14	1.79
Manor (subsoil)	Md.	101	12	40.36	31.20	10.29	0.38	1.35	1.28	1.81
Orangeburg (surface)	Miss.	134	17	40.35	31.04	10.11	0.51	0.73	0.81	1.83
Norfolk (subsoil)	N.C.	127	16	41.60	31.10	11.28	0.34	0.49	0.46	1.84
Sassafras (surface)	Md.	104	13	39.24	28.64	10.19	0.75	1.16	1.17	1.85
Huntington (surface)	Md.	89	12	37.79	27.16	11.28	0.70	1.30	2.67	1.86
Huntington (subsoil)	Md.	89	12	39.54	26.77	13.33	0.54	1.32	2.67	1.89
Sassafras (subsoil)	Md.	104	13	41.14	29.26	12.73	0.53	1.07	1.35	1.89
Hagerstown (subsoil)	Md.	101	12	41.63	29.86	11.31	0.74	1.42	1.84	1.89
Hagerstown (surface)	Md.	101	12	39.93	29.02	9.45	1.25	1.53	1.61	1.91
Iredell (subsoil)	N.C.	132	18	42.58	34.26	4.66	0.61	1.43	0.10	1.93
Dunkirk (surface)	N.Y.	101	8	43.02	27.92	13.16	0.63	2.32	3.71	1.99
Ontario (subsoil)	N.Y.	86	8	42.40	24.71	15.27	1.18	2.59	2.39	2.08
Susquehanna (subsoil)	Md.	104	12	43.71	28.44	10.72	0.24	2.56	0.86	2.10
Clarksville (subsoil)	Ky.	124	14	43.68	27.70	11.51	0.48	1.39	1.65	2.10
Ontario (surface)	N.Y.	86	8	38.01	23.30	10.84	1.37	2.34	2.23	2.13
Clarksville (surface)	Ky.	124	14	42.40	26.49	10.05	0.62	1.22	1.86	2.18
Crowley (surface)	La.	140	20	46.95	27.35	8.77	0.99	1.47	0.85	2.40
Miami (surface)	Ind.	94	10	46.37	23.96	11.47	0.92	2.15	2.47	2.50
Carrington (subsoil)	Iowa	81	8	48.04	25.19	8.80	1.29	1.53	0.89	2.64

Table 19 (continued)
COMPOSITION OF CLAYS IN SOILS FROM VARYING CLIMATES, ARRANGED WITH INCREASING SiO$_2$/R$_2$O$_3$ RATIO

Soil type	Location	Mean ann. rainfall (cm)	Ann. mean temp. (°C)	SiO$_2$	Al$_2$O$_3$	Fe$_2$O$_3$ (%)	CaO	MgO	K$_2$O	$\frac{SiO_2}{R_2O_3}$
Miami (subsoil)	Ind.	94	10	48.48	24.12	10.44	1.39	2.64	2.88	2.72
Stockton (surface)	Calif.	36	16	48.95	21.54	13.79	1.47	2.16	0.28	2.72
Carrington (surface)	Iowa	81	8	44.89	22.59	7.75	1.48	1.44	1.36	2.75
Marshall (surface)	Neb.	76	10	45.93	21.72	8.94	1.19	2.01	2.28	2.82
Stockton (subsoil)	Calif.	71	17	48.61	22.23	10.13	1.70	3.37	0.78	2.85
Stockton (surface)	Calif.	71	17	49.50	22.51	10.57	1.96	2.68	0.26	2.87
Marshall (subsoil)	Neb.	76	10	48.18	22.00	10.03	1.36	1.62	2.07	2.87
Sharkey (surface)	Miss.	137	19	51.34	22.61	8.79	1.41	2.48	—	3.07
Wabash (surface)	Neb.	86	11	51.28	20.28	9.75	2.19	2.09	1.91	3.16
Sharkey (subsoil)	Miss.	137	19	50.63	20.74	9.07	1.92	1.80	1.96	3.23
Lufkin (subsoil)	Miss.	132	18	55.44	22.66	8.98	1.85	2.03	0.65	3.30
Wabash (subsoil)	Neb.	86	11	50.55	20.87	9.57	1.59	2.21	1.97	3.33
Yolo (surface)	Calif.	51	14	46.25	16.42	10.00	1.46	5.64	2.54	3.43
Houston Black (surface)	Tex.	97	18	53.34	20.77	7.15	5.01	1.39	—	3.56
Fallon (surface)	Nev.	13	10	52.57	16.49	10.70	2.20	5.67	1.97	3.82

[a]   SiO$_2$/R$_2$O$_3$ (R$_2$O$_3$ = Al$_2$O$_3$ + Fe$_2$O$_3$).

From Robinson, W. O. and Holmes, R. S., The chemical composition of soil colloids, Bull. 1311, U.S. Department of Agriculture, Washington, D.C. 1924.

Table 20

CATION-EXCHANGE CAPACITIES AND EXCHANGEABLE CATIONS IN SOILS FROM VARIOUS CLIMATES

Soil Names	Location	Depth of sample (cm)	Mean ann. temp (°C)	Mean ann. precip. (cm)	Soil pH	Exchangeable cations (meq/100 g)					CEC meq/100g	% Base saturation
						Ca	Mg	K	Na	H*		
—	Alaska	0—8	−10	11	5.4	4.5	4.6	0.2	—	22.0	31.3	30
		28—30			5.4	11.9	10.8	1.3	—	29.7	53.9	45
Tanana	Alaska	0—8	−2	34	5.7	24.3	11.8	0.5	0.3	27.2	64.1	58
		28—53			7.7	13.6	5.6	0.1	0.3	2.8	22.4	88
Kalifonsky	Alaska	0—15	2	43	4.1	1.2	0.6	0.1	0.1	30.3	32.3	6
		15—51			5.2	0.9	<0.1	0.1	0.1	18.6	19.7	6
Exline	N.D.	0—15	6	45	7.2	13.7	4.8	1.0	0.6	0	19.3	100 +
		43—61			9.4	17.2	12.9	0.5	6.0	0	14.7	100 +
Eakin	S. D.	0—9	8	46	6.8	20.0	5.8	1.7	—	5.9	27.5	—
		20—36			6.9	18.3	6.8	0.7	0.1	3.0	24.6	—
Marlin	N.Y.	1—2	9	89	4.0	1.8	0.6	0.2	0.1	21.7	24.4	11
		5—41			4.4	0.6	0.2	0.1	0.1	15.0	16.0	6
Scituate	N.H.	0—23	8	104	5.4	3.7	0.3	0.2	0.1	12.9	17.2	25
		33—41			5.7	0.4	0.2	0.1	0.1	4.8	5.6	14
Quillayute	Wash.	0—4	9	260	4.9	0.6	0.1	0.4	0.1	15.0	16.2	7
		11—33			5.2	0.4	0.1	0.2	0.1	24.5	25.3	3
Klaus	Wash.	0—20	9	294	5.3	3.0	2.1	0	0.3	52.5	80.8	9
		58—81			5.5	0.4	0.4	0	0.2	41.4	42.1	2
Fillmore	Neb.	0—13	11	58	5.6	12.0	3.9	1.9	0.3	5.1	23.2	78
		25—43			6.6	8.2	2.4	1.5	0.3	1.2	13.2	91
Sharpsburg	Neb.	0—15	11	71	6.1	15.9	6.0	1.0	—	4.1	27.0	85
		25—56			6.5	16.9	7.6	0.5	0.1	1.1	26.2	96
Shelby	Iowa	0—18	10	84	5.6	13.1	3.4	0.4	0.1	9.1	26.1	65
		28—43			5.4	14.1	4.0	0.3	0.1	7.0	25.5	73
Edina	Iowa	0—13	11	85	5.9	13.4	2.9	0.3	0.1	4.0	20.6	81
		51—61			5.6	22.0	12.1	0.3	1.7	7.4	43.5	83
—	Ohio	0—8	11	96	5.4	8.6	4.3	0.6	0	24.3	37.8	36
		28—43			4.6	2.0	8.2	0.4	0.1	21.8	32.6	33
Alford	Ind.	0—15	13	106	6.7	6.7	1.6	0.4	—	4.4	13.1	66
		56—82			5.6	5.4	4.3	0.4	0.1	8.2	18.4	55

Table 20 (continued)
CATION-EXCHANGE CAPACITIES AND EXCHANGEABLE CATIONS IN SOILS FROM VARIOUS CLIMATES

Soil Names	Location	Depth of sample (cm)	Mean ann. temp (°C)	Mean ann. precip. (cm)	Soil pH	Exchangeable cations (meq/100 g)					CEC meq/100g	% Base saturation
						Ca	Mg	K	Na	H*		
Wehadkee	Va.	0—56	14	114	5.3	4.2	2.4	0.1	—	9.8	16.6	41
		56—91			5.9	6.8	6.0	0.3	—	5.7	18.8	69
Teas	W. Va.	0—5	11	129	5.1	6.5	1.6	0.6	0.1	14.7	23.4	37
		33—58			4.7	0.7	1.4	0.2	<0.1	10.0	12.3	19
Mohave	Ariz.	0—10	21	25	7.2	7.7	8.0	0.9	0.1	—	8.5	100+
Redding	Calif.	0—20	17	59	5.7	3.2	1.3	0.4	—	2.6	7.5	65
		50—56			5.3	12.6	13.4	0.3	0.3	6.2	32.8	81
Windthorst	Texas	0—8	18	83	5.6	1.9	0.6	0.4	—	2.6	5.6	54
		58—84			5.0	9.1	4.2	0.5	—	7.3	21.1	65
—	Texas	0—15	18	88	7.8	59.4	3.2	1.3	1.1	—	49.2	100+
					8.2	50.9	4.0	0.9	6.7	—	46.3	100+
Waimea	Hawaii	0—5	18	104	6.2	34.4	6.6	5.5	0.4	33.9	80.8	58
		13—20			7.6	37.7	5.7	6.3	0.3	24.2	74.1	67
Leon	Ga.	0—8	21	121	4.8	1.5	0	<0.1	<0.1	4.3	5.8	26
		43—46			4.7	0.2	0.2	<0.1	<0.1	27.1	27.5	1
Shubuta	Miss.	0—13	17	140	5.5	4.6	0.8	0.3	0.1	6.1	11.9	49
		35—48			4.4	0.6	3.4	0.3	<0.1	19.7	24.0	18
Bladen	N.C.	0—15	18	142	3.9	0.4	0.4	0.3	0.2	24.9	26.2	5
		56—81			4.2	<0.1	<0.1	0.1	<0.1	12.0	12.1	1
Victoria	Texas	0—3	22	71	7.9	43.6	4.9	1.8	1.2	—	37.1	100+
		30—48			7.9	40.1	6.4	1.2	4.4	—	37.8	100+
Lakewood	Fla.	0—3	22	136	5.4	1.5	0.1	<0.1	<0.1	2.5	4.1	39
		3—30			6.4	0.3	0	<0.1	0	1.3	1.6	19
Nipe	P.R.	0—28	24	192	5.1	1.4	1.5	0.1	0.1	33.5	36.6	8
		28—46			5.0	0.1	—	—	—	21.4	21.5	<1
Hilo	Hawaii	0—41	23	356	5.8	2.0	1.8	0.1	<0.1	63.6	67.6	6
		41—53			6.1	2.2	0.6	0.1	<0.1	65.5	68.4	4
Los Guineos	P.R.	0—10	21	396	4.8	0.5	0.8	0.1	0.2	28.8	30.4	5
		18—36			5.1	0.3	0.5	<0.1	0.2	15.5	16.6	7

From Soil Survey Staff, Soil Classification, a Comprehensive System, 7th Approximation, U.S. Department of Agriculture, Washington, D.C., 1960.

ACIDITY AND CROP RESPONSE

Table 21
REACTIONS LEADING TO ACID SOILS

$$Ca\text{-Soil} + 2HCO_3^- + 2H^+ \longrightarrow H_2\text{-Soil} + Ca(HCO_3)_2$$

$$3/2\ H_2\text{-Soil} \longrightarrow Al^a - Soil + 3H_2O$$

and

$$Al^{3+} - Soil \rightleftarrows Al(OH)_2 - Soil$$
$$H_2$$

[a]　Al^{3+} ions are dissolved from the clay structure and become exchangeable.

$$(Al(OH)_{3\ lattice} + 3H^+ \longrightarrow Al^{3+} + 3\ H_2O)$$

Table 22
CATIONS PRESENT IN ACID SOILS

H +	Present in small quantities in most acid soils unless excess mineral acids are present
Al^{3+}	Present as a major cation in strongly acid soils (pH 4 to 4.5); most is derived from dissolution of the clay lattice
Mg^{2+}	Present in amounts larger than Ca^{2+} in many acid soils due to release from clay structure as soil becomes acid
Mn^{2+}	Present in small but toxic quantities in soils where there is an appreciable quantity present; not important in most sandy, acid soils
$[Al(OH)_2^{+\ to\ 0.5\ +}_{2\ to\ 2.5}]_n$	A polycation with variable charge and size. Although not exchangeable by neutral salt solutions, it neutralizes the negative charge on soils (blocks CEC); charge on cation will be changed with either OH^- or H^+ titration; this polycation is the source of most "pH dependent" charge on clays
$[Fe(OH)^{0.1\ to\ 0}_{2.9\ -\ 3.0}]_n$	A similar polycation with much less residual charge and probably much larger value of n; less important than the Al-analogue per equivalent of metal

Table 23
pH VALUES OF SPECIAL SIGNIFICANCE IN SOILS

pH[a]	Soil conditions inferred from pH
< 4	Free mineral acids present; most often H_2SO_4 from oxidation of pyrite
4	pH of soil largely saturated with exchangeable Al^{3+} and moderately low to very low in exchangeable divalent cations
5.5	pH above which there is essentially no exchangeable Al^{3+}
6.0	Soil pH suitable for a wide variety of plants
7.0	Neutral pH; has no special significance in soils
8.3	Soil in equilibrium with an excess of $CaCO_3$ at the partial pressure of CO_2 in the atmosphere (0.0003 atm)
> 8.3	Na_2CO_3 present in soil

[a]　pH in 1:1 soil:water suspension.

Table 24
OPTIMUM pH RANGES OF SELECTED PLANTS

Field Crops		Rhododendron	4.5—6.0
Alfalfa	6.2—7.8	Rose, hybrid tea	5.5—7.0
Barley	6.5—7.8	Snapdragon	6.0—7.5
Bean, field	6.0—7.5	Snowball	6.5—7.5
Beets, sugar	6.5—8.0	Sweet William	6.0—7.5
Bluegrass, Ky.	5.5—7.5	Zinnia	5.5—7.5
Clover, red	6.0—7.5	**Forest Plants**	
Clover, sweet	6.5—7.5	Ash, white	6.0—7.5
Clover, white	5.6—7.0	Aspen, American	3.8—5.5
Corn	5.5—7.5	Beech	5.0—6.7
Flax	5.0—7.0	Birch, European (white)	4.5—6.0
Oats	5.0—7.5	Cedar, white	4.5—5.0
Pea, field	6.0—7.5	Club moss	4.5—5.0
Peanut	5.3—6.6	Fir, balsam	5.0—6.0
Rice	5.0—6.5	Fir, Douglas	6.0—7.0
Rye	5.0—7.0	Heather	4.5—6.0
Sorghum	5.5—7.5	Hemlock	5.0—6.0
Soybean	6.0—7.0	Larch, European	5.0—6.5
Sugar cane	6.0—8.0	Maple, sugar	6.0—7.5
Tobacco	5.5—7.5	Moss, sphagnum	3.5—5.0
Wheat	5.5—7.5	Oak, black	6.0—7.0
Vegetable Crops		Oak, pin	5.0—6.5
Asparagus	6.0—8.0	Oak, white	5.0—6.5
Beets, table	6.0—7.5	Pine, jack	4.5—5.0
Broccoli	6.0—7.0	Pine, loblolly	5.0—6.0
Cabbage	6.0—7.5	Pine, red	5.0—6.0
Carrot	5.5—7.0	Pine, white	4.5—6.0
Cauliflower	5.5—7.5	Spruce, black	4.0—5.0
Celery	5.8—7.0	Spruce, Colorado	6.0—7.0
Cucumber	5.5—7.0	Spruce, white	5.0—6.0
Lettuce	6.0—7.0	Sycamore	6.0—7.5
Muskmelon	6.0—7.0	Tamarack	5.0—6.5
Onion	5.8—7.0	Walnut, black	6.0—8.0
Potato	4.8—6.5	Yew, Japanese	6.0—7.0
Rhubarb	5.5—7.0	**Weeds**	
Spinach	6.0—7.5	Dandelion	5.5—7.0
Tomato	5.5—7.5	Dodder	5.5—7.0
Flowers and Shrubs		Foxtail	6.0—7.5
African violet	6.0—7.0	Goldenrod	5.0—7.5
Almond, flowering	6.0—7.0	Grass, crab	6.0—7.0
Alyssum	6.0—7.5	Grass, quack	5.5—6.5
Azalea	4.5—5.0	Horse tail	4.5—6.0
Barberry, Japanese	6.0—7.5	Milkweed	4.0—5.0
Begonia	5.5—7.0	Mustard, wild	6.0—8.0
Burning bush	5.5—7.5	Thistle, Canada	5.0—7.5
Calendula	5.5—7.0	**Fruits**	
Carnation	6.0—7.5	Apple	5.0—6.5
Chrysanthemum	6.0—7.5	Apricot	6.0—7.0
Gardenia	5.0—6.0	Arbor Vitae	6.0—7.5
Geranium	6.0—8.0	Blueberry, high bush	4.0—5.0
Holly, American	5.0—6.0	Cherry, sour	6.0—7.0
Ivy, Boston	6.0—8.0	Cherry, sweet	6.0—7.5
Lilac	6.0—7.5	Crab apple	6.0—7.5
Lily, Easter	6.0—7.0	Cranberry, large	4.2—5.0
Magnolia	5.0—6.0	Peach	6.0—7.5
Orchid	4.0—5.0	Pineapple	5.0—6.0
Phlox	5.0—6.0	Raspberry, red	5.5—7.0
Poinsettia	6.0—7.0	Strawberry	5.0—6.5
Quince, flowering	6.0—7.0		

Source: Spurway, C. H., Soil Reaction Preferences of Plants, *Mich. Agric. Exp. Stn. Spec. Bull.*, 306, 1941.

Table 25
pH VALUES ABOVE WHICH NO
LIME RESPONSE OCCURS

Crop	Highest reported pH where lime response occurred
Alfalfa	6.5
Banana	>4.6
Barley	6.0
Bermudagrass	5.3
Clover	6.5
Corn	5.7
Cotton	5.8
Fescue	5.2
Lespedeza	6.0
Oats	<5.0
Peanuts	5.7
Pineapple	<4.7
Potato	5.0
Sorghum	5.7
Soybeans	6.0
Sugarcane	5.5
Sweet Potato	5.5
Vetch	5.8
Tropical Grasses	4.7
Coffee	<4.2

Adapted from Adams, F. and Pearson, R. W., Crop responses to lime in the southern United States and Puerto Rico, in *Soil Acidity and Liming,* Pearson, R. W. and Adams, F., Eds., *Am. Soc. Agron. Monogr.,* No. 12, 1967.

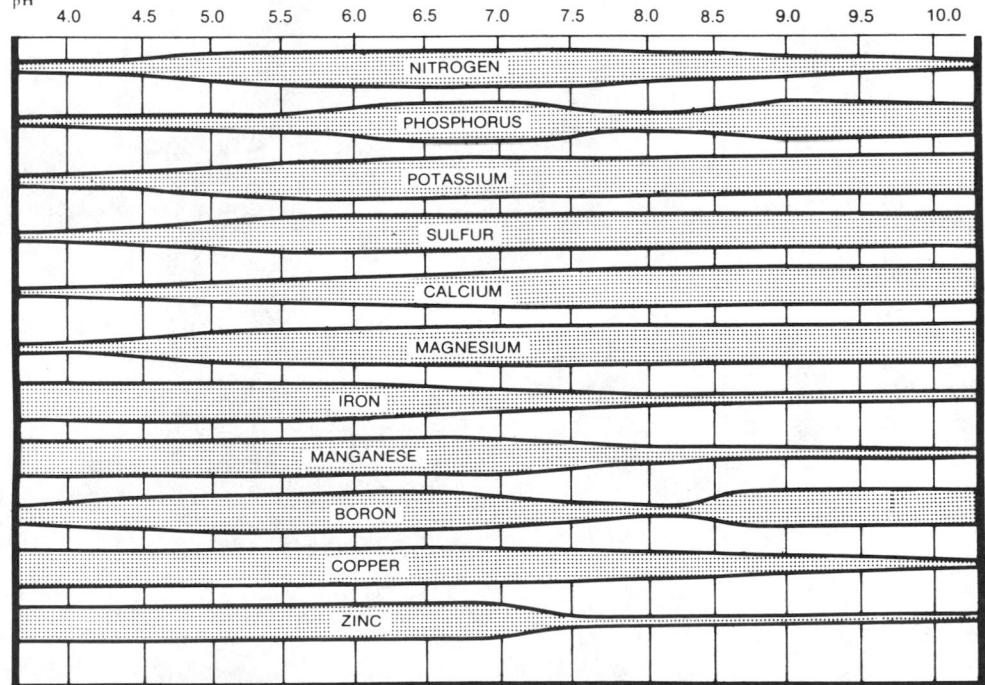

FIGURE 7.   The general relation of pH to the availability of plant nutrients in the soil; the width of the bar indicates relative availability.

Table 26
SUMMARY OF VARIOUS PHENOMENA RELATED TO CROP RESPONSES TO LIMING IN THE ACID SOIL CONTINUUM

Liming effect	Mollisols	Alfisols	Ultisols	Oxisols
Optimum pH range[a] Crop response — rel. mag.[b]	6.2—6.8 Fair to good	6.0—6.6 Excellent	5.6—6.2 Excellent	5.0—5.6 Good to excellent
Cause of favorable lime response[c]				
Chemical	pH adjustment Mobilized Ca N fixation and metabolism	pH adjustment Al inactivation Mobilized Ca N fixation and metabolism	Al (and Mn) inactivation Ca addition	Al, Mn inactivation Ca addition Slowed weathering
Physical	Improved aggreg. (Flocculation and OM coatings)	Slightly improved aggreg.	None when limed @ opt. pH	None when limed @ opt. pH
Adverse effects of over-liming[c]				
Chemical	Hardly any	Zn or Mn def. Mo toxicity	Zn, Mn (Cu, B) def. Mo toxicity Excess Ca Dispersed aggreg.	Zn, Mn, Cu, B def. Mo toxicity Excess Ca Dispersed aggreg.
Physical	Hardly any	None	Dispersed aggreg.	Dispersed aggreg.

[a] Optimum pH naturally varies with crop and even with strains of the same crop, but these pH ranges seem to approximate those where most of the crops grown on these soils do best.

[b] Assuming pH is approximately 1 unit below optimum.

[c] Arranged in decreasing order in which it most likely occurs.

From McLean, E. O., Potentially beneficial effects from liming; chemical and physical, *Soil Crop Sci. Soc. Fla.*, 31, 189, 1971. With permission.

Table 27
TONS OF LIMESTONE REQUIRED TO RAISE THE pH OF A 6⅔-IN. PLOW LAYER OF DIFFERENT TEXTURAL GROUPS TO pH 6.5

Texture of plow layer	Tons of lime recommended[a]			
	4.5—4.9 pH	5.0—5.4 pH	5.5—5.9 pH	6.0—6.4 pH
Clay and silty clay	6	5	4	2.5
Clay loams and loams	5	4	3	2
Sandy loams	4	3	2.5	1.5
Loamy sands	3	2.5	2	1
Sands	2.5	2	1.5	0.5

[a] Material of which 25% passes a 100-mesh sieve and has a neutralizing value of 90%.

Source: Christenson, D. R. and Doll, E. C., Lime for Michigan Soils, *Mich. State Coop. Ext. Serv. Ext. Bull.*, No. 471, 1973.

Table 28
CHEMICAL COMPOSITION OF 194 SAMPLES OF AGRICULTURAL LIMESTONE

Element	Range (%)	Average (%)
Ca	17.98—39.76	30.21
Mg	0.04—12.86	4.92
Si	0.03—15.53	2.36
Al	0.01—2.15	0.45
Fe	0.01—3.11	0.43
K	<0.0001—1.80	0.23
S	<0.01—1.35	0.11
Na	<0.001—0.15	0.03
	ppm	ppm
Mn	20—3000	330
F	<10—1410	230
P	10—5660	210
Zn	<1—425	31
V	<1—106	11
B	<1—21	4
Cu	<0.3—89	2.7
Mo	<0.1—92	1.1
Co	<1—6	1

Reproduced from Chichilo, P. and Whittaker, C. W., Trace elements in agricultural limestones of the United States, *Agron. J.*, 53, 139, 1961, by permission of the American Society of Agronomy, Inc.

Table 29
ANALYSES OF VARIOUS SOURCES OF LIMESTONE

Constituent	Number of samples	Range (%)	Mean (%)
P_2O_5	593	0.0—2.59	0.11
K_2O	511	tr—1.07	0.31
Ca	1892	21.53—39.83	31.70
Mg	1569	tr—14.33	3.43
C	843	8.79—12.98	11.15
S	98	0.0—1.17	0.04
Si	918	0.0—14.88	2.63
		ppm	ppm
Cu	17	4—88	42
Mn	430	tr—7900	4900
Zn	25	1—460	150
B	3	0—160	30

From *Dictionary of Plant Foods,* Meister, Willoughby, Ohio 1965, C-179. With permission.

Table 30
RANGE IN CHEMICAL COMPOSITION OF SLAGS

Element	Blast furnace (%)	Open hearth (%)	Basic (%)
Ca	26—32	16	27—42
Mg	2—7	5	1—5
Si	15—20	8	1.4—6
Al	5.4—8.5	1.5	1—2
Fe	0.2—2	24	4—17
Mn	0.2—1.2	10	0.6—4
S	1—3	0.1	0.2—11
P	<1	1	5—10

Reproduced from Barber, S. A., Liming materials and practices, in *Soil Acidity and Liming*, Pearson, R. W. and Adams, F., Eds., American Society of Agronomy, Madison, Wis., 1967, chap. 3, by permission of the American Society of Agronomy, Inc.

Table 31
ANALYSES OF 27 MARL SAMPLES SUBMITTED TO THE MICHIGAN STATE UNIVERSITY SOIL TESTING LABORATORY

Analyses	Average	Range
MgCO₃ (%)	3.06	1.40—5.18
Neutralizing value[a] (%)	84.2	32.2—102.9
Calcium carbonate equivalent (lb/yd³)[b]	1075	175—1811

[a] Neutralizing value expressed in terms of calcium carbonate.
[b] Expressed on a volume basis.

Table 32
ACIDIFYING EFFECTS OF VARIOUS FERTILIZERS

Fertilizer material	Pure lime required to neutralize acidifying effect (lb lime/1b N)
Nitrogen Sources	
Anhydrous ammonia (83% N)	1.8
Aqua ammonia (21% N)	1.8
Ammonium nitrate (33% N)	1.8
Mono ammonium phosphate (11-48-0)	5.9
Ammonium sulfate (21% N)	5.5
Diammonium phosphate (18-46-0)	3.6
Urea (45% N)	1.9
Nitrogen solutions (28—32% N)	1.8
Urea — form (38% N)	1.8
Phosphorus Sources	
None except ammoniated phosphates are acidifying	
Potassium Sources	
None are acidifying	

From *Dictionary of Plant Foods*, Meister, Willoughby, Ohio 1965, C-179. With permission.

Table 33
ALFALFA YIELD AND % STAND VS. LIME RATE, pH, AND EXCHANGEABLE Al

Lime rate mt/ha	Soil pH	Exch. Al^{3+} meq/100 g	3-Year alfalfa yield mt/ha	% Alfalfa stand after 3 years
0	4.9	0.74	0.58	0
0.56	5.1	0.42	4.30	0
1.12	5.3	0.40	8.94	20
2.24	5.7	0.14	15.55	80
4.48	5.7	0.12	17.18	90
8.96	6.2	0.10	17.49	90
17.92	6.5	0.03	17.47	99
35.84	6.8	0.02	17.92	99

From Moschler, W. W., Jones, G. D., and Thomas, G. W., Lime and soil acidity effects on alfalfa growth in a red-yellow podzolic soil, *Proc. Soil Sci. Soc. Am.,* 24, 507, 1960. With permission.

Table 34
EFFECT OF SOIL ACIDITY ON CORN GRAIN YIELD AND LEAF COMPOSITION

Soil pH	Al saturation (%)	Corn yield (ton/ha)	Composition of corn leaves				
			Ca (%)	Mg (%)	P (%)	K (%)	Mn (ppm)
Humatas clay							
3.9	66	1.15	.36	.11	.34	1.78	—
4.2	44	2.80	.42	.11	.35	1.85	—
4.5	36	4.09	.49	.12	.37	1.89	20
4.7	17	4.42	.51	.10	.36	1.95	—
5.3	3	5.34	.57	.11	.36	2.00	—
Corozal Clay							
4.3	68	0.42	.33	.10	.42	1.94	—
4.6	39	2.23	.41	.09	.49	1.66	—
4.7	18	2.91	.41	.09	.48	1.64	20
5.4	2	4.17	.48	.10	.46	1.37	—
Corozal Clay (level phase)							
4.6	35	2.35	.28	.07	.47	2.16	97
4.7	14	3.14	.28	.12	.45	2.13	98
5.2	5	3.79	.31	.10	.47	2.03	92
Corozal Clay (eroded phase)							
4.2	63	0.65	.33	.12	.41	1.85	—
4.3	52	0.95	.33	.11	.37	1.75	—
4.6	33	2.25	.40	.10	.43	1.75	20
4.8	15	2.55	.41	.10	.43	1.68	—
5.4	4	3.57	.47	.10	.43	1.35	—
Los Guineos Clay							
4.3	47	1.77	.27	.12	.43	2.87	40
4.7	27	2.65	.28	.12	.46	2.77	35
5.3	—	3.73	.30	.11	.44	2.68	31

Table 34 (continued)
EFFECT OF SOIL ACIDITY ON CORN GRAIN YIELD AND LEAF COMPOSITION

Soil pH	Al saturation (%)	Corn yield (ton/ha)	Composition of corn leaves				
			Ca (%)	Mg (%)	P (%)	K (%)	Mn (ppm)
			Coto Clay				
4.2	33	4.05	.39	.10	.19	1.89	114
4.6	19	4.15	.40	.09	.21	1.84	75
4.9	13	4.50	.39	.09	.21	1.83	85
5.3	5	4.79	.39	.09	.21	1.73	69
			Piñas Sand				
4.6	40	1.71	.41	.15	.42	1.96	121
4.7	36	2.76	.48	.14	.45	2.06	121
4.8	28	2.95	.55	.15	.42	1.90	103
5.5	—	2.79	.58	.16	.44	2.00	80
			Catalina Clay				
4.8	18	4.40	.29	.09	.42	2.12	104
5.2	—	4.95	.30	.08	.41	2.07	66
5.3	—	5.15	.33	.08	.42	2.14	86
5.6	—	4.58	.33	.10	.43	2.23	74

Source: Abruña, F., Pearson, R. W., and Perez-Escolar, R., in *Soil Management in Tropical America,* Bornemisza, E. and Alvarado, A., Eds., North Carolina State University, Raleigh, N.C., 1975, 269.

Table 35

EFFECT OF SOIL ACIDITY ON BEAN YIELD AND LEAF COMPOSITION

Soil pH	Al saturation (%)	Bean yield (ton/ha)	Composition of corn leaves				
			Ca (%)	Mg (%)	P (%)	N (%)	Mn (%)
Humatas Clay							
3.9	66	1.81	1.10	.23	.49	6.55	210
4.2	44	3.42	1.26	.21	.36	5.95	190
4.5	36	4.87	1.49	.21	.44	5.94	160
4.7	17	5.61	1.56	.25	.34	5.76	160
5.3	3	6.93	1.83	.27	.36	5.75	110
Corozal Clay							
4.3	68	4.31	.95	.27	.32	6.70	184
4.6	39	8.13	1.45	.24	.31	6.32	169
4.7	18	8.89	1.57	.29	.30	6.11	152
5.4	2	11.17	2.18	.34	.35	5.87	121
Corozal Clay (level phase)							
4.6	35	5.56	2.44	.25	.20	5.76	266
4.7	14	8.40	2.92	.25	.18	5.37	234
5.2	5	9.12	3.08	.24	.21	5.29	165
Corozal Clay (eroded phase)							
4.2	63	3.63	.88	.25	.30	6.78	157
4.3	52	5.54	1.07	.24	.30	6.25	150
4.6	33	8.76	1.34	.27	.30	6.28	132
4.8	15	9.83	1.61	.36	.32	6.14	123
5.4	4	11.70	2.00	.33	.31	5.97	122
Los Guineos Clay							
4.3	47	10.12	1.34	.46	.16	5.58	149
4.7	27	12.76	1.72	.40	.16	5.62	124
5.3	—	13.60	2.39	.31	.17	5.79	50
Coto Clay							
4.2	33	4.69	1.63	.29	.30	5.28	267
4.6	19	5.73	1.77	.33	.30	5.13	263
4.9	13	6.51	1.90	.30	.31	5.09	245
5.3	5	6.89	2.12	.30	.32	5.41	211

Source: Abruña, F., Pearson, R. W., and Perez-Escolar, R., in *Soil Management in Tropical America,* Bornemisza, E. and Alvarado, A., Eds., North Carolina State University, Raleigh, N.C., 1975, 277.

SALINITY AND CROP RESPONSE

Table 36

CHEMICAL CHARACTERISTICS OF SODIUM-AFFECTED SOILS

Soil	State	Depth (cm)	Soluble salts[a] (meq/100 g)								Exchangeable cations (meq/100 g)			% Na
			CO₃	HCO₃	Cl	SO₄	Ca	Mg	K	Na	Ca+Mg	K	Na	Na
Fresno sl[a]	Calif.	0—30	1.03	0.75	6.36	1.48	0.09	0.18	0.82	9.60	0	0.61	3.56	85
Jordan sl[a]	Utah	0—30	0.34	0.62	39.84	7.90	0.47	0.99	0.13	47.22	0	3.66	5.87	61
Lahcntan c[a]	Nev.	0—15	0.83	0.67	13.84	5.55	0.14	0.20	0.35	21.31	0	0.80	24.48	97
Hanford l[a]	Calif.	0—30	0.85	1.07	22.17	10.40	0.16	0.52	0.66	34.78	0	2.15	20.57	90
Chino l[b]	Calif.	0—30	0.55	1.22	0.10	0.17	0.06	0.14	0.10	1.76	15.88	3.63	6.43	25
Placentia[b]	Calif.	0—30	0	0.60	4.62	1.31	0.57	0.68	0.25	5.07	13.55	1.46	4.09	21
Payette[b]	Idaho	0—30	0	0.35	5.74	10.42	1.10	0.59	0.18	14.80	21.11	1.81	6.52	22
Imperial[c]	Calif.	0—30	—	0.20	77.40	10.40	24.40	12.60	2.00	48.30	15.80	0.70	1.10	6
		30—60	0	0.30	20.50	4.00	4.60	2.60	1.30	15.10	15.40	1.70	2.10	11

[a] Strongly Na-affected soils.

[b] Moderately Na-affected soils.

[c] Saline but not Na-affected.

Reproduced from Kelley, W. P., Alkali Soils, *Am. Soc. Agron. Monogr.*, No. 3, 1951, by permission of the American Society of Agronomy, Inc.

Table 37

RELATIONSHIP BETWEEN THE
ELECTRICAL RESISTANCE OF A
SATURATED PASTE AND PERCENT SALT
IN SOILS OF DIFFERENT TEXTURE

Resistance soil paste at 15—5°C (Ohms × cm)	Mixed, neutral salts in soil (%)			
	Sand	Loam	Clay loam	Clay
18	3.00	3.00	—	—
19	2.40	2.64	3.00	—
20	2.20	2.42	2.80	3.00
25	1.50	1.70	1.94	2.20
30	1.24	1.34	1.46	1.58
35	1.04	1.14	1.22	1.32
40	0.86	0.94	1.04	1.14
45	0.75	0.78	0.88	0.98
50	0.67	0.71	0.77	0.86
55	0.60	0.64	0.69	0.77
60	0.55	0.58	0.63	0.70
65	0.51	0.54	0.57	0.63
70	0.48	0.50	0.53	0.59
80	0.42	0.44	0.47	0.51
90	0.37	0.39	0.41	0.45
100	0.33	0.35	0.37	0.39
150	0.21	0.21	0.22	0.23

Source: U.S. Salinity Laboratory Staff, Diagnosis and Improvement of Saline and Alkali Soils, *U.S. Dep. Agric. Agric. Handb.*, No. 60, 1954.

Table 38

RELATION BETWEEN CONCENTRATION OF SALTS IN A
SATURATED SOIL EXTRACT, CONDUCTIVITY, AND
OSMOTIC PRESSURE OF THE SOLUTIONS

Concentration of salt (meq/ℓ)	Conductivity of saturated soil extract (EC × 10³)	Osmotic pressure of saturated soil extract (atm)
10	1	0.32
21	2	0.67
33	3	1.04
44	4	1.41
70	6	2.21
95	8	3.00
122	10	3.80
263	20	7.00
410	30	12.80
560	40	17.20

Source: U.S. Salinity Laboratory Staff, Diagnosis and Improvement of Saline and Alkali Soils, *U.S. Dep. Agric. Agric. Handb.*, No. 60, 1954.

Table 39

SALT TOLERANCE OF VEGETABLE AND CROP PLANTS

High salt tolerance	Medium salt tolerance	Low salt tolerance
$E \cdot C \times 10^3 = 12^a$	$E \cdot C \times 10^3 = 10$	$E \cdot C \times 10^3 = 4$
Table beet	Tomato	Radish
Kale	Broccoli	Celery
Asparagus	Cabbage	Green beans
Spinach	Bell pepper	
Barley	Cauliflower	$E \cdot C \times 10^3 = 3$
Sugar beet	Lettuce	
Rape	Sweet corn	$E \cdot C \times 10^3 = 4$
Cotton	Potato	
$E \cdot C \times 10^3 = 10$	Carrot	Field beans
	Onion	White clover
$E \cdot C \times 10^3 = 18$	Pea	Meadow foxtail
	Squash	Alsike clover
Bermudagrass	Cucumber	Red clover
Rhodesgrass		Ladino clover
Western wheatgrass	$E \cdot C \times 10^3 = 4$	
Birdsfoot trefoil		$E \cdot C \times 10^3 = 2$
	$E \cdot C \times 10^3 = 10$	
$E \cdot C \times 10^3 = 12$		
	Rye	
	Wheat	
	Oats	
	Rice	
	Sorghum	
	Corn	
	Flax	
	Sunflower	
	Castor bean	
	$E \cdot C \times 10^3 = 6$	
	$E \cdot C \times 10^3 = 12$	
	Sweet clover	
	Perennial ryegrass	
	Strawberry clover	
	Dallisgrass	
	Sudan grass	
	Alfalfa	
	Tall fescue	
	Orchardgrass	
	Meadow fescue	
	Reed canarygrass	
	Big trefoil	
	Smooth bromegrass	
	$E \cdot C \times 10^3 = 4$	

[a] $E \cdot C \times 10^3$ = mmhos/cm², a common measure of soil salinity. Numbers at top and bottom of each list represent the range within that list. Plants listed in order of decreasing salt tolerance.

Source: U.S. Salinity Laboratory Staff, Diagnosis and Improvement of Saline and Alkali Soils, *U.S. Dep. Agric. Agric. Handb.,* No. 60, 1954.

Table 40
THE RELATIVE PRODUCTIVITY OF SENSITIVE PLANTS WITH INCREASING SALT CONCENTRATION IN THE ROOT ZONE

Plant name	Scientific names	Relative productivity, % EC, mmhos/cm									% Productivity decrease per mmhos/cm increase	Salinity threshold (EC_t)
		1	2	3	4	5	6	7	8	9		
Algerian ivy	*Hedera canariensis*	100	81	62	35	0	0	0	0	0	—	1.0
Almond	*Prunus dulcis*	100	91	73	55	36	18	0	0	0	18	1.5
Apple[a]	*Malus sylvestris*	100	91	75	—	—	—	—	—	—	—	1.0
Apricot	*Prunus armeniaca*	100	91	68	45	23	0	0	0	—	23	1.6
Avocado	*Persea americana*	100	90	70	—	—	—	—	—	—	—	1.0
Bean	*Phaseolus vulgaris*	100	81	62	43	25	6	0	0	0	18.9	1.0
Blackberry	*Rubus* spp.	100	89	67	44	22	0	0	0	0	22.2	1.5
Boysenberry	*Rubus ursinus*	100	89	67	44	22	0	0	0	0	22.2	1.5
Burford holly	*Ilex cornuta*	100	82	59	36	14	0	0	0	0	—	1.0
Burnet	*Sanguisorba minor*	—	—	—	—	—	—	—	—	—	—	—
Carrot	*Daucus carota*	100	86	72	58	44	30	15	1	0	14.1	1.0
Celery[a]	*Apium graveolens*	100	90	75	—	—	—	—	—	—	—	1.0
Grapefruit	*Citrus paradisi*	100	97	81	65	48	32	16	0	0	16.1	1.8
Heavenly bamboo	*Nandina domestica*	100	88	75	61	47	34	20	7	0	—	1.0
Hibiscus	*Hibiscus rosa-sinensis* culti-var Brilliante	100	86	72	58	42	28	15	0	0	—	1.0
Lemon[a]	*Citrus limon*	100	91	75	—	—	—	—	—	—	—	1.0
Okra[a]	*Abelmoschus esculentus*	100	90	—	—	—	—	—	—	—	—	—
Onion	*Allium cepa*	100	87	71	55	39	23	6	0	0	16.1	1.2
Orange	*Citrus sinensis*	100	95	79	63	48	32	16	0	0	15.9	1.7
Peach	*Prunus persica*	100	94	73	52	31	10	0	0	0	18.8	3.2
Pear[a]	*Pyrus* spp.	100	91	75	—	—	—	—	—	—	—	1.0
Pineapple guava	*Feijoa sellowiana*	100	71	34	0	0	0	0	0	0	—	1.2
Plum	*Prunus domestica*	100	91	73	55	36	18	0	0	0	18.2	1.5
Prune[a]	*Prunus domestica*	100	91	75	—	—	—	—	—	—	—	1.0
Pittosporum[b]	*Pittosporum tobira*	100	89	79	69	60	50	40	30	20	—	1.0
Raspberry[a]	*Rubus idaeus*	100	80	62	—	—	—	—	—	—	—	1.0

Rose	*Rosa* spp.	100	74	36	0	0	0	0	0	0	—	1.0
Strawberry	*Fragaria*	100	67	33	0	0	0	0	0	0	33.3	1.0
Star jasmine	*Trachelospermum jasmi-noides*	100	83	61	40	18	0	0	0	0	—	1.6

Note: Salt concentration is shown as the electrical conductivity of saturated soil extracts, EC. References included in original.

[a] Tabled values are estimates based upon the EC_e for a relative yield of 90% and yield reductions for similar crops as EC_e increases. Where no productivity data are given, the plant is listed with others of similar salt tolerance.

[b] The lower part of the yield curve approaches zero asymptotically to the abscissa. Only linear data are shown.

Table 41
THE RELATIVE PRODUCTIVITY OF MODERATELY SENSITIVE PLANTS WITH INCREASING SALT CONCENTRATION IN THE ROOT ZONE

Plant name	Scientific names	Relative productivity, % EC, mmhos/cm														% Productivity decrease per mmhos/cm increase	Salinity threshold (EC$_t$)
		1	2	3	4	5	6	7	8	9	10	11	12	13	14		
Alfalfa	Medicago sativa	100	100	93	85	78	71	64	56	49	42	34	27	20	12	7.3	2.0
Arborvitae[b]	Thuja orientalis	100	100	91	81	72	62	52	43	33	24	—	—	—	—	—	2.0
Bentgrass[a]	Agrostis palustris	—	—	—	—	—	—	—	—	—	—	—	—	—	—	—	—
Bottlebrush[b]	Callistemon viminalis	100	94	85	77	68	59	50	41	33	—	—	—	—	—	—	1.5
Boxwood	Buxus microphylla var. japonica	100	96	86	76	65	54	43	32	21	11	0	0	0	0	10.8	1.7
Broadbean	Vicia faba	100	96	87	77	67	58	48	38	29	19	10	0	0	0	9.6	1.6
Cauliflower[a]	Brassica oleracea	100	100	93	85	—	—	—	—	—	—	—	—	—	—	—	2.5
Cabbage	Brassica oleracea var. capitata	100	98	88	79	69	59	50	40	30	20	11	1	0	0	9.7	1.8
Canarygrass, feed[a]	Phalaris arundinacea	—	—	—	—	—	—	—	—	—	—	—	—	—	—	—	
Clover, alsike ladino, red, strawberry	Trifolium spp.	100	94	82	70	58	40	34	22	10	0	0	0	0	0	12.0	1.5
Corn, forage	Zea mays	100	99	91	84	76	69	61	54	47	39	32	24	17	10	7.4	1.8
Corn, grain, sweet	Zea mays	100	96	84	72	60	48	36	24	12	0	0	0	0	0	12.0	1.7
Cowpea	Vigna unguiculata	100	90	76	61	47	33	19	4	0	0	0	0	0	0	14.3	1.3
Cucumber	Cucumis sativus	100	100	94	81	68	55	42	29	16	3	0	0	0	0	13.0	2.5
Dodonea	Dedonia viscosa var. atropurpurea	100	94	86	77	68	59	51	42	33	25	17	9	0	0	7.8	1.0
Flax	Linum usitatissimum	100	96	84	72	60	48	36	24	12	0	0	0	0	0	12.0	1.7
Grape	Vitis spp.	100	95	86	76	66	57	47	38	28	18	9	0	0	0	9.5	1.5
Juniper	Juniperus chinensis	100	91	81	72	63	54	45	36	27	18	9	0	0	0	9.5	1.5
Lantana	Lantana camera	100	92	82	72	62	51	41	30	20	9	0	0	0	0	—	1.8
Lettuce	Latuca sativa	100	91	78	65	52	39	26	13	0	0	0	0	0	0	13.0	1.3
Lovegrass	Eragrostis spp.	100	100	92	83	75	66	58	49	41	32	24	15	7	0	8.5	2.0

Table 41 (continued)
THE RELATIVE PRODUCTIVITY OF MODERATELY SENSITIVE PLANTS WITH INCREASING SALT CONCENTRATION IN THE ROOT ZONE

Plant name	Scientific names	Relative productivity, % EC, mmhos/cm														% Productivity decrease per mmhos/cm increase	Salinity threshold (EC_t)
		1	2	3	4	5	6	7	8	9	10	11	12	13	14		
Meadow foxtail	*Alopecurus pratensis*	100	95	85	76	66	56	47	37	27	17	8	0	0	0	9.7	1.5
Millet, foxtail	*Setaria italica*	—	—	—	—	—	—	—	—	—	—	—	—	—	—	—	
Muskmelon[a]	*Cucumis melo*	100	100	95	80	—	—	—	—	—	—	—	—	—	—	—	2.5
Oleander[b]	*Nerium oleander*	100	100	93	86	79	72	65	58	51	44	37	30	24	—	—	2.0
Pea[a]	*Pisum sativum* L.	100	100	90	—	—	—	—	—	—	—	—	—	—	—	—	2.5
Peanut	*Arachis hypogaea*	100	100	100	77	49	20	0	0	0	0	0	0	0	0	28.6	3.2
Pepper	*Capsicum annuum*	100	93	79	65	51	37	23	8	0	0	0	0	0	0	14.1	1.5
Potato	*Solanum tuberosum*	100	96	84	72	60	48	36	24	12	0	0	0	0	0	12.0	1.7
Pyracantha	*Pyracantha draperi*	100	99	90	81	72	62	53	43	34	24	14	6	0	0	9.1	2.0
Radish	*Raphanus sativus*	100	90	77	64	51	38	25	12	0	0	0	0	0	0	13.0	1.2
Reed canary	*Phalaris arundinacea*	—	—	—	—	—	—	—	—	—	—	—	—	—	—	—	
Rice, paddy	*Oryza sativa*	100	100	100	88	76	63	51	39	27	15	2	0	0	0	12.2	3.0
Sesbania	*Sesbania exaltata*	100	100	95	88	81	74	67	60	53	47	40	33	26	19	7.0	2.3
Spinach	*Spinacia oleracea*	100	100	92	85	77	70	62	55	47	39	32	24	17	9	7.6	2.0
Squash[a]	*Cucurbita maxima*	100	100	90	74	—	—	—	—	—	—	—	—	—	—	—	2.5
Sugarcane	*Saccharum officinarum*	100	98	92	86	81	75	69	63	57	51	45	39	34	28	5.9	1.7
Silverberry	*Elaeagnus pungens*	100	95	87	78	69	59	50	41	32	23	15	6	0	0	—	1.6
Sweet clover[a]	*Melilotus* spp.	—	—	—	—	—	—	—	—	—	—	—	—	—	—	—	
Sweet potato	*Ipomoea batatas*	100	95	84	73	62	51	40	29	18	7	0	0	0	0	11.0	1.5
Texas privet	*Ligustrum lucidum*	100	94	85	75	66	56	46	36	26	16	7	0	0	0	9.1	2.0
Timothy	*Phleum pratense*	—	—	—	—	—	—	—	—	—	—	—	—	—	—	—	
Tomato	*Lycopersicon lycopersicum*	100	100	95	85	75	65	55	46	36	26	16	6	0	0	9.9	2.5
Trefoil, big	*Lotus uliginosus*	100	100	87	68	49	30	11	0	0	0	0	0	0	0	18.9	2.3
Vetch, common	*Vicia sativa*	100	100	100	89	78	67	56	44	33	22	11	0	0	0	11.1	3.0

Table 41 (continued)
THE RELATIVE PRODUCTIVITY OF MODERATELY SENSITIVE PLANTS WITH INCREASING SALT CONCENTRATION IN THE ROOT ZONE

Plant name	Scientific names	Relative productivity, % EC, mmhos/cm														% Productivity decrease per mmhos/cm increase	Salinity threshold (EC,)
		1	2	3	4	5	6	7	8	9	10	11	12	13	14		
Viburnum	*Viburnum* spp.	100	90	73	58	44	32	20	10	0	0	0	0	0	0	13.2	1.4
Xylosma	*Xylosma senticosa*	100	94	81	67	54	40	27	14	0	0	0	0	0	0	13.3	1.5

Note: Salt concentration is shown as the electrical conductivity of saturated soil extracts, EC. References included in original.

[a] Tabled values are estimates based upon the EC, for a relative yield of 90% and yield reductions for similar crops as EC, increases. Where no productivity data are given, the plant is listed with others of similar salt tolerance.

[b] The lower part of the yield curve approaches zero asymptotically to the abscissa,. Only linear data are shown.

Table 42

THE RELATIVE PRODUCTIVITY OF MODERATELY TOLERANT PLANTS WITH INCREASING SALT CONCENTRATION IN THE ROOT ZONE

Plant name	Scientific names	Relative productivity, % EC, mmhos/cm																			% Productivity decrease per mmhos/cm increase	Salinity threshold (EC,)
		1	2	3	4	5	6	7	8	9	10	11	12	13	14	15	16	17	18	19		
Alkali sacaton[a]	*Sporobolus airoides*	100	100	—	—	—	—	—	—	—	—	—	—	—	0	0	0	0	0	0	—	—
Barley, forage	*Hordeum vulgare*	100	100	100	100	100	100	93	86	79	72	65	58	51	44	37	30	23	15	8	7.0	6.0
Beet, garden	*Beta vulgaris*	100	100	100	100	91	82	73	64	55	46	38	29	20	11	2	0	0	0	0	9.0	4.0
Broccoli	*Brassica oleracea var. capitata*	100	100	98	89	80	71	61	52	43	34	25	16	6	0	0	0	0	0	0	9.1	2.8
Bromegrass	*Bromis inermis*	—	—	—	—	—	—	—	—	—	—	—	—	—	—	—	—	—	—	—	—	—
Clover, berseem	*Trifolium alexandrinium*	100	97	91	86	80	74	69	63	57	51	46	40	34	29	23	17	11	6	0	5.8	1.5
Dallis grass	*Paspalum dilatatum*	—	—	—	—	—	—	—	—	—	—	—	—	—	—	—	—	—	—	—	—	—
Dracaena	*Dracaena endivisa*	100	100	100	94	85	76	67	58	49	40	31	22	13	4	0	0	0	0	0	9.1	4.0
Euonymus[b]	*Euonymus japonica var. grandiflora*	—	—	—	—	—	—	100	77	52	27	0	0	0	0	0	0	0	0	0	—	7.0
Fescue	*Festuca elatior*	100	100	100	99	94	89	84	78	73	68	62	57	52	47	41	36	31	25	20	5.3	3.9
Fig[a]	*Ficus carica*	100	100	100	100	90	85	—	—	—	—	—	—	—	—	—	—	—	—	—	—	4.2
Hardinggrass	*Phalaris tuberosa*	100	100	100	100	97	89	82	74	67	59	52	44	36	29	21	14	6	0	0	7.6	4.6

Table 42 (continued)
THE RELATIVE PRODUCTIVITY OF MODERATELY TOLERANT PLANTS WITH INCREASING SALT CONCENTRATION IN THE ROOT ZONE

Plant name	Scientific names	Relative productivity, % EC, mmhos/cm																			% Productivity decrease per mmhos/cm increase	Salinity threshold (Ec$_t$)
		1	2	3	4	5	6	7	8	9	10	11	12	13	14	15	16	17	18	19		
Kale[a]	*Brassica campestris*	100	100	100	100	100	100	90	—	—	—	—	—	—	—	—	—	—	—	—	—	6.5
Lime, rangpur[a]	*Citrus aurantifolia*	—	—	—	—	—	—	—	—	—	—	—	—	—	—	—	—	—	—	—	—	—
Mandarin, cleopatra[a]	*Citrus nobilis*	—	—	—	—	—	—	—	—	—	—	—	—	—	—	—	—	—	—	—	—	—
Milkvetch[a]	*Astragalus*	—	—	—	—	—	—	—	—	—	—	—	—	—	—	—	—	—	—	—	—	4.0
Olive	*Olea europaea*	100	100	100	100	90	85	—	—	—	—	—	—	—	—	—	—	—	—	—	—	4.0
Orchardgrass	*Dactylis glomerata*	100	97	91	84	78	72	66	60	53	47	41	35	29	22	16	10	4	0	—	6.2	1.5
Oats	*Avena sativa*	—	—	—	—	—	—	—	—	—	—	—	—	—	—	—	—	—	—	—	—	—
Pomegranate	*Punica granatum*	100	100	100	100	90	85	—	—	—	—	—	—	—	—	—	—	—	—	—	—	4.0
Rhodesgrass[a]	*Chloris gayana*	—	—	—	—	—	—	—	—	—	—	—	—	—	—	—	—	—	—	—	—	—
Rye, hay[a]	*Secale cereale*	—	—	—	—	—	—	—	—	—	—	—	—	—	—	—	—	—	—	—	—	—
Ryegrass, perennial	*Lolium perenne*	100	100	100	100	100	97	89	82	74	67	59	52	44	36	29	21	14	6	0	7.6	5.6
Safflower[c]	*Carthamus tinctorius*	100	100	100	100	100	100	97	90	85	80	75	50	—	—	—	—	—	—	—	—	6.5
Sorghum[b]	*Sorghum bicolor*	100	100	100	100	98	90	84	78	70	63	56	50	43	36	29	22	15	8	0	—	4.8
Soybean	*Glycine max*	100	100	100	100	100	80	60	40	20	0	0	0	0	0	0	0	0	0	0	20.0	5.0

Common name	Scientific name																					
Sudangrass	*Sorghum sudanense*	100	100	99	95	91	86	82	78	73	69	65	61	56	52	48	43	38	35	30	4.3	2.8
Trefoil, birdsfoot	*Lotus corniculatus tenuifolium*	100	100	100	100	100	90	80	70	60	50	40	30	20	10	0	0	0	0	0	10.0	5.0
Wheat	*Triticum aestivum*	100	100	100	100	93	86	79	71	64	57	50	43	36	29	21	14	7	—	—	7.1	6.0
Wheatgrass, slender	*Agropyron trachycaulum*	—	—	—	—	—	—	—	—	—	—	—	—	—	—	—	—	—	—	—	—	—
Wheatgrass, western[a]	*Agropyron smithi*	—	—	—	—	—	—	—	—	—	—	—	—	—	—	—	—	—	—	—	—	—
Wildrye, beardless	*Elymus triticoides*	100	100	98	92	86	80	74	68	62	56	50	44	38	32	26	20	14	8	2	6.0	2.7
Wildrye, Canada[a]	*Elymus canadensis*	—	—	—	—	—	—	—	—	—	—	—	—	—	—	—	—	—	—	—	—	—

Note: Salt concentration is shown as the electrical conductivity of saturated soil extracts, EC. References included in original.

[a] Tabled values are estimates based upon the EC, for a relative yield of 90% and yield reductions for similar crops as EC, increases. Where no productivity data are given, the plant is listed with others of similar salt tolerance.

[b] Tabled values are based upon three data points available in the literature.

[c] Tabled values based upon three data points. Productivity drops sharply towards zero for the lower 50% productivity.

Table 43
THE RELATIVE PRODUCTIVITY OF TOLERANT PLANTS WITH INCREASING SALT CONCENTRATION IN THE ROOT ZONE

Plant name	Scientific name	Relative productivity, % EC_e, mmhos/cm																					% Productivity decrease per mmhos/cm increase	Salinity threshold (EC_e)
		4	5	6	7	8	9	10	11	12	13	14	15	16	17	18	19	20	21	22	23	24		
Barley, grain	*Hordeum vulgare*	100	100	100	100	100	95	90	85	80	75	70	65	60	55	50	45	40	35	30	25	20	5.0	8.0
Bermuda grass	*Cynodon dactylon*	100	100	100	99	93	87	80	74	67	61	54	48	42	35	29	22	16	10	3	0	0	6.4	6.9
Bougainvillea[a]	*Bougainvillea spectabilis*	100	100	100	100	100	—	—	—	—	—	—	—	—	—	—	—	—	—	—	—	—	—	8.5
Cotton	*Gossypium hirsutum*	100	100	100	100	98	93	88	83	78	73	67	62	57	52	47	41	36	31	26	21	16	5.2	7.7
Date	*Phoenix dactylifera*	100	96	93	89	86	82	78	75	71	68	64	60	57	53	49	46	42	39	35	31	28	3.6	4.0
Natal plum[a]	*Carissa grandiflora*	—	—	100	—	82	—	—	—	—	—	—	—	—	—	—	—	—	—	—	—	—	—	6.0
Nutall alkali grass	*Puccinellia nuttaluana*	—	—	—	—	—	—	—	—	—	—	—	—	—	—	—	—	—	—	—	—	—	—	—
Rescue grass	*Bromus catharticus*	—	—	—	—	—	—	—	—	—	—	—	—	—	—	—	—	—	—	—	—	—	—	—
Rosemary[b,c]	*Rosmarinus lockwoodii*	100	95	85	75	68	—	—	—	—	—	—	—	—	—	—	—	—	—	—	—	—	—	4.5
Sugarbeet	*Beta vulgaris*	100	100	100	100	94	88	82	76	71	65	59	53	47	41	35	29	24	18	12	6	0	5.9	7.0
Saltgrass[a]	*Distichlis stricta*	—	—	—	—	—	—	—	—	—	—	—	—	—	—	—	—	—	—	—	—	—	—	—
Wheatgrass, crested	*Agropyron desertorum*	98	94	90	86	82	78	74	70	66	62	58	54	50	46	42	38	34	30	26	22	18	4.0	3.5
Wheatgrass, fairway	*Agropyron cristatum*	100	100	100	100	97	90	83	76	69	62	55	48	41	34	28	21	14	7	0	0	0	6.9	7.5
Wheatgrass, tall	*Agropyron elongatum*	100	100	100	100	98	94	89	85	81	77	73	68	64	60	56	52	47	43	39	35	31	4.2	7.5
Wildrye, altai	*Elymus angustus*	100	100	100	—	—	—	—	—	—	—	—	—	—	—	—	—	—	—	—	—	—	—	—
Wildrye, Russian[a]	*Elymus psathyrostachys juncea*	—	—	—	—	—	—	—	—	—	—	—	—	—	—	—	—	—	—	—	—	—	—	—

Note: Salt concentration is shown as the electrical conductivity of saturated soil extracts, EC. References included in original.

[a] Tabled values are estimates based upon the EC, for a relative yield of 90% and yield reductions for similar crops as EC, increases. Where no productivity data are given, the plant is listed with others of similar salt tolerance.

[b] The lower part of the yield curve approaches zero asymptotically to the abscissa. Only linear data are shown.

[c] Tabled values are based upon three data points available in the literature.

Table 44

QUANTITY OF SULFUR TO REDUCE SOIL pH
FOR A DEPTH OF SEVEN INCHES

Sulfur (lb/acre)

Desired pH change	Sands	Loamy sands	Sandy loams	Loams	Clay loams
7.0—6.0	300	400	500	700	1100
6.0—5.0	800	1000	1200	1400	1400
5.0—4.0	800	1000	1200	1400	1800

Source: Christenson, D. R., Lucas, R. E., and Doll, E. C., Fertilizer Recommendations for Michigan Vegetables and Field Crops, Ext. Bull. E-550, Michigan State University Cooperative Extension Service, Michigan State University, East Lansing, 1972.

Table 45

AMOUNT OF OTHER MATERIALS TO
EQUAL 1 LB OF SULFUR AS A SOIL
AMENDMENT

Amendment	Equivalent to 1 lb of sulfur (lb)
Lime-sulfur solution (CaS_2)	4.17
Sulfuric acid	3.06
Gypsum ($CaSO_4 \cdot 2H_2O$)	5.38
Iron sulfate ($FeSO_4 \cdot 7H_2O$)	8.69
Aluminum sulfate ($Al_2(SO_4)_3 \cdot 18H_2O$)	6.94

Source: Richards, L. A., Diagnosis and Improvement of Saline and Alkali Soils, *U.S. Dep. Agric. Agric. Handb.*, No. 60, 1954.

ORGANIC MATTER, CROP RESIDUES, AND N$_2$ FIXATION

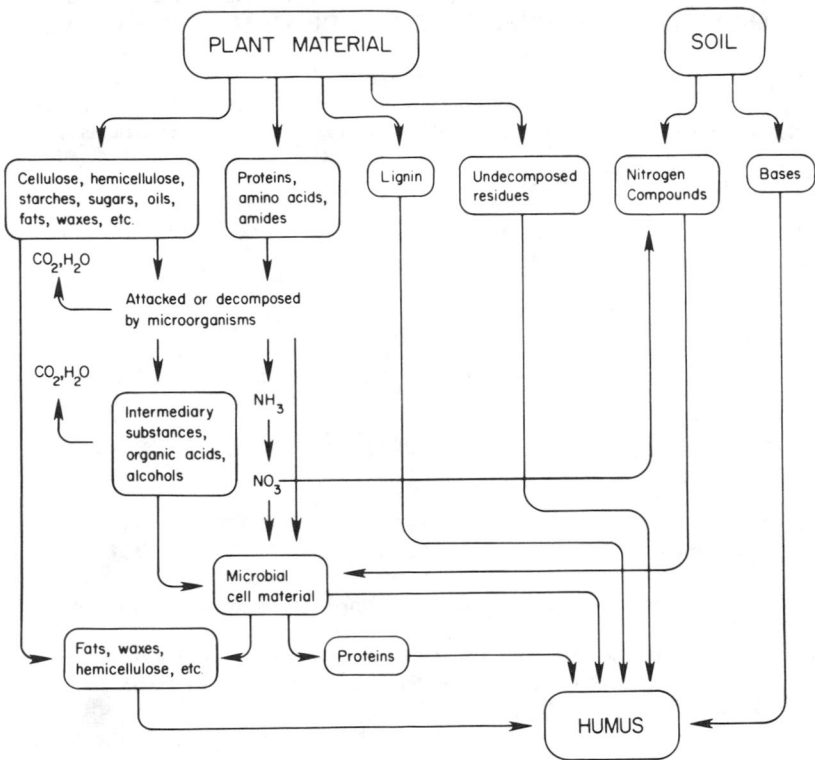

FIGURE 8.   Schematic representation of the formation of humus or organic matter from plant material in the soil.

Adapted with permission from Foth, H. D., *Fundamentals of Soil Science,* 6th ed., John Wiley & Sons, New York, N.Y. 1978. Copyright ©1978 by John Wiley & Sons, Inc.

Table 46
DISTRIBUTION OF ORGANIC MATTER IN WELL-DRAINED SILT LOAM SOIL UNDER FOREST AND PRAIRIE ECOSYSTEMS

	Forest (White oak, black oak) (Metric tons)	Prairie (Big bluestem, Indian grass) (Metric tons)
Standing crop	81.7	2.7
Litter		
Feb.	2.7	
Apr.		4.5
Oct.	1.4	1.8
Roots[a]		
0—15 cm	32.6	45.4
15—30	10.0	33.6
30—46	7.3	21.8
46—61	6.4	12.7
61—76	5.4	9.1
76—91	4.5	6.4
91—107	3.6	4.5
Total (Roots)	69.8	133.5
(Total plant biomass)	151.5	136.2

[a]   Excludes roots greater than 2.54 cm diameter, estimated to weigh an additional 14.5 t.

Source: Adapted from Nielson, G. A. and Hole, F. D., A study of the natural processes of incorporation of organic matter into soil in the University of Wisconsin Arboretum, *Wis. Acad. Sci.*, 52, 213, 1963.

Table 47
NITROGEN AND PHOSPHORUS CONTENTS OF SEVERAL SOIL CLASSES FROM IOWA AND NEW YORK AS RELATED TO SOIL TEXTURE

Soil texture	N content of soil (%)	P content of soil (%)
Sandy and loamy sand	0.02	0.03
Sandy loam	0.16	0.13
Loam	0.17	0.14
Silt loam	0.22	0.15
Silty clay loam	0.21	0.14
Clay	0.24	0.16

Adapted with permission from Foth, H. D., *Fundamentals of Soil Science,* 6th ed., John Wiley & Sons, New York, 1978. Copyright ©1978 by John Wiley & Sons, Inc.

Table 48

EQUATIONS DESCRIBING THE DECOMPOSITION OF PLANT RESIDUES

(1) $\dfrac{1}{CO_2 \text{ Evolved}} = \dfrac{1}{CH_2O} \text{ loss} = \text{lignin} - C \text{ loss} = \dfrac{\left[\left(\dfrac{C}{N}\ \text{straw}\right)\left(\%\ \text{lignin}\right)\right]}{\sqrt{\%\ \text{carbohydrates}}}$

(2) $S_o = 0.070 + 1.11\sqrt[3]{(N/C)}$
      S_o = initial proportion of easily decomposable constituents

(3) $N_t = N_o K^t$ N_t = amount of organic N per unit mass of soil remaining in soil after t years
      N_o = amount of organic N per unit mass of soil initially
      K = fraction of original organic N per unit mass of soil remaining after a year

(4) $dN/dt = A - rN$ N = amount of organic N per unit mass of soil present at time (t)
      A = rate of addition of new N per unit mass of soil
      r = annual rate of decomposition of N per unit mass of soil

(5) $dN/dt = K_1(t) \cdot N + K_2 + K_3(t) \cdot Y(t)$ t = time
      $K_1(t)$ = decomposition coefficient
      K_2 = additions to soil organic matter from noncrop sources, e.g., manure
      $Y(t)$ = crop yield
      $K_3(t)$ = coefficient related to a specific crop at time (t)

(6) $N = N_1 (\exp - r_1 t) + A_1/r_1 (1 - \exp - r_1 t) + N_2 (\exp - r_2 t) + A_2/r_2 (1 - \exp - r_2 t) + \ldots$
      Subscripts refer to humus fractions with different decomposition rates

Note: References included in the original.

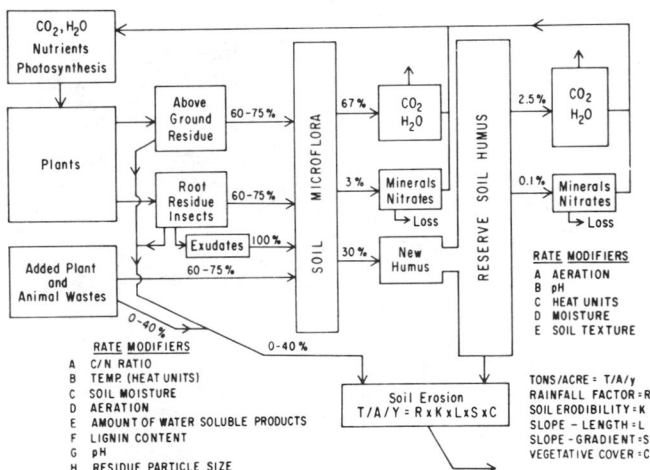

FIGURE 9.   Soil humus model, yearly change. (Reproduced from Lucas, R. E., Holtman, J. B., and Connor, L. J., *Agriculture and Energy*, Academic Press, New York, 1977, 333. With permission.)

Table 49
RESIDUE PRODUCTION (METRIC TONS) BY NINE CROPS IN THE 20 LEADING RESIDUE-PRODUCING STATES

State	Corn—1.0[a] (Zea mays L.)	Soybeans—1.5 (Glycine max [L.] Merr.)	Winter wheat—1.7[b] (Triticum aestivum L.)	Total wheat	Cotton—1.0 (Gossypium hirsutum L.)	Sorghum—1.0 (Sorghum bicolor [L.] Moench.)
Ark.	18,709	4,755,788	281,708	281,708	227,079	170,676
Calif.	628,102	—	1,427,084	1,430,630	381,096	511,391
Colo.	1,137,228	—	2,726,203	2,743,718	—	253,914
Ill.	24,986,373	11,507,778	1,808,196	1,808,196	—	93,165
Ind.	13,605,188	5,528,373	1,140,786	1,140,786	—	45,183
Iowa	30,723,167	10,779,785	50,444	50,444	—	40,881
Kan.	3,920,070	1,080,024	17,840,867	17,840,867	—	5,559,372
Mich.	3,398,497	680,415	921,716	921,716	—	—
Minn.	13,067,579	5,208,252	54,895	2,783,795	—	—
Mo.	5,824,104	5,191,479	1,182,282	1,182,282	39,349	784,014
Mont.	20,440	—	2,555,584	4,030,339	—	—
Neb.	14,117,343	1,485,033	4,348,943	4,348,943	—	3,461,880
N.C.	2,922,234	1,423,668	292,093	292,093	35,737	109,253
N.D.	256,586	209,582	108,306	8,590,915	—	—
Ohio	6,113,273	3,671,673	1,068,227	1,068,227	—	—
Okla.	199,313	188,186	7,316,239	7,316,239	93,134	779,534
S.D.	3,725,085	388,809	988,110	2,363,693	—	286,776
Tex.	1,547,664	347,735	4,571,490	4,571,490	1,023,007	10,614,735
Wash.	143,006	—	3,440,209	3,972,034	—	—
Wis.	4,415,679	246,483	25,964	37,664	—	—
Total for 20 states	130,769,640	52,698,063	52,149,346	62,775,779	1,799,402	22,710,774

State	Oats—2.0 (Avena sativa L.)	Barley—1.5 (Hordeum vulgare L.)	Rice—1.5 (Oryza sativa L.)	Rye—1.5 (Secale cereale L.)	Total
Ark.	107,200	—	1,733,357	—	7,294,517
Calif.	134,400	1,569,076	1,535,437	—	6,190,132
Colo.	56,204	403,495	—	10,881	4,605,440
Ill.	575,420	16,038	—	17,640	39,004,610
Ind.	364,452	15,121	—	11,416	20,710,519
Iowa	1,985,112	—	—	3,971	43,583,360
Kan.	116,364	107,354	—	32,073	28,656,124
Mich.	480,002	29,359	—	31,500	5,541,489
Minn.	4,154,195	1,316,728	—	120,273	26,650,822
Mo.	40,553	14,041	15,408	9,622	13,100,852
Mont.	298,474	1,963,800	—	—	6,313,053
Neb.	612,947	35,348	—	59,564	24,121,058
N.C.	109,091	93,346	—	14,509	4,999,931
N.D.	2,146,916	3,354,498	—	116,035	14,674,532
Ohio	754,039	15,318	—	7,942	11,630,472
Okla.	231,390	253,821	—	50,897	9,112,514
S.D.	2,925,973	722,842	—	303,776	10,716,954
Tex.	775,275	114,882	1,399,694	24,742	20,419,224
Wash.	69,818	465,912	—	4,582	4,655,352
Wis.	1,634,041	25,431	—	7,216	6,366,514
Total for 20 states	17,571,866	10,516,410	4,683,896	826,639	308,347,469

[a] Dry weight ratio of residue to grain.
[b] Spring wheat ratio — 1.3.

Reproduced from Larson, W. E., Holt, R. F., and Carlson, C. W., *Crop Management Systems*, Oschwald, W. R., Ed., American Society of Agronomy, Special Publication No. 31, Madison, Wis., 1978, chap. 1, by permission of the American Society of Agronomy, Inc.

Table 50

METRIC TONS OF N, P, AND K IN CROP RESIDUE AND PERCENTAGES OF
COMMERCIAL FERTILIZER USED BY STATES THAT THESE AMOUNTS
REPRESENT

State	N		P		K	
	Metric tons	% of comm. fert. N	Metric tons	% of comm. fert. P	Metric tons	% of comm. fert. K
Ark.	125,992	91	13,182	29	80,338	112
Calif.	50,780	10	7,090	10	74,489	139
Colo.	37,182	35	4,896	28	51,136	672
Ill.	553,228	84	72,479	38	481,738	106
Ind.	286,003	87	38,132	33	256,939	84
Iowa	596,878	88	82,295	52	555,617	160
Kan.	249,087	49	30,602	43	313,403	671
Mich.	62,609	44	9,097	15	69,792	52
Minn.	317,534	85	45,168	43	341,330	140
Mo.	199,033	62	24,101	32	155,447	95
Mont.	43,839	105	5,495	23	68,838	2,163
Neb.	261,067	49	38,007	65	266,862	618
N.C.	69,692	35	9,134	15	61,676	40
N.D.	104,388	87	14,201	33	167,114	1,258
Ohio	162,532	62	21,072	20	142,909	61
Okla.	69,130	36	7,978	19	94,583	304
S.D.	94,404	100	15,487	63	139,780	1,153
Tex.	202,442	28	27,580	23	254,164	253
Wash.	32,157	17	3,667	13	47,438	235
Wis.	65,333	51	11,167	20	89,009	44
Total for 20 states	3,583,310		480,830		3,712,602	

Reproduced from Larson, W. E., Holt, R. F., and Carlson, C. W., *Crop Management Systems,* Oschwald, W. R., Ed., American Society of Agronomy, Special Publication No. 31, Madison, Wis., 1978, chap. 1, by permission of the American Society of Agronomy, Inc.

Table 51
CONVERSION FACTORS
FIELD CROP RESIDUES (dry basis)

Commodity	lb/bu (kg/bu)	lb field residue/lb of commodity	Tons field residue/1000 bu (metric tons)	lb hulls/lb commodity	Tons hulls/ 1000 bu (metric tons)	lb/ton sugar cane (kg/ metric ton)	Tons field residue and gin trash/ 480-lb bale (metric tons/bale)
Flax	60 (27.2)	2.0	60.0 (54.4)	0.375	11.2 (10.2)	—	—
Wheat	60 (27.2)	1.5	45.0 (40.8)	0.145	3.45 (3.13)	—	—
Rye	56 (25.4)	2.0	56.0 (50.8)	0.100	2.8 (2.5)	—	—
Rice	45 (20.4)	1.5	33.7 (30.6)	0.200	4.5 (4.08)	—	—
Oats	32 (14.5)	2.0	32.0 (29.0)	0.300	4.8 (4.35)	—	—
Barley	48 (21.8)	1.5	36.0 (32.7)	0.200	4.8 (4.35)	—	—
Soybeans	60 (27.2)	1.4	42.0 (38.1)	0.080	2.4 (2.18)	—	—
Cotton	—	—	—	—	—	—	0.30 (0.27)
Cottonseed	—	—	—	0.210	—	—	—
Bagasse	—	—	—	—	—	300 (150)	—
Corn	56 (25.4)	1.0	—	0.055	1.54 (1.40)	—	—
Grain sorghum	56 (25.4)	1.0	—	0.080	2.24 (2.03)	—	—

Table 52
CONVERSION FACTORS
NUT SHELLS AND FRUIT PITS (dry basis)

Commodity	lb Shells/ton nuts (kg/metric ton)	lb Fruit pits/ton fruit (kg/metric ton)
Pecans	1200 (600)	—
Walnuts	1160 (580)	—
Almonds	1000 (500)	—
Filberts	1200 (600)	—
Cherries	—	80 (40)
Peaches	—	80 (40)
Apricots	—	80 (40)
Plums	—	80 (40)
Peanuts	600 (300)	—

Table 53
AVERAGE CONCENTRATIONS (%) OF NUTRIENTS CONTAINED IN THE ABOVE GROUND PORTIONS OF CROPS

Crop	Annual yield (MT/ha)	N	P	K	Ca	Mg	S	Cu	Mn	Zn
Barley										
Grain	2.15	1.822	0.341	0.433	0.052	0.104	0.156	0.0016	0.0016	0.0031
Straw	2.24	0.75	0.109	1.245	0.40	0.10	0.20	0.0005	0.016	0.0025
Corn										
Grain	9.41	1.607	0.275	0.395	0.024	0.095	0.119	0.0007	0.0011	0.0018
Stover	10.08	1.111	0.179	1.337	0.289	0.222	0.156	0.0005	0.0166	0.0033
Oats										
Grain	2.87	1.953	0.341	0.486	0.078	0.117	0.195	0.0012	0.0047	0.002
Straw	4.48	0.625	0.164	1.660	0.20	0.20	0.225	0.0008	—	0.0072
Rice										
Grain	4.03	1.388	0.242	0.231	0.083	0.111	0.083	0.0003	0.0022	0.0019
Straw	5.60	0.60	0.087	1.162	0.18	0.10	—	—	0.0316	—
Rye										
Grain	1.88	2.083	0.259	0.494	0.119	0.179	0.417	0.0012	0.0131	0.0018
Straw	3.36	0.50	0.116	0.692	0.267	0.067	0.10	0.003	0.0047	0.0023
Sorghum										
Grain	3.36	1.666	0.363	0.415	0.133	0.167	0.167	0.0003	0.0013	0.0013
Stover	6.72	1.083	0.145	1.314	0.483	0.30	—	—	—	—
Wheat										
Grain	2.68	2.083	0.615	0.519	0.042	0.25	0.125	0.0013	0.0038	0.0058
Straw	3.36	0.667	0.073	0.968	0.20	0.10	0.167	0.0003	0.0053	0.0017
Hay										
Alfalfa	8.96	2.25	0.218	1.868	1.40	0.263	0.238	0.0008	0.0055	0.0053
Bluegrass	4.48	1.5	0.218	1.245	0.40	0.175	0.125	0.0005	0.0075	0.002
Coastal Ber- muda	17.92	1.875	0.191	1.401	0.369	0.15	0.219	0.0013	—	—

Table 53 (continued)

AVERAGE CONCENTRATIONS (%) OF NUTRIENTS CONTAINED IN THE ABOVE GROUND PORTIONS OF CROPS

Crop	Annual yield (MT/ha)	N	P	K	Ca	Mg	S	Cu	Mn	Zn
Cowpea	4.48	3.0	0.273	1.66	1.375	0.375	0.325	—	0.0163	—
Peanut	5.04	2.333	0.242	1.752	1.0	0.378	0.356	—	0.0051	—
Red Clover	5.60	2.0	0.218	1.66	1.38	0.34	0.14	0.0008	0.0108	0.0072
Soybean	4.48	2.25	0.218	1.038	1.0	0.45	0.25	0.001	0.0115	0.0038
Timothy	5.60	1.2	0.218	1.577	0.36	0.12	0.10	0.0006	0.0062	0.004
Fruits & vegetables										
Apples	26.88	0.125	0.018	0.156	0.033	0.021	0.042	0.0001	0.0001	0.0001
Beans, dry	1.01	3.125	0.454	0.864	0.083	0.083	0.208	0.0008	0.0013	0.0025
Cabbage	44.80	0.325	0.038	0.270	0.05	0.02	0.11	0.0001	0.0003	0.0002
Onions	16.80	0.30	0.058	0.222	0.073	0.013	0.12	0.0002	0.005	0.0021
Oranges	62.72	0.002	0.024	0.208	0.059	0.021	0.016	0.0004	0.0001	0.004
Peaches	32.26	0.122	0.030	0.188	0.014	0.028	0.007	—	—	0.001
Potatoes	26.88	0.333	0.055	0.519	0.013	0.025	0.025	0.0002	0.0004	0.0002
Spinach	11.20	0.50	0.065	0.249	0.12	0.05	0.04	0.0002	0.001	0.001
Sweet potatoes	10.80	0.30	0.044	0.415	0.027	0.06	0.04	0.0002	0.0004	0.0002
Tomatoes	44.80	0.30	0.044	0.332	0.018	0.028	0.035	0.0002	0.0003	0.0001
Turnips	22.40	0.225	0.044	0.374	0.06	0.03	—	—	—	—
Other crops										
Cotton seed and lint	1.68	2.666	0.581	0.830	0.133	0.267	0.20	0.004	0.0073	0.0213
Stalks, burs and leaves	2.24	1.75	0.218	1.453	1.4	0.40	0.75	—	—	—
Peanuts (nuts)	2.80	3.6	0.174	0.498	0.04	0.12	0.24	0.0008	0.0004	—
Soybeans (grain)	2.69	6.25	0.636	1.902	0.292	0.292	0.167	0.0017	0.0021	0.0017
Sugar beets (roots)	33.60	0.20	0.029	0.139	0.11	0.08	0.033	0.0001	0.0025	—
Sugar cane	67.20	0.16	0.039	0.374	0.047	0.04	0.04	—	—	—
Tobacco (leaves)	2.25	3.75	0.327	4.981	3.75	0.90	0.70	0.0015	0.0275	0.0035

Source: *Our Land and Its Care,* 5th ed., National Plant Food Institute, Washington, D.C., 1967.

Table 54
CONCENTRATIONS (%) OF SOME ELEMENTS IN SELECTED ROOT MATERIAL

Plant	N	P	K	S	Ca	Mg
White clover	3.77	0.22	—	0.39	—	—
Red clover	2.79	0.26	—	0.47	—	—
Perennial grass	1.75	0.13	—	0.21	—	—
Prairie grassland						
Root	0.90	0.09	—	—	—	—
Rhizomes	—	0.15	—	—	—	—
Shoot bases	—	0.12	—	—	—	—
Wheat-maturity	1.01	0.097	—	—	—	—
Corn root						
18 days	3.06	0.79	4.82	—	0.30	0.30
32 days	3.01	0.81	4.31	—	0.30	0.26
46 days	2.61	0.41	2.90	—	0.36	0.41
60 days	2.89	0.61	3.40	—	0.33	0.36
74 days	2.52	0.59	3.69	—	0.35	0.45

Table 55
THE CARBON-NITROGEN RATIO OF SOME ORGANIC MATERIALS

Material	C:N ratio[a]
Microbial tissue	8:1
Soil humus	10:1
Sweet clover (young)	12:1
Alfalfa hay	13:1
Tomato leaves	13:1
Bean stems	15:1
Manure (rotted)	20:1
Clover residues	23:1
Green rye	36:1
Cane trash	50:1
Corn stalks	60:1
Oak leaves	65:1
Oat straw	80:1
Timothy	80:1
Pine needles	225:1
Mature rye straw	350:1
Saw dust	400:1

[a] Values are approximate; in any particular material the ratio may vary considerably.

Table 56
DISTRIBUTION OF C:N RATIOS IN HORIZON SAMPLES OF SOME PACIFIC NORTHWEST AND NORTH CENTRAL U.S. SOILS

Number of horizon samples in various ranges of C:N ratios

Soil horizon	Below 5.5 a	Below 5.5 b	Between 5.5 and 8.4 a	Between 5.5 and 8.4 b	Between 8.5 and 11.4 a	Between 8.5 and 11.4 b	Between 11.5 and 14.4 a	Between 11.5 and 14.4 b	Above 14.4 a	Above 14.4 b
A	0 (0)	0 (0)	3 (1)	0 (1)	8 (12)	7 (7)	16 (4)	18 (9)	3 (0)	5 (0)
B	0 (1)	0 (1)	8 (14)	2 (11)	10 (4)	8 (7)	2 (0)	9 (0)	1 (0)	2 (0)
C	3 (3)	1 (0)	5 (4)	2 (5)	7 (0)	10 (1)	0 (1)	2 (1)	0 (0)	0 (0)
Total	3 (4)	1 (1)	16 (19)	4 (17)	25 (16)	25 (15)	18 (4)	29 (10)	4 (0)	7 (0)
Combined total	7	2	35	21	41	40	22	39	4	7

Note: a = organic C:total N ratio; b = organic C:organic N ratio. Figures in parentheses are comparative values for some north central U.S. soils.

From Young, J. L., Inorganic soil nitrogen and carbon nitrogen ratios of some Pacific Northwest soils, *Soil Sci.*, 93, 397, 1962, © by Williams & Wilkins, 1962. With permission.

Table 57
C, N, S CONTENTS AND NET AMOUNTS OF N AND S RELEASED BY MINERALIZATION FROM SELECTED SOILS

	Total (%) C	Total (%) N	Total (%) S	C:N	C:S	Mineralized Total NO_3^- + NH_4^+ (μg N/g soil)	Mineralized S (μg S/g soil)
Pasture soils	5.0	0.38	0.060	13.2	85	111	6.3
	±0.3	±0.01	±0.004	±1.0	±8	±10	±0.9
Arable soils	2.1	0.17	0.034	12.8	63	55	3.1
	±0.1	±0.01	±0.003	±0.4	±4	±4	±1.0
All soils	3.6	0.27	0.047	13.0	74	83	4.7
	±0.4	±0.03	±0.004	±0.5	±5	±9	±0.8

Source: Swift, R. S., *Soil Organic Matter Studies,* Proc. Symp. FAO/IAEA, 1976, Braunschweig, Vol. 1, 1977, 275.

Table 58

DENSITY AND BIOMASS VALUES OF DIFFERENT GROUPS OF SOIL ORGANISMS IN CULTIVATED TEMPERATE SOILS

Group of organisms	Dry biomass[a] (g/m²)[e]	Dry biomass[b] (kg/ha)	Abundance[c] (No./m²)	Live biomass[c] (g/m²)	Abundance[d] (No./g)	Live biomass[d] (kg/ha)
Bacteria	32.76	36.9	3×10^{14}	300	10^4—10^9	300—3000
Actinomycetes		0.2			10^5—10^8	
Fungi	84—144	454.0		400	2×10^4—10^{6f}	500—5000
Algae					10^2—5×10^4	7—300
Protozoa		1.0	5×10^8	38	10^4—10^5	50—200
Nematodes		2.0	10^7	12		
Earthworms		12.0	10^5	132		
Enchytraeidae		4.0				
Molluscs		5.0	g	g		
Acari		1.0	2×10^5	3		
Collembola		2.0	2×10^4	5		
Diptera		3.0	g	g		
Other arthropods		6.0	2×10^{3g}	36g		

a   From Clark, F. E. and Paul, E. A., *Adv. in Agron.*, Vol. 22, Brady, N. C., Ed., Academic Press, New York, 1970, 405.

b   Lynch, J. M. and Poole, N. J., *Microbial Ecology, a Conceptual Approach*, Halsted Press, New York, 1979, 78.

c   From Richards, B. N., *Introduction to the Soil Ecosystem*, Longman Group Limited, New York, 1974, 3; and Macfadyen, A., *Animal Ecology: Aims and Methods*, 2nd ed., Putnam, London, 1963, 234.

d   From Alexander, M., *Introduction to Soil Microbiology*, 2nd ed., John Wiley & Sons, New York, 1977, chap. 2—6.

e   Range over non-winter months for this Saskatchewan, Canada site; most precisely measured data of that presented.

f   Shown as number of propagules/g, which is not a uniform indicator of fungal density; the more accepted approach is hyphal length for which the typical range is 10 to 100 m/g. The biomass estimate is based on the length value.

g   Organisms of these groups are grouped as other fauna in this case.

Table 59
ESTIMATED ANNUAL AMOUNTS OF N_2 FIXATION in 1975

Land use	ha × 10^{-6}	kg N_2 Fixed/ha year[a]	Metric tons × 10^{-6} N_2 fixed/year
Biological N_2 Fixation			
Agricultural	4,400		
Arable under crop	1,400		
Legumes	250	140	35
Non-legumes	1,150		
Rice	135	30	4
Other	1,015	5	5
Permanent meadows	3,000	15	45
Forest and woodland	4,100	10	40
Unused	4,900	2	10
Ice covered	1,500	0	0
Total land	14,900	—	139
Sea	36,100	1	36
Total	51,000	—	175
Abiological N_2 Fixation			
Lightning			10
Combustion			20
Industrial			
Fertilizer			40
Industrial uses			10
Total			80
Total N_2 Fixation			255

[a]   Another recent report containing estimates of N_2 fixation[1] is based on the preceding estimates with the following changes: forest and woodland, 50 × 10^6 tons; ocean, 1 × 10^6; industrial—used as fertilizer, 39 × 10^6; industrial—industrial uses, 11 × 10^6; industrial—inefficiencies (losses), 7 × 10^6, and a total of 237 × 10^6 tons. Another,[2] gives the following estimates for N_2 fixation: land, 99 × 10^6 tons; ocean, 30 × 10^6; atmospheric, 7.4 × 10^6; industrial, 40 × 10^6; combustion, 18 × 10^6. Confidence limits are about ±50% for all values except that of industrial fixation.

REFERENCES
1. Effect of Increased Nitrogen Fixation of Stratospheric Ozone, Rep. No. 53, Council for Agricultural Science and Technology, Iowa State University, Ames, 1976.
2. **Delwiche, C. C.,** Energy relations in the global nitrogen cycle, *Ambio,* 6, 106, 1977.

From Burns, R. C. and Hardy, R. W. F., *Nitrogen Fixation in Bacteria and Higher Plants,* Springer-Verlag, Berlin, 1975. Copyright© 1975 by Springer-Verlag. With permission.

Table 60
REPORTED RATES OF N₂ FIXATION BY SPECIFIC SYSTEMS

N₂-fixing system	kg N₂ fixed/ ha × yr	N₂-fixing system	kg N₂ fixed/ ha × yr
Soybean	94	Soil, Canadian	2—200
	84	Soil, algal film	To 6
	57	Under rye grass	55
Soybean, groundnut	91	Natural grassland	1—2
Pulses	50—60	Irrigated lawn	5
Peas, lentils, vetches	85	Grass	112—145
Cowpeas	84	Sand dune vegetation	0.1
Groundnuts	47		
		Beech forest	0.4
Cereal legumes	50	Coniferous forest	50
Clovers	105—220	Pine forest	0—23
	137		13—39
	150—160	Soil under Douglas fir	13—38
	104—149	Douglas fir, phylloplane	7—23
Lucerne	128	*Alnus*	156
	208		85
	158		150
	300		56
Lupin	150	*Hippophae*	179
	169		2—15
Mixed legumes	125	*Acacia*	270
Pasture legumes	118	*Casuarina*	58
Rice paddy	10—55	*Ceanothus*	60
	27	*Myrica*	9
	45		
		Gunnera-Nostoc	To 72
Wheatfield soil	4	Desert crust, Arizona	41
Cornfield soil	90	Desert crust, Australia	2—3
Soil, cultivated layer, U.S.S.R.	5	Lichens, mountains	39—84
Soil, Nile valley	15	Lake, Wisconsin	4
Fallow soil, Calif.	3.5		2.4
Under native vegetation	2	Lake, England	0.4—3
Under grasses, iris	45—66		8
Soil, Quebec (anaerobic)	To 73	Lake, Tropical	44
Under mustard	24	Marine angiosperms, rhizosphere	700

Table 61

SELECTED EXAMPLES OF SEASONAL AMOUNTS AND PERCENTAGES OF TOTAL N FIXED SYMBIOTICALLY BY FIELD-GROWN SOYBEANS

Method or conditions	kg N_2 Fixed/ha × Season		N Fixed (%)	
	Untreated	Soil amended	Untreated	Soil amended
Inoculated vs. uninoculated soybeans	29	—	—	—
Nodulated vs. nonnodulated soybean isolines				
0 kg N/ha (good growth conditions)	155	—	40	—
56 kg N/ha as NH_4NO_3	—	110	—	29
112 kg N/ha as NH_4NO_3	—	77	—	20
168 kg N/ha as NH_4NO_3	—	69	—	16
45 ton/ha as corn cobs	—	280	—	72
Nodulated vs. nonnodulated soybean isolines				
0 kg N/ha	51	—	34	—
225 kg N/ha as NH_4NO_3	—	18	—	9
Nodulated vs. nonnodulated soybean isolines				
0 kg N/ha	103	—	48	—
112 kg N/a	—	62	—	27
224 kg N/ha	—	21	—	10
449 kg N/ha	—	22	—	10
Acetylene reduction	100	—	—	—
Acetylene reduction	164	—	35	—
Acetylene reduction				
0 kg N/ha (first-year soybeans)	8	—	—	—
280 kg N/ha as NH_4NO_3	—	<1	—	—
12.7 ton/ha as corn cobs	—	57	—	—
Acetylene reduction				
0 kg N/ha	—	—	6	—
200 kg N/ha as NH_4NO_3	—	—	—	<1
Acetylene reduction				
Field-grown soybeans	84	—	25	—
Outdoor chamber-grown soybeans	76	—	26	—
Outdoor chamber-grown soybeans + CO_2	—	427[a]	—	84
Average	86		31	

[a]   No soil amendment was used. The CO_2 level in canopy varied between 800—1200 ppm between 0800—1700 hr vs. ambient CO_2 for the untreated chamber-grown soybeans.

From Criswell, J. G., Hardy, R. W. F., and Havelka, U. D., in *World Soybean Research: Proceedings of the World Soybean Research Conference*, Hill, L. D., Ed., Interstate Printers & Publishers, Danville, Ill.. 1976, 118. With permission.

ANIMAL MANURES AND SEWAGE SLUDGE

Table 62
EXCRETA PRODUCTION BY AGRICULTURAL STOCK

	Amount per week			
	Dung		Urine	
Stock	ft³	cwt	ft³	gal
Mature cattle	6	4	24	4
Young cattle	3	2	12	2
Bacon pigs	0.5	0.3	3.5	0.5
100 hens	0.75	0.4	—	—

Table 63
CONVERSION FACTORS ANIMAL RESIDUES (dry basis)

Animal	Tons/year (metric tons/year)
Beef cattle	1.33 (1.21)
Milk cows	2.01 (1.82)
Heifers (500 lb and over)	0.95 (0.86)
Steers (500 lb and over)	0.95 (0.86)
Bulls (500 lb and over)	1.60 (1.45)
Heifers, steers, bulls (under 500 lb)	0.55 (0.50)
Pigs and hogs	0.13 (0.12)
Sheep and lambs	0.18 (0.17)
Horses, mules, donkeys	1.64 (1.49)
Chickens (layers)	0.009 (0.008)
Chickens (broilers)	0.0066 (0.0060)
Turkeys	0.060 (0.054)

Table 64
AVERAGE AMOUNTS (LB/TON) OF VARIOUS NUTRIENTS IN MANURES FROM DIFFERENT FARM ANIMALS

Nutrient	Dairy cattle	Beef cattle	Poultry	Swine	Sheep
Nitrogen (N)	10.0	14.0	25.0	10.0	28.0
Phosphorus (P)	2.0	4.0	11.0	2.8	4.2
Potassium (K)	8.0	9.0	10.0	7.6	20.0
Sulfur (S)	1.5	1.7	3.2	2.7	1.8
Calcium (Ca)	5.0	2.4	36.0	11.4	11.7
Iron (Fe)	0.1	0.1	2.3	0.6	0.3
Magnesium (Mg)	2.0	2.0	6.0	1.6	3.7
Boron (B)	0.01	0.03	0.01	0.09	0.02
Copper (Cu)	0.01	0.01	0.01	0.04	0.01
Manganese (Mn)	0.03	0.01	0.18	0.04	0.02
Zinc (Zn)	0.04	0.03	0.01	0.12	0.05
% moisture	85	85	72	82	77

REFERENCES
1. **Walsh, L. M. and Hensler, R. F.,** Manage Manure for Its Value, *Wisc. Ext. Circ.,* No. 550, 1971.
2. **Miller, H. F., Ross, I. J., and Thomas, G. W.,** Farm Manure: Production-Value-Use, ID-19, University of Kentucky Cooperative Extension Service, Lexington, KY., 1976.

Table 65
TYPICAL COMPOSITION OF MANURE

Fresh manure[a]	Moisture %	Approx. composition (lb/ton)		
		N	P_2O_5	K_2O
Fresh manure				
Cow	86	11	3	10
Hog	87	11	6	9
Horse	80	13	5	10
Sheep	68	20	15	8
Steer or feed yard	75	12	7	11
Duck	61	22	29	10
Goose	67	22	11	10
Hen	73	22	18	10
Turkey	74	26	14	10
Dried commercial products				
Cow, east	10	42	63	61
Cow, west	16	18	15	31
Hen, east[b]	16	56	57	30
Hen, west[b]	13	41	37	23
Hen, west[c]	8	83	63	31

[a] With normal quantity of bedding or litter.
[b] With litter.
[c] Droppings only.

Adapted from Knott, J. E., *Handbook for Vegetable Growers,* 5th printing, John Wiley & Sons, New York, 1966. Copyright © 1966 by John Wiley & Sons.

Table 66
TYPICAL COMPOSITION OF ORGANIC FERTILIZING MATERIALS

	Percentage on a dry-weight basis		
Organic materials	Total nitrogen, N	Available phosphoric acid, P_2O_5	Water-soluble potash, K_2O
Bat guano	10.0	4.0	2.0
Blood	13.0	2.0	1.0
Blood and bone	6.5	7.0	—
Bone black	1.3	15.0	—
Bone meal, raw	3.0	15.0	—
Steamed	2.0	15.0	—
Castor bean meal	5.5	2.0	1.0
Cotton seed meal	6.0	3.0	1.0
Fish meal	10.0	4.0	—
Fish solutions	10.0	3.0	1.0
	5.0	2.0	2.0
Garbage tankage	1.5	2.0	0.7
Horn and hoof meal	12.0	2.0	—
Sewage sludge	1.5	1.3	0.4
Activated	6.0	3.0	0.1
Tankage	9.0	6.0	—

From Knott, J. E., *Handbook for Vegetable Growers,* John Wiley & Sons, New York, 1966. Copyright © 1966 by John Wiley & Sons. With permission.

Table 67
APPROXIMATE NITROGEN LOSSES FROM SWINE MANURE AS AFFECTED BY HANDLING AND STORING

Handling and storing methods	Nitrogen loss (%)[a]
Solid systems	
Manure pack	20—40
Open lot	40—60
Liquid systems	
Anaerobic storage[b]	15—30
Lagoon[c]	70—80

[a] Based on composition of manure applied to land vs. composition of freshly excreted manure.
[b] Concentrated manure with little water added.
[c] Manure plus dilution water added for biological treatment and odor control.

Source: Sutton, A. L., Melvin, S. W., and Vanderholm, D. H., Fertilizer value of swine manure, in *Pork Industry Handbook,* P1H-25, rev., Cooperative Extension Service, Purdue University, West Lafayette, Ind., 1983.

Table 68
NITROGEN LOSSES TO THE AIR AS AFFECTED BY
APPLICATION METHOD IN SPREADING MANURE

Application method	Type of manure	Nitrogen loss (%)[a]
Broadcast without cultivation	Solid	15—30
	Liquid	10—25
Broadcast with cultivation[b]	Solid	1—5
	Liquid	1—5
Injection	Liquid	0—2
Irrigation (sprinkle)	Liquid	15—35

[a] Percent of total nitrogen in manure applied that is lost within 4 d after application.
[b] Cultivation within a few hours after application.

Source: Sutton, A. L., Melvin, S. W., and Vanderholdm, D. H., Fertilizer value of swine manure, in *Pork Industry Handbook*, P1H-25, rev., Cooperative Extension Service, Purdue University, West Lafayette, Ind., 1983.

Table 69
ESTIMATED PRODUCTION OF METHANE AND FERTILIZER ELEMENTS
FROM SWINE MANURE

Size operation[a]	Gross yield (ft³/d)[b]		Net methane[c] yield (ft³/d)	Fertilizer elements (lb/d)[d]		
	Biogas	Methane		N	P_2O_5	K_2O
50	240	144	100	4	3	3
100	480	290	200	7	5	5
200	960	580	400	14	10	11
500	2,400	1.440	1,000	35	25	25
1,000	4,800	2,900	2,000	70	50	50
2,000	9,600	5,800	4,000	140	100	100
5,000	24,000	14,400	10,000	350	250	250
10,000	48,000	29,000	20,000	700	500	500
15,000	72,000	43,000	30,000	1,100	750	750

[a] 150 lb hog or equivalent live weight, one-time capacity.
[b] Assumes volatile solids destruction = 50%; gross methane yield is 8.0 ft³ CH_4 per lb volatile solids destroyed; and biogas contains 60% methane and 40% carbon dioxide (volume basis).
[c] Assumes energy requirement for heating and mixing the digester is 30% of gross biogas output.
[d] Assumes recoverable fertilizer amounts are 0.07 lb N, 0.05 lb P_2O_5, and 0.05 lb K_2O per head per day, not including losses during slurry storage.

Source: Sweeten, J. M., Fulhage, C., Humenik, F. J., Methane gas in swine manure, *Pork Industry Handbook*, P1H-76, Cooperative Extension Service, Purdue University, West Lafayette, Ind., 1981.

Table 70

TOTAL ELEMENTAL COMPOSITION DRY WEIGHT BASIS OF OVER 200 SEWAGE SLUDGE SAMPLES FROM 8 STATES

Component	Concentration (%)		
	Minimum	Maximum	Median
Organic C	6.5	48.0	30.4
Inorganic C	0.3	543.0	1.4
Total N	<0.1	17.6	3.3
NH_4-N	<0.1	6.7	1.0
NO_3-N	<0.1	0.5	<0.1
Total P	<0.1	14.3	2.3
Inorganic P	<0.1	2.4	1.6
Total S	0.6	1.5	1.1
Ca	0.10	25.0	3.9
Fe	<0.10	15.3	1.1
Al	0.10	13.5	0.4
Na	0.01	3.1	0.2
K	0.02	2.6	0.3
Mg	0.03	2.0	0.4

	Concentration (ppm)		
Zn	101	27,800	1,740
Cu	84	10,400	850
Ni	2	3,515	82
Cr	10	99,000	890
Mn	18	7,100	260
Cd	3	3,410	16
Pb	13	19,730	500
Hg	<1	10,600	5
Co	1	18	4
Mo	5	39	30
Ba	21	8,980	162
As	6	230	10
B	4	757	33

From Dowdy, R. H., Larson, R. E., and Epstein, E., Sewage sludge and effluent use in agriculture, *Proc. Conf. Land Appl. Waste Mater.*, Soil Conservation Society of America, Ankeny, Iowa, 1976, 140. With permission.

Table 71
THREE ESTIMATES OF THE PROPERTIES OF SECONDARY EFFLUENTS

	Concentration (mg/*l*)			
Constituent	Range[54]	Typical value or range[55]	Typical value[56]	Range[56]
Suspended solids	13—62	25	—	—
Biochemical oxygen demand	13—75	25	—	—
Chemical oxygen demand	50—160	70	—	—
Nitrogen, total	—	20	25	15—40
Phosphorus, total	7—10	10	10	0.5—40
Trace metals				
Cadmium	—	0.015	<0.005	<0.005—6.4
Chromium	0.01—2.5	0.02—0.14	0.025	<0.05—6.8
Cobalt	—	—	<0.05	<0.05—0.05
Copper	0.10—1.4	0.07—0.14	0.10	<0.02—5.9
Iron	0.10—3.0	0.10—4.3	—	—
Lead	0.01—1.0	0.01—0.03	0.05	<0.02—6.0
Manganese	—	0.20	—	—
Mercury	<0.005	0.01	0.001	<0.0001—0.125
Nickel	0.02—2.0	0.03—0.20	0.02	<0.02—6.0
Zinc	0.10—1.10	0.20—0.44	0.15	<0.02—20
Other Parameters				
Boron	0.5—1.0	1.0	—	—
Calcium	1—40	20	—	—
Magnesium	1—10	17	—	—
Potassium	7—10	14	—	—
Sodium	40—100	50	—	—
Chloride	40—100	45	—	—
Sulfate	12—52	—	—	—
Oil	0—10	—	—	—
Phenol	0—1	—	—	—

Reproduced from Elliott, L. F. and Stevenson, F. J., *Soils for Management of Organic Wastes and Waste Waters*, American Society of Agronomy, Madison, Wis., 1977, 47—72, by permission of the American Society of Agronomy, Inc.

Table 72
TOTAL AMOUNT OF
SLUDGE METALS
ALLOWED ON
AGRICULTURAL LAND

Metal	Soil cation exchange capacity (me/100g)[a]		
	<5	5—15	>15
	Maximum metal load (lb/acre)		
Lead	500	1000	2000
Zinc	250	500	1,000
Copper	125	250	500
Nickel	50	100	200
Cadmium	5	10	20

[a] Determined by the pH 7 ammonium acetate procedure.

Source: Galloway, H. M. and Jacobs, L. W., Sewage sludges — characteristics and management, in Utilizing Municipal Sewage Wastewaters and Sludges on Land for Agricultural Production, Jacobs, L. W., Ed., *North Central Reg. Ext.,* Pub. No. 52, 1977, 3.

PLANT NUTRIENT UPTAKE

Table 73
CHEMICAL ELEMENTS ESSENTIAL FOR MOST PLANTS OR CROPS, SYMBOL, AND FORM UTILIZED

Major or macronutrients	Secondary nutrients	Trace or micronutrients
Carbon (C), CO_2	Calcium (Ca), Ca^{2+}	Boron (B), $H_2BO_3^-$
Oxygen (O), CO_2	Magnesium (Mg), Mg^{2+}	Chlorine (Cl), Cl^-, Cl_2
Hydrogen (H), H_2O	Sulfur (S), SO_4^{2-}, SO_2	Cobalt (Co), Co^{2+}
Nitrogen (N), NH_4^+,		Copper (Cu), Cu^{2+}
NO_3^-, N_2 (Legumes)		Iron (Fe), Fe^{2+}
Phosphorus (P),		Manganese (Mn), Mn^{2+}
$H_2PO_4^-$, HPO_4^{2-}		Molybdenum (Mo), MoO_4^{2-}
Potassium (K), K^+		Zinc (Zn), Zn^{2+}

Table 74
NUTRIENT ELEMENT CONCENTRATIONS OF SPECIFIC PLANT PARTS IN RELATION TO SUFFICIENCY FOR PLANT PRODUCTION[a]

Element	Alfalfa[b] Deficient	Alfalfa[b] Low	Alfalfa[b] Sufficient	Corn[c] Deficient	Corn[c] Low	Corn[c] Sufficient	Soybeans[d] Deficient	Soybeans[d] Low	Soybeans[d] Sufficient	Wheat[e] Deficient	Wheat[e] Low	Wheat[e] Sufficient
N (%)		Legume Plant	4.50	<2.0	2.0—2.4	>2.4		Legume Plant	3.50	<2.5	2.5—3.0	>3.0
P (%)	<0.20	0.20—0.25	>0.25	<0.15	0.15—0.20	>0.20	<0.10	0.10—0.25	>0.25	<0.15	0.15—0.25	>0.25
K (%)	<1.75	1.75—2.00	>2.00	<1.4	1.4—1.6	>1.6	<1.25	1.25—1.75	>1.75	<1.00	1.00—1.50	>1.50
Ca (%)	<1.00	1.00—1.80	>1.80	<0.10	0.10—0.20	>0.20	<0.80	0.80—1.20	>1.20	<0.15	0.15—0.30	>0.30
Mg (%)	<0.20	0.20—0.30	>0.30	<0.10	0.10—0.15	>0.15	<0.15	0.15—0.25	>0.25	<0.10	0.10—0.15	>0.15
S (%)	<0.20	0.20—0.22	>0.22	<0.15	0.15—0.20	>0.20	<0.20	0.20—0.25	>0.25	<0.15	0.15—0.20	>0.20
Fe (ppm)	<20	20—30	>30	<10	10—20	>20	<20	20—25	>25	<5	5—10	>10
Mn (ppm)	<15	15—20	>20	<9	9—15	>15	<14	14—20	>20	<10	10—15	>15
Cu (ppm)	<2	2—4	>4	<1	1—2	>2	<2	2—4	>4	<2	2—5	>5
Zn (ppm)	<10	10—20	>20	<15	15—18	>18	<15	15—20	>20	<10	10—15	>15
B (ppm)	<15	15—20	>20	<2	2—5	>5	<10	10—20	>20	<2	2—5	>5
Mo (ppm)	<0.5	0.5—1.0	>1.0	<0.5	0.5—1	>1.0	<0.5	0.5—1.0	>1.0	<0.02	0.02—0.5	>0.5

[a] An amalgamation of calibration data from the Ohio and Nebraska Agricultural Experiment Stations. Chlorine is omitted since it is unlikely to be found deficient in field soils for crop production anywhere in the U.S.

[b] Plant tops shortly before bloom stage.

[c] Ear leaf at early silk stage.

[d] Fully emerged leaves before flowering stage.

[e] Tops during rapid vegetative stage.

Table 75
GENERAL GUIDELINES FOR PLANT FOOD UPTAKE AT VARIOUS LEVELS OF CROP YIELD

Crop	Yield level	Nutrient uptake in lb/A				
		N	P_2O_5	K_2O	Mg	S
Alfalfa[a]	4 tons	225	60	240	20	20
	6	338	90	360	30	30
	8	450	120	480	40	40
Clovergrass[a]	3 tons	150	45	180	15	15
	6	300	90	360	30	30
	7	350	105	420	35	35
Bermudagrass (hybrid)	6 tons	258	60	288	18	30
	8	368	96	400	26	44
	10	460	120	500	34	58
Corn	120 bu	160	68	160	39	20
	160	213	91	213	52	26
	200	266	114	266	65	33
Cotton (lint)	750 lb	105	45	65	17	15
	1,125	143	54	96	26	23
	1,500	180	63	126	35	30
Potatoes	300 cwt	150	48	270	24	14
	600	300	96	540	40	27
	900	450	144	810	72	41
Sorghum (grain)	6,000 lb	178	63	180	30	28
	8,000	238	84	240	40	38
	10,000	297	105	300	50	47
Soybeans[a]	40 bu	224	38	144	16	14
	60	315	58	205	24	20
	80	416	78	250	32	26
Wheat	40 bu	75	27	81	12	10
	80	166	54	184	24	20
	100	188	68	203	30	25

[a] Legumes obtain most of their nitrogen through nitrogen fixation.

From Potash and Phosphate Institute, *Better Crops with Plant Food,* Vol. 72(4), 1988. With permission.

Table 76

THE N, P_2O_5, K_2O, Mg, AND S UPTAKE OF HIGH YIELDING CROPS

Crop	Yield per hectare plant part	kg/ha				
		N	P_2O_5	K_2O	Mg	S
Grain crops						
Barley	5,376 kg grain	123	45	39	9	11
	Straw	45	17	129	10	11
	Total	168	62	168	19	22
Buckwheat	1,613 kg grain	34	17	11	—	6
	Straw	13	6	28	—	3
	Total	47	23	39	—	9
Corn	12,544 kg grain	168	97	64	20	17
	Stover	130	30	234	53	20
	Total	298	127	298	73	37
Oats	3,584 kg	90	28	22	6	9
	Straw	39	17	140	17	12
	Total	129	45	162	23	21
Rice	7,840 kg grain	86	52	31	9	6
	Straw	39	16	134	7	8
	Total	125	68	165	16	14
Sorghum	8,960 kg grain	134	67	34	16	25
	Stover	146	34	190	34	18
	Total	280	101	224	50	43
Wheat	5,376 kg grain	106	49	30	13	6
	Straw	47	11	151	13	17
	Total	153	60	181	26	23
Oil crops						
Flax	1,344 kg seed	45	19	17	—	3
	Straw	16	6	34	—	3
	Total	61	25	51	—	6
Oil Palm	24,640 kg	73	28	111	21	—
	Fronds, stems	120	55	189	40	—
	Total	193	83	300	61	—
Peanuts	4,480 kg nuts	157	25	39	4	11
	Vines	112	19	168	22	12
	Total	269	44	207	28	22
Rapeseed	1,960 kg seed	74	36	18	—	13
	Straw	44	16	75	—	10
	Total	118	52	97	—	23
Soybeans	4,032 kg beans	269	54	94	19	13
	Stover	94	18	65	11	15
	Total	363	72	159	30	28
Sunflower	3,920 kg seed	140	67	44	13	7
	Stover	57	11	100	34	11
	Total	197	78	144	47	18
Fiber crops						
Cotton	1,680 kg lint	105	43	49	12	8
	Stalks	96	28	92	27	26
	Total	201	71	141	39	34
Pulp Wood	98 cords	168	34	90	22	—
(Slash pine)	Bark, branches	212	8	67	34	—
	Total	380	42	157	56	—
Sugar						
Sugar Cane	224 T cane	179	101	375	45	60
	Trash	224	74	308	67	36
	Total	403	175	683	112	96
Sugar Beets	67 T	140	17	280	30	11
	Tops	146	28	336	59	39
	Total	286	45	616	89	50

Table 76 (continued)
THE N, P₂O₅, K₂O, Mg, AND S UPTAKE OF HIGH YIELDING CROPS

Crop	Yield per hectare plant part	kg/ha				
		N	P₂O₅	K₂O	Mg	S
Silage crops						
Corn silage	72 T	298	128	298	73	37
Forage sorghum	18 T	222	75	320	39	20
Oat silage	18 T	123	34	263	22	17
Sorghum-Sudan	18 T	357	137	523	53	—
Forage crops						
Alfalfa	22 T	672	134	672	59	57
Big Bluestem	7 T	71	57	118	13	—
Birdsfoot Trefoil	9 T	215	94	305	36	—
Bromegrass	11 T	246	73	353	11	22
Clover-grass	13 T	336	101	403	34	34
Guineagrass	26 T	323	113	488	111	52
Hybrid Bermudagrass	22 T	560	157	470	56	56
Indiangrass	7 T	67	57	81	10	—
Lespedeza	7 T	168	56	168	28	22
Little Bluestem	7 T	74	57	97	13	—
Napiergrass	28 T	340	165	678	71	84
Orchard grass	13 T	336	112	420	28	39
Pangola grass	26 T	335	121	482	75	52
Paragrass	27 T	345	110	515	88	46
Pensacola Bahiagrass	16 T	339	97	271	39	30
Ryegrass	11 T	241	95	269	45	—
Switchgrass	7 T	77	67	128	17	—
Tall fescue	7.8 T	151	73	207	15	—
Timothy	9 T	168	62	280	11	18
Turf Grass						
Bluegrass (turf)	7 T	224	62	202	22	28
Bentgrass (turf)	6 T	291	74	164	15	11
Bermudagrass (turf)	9 T	252	45	179	22	17
Vegetable crops						
Bell peppers	20,160 kg	81	24	99	8	—
	Vines	73	35	144	40	—
	Total	154	59	243	48	—
Cabbage	78,400 kg	157	39	143	10	72
	Stems and leaves	146	31	136	30	—
	Total	303	70	279	40	72
Celery	168 T	286	146	762	—	—
	Roots	28	39	78	—	—
	Total	314	185	840	—	—
Cucumbers	22 T	45	16	74	4	—
	Vines	56	16	121	24	—
	Total	101	32	195	28	—
Lettuce	44.8 T heads	101	34	207	—	—
Onions	67.2 T	202	90	179	20	41
Peas	2,800 kg	103	27	35	7	7
	Vines	81	12	83	13	4
	Total	184	39	118	20	11
Potatoes	56 T	194	82	315	17	17
	Vines	108	19	297	40	8
	Total	302	101	612	57	25
Snap beans	9 T	78	24	86	9	—
	Vines	76	13	96	10	—
	Total	154	37	182	19	—

Table 76 (continued)
THE N, P₂O₅, K₂O, Mg, AND S UPTAKE OF HIGH YIELDING CROPS

Crop	Yield per hectare plant part	kg/ha				
		N	P₂O₅	K₂O	Mg	S
Sweet corn	10,080 kg	90	25	58	7	6
	Stover	67	28	94	16	7
	Total	157	53	152	23	13
Sweet potato	33.6 T	82	44	189	9	—
	Vines	93	34	161	11	—
	Total	175	78	350	20	—
Table beets	56 T	190	34	235	34	15
	Tops	213	15	414	83	31
	Total	403	49	649	117	46
Tomatoes	67,200 kg	112	26	242	9	24
	Vines	90	28	134	22	22
	Total	202	54	376	31	46
Fruits						
Apples	28 T	22	9	56	2	—
	Leaves and stems	90	43	146	25	—
	Total	112	52	202	27	—
Cantaloupe	19,600 kg	44	18	91	7	—
	Vines	29	6	40	7	—
	Total	73	24	131	14	—
Grapes	27 T	74	26	134	11	—
	Vines	40	13	40	9	—
	Total	114	39	174	20	—
Oranges	605 q	101	26	181	11	8
	Leaves and stems	196	36	188	31	24
	Total	297	62	369	42	32
Peaches	322 q	35	10	65	12	—
	Leaves and stems	60	30	55	10	—
	Total	95	40	120	22	—
Tropical crops						
Bananas	1,200 plants	448	448	1,680	175	—
Cocoa	1,008 kg beans	20	8	12	2	—
	Husks, stems, and leaves	446	113	809	131	—
	Total	446	131	821	133	—
Coconuts	1200 nuts	108	35	231	15	9
Oil palm	24,640 kg	73	28	111	21	—
	Fronds and stems	120	55	189	40	—
	Total	193	83	300	61	—
Pineapple	399.8 q	171	140	668	72	16
Rubber	2,262 kg	19	11	19	4	—
	Leaves and stems	65	30	104	15	—
	Total	84	41	123	19	—
Sugar cane	244 T	179	101	375	45	60
	Trash	224	74	308	67	36
	Total	403	175	683	112	96
Tobacco						
Flue cured	3,360 kg	95	17	174	17	13
	Stalks, etc.	46	12	114	10	8
	Total	141	29	288	27	21
Burley	4,480 kg	165	15	166	22	19
	Stalks, etc.	160	27	194	15	8
	Total	325	42	360	37	27

From Agerton, R. and Martin, S., Eds., *Better Crops with Plant Food,* 63, 4, 1979. With permission.

Table 77

NUTRIENTS CONTAINED IN THE TOTAL
ABOVEGROUND PLANT MATERIAL IN AN
ACRE OF 150-BUSHEL CORN OF THE CORN
BELT[a]

	Pounds of element		
Element	Grain	Stover[b]	Total plant above ground
N	115	55	170
P	28	7	35
K	35	140	175
Ca	1	34	35
Mg	10	30	40
S	11	9	20
Cl	4	68	72
Fe	0.1	1.8	1.9
Mn	0.05	0.25	0.3
Cu	0.02	0.08	0.1
Zn	0.17	0.17	0.3
B	0.04	0.12	0.2
Mo	0.005	0.003	0.01

[a] Average values derived from analyses recorded by the Iowa, Indiana, Michigan, and Nebraska Experiment Stations of the North Central U. S.

[b] The quantities of Ca, Mg, Cl, and probably Fe in the stover represent luxury consumption of these nutrients.

Reproduced from Barbar, S. A. and Olson, R. A., *Changing Patterns of Fertilizer Use,* Nelson, L. B., Ed., Soil Science Society of America, Madison, Wisconsin, 1968, 169, by permission of the Soil Science Society of America, Inc.

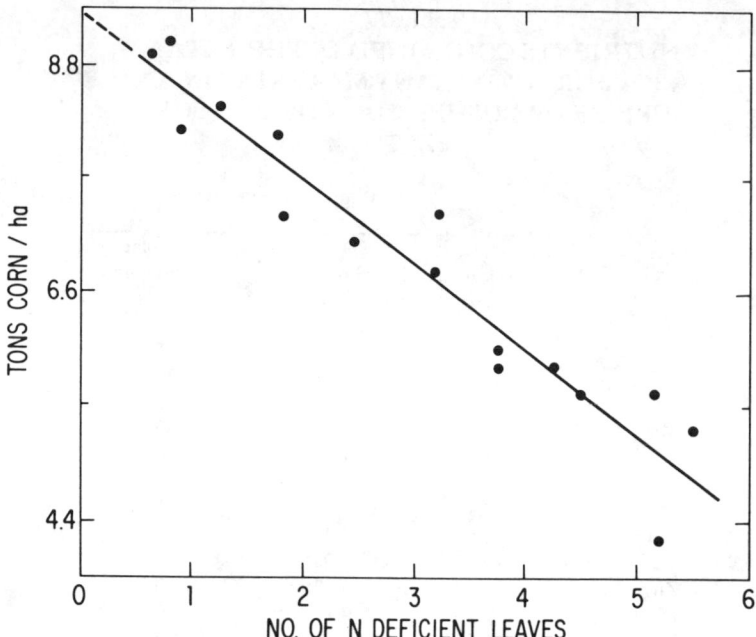

FIGURE 10.   Relationship between number of N deficient corn leaves showing characteristic tip burn at tasseling and yield. Plants without visible evidence of N deficiency on the leaves at tasseling may yield 3 tons less corn per hectare than plants with sufficient N. (Reproduced from Viets, F. G., Jr., The plant's need for and use of nitrogen, in *Soil Nitrogen,* Bartholomew, W. V. and Clark, F. E., Eds., *Agronomy J.,* 10, 503, 1965, by permission of the American Society of Agronomy, Inc.)

Plant Nutrient Uptake **95**

Table 78
FORMS OF PHOSPHORUS FOUND IN SOILS FROM DIFFERENT ORDERS

Soil, horizon, and depth (cm)	Location	Soil order	pH	% Free Fe_2O_3[a]	Ca-P	Al-P	Fe-P (ppm)	Occluded Fe-P	Al-P
Wahiawa A, 0—46	Hawaii	Oxisol	6.7	10.1	10	20	123	504	97
Catalina A, 30—90	Puerto Rico	Oxisol	6.2	18.2	6	0	42	320	40
Miami A_3, 30—40	Wis.	Alfisol	5.8	1.4	45	48	100	73	6
Miami C_1, 107—147	Wis.	Alfisol	7.8	1.2	525	21	10	7	tr
Scott B, 18—71	Neb.	Mollisols	6.7	0.75	162	85	48	16	12
Dewey A_2, 5—30	Tenn.	Ultisol	5.5	3.3	32	2	22	26	19
Rosebud B_2, 25—45	Neb.	Mollisols	7.1	0.82	122	11	8	9	6
Barnes, 8—30	N.D.	Mollisols	7.5	0.77	155	18	8	tr	tr
Barnes, 71—137	N.D.	Mollisols	8.0	0.55	190	11	11	tr	tr
Barnes, 137—162	N.D.	Mollisols	8.0	0.93	185	6	3	tr	tr

From Chang, S. C. and Jackson, M. L., Soil phosphorus fractions in some representative soils, *J. Soil Sci.*, 9, 109, 1958. Copyright© by Williams & Wilkins, 1958. With permission.

Table 79
YIELDS AND PHOSPHORUS CONTENTS OF WHEAT, CORN, AND RED CLOVER AS AFFECTED BY SOIL PHOSPHORUS LEVEL (TATUM SIL.)

Extractable P kg/ha[a]	Wheat—11 year Yield (kg/ha)	%P	Corn—13 year Yield (kg/ha)	%P	Red Clover—10 year Yield (kg/ha)	%P
6.2	1297	0.16	2465	0.14	3566	0.13
9.9	2056	0.18	3506	0.19	6485	0.15
18.8	2271	0.19	3870	0.23	6987	0.18
44.8	2500	0.23	4045	0.30	7206	0.25
165.3	2419	0.28	4083	0.28	7176	0.28

Method of P Extraction: Mehlich $HCl-H_2SO_4$

Table 80

EFFECTS OF EXCHANGEABLE K, SOIL SOLUTION K, AND CEC ON K
CONTENT OF SORGHUM

Soil	Predominant clay minerals	Soil order	Exchangeable K(meq/100g)	Soil solution K (meq/*l*)	CEC (meq/100 g)	% K in sorghum
Miller c	Mica-Mont	Mollisols	1.17	0.34	39	2.2
San Saba c	Mont	Vertisols	0.94	0.12	44	1.6
Houston Black c	Mont	Vertisols	0.93	0.09	42	1.7
Norwood sil	Mica-Mont	Entisols	0.92	0.30	19	1.9
Clareville cl	Mont	Mollisols	0.78	0.20	29	1.8
Lake Charles c	Mont	Vertisols	0.48	0.10	45	1.4
Nacogdoches fsl	Kaol	Ultisols	0.20	0.34	3.5	1.6
Lufkin fsl	Mont	Alfisols	0.05	0.12	5	1.0

Reproduced from Hipp, B. W. and Thomas, G. W., Method for predicting potassium uptake by grain sorghum, *Agron, J.,* 60, 467, 1968, by permission of the American Society of Agronomy, Inc.

Table 81

INCREASE IN GRAIN YIELD OF CORN WITH TOTAL UPTAKE
OF N, P, AND K IN CROP 87 DAYS AFTER PLANTING

Corn yield level		Nutrient uptake, excluding roots		
		N	P	K
(q/ha)	(bu/A)	(kg/ha)		
69	110	92	24	131
100	159	162	34	185
127	202	204	43	250
145	231	239	45	290

From Rhoads, F. M. and Stanley, R. L., Jr., Effect of population and fertility on nutrient uptake and yield components of irrigated corn, *Soil Crop Sci. Soc. Fla.,* 38, 78, 1978. With permission.

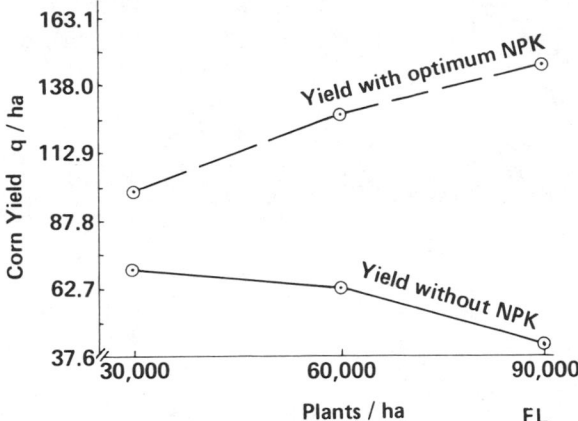

FIGURE 11.   Improved cultural practices without adequate nutrients can be unproductive. Response with optimum rates of N, P, K may be linear. (From Rhoads, F. M. and Stanley, R. L., Jr., Effect of population and fertility on nutrient uptake and yield components of irrigated corn, *Soil and Crop Sci. Soc. Fla.*, 38, 78, 1978. With permission.)

Table 82
INCREASE IN NUTRIENT UPTAKE AND YIELD LEVEL IN SOYBEANS

Soybean yield level		Total dry matter	Maximum nutrient uptake excluding roots				
			N	P	K	Ca	Mg
(kg/ha)	(bu/A)	(kg/ha)			(kg/ha)		
2351	35	8,549	148.8	16.1	121.3	63.2	30.0
3360	50	10,204	287.8	23.3	175.0	54.9	21.5
5376	80	13,885	460.5	37.8	232.5	80.5	35.8

Source: Henderson, J. B. and Kamprath, E. J., Nutrient and dry matter accumulation by soybeans, *N.C. Agric. Exp. Stn. Tech. Bull.*, No. 197, 1970.

Table 83
INCREASE IN NUTRIENT UPTAKE AND YIELD LEVEL IN ALFALFA

Tons /ha	Nutrient (kg per unit)											
	N	P	K	Ca	Mg	S	Fe	Cu	Zn	Mn	B	Mo
1	29.1	2.49	23.0	12.5	2.6	2.4	0.16	0.006	0.020	0.051	0.026	0.004
11.2	326	28	258	140	29	27	1.8	0.07	0.22	0.57	0.29	0.05
22.4	652	56	516	280	58	54	3.6	0.14	0.44	1.14	0.58	0.10

COMMERCIAL FERTILIZERS

Table 84
COMMERCIAL AND EXPERIMENTAL FERTILIZERS — MANUFACTURE AND ANALYSES

Fertilizer and common abbreviations	Typical grades ($N-P_2O_5-K_2O$)	Water solubility of N, P or K[a]	Main reactants or process used	Major fertilizer compounds present	N	P	K	Ca	Mg	S
N Sources										
Ammonia	82-0-0	100	$N_2 + H_2$	NH_3	82.0	—	—	—	—	—
Anhydrous aqua	20-0-0	100	$NH_3 + H_2O$	NH_3	15—30	—	—	—	—	—
Ammonium chloride	28-0-0	100	$NH_3 + HCl$	NH_4Cl	28.0	—	—	—	—	—
Ammonium nitrate (AN)	33.5-0-0	100	$NH_4 + NHO_3$	NH_4NO_3	33—34	—	—	—	—	—
Ammonium nitrate limestone (ANL)	20.5-0-0	100	NH_4NO_3 + lime	NH_4NO_3	20.5	—	—	8.2	4.4	0.4
Ammonium nitrate sulfate (ANS)	30-0-0	100	$NH_3 + HNO_3 + H_2SO_4$	H_4NO_3, $(NH_4)_2SO_4$	30.0	—	—	—	—	12.0
Ammonium sulfate (AS)	21-0-0	100	$NH_3 + H_2SO_4$	$(NH_4)_2SO_4$	21.0	—	—	0.3	—	24.0
Calcium cyanamid	22-0-0	>90	$CaC_2 + N_2$	$CaCN_2$	22.0	—	—	40.0	0.1	0.3
Calcium nitrate	15.5-0-0	100	$CaCO_3 + HNO_3$	$Ca(NO_3)_2$	15.5	—	—	19.0	1.5	—
Nitrogen solutions	32-0-0	100	NH_4NO_3 + urea	NH_4NO_3, $CO(NH_2)_2$	21—36	—	—	—	—	—
Sodium nitrate	16-0-0	100	Beneficiation	$NaNO_3$	16.0	—	0.2	0.1	0.1	0.1
Chilean synthetic	16-0-0	100	$Na_2CO_3 + HNO_3$	$NaNO_3$	16.0	—	—	—	—	—
Urea	46-0-0	100	$NH_3 + CO_2$	$CO(NH_2)_2$	46.0	—	—	—	—	—
Urea, S-coated (SCU)	35-0-0	v	$CO(NH_2)_2 + S$	$CO(NH_2)_2$, S	35.0	—	—	—	—	7—10
Urea-form (UF)	38-0-0	v	$CO(NH_2)_2 + CH_2O$	$CO(NH_2)_2 : CH_2O$	38.0	—	—	—	—	—
P Sources										
Bone meal	3-25-0	<1	Grinding + steam		2—4	9—12	0.2	20—25	0.4	0.1
Phosphate rocks (PR)	0-32-0	<1	Beneficiation	Apatites	—	11—17	—	33—36	—	—
Phosphoric acid										
Wet process	0-55-0	100	$PR + H_2SO_4$	H_3PO_4	—	22—26	—	—	—	—
Furnace grade	0-55-0	100	Elemental P + H_2O	H_3PO_4	—	22—26	—	—	—	—
Super acid	0-78-0	100	Elemental P + H_2O	H_3PO_4, poly P acids	—	32—34	—	—	—	—
Superphosphates										
Ordinary (OSP)	0-20-0	85	$PR + H_2SO_4$	$Ca(H_2PO_4)_2 \cdot H_2O$, $CaSO_4 \cdot _2H_2O$	—	7—9	0.2	20	0.2	12.0
Triple (TSP)	0-45-0	87	$PR + H_3PO_4$	$Ca(H_2PO_4)_2 \cdot 2H_2O$	—	18—22	0.3	14	0.3	1.4
Concentrated (CSP)	0-48-0	90	$PR + H_3PO_4$ (furnace grade)	$Ca(H_2PO_4)_2 \cdot H_2O$	—	21	—	12—14	—	<1.0
Dicalcium phosphate										
HCl process	0-40-0	4	$H_3PO_4 + CaCO_3$	$CaHPO_4 \cdot 2H_2O$, $CaHPO_4$	—	17	—	28	—	—
Furnace H_3PO_4	0-48-0	3	$H_3PO_4 + CaCO_3$	$CaHPO_4$, $CaHPO_4 \cdot 2H_2O$	—	21	—	19	—	—

Material	Grade	%	Preparation	Principal compounds	N	P	K	Ca	Mg	S
Calcium metaphosphate (CMP)	0-62-0	5	PR + P (heat)	Vitreous Ca polyphosphate, $Ca_2P_2O_7$	—	27	—	19	—	—
Fused tricalcium phosphate (FTP)	0-62-0	<2	PR + heat	$Ca_3(PO_4)_2$	—	12	—	20	—	—
Rhenania phosphate	0-33-0	<2	PR + Na_2CO_3 + SiO_2 + heat	Silico phosphates	—	14	—	30	0.3	—
Serpentine phosphate	0-22-0	<2	PR + serpentine	Ca, Mg silico phosphates	—	9	—	20	8	—
Basic slag	0-9-0	<2	Iron ore + lime + heat	Ca silico carnotite	—	3.5	—	29	3	0.3
Colloidal "clay" phosphate	0-22-0	<1	Tailings from PR beneficiation	Apatites, Al, phosphates	—	9	—	—	—	—
N, NK, & PK Sources										
Muriate of potash	0-0-60	100	Beneficiation	KCl	—	—	50—52	0.1	0.1	—
Sulfate of potash	0-0-50	100	KCl + H_2SO_4, beneficiation	K_2SO_4	—	—	40—43	—	0.1	18
K, Mg sulfate (sulpomag)	0-0-23	100	Beneficiation	K_2SO_4, $2MgSO_4$	—	—	18	0.1	11	18
Nitrate of potash	13-0-44	100	KCl + HNO_3 or $NaNO_3$	KNO_3	13	—	37	0.3	0.3	0.2
K polyphosphate (KPP)	0-59-39	5	KCl + H_3PO_4	K polyphosphates	—	26	32	—	—	—
NP Sources										
Ammoniated OSP	4-14-0	35	OSP + NH_3	$NH_4H_2PO_4$, $CaHPO_4$, apatite	4	6	—	16	0.3	10
Ammoniated CSP	5-47-0	50	CSP + NH_3	$NH_4H_2PO_4$, $CaHPO_4$	5	20	—	12—14	—	<1
Ammoniated TSP		50	TSP + NH_3	$(NH_4)_2HPO_4$, apatite	9	—	—	—	—	—
Ammonium phosphate nitrate (APN)	30-10-0	100	H_3PO_4 + NH_3 + HNO_3	$NH_4H_2PO_4$,	30	4	—	—	—	—
	25-25-0	100		NH_4PO_3,	25	11	—	—	—	—
	28-14-0	100		$(NH_4)_2HPO_4$	28	6	—	—	—	—
Monoammonium phosphate (MAP)	11-48-0	>90	NH_3 + H_3PO_4	$NH_4H_2PO_3$	11	21	—	1.5	0.2	4.5
Ammonium phosphate sulfate (APS)	13-39-0	>90	H_3PO_4 + NH_3 + H_2SO_4	$NH_4H_2PO_4$	13	17	—	—	—	2
	16-20-0	>90		$(NH_4)_2HPO_4$, $(NH_4)_2SO_4$	16	9	—	—	—	15
	16-48-0	>90		$(NH_4)_2HPO_4$	16	21	—	—	—	2
Diammonium phosphate (DAP)	21-53-0	100	H_3PO_4 + NH_3	$(NH_4)_2HPO_4$	21	23	—	—	—	—
	18-55-0	>95		$NH_4H_2PO_4$	18	24	—	—	—	—
Ammonium polyphosphate (APP)	15-60-0	100	Poly P acid + NH_3	$NH_4H_2PO_4$, $(NH_4)_2HP_2O_7$,	15	26	—	—	—	—
	12-55-0	>90		$NH_4H_2PO_4$	12	24	—	—	—	—
Urea ammonium phosphate (UAP)	34-17-0	100	$CO(NH_2)_2$ + $(NH_4)_2HPO_4$	$CO(NH_2)_2$, $(NH_4)_2HPO_4$	34	7	—	—	—	—
	29-29-0	100			29	13	—	—	—	—
	25-35-0	100			25	15	—	—	—	—
Urea ammonium polyphosphate (UAPP)	36-18-0	100	$CO(NH_2)_2$ + poly P acids	$CO(NH_2)_2$, $(NH_4)_2HP_2O_7$, $NH_4H_2PO_4$	36	8	—	—	—	—
	30-30-0	100			30	13	—	—	—	—
	22-44-0	100			22	19	—	—	—	—

Table 84 (continued)
COMMERCIAL AND EXPERIMENTAL FERTILIZERS — MANUFACTURE AND ANALYSES

Fertilizer and common abbreviations	Typical grades (N-P_2O_5-K_2O)	Water solubility of N, P or K[a]	Main reactants or process used	Major fertilizer compounds present	Macronutrient content (%)					
					N	P	K	Ca	Mg	S
Urea phosphate (UP)	18-45-0	100	$CO(NH_2)_2$ + H_3PO_4	$CO(NH_2)_2 \cdot H_3PO_4$	18	20	—	—	—	—
Nitric phosphate (NP)	20-20-0	40	PR + HNO_3 (+ H_3PO_4 + H_2SO_4)	$CaHPO_4$, $Ca(NO_3)_2$, $NH_4H_2PO_4$, apatite	20	9	—	—	—	—

[a] v = variable.

REFERENCES

1. **Terman, G. L., Bouldin, D. R., and Webb, J. R.,** Evaluation of fertilizers by biological methods, *Advan. Agron.*, 14, 265, 1962.
2. **Tisdale, S. L. and Nelson, W. L.,** *Soil Fertility and Fertilizers*, 2nd ed., Macmillan, New York, 1966.

Table 85

SALT INDICES OF SELECTED FERTILIZERS

Fertilizer N sources	Fertilizer grade	Salt index	
		On a material weight basis	Per unit (20 lb) of plant nutrients
$NaNO_3$	16.5-0-0	100	6.06
Anhydrous NH_3	82-0-0	47	0.57
NH_4NO_3	33.5-0-0	100	3.00
$(NH_4)_2SO_4$	21-0-0	69	3.25
$Ca(NO_3)_2$	15.5-0-0	65	4.20
$CaCN_2$	21-0-0	31	1.48
Urea	46-0-0	75	1.62
UAN solution	40-0-0	78	1.93
Organic N	5—13-0-0	4	0.3 to 0.7
P sources			
OSP	0-20-0	8	0.39
CSP, TSP	0-45-0	10	0.22
K sources			
KCl	0-0-60	116	1.94
K_2SO_4	0-0-50	46	0.85
$K_2SO_4 \cdot 2MgSO_4$	0-0-23	43	1.97
KNO_3	0-0-47	74	1.22
NP, NK, PK, or NPK sources			
MAP	11-48-0	30	0.41
DAP	18-46-0	34	0.46
KNO_3	13-0-44	74	1.22
APS	16-20-0	45	—
NPK	5-10-10	38	—
	10-20-20	57	—
	15-15-15	71	—

Note: References included in the original.

Table 86
EQUIVALENT ACIDITY (A) OR BASICITY (B) OF SELECTED FERTILIZERS

Fertilizer	Grade	Acidic or basic	CaCO₃ (lb) Per unit (20 lb) of N	CaCO₃ (lb) Per 100 lb of fertilizer
Anhydrous NH₃	82-0-0	A	36	146
NH₄NO₃	33.5-0-0	A	36	60
(NH₄)₂SO₄	20.5-0-0	A	107	110
Urea	46-0-0	A	36	84
N solution	32-0-0	A	36	57
Organic sources	2 to 13-0-0	A	30—50	7—20
Ca(NO₃)₂	16-0-0	B	22	20
NaNO₃	16-0-0	B	36	29
CaCN₂	22-0-0	B	57	63
KNO₃	13-0-44	B	40	26
MAP	11-48-0	A	36	20
DAP	18-46-0	A	36	32
Superphosphates	0-16 to 45-0	B	—	5—12
KCl or K₂SO₄	0-0-50 to 60	—	—	0

[a] AOAC method, equivalent pounds of $CaCO_3$/ton of fertilizer:

Equivalent acidity due to N = % N × 35.7

Basicity or acidity of ignited residue = net ml of 0.5 N NaOH; titration × 50 (1 g of fertilizer sample);

Basicity due to citrate insoluble P_2O_5 = % citrate insoluble P_2O_5 × 28.2

Net = equivalent acidity or basicity

REFERENCES

1. **Tisdale, S. L. and Nelson, W. L.,** *Soil Fertility and Fertilizers,* 2nd ed., Macmillan, New York, 1966.
2. *Farm Chemicals Handbook,* Meister, Willoughby, Ohio, 1978.

Table 87
BULK DENSITIES OF COMMERCIAL
FERTILIZERS

Product	Grade	Average bulk density (lb/ft^3)
Urea		
Prilled	46-0-0	45
Granular	46-0-0	46
Ammonium nitrate		
Prilled		
High density	33.5-0-0	58
Low density	33.5-0-0	45
Granular	33.5-0-0	46
Urea ammonium phosphate	28-28-0	48
	34-17-0	
Urea ammonium sulfate	40-0-0	45
Ammonium sulfate	21-0-0	60
Sulfur-coated urea	36-0-0	45
Diammonium phosphate	18-46-0	55
Monoammonium phosphate	11-52-0	56
Normal superphosphate	0-19-0	68
Triple superphosphate	0-45-0	66
Potassium chloride		
Granular	0-0-60	67
Coarse	0-0-60	68

Source: Balay, H. V., National Fertilizer Development Center, TVA, Muscle Shoals, Ala., unpublished data.

Table 88
CHEMICAL COMPOSITION AND PHYSICAL PROPERTIES OF LOW-PRESSURE LIQUID NITROGEN MATERIALS

Composition (%)				Nitrogen content (%)					Salting out temperature (°F)	Vapor pressure (psi) 60° F	Vapor pressure (psi) 104° F	Specific gravity at 60°F	Weight of solution per gal at 60°F (lb)	Weight of nitrogen per gal at 60°F (lb)	Vol. per 100 lb of nitrogen at 60°F (gal)
Ammonia	Ammonium nitrate	Urea	Water	Ammonia N Free	Ammonia N Combined	Nitrate N	Urea N	Total N							
Aqua ammonia															
25.0	—	—	75.0	20.6	—	—	—	20.6	−103	−9	2	0.911	7.60	1.56	64.1
29.4	—	—	70.6	24.2	—	—	—	24.2			10	0.897	7.48	1.84	54.3
Ammonia-ammonium nitrate solution															
8.0	72.0	—	20.0	6.6	12.6	12.6	—	31.8	89	—	1	1.257	10.48	3.33	30.0
10.0	72.0	—	18.0	8.2	12.6	12.6	—	33.4	85	—	4	1.230	10.26	3.43	29.2
18.5	54.2	—	27.3	15.2	9.5	9.5	—	34.2	12	—	1	1.118	9.32	3.19	31.3
16.6	66.8	—	16.6	13.7	11.7	11.7	—	37.1	50	−9	2	1.184	9.87	3.66	27.3
19.0	72.5	—	8.5	15.6	12.7	12.7	—	41.0	61	−6	7	1.194	9.96	4.08	24.5
22.2	65.0	—	12.8	18.3	11.4	11.4	—	41.1	21	−5	10	1.138	9.49	3.90	25.6
21.9	65.7	—	12.4	18.0	11.5	11.5	—	41.0	21	−5	10	1.139	9.50	3.90	25.6
24.3	60.0	—	15.7	20.0	10.5	10.5	—	41.0	1	—	13	1.105	9.22	3.78	26.5
26.5	55.0	—	18.5	21.8	9.6	9.6	—	41.0	−23	—	16	1.077	8.98	3.68	27.2
26.3	55.5	—	18.2	21.6	9.7	9.7	—	41.0	−25	−3	17	1.078	8.99	3.69	27.1
19.0	74.0	—	7.0	15.6	13.0	13.0	—	41.6	64	−3	8	1.186	9.89	4.11	24.3
23.8	69.8	—	6.4	19.6	12.2	12.2	—	44.0	25	−2	18	1.147	9.57	4.21	23.8
25.0	69.0	—	6.0	20.6	12.1	12.1	—	44.8	6	0	17	1.124	9.37	4.20	23.8
Ammonia-urea															
4.3	—	37.5	58.2	3.5	—	—	17.5	21.0	33	—	—	1.085	9.05	1.90	52.6

Ammonia-ammonium nitrate-urea solutions

13.3	53.4	15.9	9.4	10.9	9.3	9.3	7.4	36.9	36	0	—	1.235	10.30	3.80	26.3
16.0	70.0	6.0	8.0	13.2	12.3	12.3	2.8	40.6	66	—	4	1.228	10.24	4.16	24.0
19.0	58.0	11.0	12.0	15.6	10.2	10.2	5.1	41.1	7	-6	10	1.162	9.69	3.98	25.1
19.0	66.0	5.0	10.0	15.6	11.6	11.6	2.3	41.1	44	—	8	1.182	9.86	4.05	24.7
19.0	65.6	6.0	9.4	15.6	11.5	11.5	2.8	41.4	35	-7	11	1.178	9.82	4.07	24.6
19.5	66.3	6.0	8.2	16.0	11.6	11.6	2.8	42.0	34	—	10	1.179	9.83	4.13	24.2
20.0	68.0	6.0	6.0	16.4	11.9	11.9	2.8	43.0	33	-3	15	1.176	9.81	4.22	23.7
21.8	62.4	7.0	8.8	17.9	10.9	10.9	3.3	43.0	12	—	15	1.165	9.72	4.18	23.9
22.0	66.0	6.0	6.0	18.1	11.6	11.6	2.8	44.1	13	-2	19	1.155	9.63	4.25	23.5
24.0	56.0	10.0	10.0	19.7	9.8	9.8	4.7	44.0	-13	—	19	1.119	9.33	4.11	24.3

From National Fertilizer Solutions Associations, *Fluid Fertilizer Manual*, Vol. 2, 9-9, 1985. With permission.

Table 89
SOME CONVENIENT CONVERSION FACTORS FOR FERTILIZER MATERIALS

Multiply	By	To get
Ammonia — NH_3	4.700	Ammonium nitrate — NH_4NO_3
Ammonia — NH_3	3.888	Ammonium sulfate — $(NH_4)_2SO_4$
Ammonia — NH_3	0.823	Nitrogen — N
Ammonium nitrate — NH_4NO_3	0.350	Nitrogen — N
Ammonium sulfate — $(NH_4)_2SO_4$	0.212	Nitrogen — N
Borax — $Na_2B_4O_7$, 10 H_2O	0.114	Boron — B
Boric acid — H_3BO_3	0.177	Boron — B
Boron — B	8.807	Borax — $Na_2B_4O_7$, 10 H_2O
Boron — B	5.636	Boric acid — H_3BO_3
Calcium — Ca	1.399	Calcium oxide — CaO
Calcium — Ca	2.498	Calcium carbonate — $CaCO_3$
Calcium — Ca	1.849	Calcium hydroxide — $Ca(OH)_2$
Calcium — Ca	4.295	Calcium sulfate — $CaSO_4$, 2 H_2O (gypsum)
Calcium carbonate — $CaCO_3$	0.400	Calcium — Ca
Calcium carbonate — $CaCO_3$	1.351	Calcium hydroxide — $Ca(OH)_2$
Calcium carbonate — $CaCO_3$	0.560	Calcium oxide — CaO
Calcium carbonate — $CaCO_3$	0.403	Magnesia — MgO
Calcium carbonate — $CaCO_3$	0.842	Magnesium carbonate — $MgCO_3$
Calcium hydroxide — $Ca(OH)_2$	0.541	Calcium — Ca
Calcium hydroxide — $Ca(OH)_2$	0.741	Calcium carbonate — $CaCO_3$
Calcium hydroxide — $Ca(OH)_2$	0.756	Calcium oxide — CaO
Calcium oxide — CaO	0.715	Calcium — Ca
Calcium oxide — CaO	1.785	Calcium carbonate — $CaCO_3$
Calcium oxide — CaO	1.323	Calcium hydroxide — $Ca(OH)_2$
Calcium oxide — CaO	3.071	Calcium sulfate — $CaSO_4$, 2 H_2O (gypsum)
Gypsum — $CaSO_4$, 2 H_2O	0.326	Calcium oxide — CaO
Gypsum — $CaSO_4$, 2 H_2O	0.186	Sulfur — S
Magnesia — MgO	2.480	Calcium carbonate — $CaCO_3$
Magnesia — MgO	0.603	Magnesium — Mg
Magnesia — MgO	2.092	Magnesium carbonate — $MgCO_3$
Magnesia — MgO	2.986	Magnesium sulfate — $MgSO_4$
Magnesia — MgO	6.114	Magnesium sulfate — $MgSO_4$, 7 H_2O (Epsom salts)
Magnesium — Mg	4.112	Calcium carbonate — $CaCO_3$
Magnesium — Mg	1.658	Magnesia — MgO
Magnesium — Mg	3.466	Magnesium carbonate — $MgCO_3$
Magnesium — Mg	4.951	Magnesium sulfate — $MgSO_4$
Magnesium — Mg	10.136	Magnesium sulfate — $MgSO_4$, 7 H_2O (Epsom salts)
Magnesium carbonate — $MgCO_3$	1.187	Calcium carbonate — $CaCO_3$
Magnesium carbonate — $MgCO_3$	0.478	Magnesia — MgO
Magnesium carbonate — $MgCO_3$	0.289	Magnesium — Mg
Magnesium sulfate — $MgSO_4$	0.335	Magnesia — MgO
Magnesium sulfate — $MgSO_4$	0.202	Magnesium — Mg
Magnesium sulfate — $MgSO_4$, 7 H_2O (Epsom salts)	0.164	Magnesia — MgO
Magnesium sulfate — $MgSO_4$, 7 H_2O (Epsom salts)	0.099	Magnesium — Mg
Manganese — Mn	2,749	Manganese(ous) sulfate — $MnSO_4$
Manganese — Mn	4.060	Manganese(ous) sulfate — $MnSO_4$, 4 H_2O
Manganese(ous) sulfate — $MnSO_4$	0.364	Manganese — Mn
Manganese(ous) sulfate — $MnSO_4$, 4 H_2O	0.246	Manganese — Mn
Muriate of potash — KCl	0.632	Potash — K_2O
Muriate of potash — KCl	0.524	Potassium — K
Nitrate — NO_3	0.226	Nitrogen — N
Nitrate of potash — KNO_3	0.466	Potash — K_2O
Nitrate of potash — KNO_3	0.387	Potassium — K

Table 89 (continued)
SOME CONVENIENT CONVERSION FACTORS FOR FERTILIZER
MATERIALS

Nitrate of soda — $NaNO_3$	0.165	Nitrogen — N
Nitrogen — N	1.216	Ammonia — NH_3
Nitrogen — N	2.856	Ammonium nitrate — NH_4NO_3

From Knott, J. E., *Handbook for Vegetable Growers,* John Wiley & Sons, New York, 1966, 74. Copyright © by John Wiley & Sons, 1966. With permission.

PLANTS

PLANT NAMES AND DISTRIBUTION

Table 90
SELECTED ECONOMIC PLANTS, THEIR COMMON NAMES, SCIENTIFIC NAMES, AUTHORITIES, CHROMOSOME NUMBER, AND TARGET YIELD

Common name[a]	Scientific name[b]	Chromosomes (2n)[c]	Target yield[d]
Alfalfa	*Medicago sativa* L.	16, 32, 64	74,000 hay
Almond	*Prunus dulcis* (Mill.) D.A. Webb	16	3,000 sd
Aloe, Barbados	*Aloe barbadensis* Mill.	14, 10	250 dr
Amaranth, Chinese	*Amaranthus tricolor* L.	32, 34	
Apple	*Malus sylvestris* Mill.	34, 51	40,000 fr
Apricot	*Prunus armeniaca* L.	16	75 fr/
Artichoke, globe	*Cynara scolymus* L.	34	9,000 fl
Artichoke, Jerusalem	*Helianthus tuberosus* L.	102	20,000 rt
Asparagus, garden	*Asparagus officinalis* L.	20, 40	4,500 sh
Avocado	*Persea americana* Mill.	24	13,500 fr
Bahiagrass	*Paspalum notatum* Fluegge	20, 30, 40	24,000 dm
Banana	*Musa* × *paradisiaca* L.	22, 32—35	18,000 fr
Barley	*Hordeum vulgare* L.	14, 28	3,300 sd
Basil	*Ocimum basilicum* L.	48	6,800 lf
Bay rum tree	*Pimenta racemosa* (Mill.) J.W. Moore	22	30,000 lf
Beachgrass, American	*Ammophila breviligulata* Fern.	28	
Beachgrass, European	*Ammophila arenaria* (L.) Link	28, 14, 56	
Bean, African locust	*Parkia filicoidea* Welw. ex Oliv.		5,000 sd
Bean, broad	*Vicia faba* L.	12, 14	6,600 sd
Bean, green	*Phaseolus vulgaris* L.	22	2,500 sd
Bean, lablab	*Dolichos lablab* L.	22, 24	1,400 sd
Bean, lima	*Phaseolus lunatus* L.	22	2,500 sd
Bean, mung	*Vigna radiata* (L.) Wilczek var. radiata	22	1,100 sd
Bean, scarlet runner	*Phaseolus coccineus* L.	22	1,700 sd
Bean, tepary	*Phaseolus acutifolius* A. Gray	22	1,700 sd
Bean, winged	*Psophocarpus tetragonolobus* (L.) DC.	26	4,000 rt
Bean, yard-long	*Vigna unguiculata* subsp. *sesquipe-dalis* (L.) Verdc.	22, 24	10,000 fr
Beet, garden	*Beta vulgaris* L.	18, 27, 36	42,000 rt
Bentgrass, colonial	*Agrostis tenuis* Sibth.	28—35	400 sd
Bentgrass, creeping	*Agrostis stolonifera* L.	28, 35±	400 sd
Bentgrass, velvet	*Agrostis canina* L.	14, 28, 35	2,000 fo
Bermudagrass	*Cynodon dactylon* (L.) Pers.	18, 27, 36	15,000 dm
Bermudagrass, African	*Cynodon transvaalensis* Burtt Davy	18, 20	
Blackberry	*Rubus* spp.		7,000 fr
Blueberry, highbush	*Vaccinium corymbosum* L.	48, 72	6,000 fr
Blueberry, lowbush	*Vaccinium angustifolium* Ait.	24, 48	1,100 fr
Blueberry, rabbiteye	*Vaccinium ashei* Reade	72	11,000 fr
Bluegrass, Kentucky	*Poa pratensis* L.	25—96	6,400 dm
Bluegrass, roughstalk	*Poa trivialis* L.	14, 15, 28	1,000 sd
Bluestem, Australian	*Bothriochloa intermedia* (R. Br.) A. Camus	40, 50, 60	5,000 hay
Bluestem, big	*Andropogon gerardii* Vitm.	20, 60, 80	200 sd
Bluestem, Caucasian	*Bothriochloa caucasica* (Trin.) C.E. Hubb.	40	15,000 hay
Bluestem, diaz	*Dichanthium annulatum* (Forsk.) Stapf	20, 40, 60	8,000 hay

Table 90 (continued)
SELECTED ECONOMIC PLANTS, THEIR COMMON NAMES, SCIENTIFIC NAMES, AUTHORITIES, CHROMOSOME NUMBER, AND TARGET YIELD

Common name[a]	Scientific name[b]	Chromosomes (2n)[c]	Target yield[d]
Bluestem, little	*Schizachyrium scoparium* (Michx.) Nash	40	
Bluestem, sand	*Andropogon hallii* Hack.	60, 70	150 sd
Bluestem, yellow	*Bothriochloa ischaemum* (L.) Keng var. *ischaemum*	40, 50, 60	12,000 hay
Breadfruit	*Artocarpus altilis* (Parkins.) Fosb.	56, 54, 81	9,000 fr
Broccoli	*Brassica oleracea* var. *botrytis* L.	18	13,000 lf
Brome, smooth	*Bromus inermis* Leyss.	28, 42, 56	12,500 hay
Brussels sprouts	*Brassica oleracea* var. *gemmifera* DC.	18	25,000 sh
Buckwheat	*Fagopyrum esculentum* Moench	16, 32	4,000 sd
Buffalograss	*Buchloe dactyloides* (Nutt.) Engelm.	40, 56, 60	1,100 sd
Buffelgrass	*Cenchrus ciliaris* L.	32, 34, 36	50,000 wm
Burclover, California	*Medicago polymorpha* L.	14, 16	7,500 hay
Burnet	*Sanguisorba minor* Scop.	28, 54, 56	
Button clover	*Medicago orbicularis* (L.) Bartal.	16, 32	
Cabbage	*Brassica oleracea* var. *capitata* L.	18, 36	45,000 lf
Cabbage, Chinese	*Brassica rapa* L.	20	40,000 lf
Cacao	*Theobroma cacao* L.	16, 20, 26	3,300 sd
Camphor tree	*Cinnamomum camphora* (L.) J.S. Presl	24	
Canarygrass	*Phalaris canariensis* L.	12, 28	3,000 sd
Canarygrass, reed	*Phalaris arundinacea* L.	14, 28, 42	18,000 dm
Canna, edible	*Canna edulis* Ker.	18, 27	85,000 rt
Cantaloupe	*Cucumis melo* L.	24	13,500 fr
Caraway	*Carum carvi* L.	20, 22	800 sd
Carpetgrass	*Axonopus affinis* Chase	54, 80	50,000 hay
Carrot	*Daucus carota* L. subsp. *sativus* (Hoffm.) Arcang.	18, 22	42,000 rt
Cashew	*Anacardium occidentale* L.	40, 42	1,000 sd
Cassava	*Manihot esculenta* Crantz	36	56,000 rt
Castorbean	*Ricinus communis* L.	20	5,000 sd
Cauliflower	*Brassica oleracea* var. *botrytis* L.	18	25,000 lf
Celery	*Apium graveolens* L. var. *dulce* (Mill.) Pers.	22	43,000 st
Chard, Swiss	*Beta vulgaris* L.	18	11,000 lf
Cherry, sour	*Prunus cerasus* L.	32	
Cherry, sweet	*Prunus avium* (L.) L.	16—36	18 fr/
Chestnut, American	*Castanea dentata* (Marsh.) Borkh.	24	
Chestnut, Chinese	*Castanea mollissima* Blume	24	1,000 fr
Chestnut, European	*Castanea sativa* Mill.	22, 24	
Chickpea	*Cicer arietinum* L.	14, 16, 24	2,000 sd
Chicory	*Cichorium intybus* L.	18	27,000 rt
Chives	*Allium schoenoprasum* L.	16, 24, 32	
Cinnamon	*Cinnamomum verum* J.S. Presl	24	
Citron	*Citrus medica* L.	18, 28	
Clove	*Syzygium aromaticum* (L.) Merr. & Perry		
Clover, alsike	*Trifolium hybridum* L.	16	7,000 dm
Clover, alyce	*Alysicarpus vaginalis* (L.) DC.	16, 20	45,000 wm
Clover, arrowleaf	*Trifolium vesiculosum* Savi	16	
Clover, ball	*Trifolium nigrescens* Viv.	16, 32	
Clover, berseem	*Trifolium alexandrinum* L.	16	37,500 wm
Clover, crimson	*Trifolium incarnatum* L.	14, 16	12,000 dm

Table 90 (continued)
SELECTED ECONOMIC PLANTS, THEIR COMMON NAMES, SCIENTIFIC NAMES, AUTHORITIES, CHROMOSOME NUMBER, AND TARGET YIELD

Common name[a]	Scientific name[b]	Chromosomes (2n)[c]	Target yield[d]
Clover, kura	*Trifolium ambiguum* Bieb.	16, 32, 48	
Clover, red	*Trifolium pratense* L.	14, 28	7,100 dm
Clover, sub	*Trifolium subterraneum* L.	16, 12	
Clover, white	*Trifolium repens* L.	32, 48, 64	7,000 dm
Coconut	*Cocos nucifera* L.	32	5,000 copra
Coffee	*Coffea arabica* L.	22, 44, 66	6,600 fr
Coffee, robusta	*Coffea canephora* Pierre ex Froehner	22, 44	2,000 fr
Collard	*Brassica oleracea* var. *acephala* DC.		100,000 lf
Comfrey	*Symphytum* sp.	36	24,000 hay
Corn	*Zea mays* L. subsp. *mays*	20	140,000 wm
Cotton, Sea Island	*Gossypium barbadense* L.	52	700 fib
Cotton, tree	*Gossypium arboreum* L.	26	700 fib
Cotton, upland	*Gossypium hirsutum* L.	52	3,400 fib
Cotton, wild	*Gossypium anomalum* Wawra ex Wawra & Peyr.	26	
Cow pea	*Vigna unguiculata* (L.) Walp. subsp. *unguiculata*	22	2,500 sd
Crambe	*Crambe abyssinica* Hochst. ex. R.E. Fries	90	5,000 sd
Cranberry	*Vaccinium macrocarpon* Ait.	24	6,000 fr
Cress, garden	*Lepidium sativum* L.	16, 24, 32	
Crotalaria, showy	*Crotalaria spectabilis* Roth	16	50,000 hay
Crownvetch	*Coronilla varia* L.	24	12,400 hay
Cucumber	*Cucumis sativus* L.	14	90,000 fr
Current, black	*Ribes nigrum* L.	16	2,900 fr
Current, red	*Ribes rubrum* L.	16	2,900 fr
Dallisgrass	*Paspalum dilatatum* Poir.	40, 50, 60	24,000 dm
Date, desert	*Balanites aegyptiaca* (L.) Del.	16, 18	
Dill	*Anethum graveolens* L.	22	750 sd
Eggplant	*Solanum melongena* L.	24, 36, 48	24,000 fr
Einkorn	*Triticum monococcum* L.	14	
Elder, American	*Sambucus canadensis* L.		10,000 fr
Emmer	*Triticum dicoccon* Schrank	28	
Endive	*Cichorium endivia* L.	18, 36	50,000 lf
Esparto	*Stipa tenacissima* L.	32, 40	1,500 hay
Eucalyptus	*Eucalyptus* sp.	22	
Fescue, meadow	*Festuca pratensis* Huds.	14, 16, 42	4,000 dm
Fescue, red	*Festuca rubra* L.	14, 28, 42	
Fescue, sheep	*Festuca ovina* L.	14, 21, 28	
Fescue, tall	*Festuca arundinacea* Schreb.	42, 28	18,000 dm
Fig, common	*Ficus carica* L.	26	10,000 fr
Filbert, European	*Corylus avellana* L.	22, 28	
Flax	*Linum usitatissimum* L.	30, 32	1,000 sd
Foxtail, creeping	*Alopecurus arundinaceus* Poir.	28	
Foxtail, meadow	*Alopecurus pratensis* L.	28, 42	6,000 dm
Frankincense	*Boswellia carteri* Birdw.		
Garlic	*Allium sativum* L.	16, 48	10,000 rt
Ginger	*Zingiber officinale* Roscoe	22, 24	30,000 rt
Gooseberry, European	*Ribes uva-crispa* L.	16	
Grama, blue	*Bouteloua gracilis* Willd. ex H.B.K. Lag. ex Griffiths	20, 21, 28	200 sd
Grama, side-oats	*Bouteloua curtipendula* (Michx.) Torr.	20, 103	

Table 90 (continued)
SELECTED ECONOMIC PLANTS, THEIR COMMON NAMES, SCIENTIFIC
NAMES, AUTHORITIES, CHROMOSOME NUMBER, AND TARGET YIELD

Common name[a]	Scientific name[b]	Chromosomes (2n)[c]	Target yield[d]
Grape, muscadine	*Vitis rotundifolia* Michx.	40	20,000 fr
Grape, wine	*Vitis vinifera* L.	38, 40, 57	30,000 fr
Grapefruit	*Citrus paradisi* Macfad.	18, 27, 36	45,000 fr
Grass, centipede	*Eremochloa ophiuroides* (Munro) Hack.	18	
Grass, elephant	*Pennisetum purpureum* Schumach.	28, 27, 56	160,000 wm
Grass, Harding	*Phalaris stenoptera* Hack.	14, 28	
Grass, jaragua	*Hyparrhenia rufa* (Nees) Stapf	20, 30, 40	13,000 dm
Grass, limpo	*Hemarthria altissima* (Poir.) Stapf & C.E. Hubb.	20	12,500 dm
Grass, molasses	*Melinis minutiflora* Beauv.	36	43,000 hay
Grass, Para	*Brachiaria mutica* (Forsk.) Stapf	36	20,000 dm
Grass, St. Augustine	*Stenotaphrum secundatum* (Walt.) Ktze.	18, 36, 54	
Groundnut, Bambarra	*Voandzeia subterranea* (L.) Thou.	22	4,200 sd
Guar	*Cyamopsis tetragonoloba* (L.) Taub.	14	2,000 sd
Guayule	*Parthenium argentatum* A. Gray	36—111	4,500 rubber
Guineagrass	*Panicum maximum* Jacq.	18, 36, 32	250,000 wm
Gum arabic	*Acacia senegal* (L.) Willd.	26	
Gutta percha	*Palaquium gutta* (Hook.) Baill.	24	1 latex/
Hazel, beaked	*Corylus cornuta* Marsh.	28	
Hazel, Chinese	*Corylus chinensis* Franch.		
Hemp, Mauritius	*Furcraea foetida* (L.) Haw.		3,700 fib
Hickory	*Carya* spp.		
Hops	*Humulus lupulus* L.	20	4,000 cones
Horseradish	*Armoracia rusticana* Gaertn., Mey. & Scherb.	28, 32	9,000 rt
Indiangrass	*Sorghastrum nutans* (L.) Nash		
Indigo	*Indigofera tinctoria* L.	16	13,000 wm
Indigo, hairy	*Indigofera hirsuta* L.	16	17,000 wm
Jackfruit	*Artocarpus heterophyllus* Lam.	56	9,000 fr
Johnsongrass	*Sorghum halepense* (L.) Pers.	20, 40	19,000 hay
Jojoba	*Simmondsia chinensis* (Link) C. Schneid.	56, ±100	2 fr/
Jute, China	*Abutilon theophrasti* (Medik.)	42	2,500 fib
Jute, tussa	*Corchorus olitorius* L.	14, 28	1,900 fib
Kale	*Brassica oleracea* L. var *acephala* DC.		100,000 lf
Kenaf	*Hibiscus cannabinus* L.	36, 72	7,200 fib
Kikuyugrass	*Pennisetum clandestinum* Hochst. ex Chiov.	36	3,800 wm
Kiwi	*Actinidia chinensis* Planch.	±116, 160	40,000 fr
Kleingrass	*Panicum coloratum* L.	18, 36, 54	13,000 dm
Kochia	*Kochia scoparia* (L.) Schrad.	18	11,000 dm
Kohlrabi	*Brassica oleracea* var. *gongylodes* L.	18	11,000 sh
Kudzu	*Pueraria lobata* (Willd.) Ohwi	24	40,000 wm
Kudzu, tropical	*Pueraria phaseoloides* (Roxb.) Benth.	22	50,000 wm
Kumquat	*Fortunella* sp.	18, 36	
Lablab	*Dolichos lablab* L.	22, 24	1,400 sd
Lawngrass, Japanese	*Zoysia japonica* Steud.	40	
Leek	*Allium ampeloprasum* L.	16, 32, 48	
Lemon	*Citrus limon* (L.) Burm.	18, 36	100,000 fr

Table 90 (continued)
SELECTED ECONOMIC PLANTS, THEIR COMMON NAMES, SCIENTIFIC NAMES, AUTHORITIES, CHROMOSOME NUMBER, AND TARGET YIELD

Common name[a]	Scientific name[b]	Chromosomes (2n)[c]	Target yield[d]
Lentil	*Lens culinaris* Medik.	14	1,700 sd
Lespedeza, common	*Lespedeza striata* (Thunb. ex Murr.) Hook. & Arn.	22	7,500 hay
Lespedeza, Korean	*Lespedeza stipulacea* Maxim.	20	7,500 hay
Lespedeza, sericea	*Lespedeza cuneata* (Dum.) G. Don	20, 18	9,000 wm
Lettuce	*Lactuca sativa* L.	18, 36	25,000 lf
Licorice, common	*Glycyrrhiza glabra* L.	16	22,000 rt
Lime	*Citrus aurantiifolia* (Christm.) Swingle	18, 27	18,000 fr
Litchi	*Litchi chinensis* Sonn.	28, 30	10,000 fr
Lovegrass, Lehmann	*Eragrostis lehmanniana* Nees	40, 60	19,000 wm
Lovegrass, weeping	*Eragrostis curvula* (Schrad.) Nees	20, 40, 50	9,000 hay
Lupine, European blue	*Lupinus angustifolius* L.	40, 48	7,500 hay
Lupine, European yellow	*Lupinus luteus* L.	52, 104	7,500 hay
Lupine, white	*Lupinus albus* L.	26, 22	1,000 sd
Macadamia	*Macadamia* spp.	28, 36, 56	3,750 fr
Mango	*Mangifera indica* L.	40	50,000 fr
Maple, sugar	*Acer saccharum* Marsh.	26	
Marihuana	*Cannabis sativa* L.	20, 10, 40	1,700 fib
Marrow, Boston	*Cucurbita maxima* Duch.	40	23,000 fr
Marrow, vegetable	*Cucurbita pepo* L.	24, 28, 40	23,000 fr
Mate	*Ilex paraguariensis* St. Hil.	40	450 lf
Medic, black	*Medicago lupulina* L.	16, 32	
Mesquite, curly	*Hilaria belangeri* (Steud.) Nash	36, 72, 74	
Millet, browntop	*Brachiaria ramosa* (L.) Staph	32, 36, 42	
Millet, finger	*Eleusine coracana* (L.) Gaertn.	36	5,000 sd
Millet, Italian	*Setaria italica* (L.) Beauv.	18	4,500 sd
Millet, pearl	*Pennisetum americanum* (L.) Leeke	14	3,500 sd
Millet, proso	*Panicum miliaceum* L.	36, 54, 72	3,400 sd
Mint, field	*Mentha arvensis* L.	12, 60—90	27,500 wm
Mulberry, red	*Morus rubra* L.	28	14 fr/
Mustard, black	*Brassica nigra* (L.) Koch	8—22	1,100 sd
Mustard, white	*Sinapis alba* L.	24	8,000 sd
Nut, Brazil	*Bertholletia excelsa* Humb. & Bonpl.	34	
Nut, macadamia	*Macadamia* spp.	28, 36, 56	4,000 sd
Nut, pistachio	*Pistacia vera* L.	30	7 fr/
Nutmeg	*Myristica fragrans* Houtt.	42, 44	1,200 nutmeg
Oak, cork	*Quercus suber* L.	24	
Oat (Oats)	*Avena sativa* L.	42, 48, 63	3,000 sd
Okra	*Abelmoschus esculentus* (L.) Moench	72, 144	23,000 fr
Olive	*Olea europaea* L.	46	2,200 fr
Onion	*Allium cepa* L.	16, 32	29,000 rt
Orange, sour	*Citrus aurantium* L.	18	50,000 fr
Orange, sweet	*Citrus sinensis* (L.) Osb.	18, 27, 36	50,000 fr
Orchardgrass	*Dactylis glomerata* L.	28, 14	30,000 dm
Palm, African oil	*Elaeis guineensis* Jacq.	32, 36	2,200 oil
Palm, date	*Phoenix dactylifera* L.	28, 36	10,000 fr
Palm, sugar	*Arenga pinnata* (Wurmb) Merr.	26, 32	20,000 sugar
Pangolagrass	*Digitaria decumbens* Stent	27, 30, 36	31,000 dm
Panicgrass, blue	*Panicum antidotale* Retz.	18, 36	100,000 wm
Papaya	*Carica papaya* L.	18, 36	100,000 fr

Table 90 (continued)
SELECTED ECONOMIC PLANTS, THEIR COMMON NAMES, SCIENTIFIC NAMES, AUTHORITIES, CHROMOSOME NUMBER, AND TARGET YIELD

Common name[a]	Scientific name[b]	Chromosomes (2n)[c]	Target yield[d]
Parsley	*Petroselinum crispum* (Mill.) Nym. ex A.W. Hill	22	1,500 sd
Parsnip	*Pastinaca sativa* L.	22	150,000 ft
Pea, cow	*Vigna unguiculata* (L.) Walp. subsp. *unguiculata*	22	2,500 sd
Pea, garden	*Pisum sativum* L.	14, 28, 30	1,800 sd
Pea, pigeon	*Cajanus cajan* L. Huth.	22, 44, 66	12,800 fr
Pea, rough	*Lathyrus hirsutus* L.	14	
Peach	*Prunus persica* (L.) Batsch	16	80 fr/
Peanut	*Arachis hypogaea* L.	40	5,000 sd
Pear	*Pyrus communis* L.	34, 51	87,000 fr
Pecan	*Carya illinoensis* (Wangenh.) K. Koch	32	10 fr/
Pepper, black	*Piper nigrum* L.	48, 52, 104	6,000 fr
Pepper, Indian long	*Piper longum* L.	24, 48, 96	1,700 fr
Pepper, tabasca	*Capsicum frutescens* L.	24	1,000 dry fr
Peppermint	*Mentha* × *piperita* L.	36, 48, 64	70,000 wm
Persimmon, common	*Diospyros virginiana* L.	60, 90	
Persimmon, Japanese	*Diospyros kaki* L.	54, 56	227 fr/
Pineapple	*Ananas comosus* (L.) Merr.	50, 75	40,000 fr
Pistachio	*Pistacia vera* L.	30	7 fr/
Plum, common	*Prunus domestica* L.	16, 48	14,000 fr
Plum, damson	*Prunus insititia* L.	48	11,500 fr
Plum, Japanese	*Prunus salicina* Lindl.	16	14,000 fr
Plum, Klamath	*Prunus subcordata* Benth.		
Pomegranate	*Punica granatum* L.	16, 18, 19	50 fr/
Poppy, opium	*Papaver somniferum* L.	22, 20	900 sd
Potato	*Solanum tuberosum* L.	24, 36, 48	32,000 rt
Potato, sweet	*Ipomoea batatas* (L.) Lam.	84, 90	36,000 rt
Pumpkin	*Cucurbita moschata* (Duch.) Duch. ex Poir.	24, 40, 48	23,000 fr
Quince	*Cydonia oblonga* Mill.	34	
Quinine	*Cinchona* spp.	34	600 bark
Radish	*Raphanus sativus* L.	18	45,000 rt
Ramie	*Boehmeria nivea* (L.) Gaudich.	22, 28, 42	2,000 fib
Rape	*Brassica napus* L.	38, 32	3,000 sd
Raspberry, American red	*Rubus idaeus* L. var. *strigosus* (Michx.)	14, 21	13,000 fr
Raspberry, black	*Rubus occidentalis* L.	14	3,000 fr
Reed, common	*Phragmites australis* (Cav.) Trin. ex Steud.	48, 36, 54	40,000 dm
Rhodesgrass	*Chloris gayana* Kunth	20, 30, 40	17,200 dm
Rhubarb	*Rheum rhabarbarum* L.	44	30,000 petiole
Rhubarb, Chinese	*Rheum palmatum* L.	22, 44	
Rice	*Oryza sativa* L.	24	6,000 sd
Rice, northern wild	*Zizania palustris* L.	30	500 sd
Ricegrass, Indian	*Oryzopsis hymenoides* (Roem. & Schult.) Ricker	48	
Rubber tree	*Hevea brasiliensis* (Willd. ex A. Juss.) Muell. Arg.	36, 34, 72	2,300 latex
Rutabaga	*Brassica napus* var. *napobrassica* (L.) Reichenb.		100,000 rt
Rye	*Secale cereale* L.	14—29	3,800 sd
Ryegrass, annual	*Lolium multiflorum* Lam.	14	24,000 wm
Ryegrass, perennial	*Lolium perenne* L.	14, 28	

Table 90 (continued)
SELECTED ECONOMIC PLANTS, THEIR COMMON NAMES, SCIENTIFIC NAMES, AUTHORITIES, CHROMOSOME NUMBER, AND TARGET YIELD

Common name[a]	Scientific name[b]	Chromosomes (2n)[c]	Target yield[d]
Safflower	*Carthamus tinctorius* L.	24, 32	4,500 sd
Sage	*Salvia officinalis* L.	14, 16	1,000 dry lf
Sainfoin	*Onobrychis viciifolia* Scop.	28, 22, 29	500 sd
Savory, summer	*Satureja hortensis* L.	45—48	6,000 dm
Savory, winter	*Satureja montana* L.	12, 30	
Serradella	*Ornithopus sativus* Brot.	14, 16	8,000 dm
Sesame	*Sesamum indicum* L.	26, 52, 58	5,900 sd
Sisal	*Agave sisalana* Perr.	138—150	2,250 fib
Smilograss	*Oryzopsis miliacea* (L.) Aschers. & Schweinf.	24	35,000 wm
Sorghum, almum	*Sorghum* × *almum* Parodi	40	14,000 hay
Sorghum, grain	*Sorghum bicolor* (L.) Moench	20	70,000 hay
Soybean	*Glycine max* (L.) Merr.	40, 38	3,100 sd
Soybean, perennial	*Glycine wightii* (Grah. ex Wight & Arn.) Verdc.	22, 44, 20	7,000 hay
Spearmint	*Mentha spicata* L.	36, 48, 64	2,500 dm
Spelt	*Triticum spelta* L.	42	
Spinach	*Spinacea oleracea* L.	12	14,000 lf
Squash, mixta	*Cucurbita mixta* Pang.	40	23,000 fr
Stargrass	*Cynodon plectostachyus* (K. Schum.) Pilg.	18, 54	67,000 wm
Strawberry, Chilean	*Fragaria chiloensis* (L.) Duch.	56	1,600 fr
Strawberry, garden	*Fragaria* × *ananassa* Duch.	56	19,500 fr
Strawberry, Virginia	*Fragaria virginiana* Duch.	56	
Stylo, Townsville	*Stylosanthes humilis* H.B.K.	20	1,200 sd
Sudangrass	*Sorghum sudanense* (Piper) Stapf	20	45,000 hay
Sugarcane	*Saccharum officinarum* L.	80, 60, 90	73,000 dm
Sunflower	*Helianthus annuus* L.	14, 34	3,700 sd
Sweetclover, white	*Melilotus alba* Medik.	16, 24, 32	32,000 wm
Sweetclover, yellow	*Melilotus officinalis* Lam.	16, 32	8,500 hay
Switchgrass	*Panicum virgatum* L.	18—108	10,000 hay
Tangerine	*Citrus reticulata* Blanco	18, 36	50,000 fr
Taro	*Colocasia esculenta* (L.) Schott	28, 42	34,000 rt
Tea	*Camellia sinensis* (L.) Ktze.	30, 45, 60	6,700 dry lf
Teff	*Eragrostis tef* (Zuccagni) Trotter	40	3,000 sd
Teosinte	*Zea mays* subsp. *mexicana* (Schrad.) Iltis	20	70,000 wm
Thyme, common	*Thymus vulgaris* L.	30	900 lf
Timothy	*Phleum pratense* L.	14, 21, 35	600 sd
Tobacco	*Nicotiana tabacum* L.	24, 48, 72	2,300 dry lf
Tomato	*Lycopersicon esculentum* Mill.	12, 24, 48	22,500 fr
Trefoil, big	*Lotus pedunculatus* Cav.	12, 24	
Trefoil, birdsfoot	*Lotus corniculatus* L.	12, 24, 26	600 sd
Trefoil, narrowleaf	*Lotus tenuis* Waldst. & Kit. ex. Willd.	12, 24	250 sd
Triticale	× *Triticosecale (Triticum* × *Secale)*	42, 56	5,000 sd
Tung-oil tree	*Aleurites fordii* Hemsl.	22	5,000 fr
Turmeric	*Curcuma domestica* Val.	32, 64, 63	35,000 rt
Turnip	*Brassica rapa* L.	20	65,000 rt
Vanilla	*Vanilla planifolia* Andr.	28—32	800 fr
Velvetbean	*Mucuna deeringiana* (Bort) Merr.	22	18,000 hay
Vetch, common	*Vicia sativa* L. subsp. *sativa*	14, 10, 24	6,800 hay
Vetch, winter (hairy)	*Vicia villosa* Roth	14	26,000 wm
Walnut, black	*Juglans nigra* L.	32	7,500 sd

Table 90 (continued)
SELECTED ECONOMIC PLANTS, THEIR COMMON NAMES, SCIENTIFIC NAMES, AUTHORITIES, CHROMOSOME NUMBER, AND TARGET YIELD

Common name[a]	Scientific name[b]	Chromosomes (2n)[c]	Target yield[d]
Watercress	*Nasturtium officinale* R. Br.	14, 32, 48	
Watermelon	*Citrullus lanatus* (Thunb.) Matsum. & Nakai	22	20,000 fr
Wheat, bread	*Triticum aestivum* L.	42	4,000 sd
Wheat, club	*Triticum compactum* Host	42	
Wheat, durum	*Triticum durum* Desf.	28	
Wheatgrass, bluebunch	*Agropyron spicatum* (Pursh) Scribn. & Smith	14, 21, 28	
Wheatgrass, crested	*Agropyron desertorum* Fisch. ex Link	28, 29, 32	2,000 hay
Wheatgrass, Fairway	*Agropyron cristatum* (L.) Gaertn.	14, 28, 42	
Wheatgrass, intermediate	*Agropyron intermedium* (Host) Beauv. var. *intermedium*	42, 28	2,400 hay
Wheatgrass, tall	*Agropyron elongatum* (Host) Beauv.	14, 56, 70	15,000 hay
Wheatgrass, western	*Agropyron smithii* Rydb.	28, 42, 56	2,300 hay
Wildrye, Canada	*Elymus canadensis* L.	28, 42	
Wildrye, giant	*Elymus condensatus* Presl	28, 56	
Wildrye, Russian	*Psathyrostachys juncea* (Fisch.) Nevski	14	2,800 dm
Wintergreen	*Gaultheria procumbens* L.	24	
Yam, composite	*Dioscorea composita* Hemsl.	36, 54	185,000 rt
Yam, eboe	*Dioscorea rotundata* Poir.	40	16,000 rt
Yam, winged	*Dioscorea alata* L.	20, 30, 40	30,000 rt

[a]   After Duke, J. A., Hurst, S. J., and Terrell, E. E., Ecological distribution of 1000 economic plants, Information al Dia Alerta, IICA-Tropicos, Agronomia No. 1, Turrialba, Costa Rica, 1975.

[b]   After Terrell, E. E., A checklist of names for 3000 vascular plants of economic importance, *U.S. Dep. Agric. Agric. Handb.,* No. 505, 1977.

[c]   Duke lists up to three counts for diploid chromosome numbers, recognizing that many more counts may have been reported.

[d]   Duke provides information on yield in kilograms, recognizing that yield data from different countries or different technologies may vary by a factor of 10. He takes the higher reported figures, where credible, on the assumption that this could represent a good target yield. The symbols used are as follows: dm = dry matter, dr = drug, fib = fiber, fl = flower, fo = fodder, fr = fruit, hay = hay, lf = leaf, rt = root or rhizome, sd = seed, sh = shoot, st = stem, wm = wet matter or green biomass. Where hectare yields were not available, per-plant yield is indicated by the slash symbol (/) following the item for which yield was recorded.

Adapted from Duke, J. A., The quest for tolerant germplasm, in *Crop Tolerance to Suboptimal Land Conditions,* Jung, G. A., Ed., Spec. Pub. No. 32, American Society of Agronomy, Madison, Wis., 1978. With permission.

Table 91
**SELECTED ECONOMIC PLANTS AND FACTORS AFFECTING THEIR
DISTRIBUTION**

Common name	Growth type[a]	pH	Annual precipitation (dm)	Annual temperature (°C)[b]	Center of diversity[c]
Alfalfa	P, H	4.3—8.7	2—27	5—28	CE, NE, ME
Almond	P, T	5.3—8.6	5—15	11—17	CE, NE
Aloe, Barbados	P, H	5.0—8.0	6—40	19—27	AF, ME
Amaranth, Chinese	A, H	4.3—7.8	3—27	7—32	II, CJ
Apple	P, T	4.5—8.3	3—46	6—24	ES, CE
Apricot	P, S, T	4.9—8.6	3—15	8—19	CE, NE, CJ
Artichoke, globe	P, H	5.0—8.3	3—26	9—21	NE
Artichoke, Jerusalem	P/A, H	4.5—8.3	3—28	7—27	NA
Asparagus, garden	P, H	4.3—8.2	3—40	6—27	ME, ES
Avocado	P, T	4.3—8.3	3—41	13—27	MA
Bahiagrass	P, G	4.3—8.3	3—41	13—27	SA
Banana	P, H, S	4.3—8.3	6—42	16—29	II, HI, AF
Barley	A, G	4.3—8.6	2—27	5—32	NE, ME, CJ
Basil	A, H	4.3—8.2	6—42	7—27	II
Bay rum tree	P, T	4.8—7.1	7—28	21—27	MA
Beachgrass, American	P, G	4.9—5.6	11—12	7—12	NA
Beachgrass, European	P, G	4.5—6.8	5—12	7—20	ES
Bean, broad	A, H	4.2—8.6	2—26	5—28	CE, ME, SA
Bean, green	A, H	4.2—8.7	3—46	5—28	MA, SA, CJ
Bean, lablab	A, P, H	5.0—7.8	2—40	9—27	AF
Bean, lima	B/A, H	4.3—8.4	3—42	8—27	MA, SA
Bean, mung	A, H	4.3—8.3	4—41	12—27	II, CJ, HI
Bean, scarlet runner	P/A, H	4.8—8.2	4—17	6—27	MA
Bean, tepary	A, H	5.0—7.3	7—17	17—26	MA
Bean, winged	P/A, H	4.3—7.5	7—41	21—32	II
Bean, yard-long	A, H	4.3—7.3	6—41	13—27	AF, HI
Beet, garden	A/B, H	4.2—8.3	4—46	5—27	ES, CE, ME
Bentgrass, colonial	P, G	4.5—7.5	3—26	5—23	ES, NE, ME
Bentgrass, creeping	P, G	4.5—8.3	3—26	6—18	ES, NA
Bentgrass, velvet	P, G	4.5—7.5	4—11	6—19	ES, NA
Bermudagrass	P, G	4.3—8.3	2—42	6—29	HI
Bermudagrass, African	P, G	5.6—7.3	7—15	13—26	AF
Blackberry	P, H	4.5—8.0	5—17	7—27	NA
Blueberry, highbush	P, S	4.8—7.3	7—13	6—19	NA
Blueberry, lowbush	P, S	4.8—7.5	4—11	6—19	NA
Blueberry, rabbiteye	P, S	4.8—6.5	11—15	9—19	NA
Bluegrass, Kentucky	P, G	4.5—8.3	3—17	5—19	ES, NA
Bluegrass, roughstalk	P, G	4.5—8.0	3—17	5—19	ES
Bluestem, Australian	P, G	5.9—7.5	7—19	10—27	HI, II, AU
Bluestem, big	P, G	4.5—7.6	4—28	7—28	NA
Bluestem, Caucasian	P, G	5.2—5.6	11—12	12—16	ES
Bluestem, diaz	P, G	5.5—8.3	2—19	19—28	AF, HI
Bluestem, little	P, G	4.5—7.1	9—15	10—20	NA
Bluestem, sand	P, G	4.9—7.5	5—12	7—28	NA
Bluestem, yellow	P, G	5.6—7.8	5—12	8—26	ES
Breadfruit	P, T	5.0—8.0	7—40	17—29	II
Broccoli	B, H	4.2—8.3	3—46	4—27	ME
Brome, smooth	P, G	4.9—8.2	3—17	5—20	ES
Brussels sprouts	B/A, H	4.2—7.8	5—19	6—27	ME, ES
Buckwheat	A, H	4.8—8.2	4—13	6—25	NE
Buffalograss	P, G	6.3—7.8	3—12	8—19	NA
Buffelgrass	P, G	4.3—8.3	3—27	13—32	HI, II
Burclover, California	A, H	5.3—8.2	3—19	10—27	ES, HI

Table 91 (continued)
SELECTED ECONOMIC PLANTS AND FACTORS AFFECTING THEIR DISTRIBUTION

Common name	Growth type[a]	pH	Annual precipitation (dm)	Annual temperature (°C)[b]	Center of diversity[c]
Burnet	P, H	4.8—8.2	3—13	7—19	AF, ES
Button clover	A, H	6.3—8.0	4—13	14—21	ME, CE
Cabbage	B/A, H	4.2—8.3	3—46	4—27	ME, NE
Cabbage, Chinese	B/A, H	4.3—7.0	7—41	7—27	CJ
Cacao	P, T	4.3—8.7	5—42	18—28	SA, MA
Camphor tree	P, T	4.3—8.0	7—40	15—27	CJ
Canarygrass	A, G	4.5—8.3	3—21	6—21	ME
Canarygrass, reed	P, G	4.5—8.2	3—26	5—23	ES
Canna, edible	P, H	4.5—8.0	7—40	7—26	SA
Cantaloupe	A, V	4.3—8.7	2—40	7—28	CJ, CE, NE
Caraway	P/A, H	4.8—7.8	4—13	6—19	ES
Carpetgrass	P, G	4.3—7.0	8—41	16—27	MA
Carrot	A, B, H	4.2—8.7	3—46	3—27	ME, CE, ES
Cashew	P, T	4.3—8.7	7—42	15—28	SA, MA
Cassava	P, S, T	4.3—8.0	5—40	8—29	SA, MA, II
Castorbean	P/A, H	4.3—8.3	2—42	5—29	AF
Cauliflower	B, H	4.2—8.3	3—46	4—27	ME
Celery	P/A, H	4.2—8.3	3—46	5—27	ME
Chard, Swiss	A, H	4.2—8.3	4—46	5—27	ES
Cherry, sour	P, S, T	4.5—8.3	3—15	5—20	ES, NE, CJ
Cherry, sweet	P, T	4.5—8.3	3—15	5—17	NE
Chestnut, American	P, T	5.6—7.3	5—11	9—15	NA
Chestnut, Chinese	P, T	5.0—7.3	8—12	10—19	CJ
Chestnut, European	P, T	4.5—7.5	5—13	7—16	NE
Chickpea	A, H	5.0—8.6	3—25	6—27	HI, CE, NE
Chicory	P, H	4.5—8.3	3—40	6—27	ES
Chives	P, H	5.0—8.3	3—40	7—26	CJ
Cinnamon	P, T	5.3—8.0	7—39	21—27	CJ
Citron	P, T	4.3—8.3	3—41	13—27	II, NE
Clove	P, T	4.3—7.3	7—41	21—27	II
Clover, alsike	P, B, H	4.8—7.8	3—26	5—27	ES
Clover, alyce	P, H	4.3—8.7	10—42	19—29	AF, HI, II
Clover, arrowleaf	A, H	5.5—8.2	6—14	11—20	ME
Clover, ball	H	5.6—8.2	6—15	12—21	NE, ES
Clover, berseem	A, H	4.8—8.2	4—17	5—26	ME
Clover, crimson	A, H	4.8—8.3	3—15	5—21	ME, ES
Clover, kura	H	4.5—7.3	5—11	9—12	NE
Clover, red	B, P, H	4.5—8.3	3—21	5—20	ES
Clover, sub	A, H	4.5—8.2	4—16	6—21	ME, AU
Clover, white	P, H	4.5—8.3	3—26	5—21	ME, NE, ES
Coconut	P, T	4.3—8.3	7—42	19—29	II, HI
Coffee	P, S, T	4.3—8.0	7—46	16—27	AF, NE, ME
Coffee, robusta	P, S, T	4.3—8.0	8—41	20—27	AF
Collard	B/A, H	4.2—8.3	3—46	4—31	
Comfrey	P, H, S	5.3—8.7	5—27	6—25	ES, NE
Corn	A, G	4.3—8.3	3—40	5—29	MA, SA, CJ
Cotton, Sea Island	P, S	4.3—8.4	5—40	9—27	SA, MA, NA
Cotton, tree	A, P, S	4.8—8.4	5—42	11—27	CJ, II, HI
Cotton, upland	A, H, S	4.3—8.4	3—27	7—32	MA, SA, AF
Cotton, wild		5.8—8.3	5—17	13—26	AF
Cowpea	A, H	4.3—8.3	3—41	13—27	AF, HI
Crambe	A, H	5.0—7.8	4—11	6—19	AF

Table 91 (continued)
SELECTED ECONOMIC PLANTS AND FACTORS AFFECTING THEIR
DISTRIBUTION

Common name	Growth type[a]	pH	Annual precipitation (dm)	Annual temperature (°C)[b]	Center of diversity[c]
Cranberry	P, L	4.5—7.5	4—13	6—13	NA
Cress, garden	A, B, H	4.9—8.0	5—26	6—27	AF
Crotalaria, showy	A, H	4.9—8.0	9—28	11—27	AF
Crownvetch	P, H	4.8—7.8	5—40	7—15	ES
Cucumber	A, H	4.3—8.7	2—46	6—28	CJ, HI, NE
Currant, black	P, S	4.5—7.5	5—17	5—15	ES
Currant, red	P, S	4.5—7.5	5—17	5—15	ES
Dallisgrass	P, G	4.3—8.3	3—41	11—27	SA
Date, desert	P, T	5.1—8.3	2—17	19—28	ME
Dill	A, B, H	5.0—7.8	5—17	6—26	ME, HI
Eggplant	P/A, H	4.3—8.7	2—42	7—28	CJ, HI
Einkorn	A, G	5.8—8.3	3—16	7—19	NE
Elder, American	P, S	4.9—8.7	3—40	6—28	NA
Emmer	A, G	5.3—8.3	3—16	7—19	NE, ES
Endive	A, B, H	4.5—8.3	3—27	7—27	ES
Esparto	P, G	5.6—8.0	3—11	13—23	ME
Eucalyptus	P, T	4.3—8.3	2—42	9—28	AU, ME
Fescue, meadow	P, G	4.5—8.2	3—26	5—21	ES
Fescue, red	P, G	4.5—8.3	3—17	5—19	ES
Fescue, sheep	P, G	4.5—8.2	3—21	5—19	ES
Fescue, tall	P, G	4.8—8.5	3—21	5—23	ES
Fig, common	P, S, T	4.3—8.6	2—41	9—32	NE
Filbert, European	P, S, T	4.5—8.3	3—13	5—19	NE, ME
Flax	A, H	4.8—8.3	3—13	6—27	CE, NE, ME
Foxtail, creeping	P, G	5.5—7.8	3—11	6—19	ES, HI
Foxtail, meadow	P, G	4.5—7.8	3—17	5—18	ES
Garlic	P/A, H	4.5—8.3	3—26	6—27	CJ, CE, NE
Ginger	P/A, H	4.3—7.5	7—42	14—27	HI
Gooseberry, European	P, S	4.5—7.0	5—17	5—11	ES
Grama, blue	P, G	6.3—8.3	3—13	7—15	NA
Grama, side-oats	P, G	5.6—7.8	4—16	7—19	NA
Grape, muscadine	P, L, V	5.8—8.0	5—17	13—23	NA
Grape, wine	P, L, V	4.3—8.7	5—33	7—28	CEN, NE, ME
Grapefruit	P, T	4.3—8.3	3—41	13—28	II, MA
Grass, centipede	P, G	4.8—7.8	7—17	15—27	CJ
Grass, elephant	P, G	4.3—8.2	2—40	12—29	AF
Grass, Harding	P, G	5.6—8.2	3—16	11—18	ME
Grass, jaragua	P, G	4.3—6.5	8—40	16—27	AF
Grass, limpo	P, G	5.0—7.1	8—15	7—26	AF
Grass, molasses	P, G	4.3—8.4	7—27	19—32	AF
Grass, Para	P, G	4.3—7.9	8—41	19—32	AF
Grass, St. Augustine	P, G	5.0—8.3	3—26	13—27	MA
Groundnut, Bambarra	A, H	4.3—7.1	5—41	19—27	AF
Guar	A, H	4.8—8.3	4—24	7—27	HI, AF
Guayule	P, H, S	5.6—7.8	3—21	11—25	MA
Guineagrass	P, G	4.3—8.4	5—42	13—27	AF
Gum arabic	P, S, T	5.0—8.0	3—22	17—27	AF
Gutta percha	P, T	5.0—5.3	15—40	20—27	II
Hazel, beaked	P, S, T	5.0—7.5	4—11	6—15	NA
Hazel, Chinese	P, T	4.9—6.8	7—11	14—15	CJ
Hemp, Mauritius	P, S	5.8—	21—25	19—27	SA, MA
Hickory	P, T	4.5—7.3	7—13	8—19	NA

Table 91 (continued)
SELECTED ECONOMIC PLANTS AND FACTORS AFFECTING THEIR DISTRIBUTION

Common name	Growth type[a]	pH	Annual precipitation (dm)	Annual temperature (°C)[b]	Center of diversity[c]
Hops	P, V	4.5—8.3	3—13	5—21	ES
Horseradish	P, H	5.0—7.5	5—17	5—19	ES
Indiangrass	P, G	5.6—7.1	11—17	12—26	NA
Indigo	P, S	5.0—7.3	7—42	16—27	AF
Indigo, hairy	A, B, H	4.3—8.0	8—42	15—29	AF
Jackfruit	P, T	4.3—8.0	7—42	19—29	HI
Johnsongrass	P, G	4.3—8.3	3—42	9—27	ME
Jojoba	P, S, T	7.3—8.2	2—11	16—26	MA
Jute, China	A, H	4.9—8.2	7—15	10—21	CJ
Jute, tussa	A, H	4.5—8.2	2—42	17—27	AF, HI, CJ
Kale	B/A, H	4.2—8.3	3—46	4—31	
Kenaf	A, H	4.3—8.2	2—40	12—27	AF, HI
Kikuyugrass	P, G	4.3—8.2	2—40	12—27	AF
Kiwi	P, S, V	5.5—7.3	7—15	15—24	CJ
Kleingrass	P, G	5.0—7.8	2—26	13—27	AF
Kochia	P/A, H	5.0—8.3	3—25	6—19	ES, CJ
Kohlrabi	B, H	4.5—7.5	5—17	7—23	ME
Kudzu	P, H	5.0—7.1	11—40	13—26	II, CJ
Kudzu, tropical	P, V	4.3—8.0	10—42	23—27	II
Kumquat	P, S, T	5.0—7.8	7—25	15—27	CJ
Lablab	A, P, V	5.0—7.8	2—40	9—27	AF
Lawngrass, Japanese	P, G	4.3—8.0	7—41	9—27	CJ
Leek	B/A, H	4.5—8.3	4—27	6—26	HI, NE, ES
Lemon	P, T	4.8—8.3	3—41	12—28	II, ME
Lentil	A, H	4.5—8.3	3—24	6—27	NE
Lespedeza, common	A, P, H	4.9—8.8	5—17	9—26	CJ
Lespedeza, Korean	A, H	5.5—7.3	5—17	9—26	CJ
Lespedeza, sericea	P, H, S	4.9—6.8	11—17	11—19	CJ
Lettuce	A, H	4.2—8.7	3—41	5—28	ES
Licorice, common	P, H	5.5—8.2	3—11	6—25	ME, ES
Lime	P, S, T	4.3—8.3	4—41	13—28	II
Litchi	P, T	5.0—8.0	7—40	19—27	CJ
Lovegrass, Lehmann	P, A, G	5.5—7.3	3—12	13—19	AF
Lovegrass, weeping	P, G	4.3—8.3	3—16	9—23	AF
Lupine, European blue	A, H	4.8—8.3	3—26	7—26	ME
Lupine, European yellow	A, H	4.5—8.2	3—26	7—26	ME
Lupine, white	A, H	4.8—8.2	3—26	7—26	HI
Macadamia	P, T	4.5—8.0	7—26	15—25	AU
Mango	P, T	4.3—8.0	2—42	15—29	HI, II
Maple, sugar	P, T	4.5—7.3	5—15	7—21	NA
Marihuana	A, H	4.3—8.3	3—40	6—29	CE, HI, ES
Marrow, Boston	A, B, V	4.3—8.3	3—41	7—32	HI, SA
Marrow, vegetable	A, H, V	4.3—8.7	3—41	7—32	MA, SA
Mate	P, T	5.7—7.3	6—10	14—15	SA
Medic, black	A, P, H	4.5—8.3	3—17	6—21	ES
Mesquite, curly	P, G	6.8—7.8	3—21	15—16	NA, MA
Millet, browntop	A, P, G	4.3—8.0	4—27	14—32	AF
Millet, finger	A, G	4.3—8.4	3—42	12—27	HI, AF
Millet, Italian	A, G	5.0—8.3	3—42	6—27	CJ
Millet, pearl	A, G	4.5—8.3	2—26	12—27	AF
Millet, proso	A, G	4.8—8.5	3—42	6—27	CJ
Mint, field	P, H	5.0—8.3	3—42	6—27	CJ, ME, ES
Mulberry, red	P, T	5.3—8.0	5—17	6—23	NA

Table 91 (continued)
SELECTED ECONOMIC PLANTS AND FACTORS AFFECTING THEIR DISTRIBUTION

Common name	Growth type[a]	pH	Annual precipitation (dm)	Annual temperature (°C)[b]	Center of diversity[c]
Mustard, black	A, H	4.2—8.2	3—25	6—27	
Mustard, white	A, H	4.5—8.0	4—18	5—24	ME
Nut, Brazil	P, T	4.3—8.0	11—41	19—27	SA
Nut, macadamia	P, T	4.5—8.0	7—26	15—25	AU
Nut, pistachio	P, T	5.7—7.8	3—11	15—19	CE
Nutmeg	P, T	4.3—8.0	7—42	15—27	II
Oak, cork	P, T	4.9—8.3	3—13	7—27	ME
Oat, (Oats)	A, G	4.5—8.6	2—21	5—26	ME, NE, CJ
Okra	A, H	4.3—8.7	3—41	13—28	HI
Olive	P, T	5.3—8.6	3—17	13—23	ME
Onion	B/A, H	4.3—8.2	3—41	6—27	CE, ME
Orange, sour	P, T	4.5—8.4	2—40	13—28	II, ME, CJ
Orange, sweet	P, T	4.3—8.3	3—41	13—28	II, CJ, ME
Orchardgrass	P, G	4.5—8.3	3—26	5—23	ES
Palm, African oil	P, T	4.3—8.0	7—42	17—27	AF, SA
Palm, date	P, T	5.0—8.3	2—40	13—28	AF
Palm, sugar	P, T	5.0—8.0	7—40	19—27	II, HI
Pangolagrass	P, G	4.3—8.0	7—41	17—27	AF
Panicgrass, blue	P, G	5.0—8.3	4—25	13—27	HI, II, AU
Papaya	P, S, T	4.3—8.0	7—42	17—29	MA
Parsley	B, H	4.9—8.3	3—46	5—26	ME
Parsnip	B/A, H	4.2—8.3	3—15	6—21	ES
Pea, cow	A, H	4.3—8.3	3—41	13—27	AF, HI
Pea, garden	A, H	4.2—8.6	3—46	5—27	NE, ME, AF
Pea, pigeon	P/A, S	4.3—8.4	3—40	15—27	HI, AF
Pea, rough	A, H, V	5.5—8.2	3—13	5—19	CE, ME
Peach	P, T	4.5—8.6	3—21	6—23	CJ, CE, NE
Peanut	A, H	4.3—8.3	3—41	10—28	SA, AF
Pear	P, T	4.5—8.3	3—46	5—24	ES
Pecan	P, T	4.5—8.3	3—15	13—21	NA, MA
Pepper, black	P, L, V	4.3—7.4	7—42	20—27	HI
Pepper, Indian long	P, H, S		—42	—27	HI
Pepper, tabasco	P/A, S	4.3—8.7	3—41	7—29	MA
Peppermint	P, H	5.0—8.0	3—40	7—27	ES, ME
Persimmon, common	P, T	4.9—8.0	7—46	13—25	NA
Persimmon, Japanese	P, T	4.3—8.3	3—46	13—27	CJ, NA
Pineapple	P, H	3.5—7.8	7—41	16—28	SA
Pistachio	P, T	5.7—7.8	3—11	15—19	CE
Plum, common	P, S, T	4.5—8.3	3—21	5—19	ES, NE, CJ
Plum, damson	P, S, T	4.5—7.5	5—15	5—19	NE, ES, CJ
Plum, Japanese	P, T	5.0—7.5	6—15	7—21	CJ
Plum, Klamath	P, S, T	5.8—	—11	—15	NA
Pomegranate	P, S, T	4.3—8.7	3—42	13—27	NE
Poppy, opium	A, H	4.5—8.3	3—17	7—23	ME, CE, NE
Potato	A, H	4.2—8.3	3—46	4—27	SA, MA, ES
Potato, sweet	P, A, V	4.3—8.7	3—46	9—27	MA, SA, II
Pumpkin	A, V	4.3—8.4	3—28	7—32	MA, SA
Quince	P, T	5.0—7.8	4—13	9—21	NE
Quinine	P, T	4.8—6.5	12—31	16—24	SA
Radish	A, B, H	4.2—8.3	3—46	5—27	CJ, HI, ME
Ramie	P, S	4.3—7.5	7—40	15—27	CJ
Rape	A, B, H	4.2—8.3	3—28	5—27	ME, ES
Raspberry, American red	P, S	4.5—7.8	3—17	5—21	ES, NA

Table 91 (continued)
SELECTED ECONOMIC PLANTS AND FACTORS AFFECTING THEIR
DISTRIBUTION

Common name	Growth type[a]	pH	Annual precipitation (dm)	Annual temperature (°C)[b]	Center of diversity[c]
Raspberry, black	P, S	4.9—7.8	3—15	5—17	NA
Reed, common	P, G	4.8—8.3	2—40	7—27	ES
Rhodesgrass	P, G	4.3—8.3	2—40	9—28	AF
Rhubarb	P, H	4.2—7.8	3—46	5—18	ES, CE
Rhubarb, Chinese	P, H	5.3—6.8	7—13	7—16	CJ, CE
Rice	A, G	4.3—8.3	5—42	9—29	CJ, II, HI
Rice, northern wild	A, G	5.8—7.8	4—13	7—19	NA
Ricegrass, Indian	P, G	7.0—7.5	3—5	6—13	NA
Rubber tree	P, T	4.3—8.0	11—42	18—28	SA, II
Rutabaga	B/A, H	4.2—7.8	3—17	4—27	
Rye	A, G	4.5—8.3	3—17	5—21	CE, NE, CJ
Ryegrass, annual	A, B, G	4.5—8.3	3—21	5—23	ME, ES
Ryegrass, perennial	P, G	4.5—8.3	3—26	5—23	ME, ES
Safflower	A, H	5.4—8.3	2—15	6—27	CE, NE
Sage	P, H, S	4.2—8.3	3—26	5—26	ME, II
Sainfoin	P, H	4.9—8.2	4—12	3—19	NE, ES
Savory, summer	A, H	5.6—8.2	3—13	7—21	ME
Savory, winter	P, S	6.5—7.3	7—17	7—23	ME
Serradella	A, H	5.3—7.3	6—11	7—17	ME
Sesame	A, H	4.3—8.7	2—40	11—29	AF, CJ, HI
Sisal	P, H	4.5—8.3	2—32	15—27	MA
Smilograss	P, G	7.2—8.0	3—7	13—19	ME
Sorghum, almum	P, G	5.0—8.3	3—25	9—26	SA
Sorghum, grain	A, G	4.3—8.7	4—41	8—27	CJ, HI, ME
Soybean	A, H	4.3—8.4	4—41	6—29	CJ
Soybean, perennial	P, V, L	4.3—7.1	7—42	18—27	AF
Spearmint	P, H	4.5—7.5	6—27	6—26	ME, ES
Spelt	A, G	4.9—8.3	3—16	5—19	NE, ES, AF
Spinach	A, H	4.2—8.3	3—26	6—23	CE
Squash, mixta	A, V	4.0—7.5	4—28	7—32	MA, SA
Stargrass	P, G	4.3—8.0	2—41	13—27	AF
Strawberry, Chilean	P, H	4.9—7.3	5—26	5—23	SA, NA, ES
Strawberry, garden	P, H	4.5—8.3	3—26	5—25	ES, NA, SA
Strawberry, Virginia	P, H	4.9—7.5	4—17	5—27	NA, ES
Stylo, Townsville	A, H	4.5—8.0	5—25	19—27	MA, AF
Sudangrass	A, G	5.0—8.2	2—25	8—27	AF
Sugarcane	P, G	4.3—8.4	5—42	16—30	II, HI
Sunflower	A, H	4.5—8.7	2—40	6—28	NA, ES
Sweetclover, white	B, A, H	4.8—8.3	3—16	5—23	ES, ME
Sweetclover, yellow	B, A, H	4.8—8.5	3—16	5—21	ES, ME
Switchgrass	P, G	4.9—7.6	4—26	7—27	MA, NA
Tangerine	P, T	4.3—8.3	3—41	13—28	II, CJ, MA
Taro	P/A, H	4.3—7.5	7—41	11—29	II, CJ
Tea	P, T	4.5—7.3	7—31	14—27	CJ, HI
Teff	A, G	4.3—8.3	3—12	13—19	AF
Teosinte	A, G	4.8—8.2	5—25	8—27	MA, SA
Thyme, common	P, H	4.5—8.0	4—28	7—25	ME
Timothy	P, G	4.5—7.8	3—17	5—19	ES
Tobacco	P/A, H	4.3—8.7	3—40	6—28	SA, NA
Tomato	A, P, H	4.3—8.7	3—46	6—28	SA, MA
Trefoil, big	P, H	5.3—8.2	3—15	7—21	ES, CA
Trefoil, birdsfoot	P, H	4.5—8.2	3—19	5—23	ES
Trefoil, narrowleaf	P, H	4.8—8.0	4—12	7—17	ES

Table 91 (continued)
SELECTED ECONOMIC PLANTS AND FACTORS AFFECTING THEIR
DISTRIBUTION

Common name	Growth type[a]	pH	Annual precipitation (dm)	Annual temperature (°C)[b]	Center of diversity[c]
Triticale	A, G	5.3—7.5	4—25	6—26	NA, ES
Tung-oil tree	P, T	5.3—7.3	6—25	13—26	CJ, NA
Tumeric	P, H	4.3—6.8	7—42	18—27	HI
Turnip	B, H	4.2—8.3	3—41	4—27	ME, ES, CJ
Vanilla	P, V	4.3—8.0	7—42	19—28	MA
Velvetbean	A, H, V	4.5—6.8	7—31	17—27	II
Vetch, common	A, H, V	4.5—8.2	3—17	5—23	NE, ME
Vetch, purple	A, H	4.8—8.2	6—17	15—19	ME
Vetch, winter	A, B, V	4.8—8.2	4—17	5—21	NE
Walnut, black	P, T	5.5—8.3	3—13	7—19	NA
Watercress	P/A, H	4.3—8.3	3—42	6—27	ES
Watermelon	A, H, V	5.3—8.7	3—40	10—29	HI, ME, AF
Wheat, bread	A, G	4.5—8.6	3—25	5—27	CJ, HI, CE
Wheat, club	A, G	5.8—8.3	3—16	6—16	ES, NE, CE
Wheat, durum	A, G	4.8—8.6	3—21	6—27	NE
Wheatgrass, bluebunch	P, G	5.6—8.2	3—11	6—19	NA
Wheatgrass, crested	P, G	5.3—8.2	3—12	5—19	ES, NA
Wheatgrass, fairway	P, G	5.5—8.2	3—17	5—23	ES
Wheatgrass, intermediate	P, G	5.3—8.2	3—11	5—19	ES, NE
Wheatgrass, tall	P, G	5.3—9.0	3—21	5—19	ME
Wheatgrass, western	P, G	5.3—8.2	3—11	5—15	NA
Wildrye, Canada	P, G	5.0—7.8	3—13	6—20	NA
Wildrye, giant	P, G	5.6—7.8	3—11	7—13	NA
Wildrye, Russian	P, G	4.9—7.8	3—12	5—16	ES
Wintergreen	P, S	4.8—7.5	6—13	8—17	NA
Yam, composite	P, V	5.0—8.0	8—21	16—26	MA
Yam, eboe	P, V	5.1—5.8	9—28	21—27	AF
Yam, winged	P, V	4.8—8.0	7—42	15—29	II, AF

Note: Absence of a maximum or minimum indicates data were not available.

[a] A = Annual; B = biennial; P = perennial; P/A = perennial treated as an annual; H = herb; G = grass; L = woody vine; S = shrub; T = tree; V = herbaceous vine.

[b] Average of monthly means with values below 0°C treated as 0.

[c] From various sources, Duke lists up to three centers, and no more. The first symbol is the possible center of origin. Symbols are as follows: CJ = China, Japan; II = Indochina-Indonesia; AU = Australia; HI = Hindustani; CE = Central Asia; NE = Near East; ME = Mediterranean; AF = Africa; ES = Eurosiberian; SA = South America; MA = Middle America; NA = North America.

Adapted from Duke, J. A., The quest for tolerant germplasm, in *Crop Tolerance to Suboptimal Land Conditions,* Jung, G. A., Ed., Spec. Pub. No. 32, American Society of Agronomy, Madison, Wis., 1978. With permission.

SEEDS AND SEED CERTIFICATION

Table 92

LONGEVITY OF SEEDS STORED UNDER VARIOUS CONDITIONS IN OPEN AND POROUS CONTAINERS

Species	Moisture (% when stored)	Storage			Germination (%)	
		RH (%)	Time (years)	Temperature (°C)	When stored	After storage
African daisy						
Dimorphotheca sinuata de Candolle	Not given	40	16	5	64	76
D. pluvialis (L.) Moench	Not given	40	16	5	62	73
Ageratum						
Ageratum mexicanum Sims, now *A. houstonianum* Mill.	Not given	35—40	16	5	89	89
Alfalfa, lucerne						
Medicago sativa L.	Not given	Ambient	20	Ambient	99	83
	13.9	45	6.25	22—27	100	94
	13.9	60	6.25	22—27	100	12
	13.9	75	2.25	22—27	100	3
	9.4	60—70	4	0	95	93
Amaranth, feathered						
Celosia cristata L. (Plumosa group)	Not given	35—40	16	5	98	99
Apricot						
Prunus armeniaca L.	10—11	50—65	2.50	20—25	100	39
	10—11	80—90	2.50	2—10	100	0
Apple						
Malus sylvestris Mill.	10—11	50—65	2.50	20—25	95	33
	10—11	80—90	0.83	2—10	95	4
Artichoke						
Cynara scolymus L.	Not given	35—40	17	5	69	68
Ash, European						
Fraxinus excelsior L.	15.4	Not given	2	5	53	22
Asparagus						
Asparagus officinalis L.	Not given	35—40	17	5	87	78

Species						
Aster						
Callistephus chinensis (L.) Nees	Not given	Not given	16.33	-4	91	21
	Not given	35—40	14	5	82	75
Bahiagrass						
Paspalum notatum Fluegge	Not given	35—40	17	5	66	60
Bamboo						
Bamboo arundinacea Retz	18	Not given	0.6	21—32	Not given	0
Barley						
Hordeum vulgare L.	Not given	Ambient	5	18	100	5
	12.5	77—87	4	5—15	95	68
	12.5	60—70	4	0	95	94
Bean, garden						
Phaseolus vulgaris L.	7	Not given	8	30	97	6
	7	Not given	4	20	97	18
	7	Not given	15	10	97	86
	7	Not given	15	-2	97	97
Bean, lima (green seed)						
Phaseolus vulgaris L.	Not given	35—40	18	5	90	99
	Not given	90	0.50	21	97	52
	Not given	50	3	21	97	99
Bean, lima (bleached seed)	Not given	90	0.50	21	76	21
	Not given	70	1.50	21	76	30
	Not given	50	3	21	76	70
Bean, scarlet runner						
Phaseolus coccineus L.	Not given	35—40	16	5	92	95
Beet, garden						
Beta vulgaris L.	11.1	85—95	6	0—2	83	70
	11.1	70—80	6	5—20	83	78
	Not given	35—40	17	5	87	88
Beet, mangel						
Beta vulgaris L.	Not given	35—40	17	5	87	86
Beet, sugar						
Beta vulgaris L.	Not given	35—40	17	5	91	90
	Not given	Ambient	22	10 to -10	84	75
Whole seed	Not given	Ambient	20	Ambient	67	65
Segmented	Not given	Ambient	20	Ambient	73	64
Swiss chard						
Beta vulgaris L.	Not given	35—40	15	5	90	94

Table 92 (continued)

LONGEVITY OF SEEDS STORED UNDER VARIOUS CONDITIONS IN OPEN AND POROUS CONTAINERS

Species	Moisture (% when stored)	RH (%)	Storage Time (years)	Temperature (°C)	Germination (%) When stored	After storage
Bentgrass, colonial						
Agrostis tenuis Sibth.	11.0	77—87	4	5—15	72	2
	11.0	60—70	4	0	82	63
Bindweed						
Convolvulus arvensis L.	Not given	Ambient	50	Ambient	Not given	62
Bluegrass, Kentucky						
Poa pratensis L.	8.7	20—40	2	21	86	3
	8.7	10—15	2.50	32	86	81
	3.8	15	2	43	75	60
	Not given	35—40	17	5	96	93
Bluegrass, wood						
P. nemoralis L.	Not given	Ambient	5	18	84	2
	Not given	35—40	10	5	99	96
Bluegrass, roughstalk						
P. trivialis L.	Not given	35—40	6	18	98	9
Bluestem, big						
Andropogon gerardii Vitm.	Not given	35—40	12	5	73	80
Bluestem, sand						
A. hallii Hack.	Not given	35—40	7	5	64	66
Bluestem, little						
A. scoparius (Michx.) Nash, now *Schizachyrium scoparium* (Michx.) Nash	Not given	35—40	15	5	79	32
Broad bean, fava bean						
Vicia faba L.	Not given	35—40	17	5	91	88
Brome, field						
Bromus arvensis L.	Not given	Ambient	4	18	95	8
	Not given	35—40	15	5	75	58

Species						
Brome, mountain						
B. marginatus Nees ex Steud.	7—9	Not given	5 –	15—23	96	93
	Not given	35—40	10	5	89	90
B. polyanthus Scribn.	7—9	Not given	5 –	15—23	94	93
Brome, smooth						
B. inermis Leyss.	8	80	0.50	30	91	1
	8	50	2 +	10	91	91
	Not given	25—40	13	5	91	57
Brussels sprouts						
Brassica oleracea L. var. *gemminifera*	6—8	58—83	6.50	11—19	71	13
	Not given	35—40	19	5	92	94
Buckwheat						
Fagopyrum esculentum Moench	Not given	35—40	15	5	97	92
Buffel grass						
Pennisetum ciliare (L.) Link	Not given	15	6.25	22—27	46	26
	Not given	60	4.25	22—27	46	1
now *Cenchrus ciliaris* L.	Not given	35—40	15	5	85	15
Buttercup, creeping						
Ranunculus repens L.	11.2	77—87	4	5—15	96	100
	11.2	60—70	4	0	96	100
Cabbage						
Brassica oleracea L. var. *capitata*	3.1	50—60	2	1	96	95
	3.1	80—95	1.50	10	96	10
	2.9	15	2	43	94	67
B. oleracea L. *capitata alba*	Not given	35—40	18	5	84	82
B. oleracea L. var. *quercifolia*	Not given	Ambient	8	18	94	11
	Not given	Ambient	9	18	92	18
Campion, white						
Lychnis alba Mill.	11.5	77—87	4	5—15	95	100
now *Silene alba* (Mill.) Krause	11.5	60—70	4	0	95	92
Canarygrass						
Phalaris canariensis L.	Not given	Ambient	15	18	99	3
		35—40	10	5	75	95
Canarygrass, reed						
Phalaris arundinacea L.	8.2	40	12	—1	89	90
	8.2	35	12	5	89	83
	8.2	70	1	21	89	33
	8.2	50	1.50	32	89	8
	Not given	35—40	17	5	68	65

Table 92 (continued)
LONGEVITY OF SEEDS STORED UNDER VARIOUS CONDITIONS IN OPEN AND POROUS CONTAINERS

Species	Moisture (% when stored)	RH (%)	Storage Time (years)	Temperature (°C)	Germination (%) When stored	Germination (%) After storage
Cantaloupe (muskmelon)						
Cucumis melo L.	Not given	15	12	32	91	64
	Not given	60	12	10	91	84
	Not given	35—40	18	5	90	88
Caraway						
Carum carvi L.	Not given	Ambient	2	18	77	13
	Not given	35—40	17	5	66	65
Carrot						
Daucus carota L.	10.2	85—95	6	−20	89	76
	10.2	85—95	6	0—2	89	41
	Not given	35—40	17	5	91	88
Castorbean						
Ricinus communus L.	Not given	35—40	17	5	96	90
Cedar, Port-Orford						
Chamaecyparis lawsoniana (A. Murr.) Parl.	8.0	Not given	7	−18	56	43
	8.0	Not given	7	0	56	43
Cedar, western red						
Thuja plicata Donn.	7.9	Not given	7	−18	68	34
	7.9	Not given	3	0	68	6
	7.9	Not given	3	Ambient	68	0
Celery						
Apium graveolens L. var. *dulce* (Mill.) Pers.	Not given	Ambient	7	18	67	1
	8.3	70—80	6	5—20	79	61
	Not given	35—40	17	5	77	80
Chess, soft						
Bromus mollis L.	12.5	60—70	4	0	88	82
	Not given	35—40	4	18	73	12
	Not given	35—40	10	5	89	90

Species						
Chives						
Allium schoenoprasum L.	11	85—95	6	−20	98	86
	11	85—95	6	0—2	98	1
	11	70—80	6	5—20	98	0
	Not given	35—40	16	5	92	83
Chrysanthemum						
Chrysanthemum coronarium L.	Not given	Ambient	8	18	48	4
Clover, alsike						
Trifolium hybridum L.	Not given	Ambient	25	18	99	3
	Not given	Not given	10	Ambient	93	45
	Not given	35—40	11	5	90	90
Clover, crimson						
Trifolium incarnatum L.	7	70	16	−12	91	81
	10	40	16	−1	91	74
	10	60	16	10	91	12
	10	30	16	21	91	25
Clover, ladino						
Trifolium repens L.	Not given	Ambient	26	18	98	8
Clover, red						
Trifolium pratense L.	10.8	77—87	4	5—15	80	2
	10.8	60—70	4	0	80	76
	Not given	Ambient	20	Ambient	97	33
Continuously frozen	Not given	60—90	13	−5 to −15	96	97
Thawed weekly	Not given	60—90	13	−5 to −15	96	96
Clover, small hop						
Trifolium dubium Sibth.	10.2	77—87	4	5—15	90	10
	10.2	60—70	4	0	90	93
Clover, white						
Trifolium repens L.	9.8	Ambient	16	0	92	92
	9.8	Ambient	16	Ambient	92	2
	Not given	35—40	15	5	99	88
	10	77—87	4	5—15	98	5
	10	60—70	4	0	98	84
Cocoa (pods)						
Theobroma cacao L.	In moist steam-sterilized charcoal	Not given	0.09	Ambient	100	100

Table 92 (continued)
LONGEVITY OF SEEDS STORED UNDER VARIOUS CONDITIONS IN OPEN AND POROUS CONTAINERS

Species	Moisture (% when stored)	RH (%)	Storage Time (years)	Temperature (°C)	Germination (%) When stored	Germination (%) After storage
Corn, field inbred lines						
Zea mays L.	10	Ambient	7	13—21	97	55
	10	Ambient	7	4	97	90
Open pollinated	Not given	35—40	15	5	94	93
	Not given	35—40	17	5	97	98
	Not given	15	6	22—27	95	83
	Not given	30	6.25	22—27	95	80
	Not given	60	4.75	22—27	95	21
	Not given	90	0.83	22—27	95	3
Cotton, upland						
Gossypium hirsutum L.	8—11	Ambient	11	1—32	96	82
	Not given	35—40	15	5	82	84
Cotton, sea island						
G. barbadense L.	Not given	35—40	13	5	80	80
Cowpea (southern pea)						
Vigna unguiculata (L.) Walp	Not given	Not given	4	5	84	84
subsp. *unguiculata*	Not given	Humid	4.50	Warm	84	45
	Not given	35—40	17	5	92	96
Crambe						
Crambe abyssinica Hochst.	Not given	50	4	32	93	51
ex R. E. Fries	Not given	70	1.50	21	93	33
	Not given	70	8	10	93	67
	Not given	50	8	10	93	82
	Not given	40	8	−1	93	88
Cress, garden						
Lepidium sativum L.	Not given	35—40	14	5	74	82
	Not given	35—40	17	5	94	31

Species						
Crotalaria						
Crotalaria juncea L.	Not given	35—40	14	5	72	70
Cucumber						
Cucumis sativus L.	Not given	70	1	32	96	25
	Not given	70	1.25	21	96	83
	Not given	50	5	10	96	94
	Not given	35—40	17	5	97	98
Datura						
Datura ferox L.	Not given	35—40	10	5	99	96
Datura						
D. innoxia Mill.	Not given	35—40	10	5	84	98
Datura						
D. metal L.	Not given	35—40	10	5	90	74
Datura, jimson weed						
D. stramonium L.	Not given	35—40	10	5	98	98
Delphinium, annual						
Delphinium spp.	Not given	Not given	5	5	57	0
	Not given	Not given	10	-4	57	6
Delphinium, perennial						
Delphinium spp.	Not given	Not given	5	5	43	0
	Not given	Not given	10	-4	43	6
	Not given	35—40	16	5	87	88
Dill						
Anethum graveolens L.	Not given	35—40	18	5	92	64
Dock, curly						
Rumex crispus L.	9.8	77—87	4	5—15	63	98
	9.8	60—70	4	0	63	88
Dogtail, crested						
Cynosurus cristatus L.	Not given	Ambient	4	18	63	5
Eggplant						
Solanum melongena L.	Not given	Ambient	2.16	Ambient	73	25
	Not given	35—40	17	5	99	97
Elm, American						
Ulmus americana L.	Not given	Not given	10	-4	96	14
Evening primrose						
Oenothera biennis L.	Not given	35—40	11	5	99	99
	Not given	35—40	11	5	99	14

Table 92 (continued)
LONGEVITY OF SEEDS STORED UNDER VARIOUS CONDITIONS IN OPEN AND POROUS CONTAINERS

Species	Moisture (% when stored)	RH (%)	Storage Time (years)	Temperature (°C)	Germination (%) When stored	After storage
Fescue, creeping red						
Festuca rubra L. subsp. *rubra*	3.6	50—60	2	1	95	91
	3.6	80—95	2	10	95	0
	3.6	25—30	2	27	95	83
	Not given	35—40	17	5	90	72
	13.6	60—70	4	0	87	80
Fescue, hard						
Festuca ovina var. *duriuscula* (L.) Koch now *F. longifolia* Thuill.	Not given	Ambient	5	18	89	1
	Not given	35—40	13	5	94	85
Fescue, meadow						
Festuca pratensis Huds.	Not given	35—40	5	18	98	6
	Not given	35—40	10	5	95	85
	12.5	60—70	4	0	90	89
Fescue, tall						
Festuca arundinacea Schreb.	12.8	77—87	2	5—15	76	15
	12.8	60—70	4	0	76	63
	Not given	35—40	5	18	95	14
	Not given	35—40	17	5	89	66
Fir, balsam						
Abies balsamea (L.) Mill.	Not given	Ambient	5	18	29	2
Fir, Douglas						
Pseudotsuga menziesii (Mirbel) Franco	15	Not given	3	−4	88	7
	10	Not given	3	−18	88	78
	8.3	Not given	7	−18	43	27
	8.3	Not given	7	0	43	14
Fir, noble						
Abies nobilis (Dougl. ex D. Don) Lindl. now *A. procena* Rehd.	9.0	Not given	7	−18	43	33
	9.0	Not given	7	0	43	32
	9.0	Not given	7	Ambient	43	2

Species						
Fir, white						
A. concolor (Gord. & Glend.) Lindl. ex Hildebr.	6.3	Not given	7	−18	72	62
	6.3	Not given	7	0	72	55
	6.3	Not given	3	Ambient	72	20
Flax						
Linum usitatissimum L.	Not given	Ambient	10	18	72	1
	Not given	35—40	17	5	99	95
Foxtail, black grass						
Alopecurus myosuroides Huds.	13.6	77—87	4	5—15	44	7
	13.6	60—70	4	0	44	40
Foxtail, creeping						
A. arundinaceus Poir.	Not given	35—40	14	5	87	88
Foxtail, meadow						
A. pratensis L.	Not given	Ambient	9	18	89	1
	Not given	40	10	5	84	87
	Not given	30—40	10	21	84	82
Geranium, cutleaf						
Geranium dissectum L.	9.1	77—87	4	5—15	99	90
	9.1	60—70	4	0	99	99
Geranium, dovefoot						
G. molle L.	9.1	77—87	4	5—15	99	83
	9.1	60—70	4	0	99	98
Gherkin, West Indian						
Cucumis anguria L.	Not given	35—40	17	5	85	90
Gladiolus						
Gladiolus spp.	Not given	Not given	5	5	82	0
	Not given	Not given	10	−4	82	4
Grapefruit						
Citrus paradisi Macfad.	60	Not given	1	5	94	88
	100	Not given	1	5	96	88
	123	Not given	1	5	32	16
Guar						
Cyamopsis tetragonoloba (L.) Taub.	Not given	35—40	16	5	96	96
Guayule						
Parthenium argentatum A. Gray	Not given	35—40	16	5	66	62

Table 92 (continued)
LONGEVITY OF SEEDS STORED UNDER VARIOUS CONDITIONS IN OPEN AND POROUS CONTAINERS

Species	Moisture (% when stored)	RH (%)	Storage Time (years)	Temperature (°C)	Germination (%) When stored	Germination (%) After storage
Hemlock, western						
Tsuga heterophylla (Raf.) Sarg.	8	Not given	3	−18	12	13
	7	Not given	7	−18	70	70
	7	Not given	7	0	70	56
	7	Not given	7	Ambient	70	0
Hemp						
Cannabis sativa L.	5.7	50	8.25	10	98	97
	5.7	50	15	10	98	86
Indiangrass						
Sorghastrum nutans (L.) Nash	Not given	35—40	16	5	66	41
Juniper, alligator						
Juniperus deppeana Steud.	Not given	Ambient	9	Ambient	Not known	16
Juniper, one-seed						
J. monosperma (Engelm.) Sarg.	Not given	Ambient	21	Ambient	Not known	54
J. osteosperma (Torr.) Little	Not given	Ambient	45	Ambient	Not known	17
Kenaf						
Hibiscus cannabinus L.	Not given	35—40	13	5	78	77
Ladysthumb						
Polygonum persicaria L.	10.5	77—87	4	5—15	95	73
	10.5	60—70	4	0	95	99
Lambsquarters						
Chenopodium album L.	8.6	77—87	4	5—15	86	98
	8.6	60—70	4	0	86	98
Larch						
Larix spp.	Not given	Ambient	3	18	32	4
Lemon, rough						
Citrus limon (L.) Burm. f.	56	Ambient	1.50	5	84	14

Species						
Leek *Allium ampeloprasum* L.	Not given	35—40	17	5	97	99
Lentil *Lens culinaris* Medik	Not given	35—40	5	5	99	99
Lespedeza, Kobe *Lespedeza striata* (Thkunb. ex Murr.) Hook & Arn.	Not given	Not given	4	Ambient	87	39
	Not given	Not given	4	−7 to −11	87	90
	Not given	Not given	18	−7 to −11	87	64
	Not given	35—40	10	5	95	96
Lespedeza, Korean *L. stipulacea* Maxim.	Not given	Not given	4	Ambient	97	54
	Not given	Not given	4	−7 to −11	97	95
	Not given	Not given	18	Ambient	97	4
	Not given	Not given	18	−7 to −11	97	77
Lespedeza, Sericea *L. cuneata* (Dum. G. Don)	Not given	Not given	4	Ambient	89	46
	Not given	Not given	4	−7 to −11	89	89
	Not given	Not given	18	Ambient	89	4
	Not given	Not given	18	−7 to −11	89	72
Lettuce *Lactuca sativa* L.	Not given	75	1.25	22—27	96	2
	7	70	19	−12	97	93
	7	40	19	−1	97	17
	7	15	8	32	97	2
	10	70	19	−12	95	94
Lily, regal *Lilium regale* Wilson	Not given	Not given	5	5	90	0
	9.9	Not given	15	−5	86	74
	9.9	Not given	15	5	86	8
Lobelia, cardinal flower *Lobelia cardinalis* L.	6.9	Not given	3	Ambient	63	13
	6.9	Not given	2	5	63	6
	6.2	Not given	16	−5	63	14
Trefoil, birdsfoot[a] *Lotus angustissimus* L.	Not given	35—40	11	5	68	64
L. arabicus L.	Not given	35—40	11	5	96	95

Table 92 (continued)

LONGEVITY OF SEEDS STORED UNDER VARIOUS CONDITIONS IN OPEN AND POROUS CONTAINERS

Species	Moisture (% when stored)	RH (%)	Storage Time (years)	Temperature (°C)	Germination (%) When stored	Germination (%) After storage
L. arenarius Brot.	Not given	35—40	5	5	94	94
L. caucasicus Kuprian	Not given	35—40	5	5	92	89
L. collinus (Boiss.) Heldr.	Not given	35—40	6	5	97	88
L. conimbricensis Brot.	Not given	35—40	6	5	98	99
L. conjugatus L.	Not given	35—40	11	5	91	96
L. corniculatus L.	Not given	Ambient	25	18	93	14
L. creticus L.	Not given	35—40	11	5	92	94
L. decumbens Poir.	Not given	35—40	11	5	92	91
L. divaricatus Boiss.	Not given	35—40	4		61	77
L. edulis L.	Not given	35—40	11	5	95	94
L. frondosus Freyn.	Not given	35—40	6	5	99	74
L. glinoides Del.	Not given	35—40	11	5	75	68
L. hispidus Desf.	Not given	35—40	11	5	91	97
L. major Scop.	Not given	35—40	11	5	91	89
L. mearnsii Britton	Not given	35—40	6	5	96	87
L. ornithopodioides L.	Not given	35—40	5	5	82	84
L. palustris Willd.	Not given	35—40	11	5	96	83
L. parviflorus Desf.	Not given	35—40	11	5	94	92
L. pedunculatus Cav. L.	Not given	35—40	6	5	87	88
L. peregrinus L.	Not given	35—40	11	5	90	98
L. purshianus (Benth.) Clements and Clements	Not given	35—40	6	5	93	93
	Not given	35—40	5	5	91	81
L. pusillus Medic	Not given	35—40	5	5	91	96
L. strictus Fisch. & C. A. Mey	Not given	35—40	5	5	72	39
L. tenuis Waldst. & Kit. ex Willd.	Not given	35—40	11	5	96	93
L. tetragonolobus L. now *Tetragonolobus purpureus* L.	Not given	35—40	6	5	90	97
L. weilleri Marie	Not given	35—40	5	5	87	90

Species						
Luffa, angled						
L. acuangula (L.)	Not given	35—40	12	5	90	86
Luffa, vegetable sponge						
L. aegyptiaca Mill.	Not given	35—40	11	5	69	42
Lupine, yellow						
Lupinus luteus L.	Not given	Ambient	8	18	85	10
	Not given	35—40	15	5	86	76
Macadamia nut						
Macadamia spp.	Not given	Ambient	0.91	18—27	35	8
	Not given	40—45	6	5	52	53
Marigold						
Tagetes erecta L.	Not given	35—40	16	5	97	94
Marigold, French						
T. patula L.	Not given	35—40	16	5	91	82
Mayweed, scentless						
Matricaria inodora L.	12.2	77—87	4	5—15	91	96
	12.2	60—70	4	0	91	99
Medic, black						
Medicago lupulina L.	9.5	77—87	2	5—15	84	14
	9.5	60—70	4	0	84	93
Mesquitegrass, vine						
Panicum obtusum H.B.K.	Not given	Ambient	15	18	99	1
Mignonette, common						
Reseda odorata L.	Not given	Ambient	25	Ambient	98	7
Mustard						
Brassica campstris L., now *B. rapa* L.	Not given	Ambient	10	18	59	2
Mustard, white						
Sinapis alba L.	Not given	Ambient	12	18	98	3
Needlegrass, green						
Stipa viridula Trin.	Not given	Ambient	15	18	96	1
	7 to 9	Not given	5	15—23	28 (dormant)	98
Oatgrass, tall						
Avena elatior L., now *Arrhenatherum elatius* (L.) Beauv. ex J. & C. Presl.	Not given	Ambient	6	18	98	6
	Not given	40	10	5	90	92
Oats						
Avena sativa L.	Not given	Ambient	10	18	84	2
	Not given	Not given	10	Ambient	95	96
	12	77—87	4	5—15	92	22
	Not given	35—40	16	5	99	95

Table 92 (continued)
LONGEVITY OF SEEDS STORED UNDER VARIOUS CONDITIONS IN OPEN AND POROUS CONTAINERS

Species	Moisture (% when stored)	Storage RH (%)	Storage Time (years)	Storage Temperature (°C)	Germination When stored	Germination After storage
Oat, bristle						
A. strigosa Schreb.	13.5	77—87	4	5—15	92	29
	13.5	60—70	4	0	92	88
Oat, red						
A. byzantina K. Koch	Not given	35—40	16	5	99	99
Oat, wild						
A. fatua L.	14	77—87	4	5—15	34	8
	14	60—70	4	0	34	42
Okra						
Abelmoschus esculentus (L.) Moench	Not given	35—40	17	5	82	92
Onion						
Allium cepa L.	7.7	Ambient	18	-2	95	15
	7.7	Ambient	18	-18	95	87
	8	Ambient	7	Ambient	86	0
	4.1	50—60	2	1	95	93
	4.1	80—95	1	10	95	13
	Not given	35—40	17	5	93	93
Onion, Beltsville bunching						
Allium cepa L. × A. fistulosum L.	Not given	35—40	13	5	93	64
Onion, Welsh						
A. fistulosum L.	Not given	35—40	17	5	97	78
Orange, sour						
Citrus aurantium L.	80	Ambient	1.50	5	90	70
Orange, sweet						
C. sinensis (L.) Osb.	82	Ambient	1.50	5	71	15
Orchardgrass, cocksfoot						
Dactylis glomerata L.	10.9	77—87	4	5—15	90	1
	10.9	60—70	4	0	90	88
	Not given	35—40	15	5	86	85

Seeds and Seed Certification

Osteospermum						
Osteospermum ecklonis (DC.) Norl.	Not given	50	1	32	79	13
	Not given	50	6	10	79	77
Pansy						
Viola tricolor L.	Not given	35—40	16	5	90	80
Papaya						
Carica papaya L.	Not given	50	6	10	68	53
Parsley						
Petroselinum sativum Hoffm.	Not given	Ambient	6	18	65	3
now *P. crispum* (Mill.) Nym. ex A. W. Hill	Not given	35—40	18	5	74	84
Parsnip						
Pastinaca sativa L.	Not given	Ambient	4	18	92	3
	Not given	35—40	18	5	90	88
Pea, garden						
Pisum sativum L.	Not given	Ambient	10	18	87	7
	Not given	35—40	16	5	90	96
Pea, pigeon						
Cajanus cajan (L.) Huth	Not given	35—40	10	5	85	71
	Not given	35—40	15	5	85	20
Peanut, Spanish (groundnut)						
Arachis hypogaea L.	Ambient	Not given	6	21—29	85	56
	Not given	35—40	14	5	98	92
Peanut, Valencia	Ambient	Not given	6	21—29	53	47
	Not given	35—40	17	5	88	94
Pear						
Pyrus communis L.	10—11	50—65	2.50	20—25	96	61
	10—11	80—90	0.83	2—10	96	36
Peony						
Paeonia suffruticosa Andr.	Not given	Not given	5	5	50	3
	Not given	Not given	10	–4	50	29
Pepper						
Capsicum annuum L.	Not given	Not given	20	–4	73	19
	Not given	35—40	18	4	87	87
Petunia						
Petunia × *hybrida* Vilm.	Not given	35—40	16	5	68	74
Pine, Austrian						
Pinus nigra Arnold var. *austriaca* (Hoess) Baddux	Not given	Ambient	8	18	66	2

Table 92 (continued)
LONGEVITY OF SEEDS STORED UNDER VARIOUS CONDITIONS IN OPEN AND POROUS CONTAINERS

Species	Moisture (% when stored)	RH (%)	Storage Time (years)	Temperature (°C)	Germination (%) When stored	Germination (%) After storage
Pine, loblolly *P. taeda* L.	Not given	Not given	12	5	82	15
	Not given	Not given	12	−4	82	80
Pine, lodgepole *P. contorta* Dougl. ex Loud.	8.8	Not given	7	0	32	32
Pine, mugo *P. montana* Mill. now *P. mugo* Turra	Not given	Not given	14	18	96	2
Pine, Norway *P. resinosa* Ait.	6.5	Not given	30	2—10	95	8
Pine, ponderosa *P. ponderosa* Dougl. ex P. &. C. Lawson	8.7	Not given	35	10—27	Not given	16
	8.1	Not given	7	0	66	67
	8.1	Not given	7	Ambient	66	35
	15	Not given	3	−4	95	92
Pine, Scotch *P. sylvestris* L.	Not given	Ambient	11	18	69	5
Pine, shortleaf *P. echinata* Mill.	Not given	Not given	16	5	59	3
	Not given	Not given	16	−4	59	36
Plantain, buckhorn *Plantago lanceolata* L.	10.2	77—87	4	5—15	98	48
	10.2	60—70	4	0	98	98
Plum, American *Prunus americana* Marsh.	10—11	50—65	2.50	20—25	100	62
	10—11	80—90	2.50	2—10	100	23
Potato *Solanum tuberosum* L.	Not given	35—40	10	5	95	98

Crop						
Pyrethrum						
Chrysanthemum cineraiifolium (Trevir.) Vis.	6.5	Not given	15	−18.5	57	38
	6.5	Not given	4	−4	57	5
	6.5	Not given	15	5	57	29
	6.5	Not given	13	20	57	1
Quinine						
Cinchona ledgerana Moens ex Trimen	6.0	Not given	1.50	5	74	3
now *C. calisaya* Wedd.	6.0	Not given	4	−4	74	43
Radish						
Raphanus sativus L.	Not given	Ambient	11	18	71	1
	35—40	13	5	99	99
Rice						
Oryza sativa L.	11.5—13.2	60	3.50	15	96	98
	Not given	30	6	7	92	42
	Not given	60	6	7	92	12
	Not given	Ambient	6	7	92	2
Roselle						
Hibiscus sabdariffa L.	Not given	35—40	13	5	78	80
Rye						
Secale cereale L.	Not given	Ambient	4	18	93	3
	9	58—83	3 +	9—20	90	90
	Not given	35—40	17	5	82	58
Ryegrass, Italian						
Lolium multiflorum Lam.	9	35	4	Ambient	96	92
	11	53	4	4—21	96	92
	19	89	1	4—21	96	10
	Not given	35—40	14	5	93	83
Ryegrass, perennial						
L. perenne L.	Not given	35—40	13	5	98	96
	12.6	77—87	4	5—15	92	51
Safflower						
Carthamus tinctorus L.	10	40	17	−1	95	91
	10	60	17	10	95	53
	10	30	17	21	95	10
Sage						
Salvia officinalis L.	Not given	Ambient	7	18	74	3
	Not given	35—40	15	5	75	92

Table 92 (continued)
LONGEVITY OF SEEDS STORED UNDER VARIOUS CONDITIONS IN OPEN AND POROUS
CONTAINERS

Species	Moisture (% when stored)	RH (%)	Storage Time (years)	Temperature (°C)	Germination (%) When stored	After storage
Sainfoin						
Onobrychis sativa Lam.	Not given	Ambient	7	18	60	1
Sainfoin						
O. viciifolia Scop.	Not given	Ambient	14	Ambient	Not given	36
	Not given	35—40	5	5	94	84
Salsify						
Tragopogon porrifolius L.	Not given	35—40	14	5	98	99
Savory, summer						
Satureja hortensis L.	Not given	35—40	15	5	67	46
Scabiosa						
Scabiosa atropurpurea L. and *S. caucasica* Bieb.	Not given	35—40	16	5	77	68
Serradella						
Ornithopus sativus Brot.	Not given	Ambient	10	18	97	5
Sesame						
Sesamum indicum L.	10	30	17	21	88	84
	10	15	11	32	88	51
Silver bells						
Browallia viscosa H.B.K.	Not given	35—40	14	5	99	99
Snapdragon						
Antirrhinum majus L.	Not given	35—40	11	5	90	80
	Not given	35—40	16	5	90	40
Sorghum						
Sorghum bicolor (L.) Moench	10	40	16	−1	91	83
	Not given	75—80	3.50	4—7	95	85
	Not given	Ambient	3.50	19—31	95	1

Species						
Soybean						
Gycine max (L.) Merr.	11.4	15	6.25	22—27	100	3
	11.4	Ambient	1.66	22—27	100	3
	Not given	30	6	7	97	8
	Not given	30	6	10	97	6
	Not given	35—40	16	5	92	79
Spinach, New Zealand						
Tetragonia tetragonioides (Pall.) Ktze.	Not given	Ambient	10	Ambient	1	95
	Not given	35—40	17	5	74	42
Spinach						
Spinacia oleracea L.	Not given	35—40	17	5	88	88
Spruce, Engelmann						
Picea engelmannii Parry ex Engelm.	7.2	Not given	7	−18	71	46
	7.2	Not given	7	0	71	54
	7.2	Not given	7	Ambient	71	0
Spruce, Norway						
Picea excelsa (Lam.) Link now *P. abies* (L.) Karst.	Not given	Ambient	8	18	83	2
	7.3	Not given	7	−18	80	81
Spruce, Sitka						
P. sitchensis (Bong.) Carr.	7.3	Not given	7	0	80	78
	13	Not given	2	−4	56	11
Spruce, white						
Picea glauca (Moench) Voss	6.7	Not given	7	−18	52	46
	6.7	Not given	7	0	52	47
	6.7	Not given	3	Ambient	52	27
Spurrey						
Spergula maxima Weike, now *S. arvensis* L.	Not given	Ambient	13	18	83	2
S. sativa Boenn., now *S. arvensis* L.	Not given	Ambient	14	18	86	1
Stock						
Matthiola incana (L.) R. Br.	Not given	35—40	16	5	94	92
Strawberry						
Fragaria spp.	Not given	35—40	10	5	72	82
Sudangrass						
Sorghum sudanense (Piper) Stapf	Not given	35—40	17	5	88	82
Sunflower						
Helianthus annuus L.	Not given	35—40	16	5	90	90

Table 92 (continued)
LONGEVITY OF SEEDS STORED UNDER VARIOUS CONDITIONS IN OPEN AND POROUS CONTAINERS

Species	Moisture (% when stored)	Storage RH (%)	Storage Time (years)	Storage Temperature (°C)	Germination (%) When stored	Germination (%) After storage
Sweetclover, white						
Melilotus alba Medik.	Not given	Ambient	20	Ambient	95	70
	Not given	35—40	13	5	81	82
Sweetclover, yellow						
Melilotus officinalis Lam.	Not given	Ambient	14	Ambient	Not given	24
	Not given	35—40	17	5	84	64
Sweet pea						
Lathyrus odoratus L.	Not given	35—40	16	5	96	96
Switchgrass						
Panicum virgatum L.	Not given	35—40	16	5	93	97
Tampala						
Amaranthus tricolor L.	Not given	35—40	17	5	88	85
Tea						
Thea spp.	Not given	100	0.09	0	70+	50—70
	Not given	40	0.09	0	70+	0
Thyme						
Thymus spp.	Not given	Ambient	5	18	62	3
Timothy						
Phleum nodosum L., now *P. pratense* L.	10.1	77—87	2	5—15	89	13
	10.1	60—70	4	0	89	76
Phleum pratense L.	10.6	77—87	4	5—15	97	1
	Not given	30—40	10	21	98	87
Tobacco						
Nicotiana tobacum L.	9	Not given	25	5	Not given	5
	Not given	35—40	18	5	91	92
Tomato						
Lycopersicon esculentum Mill.	Not given	70	1	32	90	56
	Not given	50	5	10	90	90
	Not given	Not given	18	5	93	88

Turnip						
Brassica Rapa L. (Rapifera group)	6—8	58—83	6.50	11—19	98	98
	Not given	35—40	19	5	90	39
	Not given	35—40	19	5	99	98
Velvetgrass						
Holcus lanatus L.	12.9	77—87	4	5—15	88	6
	12.9	60—70	4	0	88	85
Verbena						
Verbena spp.	Not given	Not given	16.33	-4	45	5
	Not given	35—40	16	5	56	38
Vetch, kidney						
Anthyllis vulneraria L.	Not given	Ambient	11	18	96	1
Watermelon						
Citrullus lanatus (Thunb.) Matsum. and Nakai	Not given	90	0.25	32	90	41
	Not given	70	1	32	90	40
	Not given	70	1.25	21	90	81
	Not given	50	5	10	90	90
	Not given	35—40	17	5	95	94
Wheat						
Triticum aestivum L.						
Wheat, spring	12.5	77—87	4	5—15	99	37
	12.5	60—70	4	0	99	92
Wheat, winter	12	77—87	4	5—15	92	19
	12	60—70	4	0	92	91
Wheatgrass, fairway crested						
Agropyron cristatum (L.) Gaertn.	7—9	Not given	5 —	15—23	95	84
	Not given	35—40	15	5	85	93
Wheatgrass, intermediate						
A. intermedium (Host) Beauv. var. *intermedium*	Not given	30—40	10	21	97	55
	Not given	35—40	16	5	86	78
Wheatgrass, slender						
A. trachycaulum (Link) Malte ex H.F. Lewis	7—9	Not given	5 —	15—23	98	24
	Not given	35—40	11	5	99	33
Wheatgrass, western						
A. smithii Rydb.	7—9	Not given	5 —	15—23	93	63
	Not given	35—40	13	5	70	29

Table 92 (continued)

LONGEVITY OF SEEDS STORED UNDER VARIOUS CONDITIONS IN OPEN AND POROUS CONTAINERS

Species	Moisture (% when stored)	RH (%)	Storage Time (years)	Temperature (°C)	Germination (%) When stored	Germination (%) After storage
Wildrye, blue						
Elymus glaucus Buckl.	7—9	Not given	5 –	15—23	92	2
Wildrye, Canada						
E. canadensis L.	Not given	35—40	12	5	92	76
Wildrye, Russian						
E. junceus Fisch.	Not given	35—40	14	5	95	86
now *Psathyrostachys juncea* (Fisch.) Nevski	Not given	35—40	14	5	95	86
Zinnia						
Zinnia elegans Jacq.	Not given	35—40	16	5	84	84

[a] Birdsfoot trefoil is the common name of *Lotus corniculatus* L. Common names of other lotus species are not listed.

Adapted from Bass, L. N., in *CRC Handbook of Transportation and Marketing in Agriculture*, Vol. 2, Finney, E. E., Jr., Ed., CRC Press, Boca Raton, Fla., 1981, 239. References are included in original.

Table 93
GERMINATION REQUIREMENTS FOR INDICATED KINDS

Seeds and Seed Certification

page147

Name of seed	Substrata[a]	Temp (°C)	First count (days)	Final[b] count (days)	Additional directions[c]	
					Specific requirements and photo numbers	Fresh and dormant seed
Agricultural Seed						
Alfalfa — *Medicago sativa*	B, T, S	20	4	[1]7	Photos 2481, 2486; see par. (b)(10)	
Alfilaria — *Erodium cicutarium*	B, T	20—30	3	14	Clip seeds	
Alyceclover — *Alysicarpus vaginalis*	B, T	35	4	[1]21	See par. (b)(1) for swollen seeds	
Bahiagrass — *Paspalum notatum*						
Var. *Pensacola*	P, S	20—35	7	[2]28	Light; see par. (b)(2)	Scratch caryopses; KNO₃
All other vars	P	30—35	3	21	Light; remove glumes; see par. (b)(2)	Prechill 5 days at 5 or 10°C or predry
Barley — *Hordeum vulgare*	B, T, S	20; 15	4	7	—	
Barrelclover — *Medicago tribuloides*	B, T	20	4	[1]14	Remove seeds from bur; see par. (b)(11)	
Bean						
Adzuki — *Phaseolus angularis*	B, T, S	20—30	4	[1]10	—	
Field — *P. vulgaris*	B, T, S	20—30; 25	5	[1]8	—	
Mung — *P. aureus*	B, T, S	20—30	3	[1]7	—	
Beet, field — *Beta vulgaris*	B, T, S	20—30	3	14	Photos 19557, 19558; see par. (b)(3)	
Beet, sugar — *B. vulgaris*	B, T, S	20—30; 20	3	10	Do	
Beggarweed, Florida — *Desmodium tortuosum*	B, T	30	5	28	—	

Table 93 (continued)
GERMINATION REQUIREMENTS FOR INDICATED KINDS

Name of seed	Substrata[a]	Temp (°C)	First count (days)	Final[b] count (days)	Additional directions[c]	
					Specific requirements and photo numbers	Fresh and dormant seed
Agricultural Seed						
Bentgrass						
Colonial (including Astoria and Highland) — *Agrostis tenuis*	P	15—30; 10—30; 15—25	7	28	Light; KNO₃	Prechill at 5 or 10°C for 7 days; see par. (a)(2)
Creeping — *A. palustris*	P	15—30; 10—30; 15—25	7	28	Do	Prechill at 5 or 10°C for 7 days
Velvet — *A. canina*	P	20—30	7	21	Do	
Bermudagrass — *Cynodon dactylon*	P	20—35	7	21	Light; KNO₃; photo 2518; see par. (a)(9)	
Bermudagrass, giant — *C. dactylon* var. *aridus*	P	20—35	7	21	Do	Prechill at 10°C for 7 days and then test at 20—35°; continue tests of hulled seed for 14 days and of unhulled seed for 21 days
Bluegrass						
Bulbous — *Poa bulbosa*	P, S	10	10	35	KNO₃ or soil	Prechill all samples at 5°C for 7 days
Canada — *P. compressa*	P	15—25; 15—30	10	28	Light; KNO₃; see par. (a)(2)	10—30°C
Glaucantha — *P. glaucantha*	P	15—25; 15—30	10	28	Light; KNO₃	
Kentucky (all vars.) — *P. pratensis*	P	15—25	10	28	Do	Prechill at 10°C for 5 days

Nevada — *P. nevadensis*	P	20—30	7	21	Do	
Rough — *P. trivialis*	P	20—30	7	21	Light	
Texas — *P. arachnifera*	P	20—30	7	28	Light; KNO₃	Prechill at 5°C for 2 weeks
Wood — *P. nemoralis*	P	20—30	7	28	Light	
Bluestem						
Big — *Andropogon gerardi*	P, TS	20—30	7	28	Light; KNO₃	Prechill at 5°C for 2 weeks
Little — *A. scoparius*	P, TS	20—30	7	28	Do	Do
Sand — *A. hallii*	P, TS	20—30	7	28	Do	Do
Yellow — *A. ischaemum*	P, TS	20—30	5	21	Do	Do
Brome						
Field — *Bromus arvensis*	P, TB	15—25; 20—30	6	14	Light	Prechill at 10°C for 5 days
Mountain — *B marginatus*	P	20—30	6	14	Do	
Smooth — *B. inermis*	P, B, TB	20—30	6	14	Light optional	Prechill at 5 or 10°C for 5 days, then test at 30°C for 9 additional days
Broomcorn — *Sorghum vulgare* var. *technicum*	B, T, S	20—30	3	10	—	
Buckwheat — *Fagopyrum esculentum*	B, T	20—30	3	6	—	
Buffalograss — *Buchloe dactyloides* (Burs)	P, TB, TS	20—35	7	28	Light; KNO₃	Prechill at 5°C for 6 weeks; test 14 additional days
(Caryopses)	P	20—35	5	14	Light; KNO₃	
Buffelgrass — *Pennisetum ciliare*	S	30	7	28	Light; press fascicles into well-packed soil and prechill at 5°C for 7 days	See par. (b)(4)
Burclover, California — *Medicago hispida*	B, T	20	4	[1]14	Remove seeds from bur; see par. (b)(10)	
Burclover, spotted — *M. arabica*	B, T	20	4	[1]14	Do	
Burnet, little — *Sanguisorba minor*	B, T	15	5	14		

Table 93 (continued)
GERMINATION REQUIREMENTS FOR INDICATED KINDS

Name of seed	Substrata[a]	Temp (°C)	First count (days)	Final[b] count (days)	Additional directions[c]	
					Specific requirements and photo numbers	Fresh and dormant seed
Buttonclover — *M. orbicularis*	B, T	20	4	[c]10	See par. (b)(10)	15°C
Canarygrass — *Phalaris canariensis*	B, T	20—30	3	7		
Canarygrass, reed — *P. arundinacea*	P	20—30	5	21	Light; KNO$_3$	
Carpetgrass — *Axonopus affinis*	P	20—35	10	21	Light	KNO$_3$
Castorbean — *Ricinus communis*	T, S	20—30	7	14	Remove caruncle if mold interferes with test	
Chess, soft — *Bromus mollis*	P	20—30	7	14	Light	Prechill at 5 or 10°C for 7 days
Chickpea — *Cicer arietinum*	T, S	20—30	3	7		
Clover						
Alsike — *Trifolium hybridum*	B, T, S	20	3	[c]7	See par. (b)(10)	15°C
Berseem — *T. alexandrinum*	B, T, S	20	3	[c]7	Do	Do
Cluster — *T. glomeratum*	B, T	20	4	[c]10	Do	Do
Crimson — *T. incarnatum*	B, T, S	20	4	[c]7	See par. (b)(10); photos 2479, 2482	Do
Kenya — *T. semipilosum*	B, T, S	20	3	[c]7		15°C
Ladino — *T. repens*	B, T, S	20	3	[c]7	See par. (b)(10)	Do
Lappa — *T. lappaceum*	B, T	20	3	[c]7	Do	Do
Large hop — *T. procumbens (T. campestre)*	B, T	20	4	[c]14	Do	Do
Persian — *T. resupinatum*	B, T	20	3	[c]7	Do	Do
Red — *T. pratense*	B, T, S	20	4	[c]7	See par. (b)(10); photos 2483, 2484	Do
Rose — *T. hirtum*	B, T	20	4	[c]10	See par. (b)(10)	Do

Small hop (Suckling) — *T. dubium*	B, T	20	4	[1]14	Do	Do
Strawberry — *T. fragiferum*	B, T	20	3	[1]7	Do	Do
Sub — *T. subterraneum*	B, T	20	4	[1]14	Do	Do
White — *T. repens*	B, T, S	20	3	[1]7	Do	Do
Corn						
Field — *Zea mays*	B, T, S	20—30; 25	4	7	Photos 2510, 2511, 2512,	
Pop — *Z. mays* var. *everta*	B, T, S	20—30; 25	4	7		
Cotton — *Gossypium* spp.	B, T, S	20—30; 30	4	[1]12	Photos 19553, 19554	Test by alternate method; see par. (b)(5)
Cowpea — *Vigna sinensis*	B, T, S	20—30	5	[1]8	Photos 1989, 1990, 2377	
Crambe — *Crambe abyssinica*	T	25	3	7		
Crested dogtail — *Cynosurus cristatus*	P	20—30	10	21	Light	Prechill for 3 days at 5 or 10°C
Crotalaria						
Lance — *Crotalaria lanceolata*	B, T, S	20—30	4	[1]10		
Showy — *C. spectabilis*	B, T, S	20—30	4	[1]10	Photos 2496, 2497	
Slenderleaf — *C. intermedia*	B, T, S	20—30	4	[1]10		
Striped — *C. mucronata*	B, T, S	20—30	4	[1]10		
Sunn — *C. juncea*	B, T, S	20—30	4	[1]10		
Crownvetch — *Coronilla varia*	B, T, S	20	7	[1]14		
Dallisgrass — *Paspalum dilatatum*	P	20—35	7	21	Light; KNO$_3$	
Dichondra — *Dichondra repens*	B, T	20—30	7	[1]28		
Dropseed, sand — *Sporobolus cryptandrus*	P	5—35; 15—35	5	28	Do	Prechill at 5°C for 4 weeks
Emmer — *Tricticum dicoccum*	B, T, S	20; 15	4	7	Photos 2507, 2520—2522	Prechill at 5 or 10°C for 5 days or predry
Fescue						
Chewing — *Festuca rubra* var. *commutata*	P	15—25	7	21	Light and KNO$_3$ optional	Prechill at 5 or 10°C for 5 days
Hard — *F. ovina* var. *duriuscula*	P	15—25	7	21	Light and KNO$_3$ optional	

Table 93 (continued)
GERMINATION REQUIREMENTS FOR INDICATED KINDS

Name of seed	Substrata[a]	Temp (°C)	First count (days)	Final[b] count (days)	Specific requirements and photo numbers	Additional directions[c] Fresh and dormant seed
Hair — *F. capillata*	P	10—25	10	28	Light; KNO$_3$	
Meadow — *F. elatior*	P	15—25; 20—30	5	14	Light and KNO$_3$ optional	
Red — *F. rubra*	P	15—25	7	21	Light and KNO$_3$ optional	
Sheep — *F. ovina*	P	15—25	7	21	Light and KNO$_3$ optional	
Tall — *F. arundinacea*	P	15—25; 20—30	5	14	Light and KNO$_3$ optional	Prechill at 5 or 10°C for 5 days and test for 21 days
Flax — *Linum usitatissimum*	B, T, S	20—30	3	7	Photos 2003, 2008, 2485, 2487	
Grama						
Blue — *Bouteloua gracilis*	P, TB	20—30	7	28	Light	KNO$_3$
Side-oats — *B. curripendula*	P	15—30	7	28	Light, KNO$_3$	
Guar — *Cyamopsis tetragonoloba*	B, T	30; 20—30	5	[1]14		
Guineagrass — *Panicum maximum*	P	15—35	10	28	Light; KNO$_3$ optional	
Hardinggrass — *Phalaris tuberosa* var. *stenoptera*	P	10—30	7	28	Light	Do
Alternate method	P	15—25	7	14	Light; KNO$_3$ presoak at 15°C for 24 hr	
Hemp — *Cannabis sativa*	B, T	20—30	3	7		
Indiangrass, yellow — *Sorghastrum nutans*	P, TS	20—30	7	28	Light; KNO$_3$	Prechill at 5°C for 2 weeks
Indigo, hairy — *Indigofera hirsuta*	B, T	20—30	5	[1]14		
Japanese lawngrass — *Zoysia japonica*	P	35—20	10	28	Light; KNO$_3$	
Johnsongrass — *Sorghum halepense*	P	20—35	7	35	Light	KNO$_3$
Kudzu — *Pueraria thunbergiana*	B, T	20—30	5	[1]14		

Species	Method	Temp °C	First count	Final count	Directions	Special
Lentil — *Lens culinaris*	B, T	20	5	'10		
Lespedeza						
Korean — *Lespedeza stipulacea*	B, T, S	20—35	5	'14		
Sericea or Chinese — *L. cuneata* (*L. sericea*)	B, T, S	20—35	7	'21	Photo 2494	
Siberian — *L. hedysaroides*	B, T, S	20—35	7	'21		
Striate (Common, Kobe, Tenn. 76) — *L. striata*	B, T, S	20—35	7	'14		
Lovegrass, sand — *Eragrostis trichodes*	P	20—30	5	14	Light; KNO₃	Prechill at 5 or 10°C for 6 weeks
Lovegrass, weeping — *E. curvula*	P	20—35	5	14	Light	KNO₃
Lupine						
Blue — *Lupinus angustifolius*	B, T, S	20	4	'10	Photos 14535-14542	
White — *L. albus*	B, T	20	3	'10		
Yellow — *L. luteus*	B, T	20	7	'10		
Manilagrass — *Zoysia matrella*	P	35—20	10	28	Light; KNO₃	
Meadow foxtail — *Alopecurus pratensis*	P	20—30	7	14	Light	
Medick, black — *Medicago lupulina*	B, T, S	20	4	'7	See par. (b)(10)	
Millet						
Browntop — *Panicum ramosum*	B, P, T	20—30; 30	4	14	Light and KNO₃ optional	Predry at 35 or 40°C for 7 days; and test at 30°C
Alternate method	B, P, T	5—35	4	14		
Foxtail — Such as Common, White Wonder, German, Hungarian, Siberian, or Golden — *Setaria italica*	B, T	15—30; 20—30	4	10	Light; KNO₃	
Japanese — *Echinochloa crusgalli* var. *frumentacea*	B, T	20—30	4	10		
Pearl — *Pennisetum glaucum*	B, T	20—30	3	7		

Table 93 (continued)
GERMINATION REQUIREMENTS FOR INDICATED KINDS

Name of seed	Substrata[a]	Temp (°C)	First count (days)	Final[b] count (days)	Additional directions[c]	
					Specific requirements and photo numbers	Fresh and dormant seed
Proso — *Panicum miliaceum*	B, T	20—30	3	7		
Molassesgrass — *Melinis minutiflora*	P	20—30	7	21	Light	
Mustard						
Black — *Brassica nigra*	P	20—30	3	7	Do	KNO$_3$ and prechill at 10°C for 3 days
India — *B. juncea*	P	20—30	3	7	Light	Prechill at 10°C for 7 days and test for 5 days; KNO$_3$
White — *B. hirta*	P	20—30	3	5	Do	
Napiergrass — *Pennisetum purpureum*	B, T	20—30	3	10		
Oat — *Avena* spp.	B, T, S	20; 15	5	10	Photos 2407, 2408, 2524-2527, 19545, 19546	Prechill at 5 or 10°C for 5 days and test for 7 days or predry and test for 10 days
Oatgrass tall — *Arrhenatherum elatius*	P	20—30	6	14	Light	
Orchardgrass — *Dactylis glomerata*	P, TS	15—25	7	21	Light; germination more rapid on soil	Prechill at 5 or 10°C for 7 days and test at 15—25°C
Panicgrass, blue — *Panicum antidotale*	P, TS	20—30	7	28	Light	
Panicgrass, green — *P. maximum* var. *trichoglume*	P	15—35	10	28	Light; KNO$_3$ optional	
Peanut — *Arachis hypogaeo*	B, T, S	20—30; 25	5	10	Remove shells; photos 19541, 19542	Predry up to 14 days at 40°C
Pea, field — *Pisum sativum* var. *arvense*	B, T, S	20	3	¹8	Photos 2503, 2506, 14543-14547	

Rape						
Annual — *Brassica napus* var. *annua*	B, T	20—30	3	7		
Bird — *B. campestris*	P	20—30	3	10	Light	KNO₃
Turnip — *B. campestris* vars.	B, T	20—30	3	7		
Winter — *B. napus* var. *biennis*	B, T	20—30	3	7		
Redtop — *Agrostis alba*	P, TB	20—30	5	10	Light	Do
Rescuegrass — *Bromus catharticus*	P, S	10—30	7	28	Light, see par. (b)(7) for alternate method	In soil at 15°C
Rhodesgrass — *Chloris gayana*	P	20—30	6	14	Light; KNO₃	Presoak, see par. (b)(9)
Rice — *Oryza sativa*	T, S	20—30; 30	5	14	Photos 19549, 19950; see par. (b)(8) for alternate method	
Ricegrass, Indian — *Oryzopsis hymenoides*	S	5—15; 15—25	7	28		Dark; prechill in soil at 5°C for 4 weeks
Roughpea — *Lathyrus hirsutus*	B, T	20	7	'14		
Rye — *Secale cereale*	B, T, S	20; 15	4	7	Photos 2403, 2406, 2528-2531	Prechill at 5 or 10°C for 5 days or predry
Ryegrass						
Annual (Italian) — *Lolium multiflorum*	P, TB	15—25	5	14	Light optional; see par. (b)(9) for fluorescence test	Light; KNO₃; prechill at 5 or 10°C for 5 days and test at 15—25°C; if the seeds are still dormant rechill for 3 days and continue the test at 15—25°C an additional 4 days
Perennial — *L. perenne*	P, TB	15—25	5	14	Do	Do
Wimmera — *L. rigidum*	P, TB	15—25; 20—30	5	14	Do	Do
Safflower — *Carthamus tinctorius*	P, B, T, S	15; 20	4	14	Light at 15°C	

Table 93 (continued)
GERMINATION REQUIREMENTS FOR INDICATED KINDS

Name of seed	Substrata[a]	Temp (°C)	First count (days)	Final[b] count (days)	Additional directions[c]	
					Specific requirements and photo numbers	Fresh and dormant seed
Sainfoin — *Onobrychis viciaefolia*	B, T	20—30	4	[l]14		
Sesame — *Sesamum indicum*	B, T, TB	20—30	3	6		
Sesbania — *Sesbania exaltata*	B, T	20—30	5	[l]7		
Smilo — *Oryzopsis miliacea*	P	20—30	7	42	Light	Prechill at 5°C for 2 weeks
Sorghum Grain and Sweet — *Sorghum vulgare*	B, T, S	20—30	4	10	Photos 2413-2416	Prechill grain varieties at 5 or 10°C for 5 days; test sweet varieties at 30—45°C for 2—4 hr/day
Sorghum almum — *S. almum*	T, S	20—35; 15—35	5	21		Prechill at 5°C for 5 days. Upon the 10th day of test, clip or pierce the distal end of ungerminated seeds
Sorghum-sudangrass hybrid — *S. vulgare × S. sudanense*	B, T, S	20—30	4	10		Prechill at 5 or 10°C for 5 days
Sorgrass[d]	B, T, S	15—35; 20—35	5	21	Photos 2413—2416	Prechill at 5 or 10°C for 7 days
Sourclover — *Melilotus indica*	B, T	20	3	[l]14	See paragraph (b)(10)	
Soybean — *Glycine max*	B, T, S	20—30; 25	5	[l]8	Photos 2371, 2372, 2378	
Spelt — *Triticum spelta*	B, T, S	20; 15	4	7	Photos 2507, 2520-2522	Prechill at 5 or 10°C for 5 days or predry
Sudangrass — *Sorghum sudanense*	B, T, S	20—30; 15—30	4	10	Photos 2449-2452	Prechill at 10°C for 5 days
Sunflower (Cult.) — *Helianthus annuus*	T, B	20—30	3	7		

Sweetclover						
White — *Melilotus alba*	B, T, S	20	4	[1]7	Photos 2374, 2375, 2376, 2381; see par. (b)(10)	
Yellow — *M. officinalis*	B, T, S	20	4	[1]7	Do	
Sweet vernalgrass — *Anthoxanthum odoratum*	P	20—30	6	14	Light	
Switchgrass — *Panicum virgatum*	P, TS	15—30	7	28	Light, KNO_3	Prechill at 5°C for 2 weeks
Timothy — *Phleum pratense*	P, TB	15—25; 20—30	5	10	Light; photo 2399; see par. (a)(9)	KNO_3, and prechill at 5 or 10°C for 5 days
Tobacco — *Nicotiana tabacum*	P, TB	20—30	7	14	Light	
Trefoil						
Big — *Lotus uliginosus (L. major)*	B, T	20	5	12	Light	
Birdsfoot — *L. corniculatus*	B, T	20	5	12	Photos 19531, 19532	
Triticale — *Tritico secale*	B, T, S	20; 15	4	7		Prechill 5 days at 5 or 10°C or predry
Vaseygrass — *Paspalum urvillei*	P	20—35	7	21	Light	KNO_3
Veldtgrass — *Ehrharta calycina*	P	10—30	7	28	Do	
Velvetbean — *Stizolobium deeringianum*	B, T, S, C	20—30	3	[1]14	Photos 19539, 19540	
Velvetgrass — *Holcus lanatus*	P	20—30	6	14	Light	
Vetch						
Common — *Vicia sativa*	B, T	20	5	[1]10		
Hairy — *V. villosa*	B, T	20	5	[1]14		
Hungarian — *V. pannonica*	B, T	20	5	[1]10		
Monantha — *V. articulata V. monantha*	B, T	20	5	[1]10		
Narrowleaf — *V. angustifolia*	B, T	20	5	[1]14		
Purple — *V. benghalensis*	B, T	20	5	[1]10		
Woolypod — *V. dasycarpa*	B, T	20	5	[1]14		Prechill at 10°C for 5 days and test at 15°C

Table 93 (continued)
GERMINATION REQUIREMENTS FOR INDICATED KINDS

Name of seed	Substrata[a]	Temp (°C)	First count (days)	Final[b] count (days)	Specific requirements and photo numbers	Additional directions[c] Fresh and dormant seed
Wheat						
Common — *Triticum aestivum*	B, T, S	20; 15	4	7	Photos 2507, 2520—2522	Prechill at 5 or 10°C for 5 days, or predry
Club — *T. compactum*	B, T, S	20; 15	4	7	Do	Do
Durum — *T. durum*	B, T, S	20; 15	4	10	Do	Do
Polish — *T. polonicum*	B, T, S	20; 15	4	7	Do	Do
Poulard — *T. turgidum*	B, T, S	20; 15	4	7	Do	Do
Wheatgrass						
Beardless — *Agropyron inerme*	P, TB	15—25	7	14	Light and KNO_3 optional	KNO_3 and prechill at 5 or 10°C for 7 days
Fairway crested — *A. cristatum*	P, TB	15—25; 20—30	5	14	Do	Do
Standard crested — *A. desertorum*	P, TB	15—25; 20—30	5	14	Do	Do
Intermediate — *A. intermedium*	P	15—25	5	28	Do	Do
Pubescent — *A. tricophorum*	P	15—25	5	28	Do	Do
Alternate method	P	20—30	5	28	Light	Do
Siberian — *A. sibiricum*	P, TB	15—25	7	14	Light and KNO_3 optional	Do
Slender — *A. trachycaulum*	P, TB	15—25; 10—30	5	14	Do	Prechill at 5 or 10°C for 5 days. If still dormant on 10th day rechill 2 days
Streambank — *A. riparium*	P, TB	15—25	5	14	Do	Prechill at 5 or 10°C for 5 days
Tall — *A. elongatum*	P	15—25	5	21	Do	Do
Alternate method	P	20—30	5	21	Light	Do

Western — *A. smithii*	P, B, T	15—25	7	28	Light and KNO_3 optional	KNO_3 or soil and prechill at 5 or 10°C for 7 days
Wildrye						
Canada — *Elymus canadensis*	P	15—30	7	21	Light	Prechill at 5°C for 2 weeks
Russian — *E. junceus*	P	20—30	5	14	Do	Prechill at 5 or 10°C for 5 days
Vegetable Seed						
Artichoke — *Cynara scolymus*	B, T	20—30	7	21		
Asparagus — *Asparagus officinalis*	B, T, S	20—30	7	21	Photos 19533, 19534	
Asparagusbean — *Vigna sesquipedalis*	B, T, S	20—30	5	[1]8		
Beans						
Garden — *Phaseolus vulgaris*	B, T, S	20—30; 25	5	[1]8	Photos 1834, 1835, 1846, 1854, 1855	
Lima — *P. lunatus* var. *macrocarpus*	B, T, C, S	20—30	5	[1]9	Photos 2380, 2400, 2401	
Runner — *P. coccineus*	B, T, S	20—30	5	[1]9	See par. (b)(3); photos 19557, 19558	
Beet — *Beta vulgaris*	B, T, S	20—30	3	14	See par. (b)(10)	
Broadbean — *Vicia faba*	S, C	20	4	14		Prechill at 10°C for 3 days
Broccoli — *Brassica oleracea* var. *botrytis*	B, P, T	20—30	3	10		Prechill at 5 or 10°C for 3 days; KNO_3 and light
Burdock, great — *Arctium lappa*	B, T	20—30	7	14		
Brussels sprouts — *B. oleracea* var. *gemmifera*	B, P, T	20—30	3	10		Prechill at 5 or 10°C for 3 days; KNO_3 and light
Cabbage — *B. oleracea* var. *capitata*	B, P, T	20—30	3	10	Photos 19551, 19552	Do
Cabbage, Chinese — *B. pekinensis*	B, T	20—30	3	7		

Table 93 (continued)
GERMINATION REQUIREMENTS FOR INDICATED KINDS

Name of seed	Substrata[a]	Temp (°C)	First count (days)	Final[b] count (days)	Additional directions[c]	
					Specific requirements and photo numbers	Fresh and dormant seed
Cabbage, tronchuda — *B. oleracea* var. *tronchuda*	B, P	20—30	3	10	Photos 19551, 19552	Do
Cardoon — *Cynara cardunculus*	B, T	20—30	7	21	Photos 19547, 19548	
Carrot — *Daucus carota*	B, T	20—30	6	21	Photo 19561	
Cauliflower — *B. oleracea* var. *botrytis*	B, P, T	20—30	3	10		Prechill at 5 or 10°C for 3 days; KNO₃ and light
Celeriac — *Apium graveolens* var. *rapaceum*	P	15—25; 20	10	21	Light; see par. (a)(9)	
Celery — *A. graveolens* var. *dulce*	P	15—25; 20	10	21	Do	
Chard, Swiss — *Beta vulgaris* var. *cicia*	B, T, S	20—30	3	14	See par. (b)(13)	
Chicory — *Cichorium intybus*	P, TS	20—30	5	14	Light; KNO₃ or soil; photo 2504; see par. (a)(9)	
Chives — *Allium schoenophrasum*	B, T	20	6	14		
Citron — *Citrullus vulgaris*	B, T	20—30	7	14	Soak seeds 6 hr	Test at 30°C
Collards — *Brassica oleracea* var. *acephala*	B, P, T	20—30	3	10		Prechill at 5 or 10°C for 3 days; KNO₃ and light
Corn, sweet — *Zea mays*	B, T, S	20—30; 25	4	7	Photos 2510—2512, 2514	
Cornsalad (Fetticus) — *Valerianella locusta* var. *olitoria*	B, T	15	7	28		Test at 10°C
Cowpea — *Vigna sinensis*	B, T, S	20—30	5	'8	Photos 1989, 1990, 2377	
Cress Garden — *Lepidium sativum*	B, P, T	15	4	10		Light
Upland — *Barbarea verna*	P, TB	20—35	4	7	Light; KNO₃	

Kind	Substrata	Temperature (°C)	First count	Final count	Directions	Additional directions
Water — *Rorippa nasturtium-aquaticum*	P	20—30	4	14	Light	
Cucumber — *Cucumis sativus*	B, T, S	20—30	3	7	Keep substratum on dry side; see par. (a)(3); photos 19535, 19536	
Dandelion — *Taraxacum officinale*	P, TB	20—30	7	21	Light; see par. (a)(9)	
Eggplant — *Solanum melongena var. esculentum*	P, TB, RB	20—30	7	14		Light; KNO_3
Endive — *Cichorium endivia*	P, TS	20—30	5	14		See par. (b)(6)
Kale — *Brassica oleracea var. acephala*	B, P, T	20—30	3	10	Light; KNO_3 or soil	Prechill at 5 or 10°C for 3 days; KNO_3 and light
Kale, Chinese — *Brassica oleracea var. alboglabra*	B, P, T	20—30	3	10		Light; KNO_3; prechill at 5 or 10°C for 3 days
Kale, Siberian — *B. napus var. pabularia*	B	20—30	3	7		
Kohlrabi — *B. oleracea var. gongylodes*	B, P, T	20—30	3	10		Prechill at 5 or 10°C for 3 days; KNO_3 and light
Leek — *Allium porrum*	B, T	20	6	14		
Lettuce — *Lactuca sativa*	P	20	None	7		Prechill at 10°C for 3 days or test at 15°C
Muskmelon (cantaloupe) — *Cucumis melo*	B, T, S	20—30	4	10	Keep substratum on dry side [par. (a)(3)]	
Mustard, India — *Brassica juncea*	P	20—30	3	7	Light	Prechill at 10°C for 7 days and test for 5 additional days KNO_3
Mustard, spinach — *B. perviridis*	B, T	20—30	3	7		
Okra — *Hibiscus esculentus*	B, T	20—30	4	'14	Photos 19543, 19544	
Onion — *Allium cepa*	B, T	20	6	10	Photos 1962, 2253, 2254, 2328, 2330, 2340, 2341, 2469	

Table 93 (continued)
GERMINATION REQUIREMENTS FOR INDICATED KINDS

Name of seed	Substrata[a]	Temp (°C)	First count (days)	Final[b] count (days)	Additional directions[c] Specific requirements and photo numbers	Fresh and dormant seed
Alternate method	S	20	6	12		
Onion, Welsh — *A. fistulosum*	B, T	20	6	10		
Pak-choi — *B. chinensis*	B, T	20—30	3	7		
Parsley — *Petroselinum hortense* (*P. crispum*)	B, T, TS	20—30	11	28		
Parsnip — *Pastinaca sativa*	B, T, TS	20—30	6	28		
Pea — *Pisum sativum*	B, T, S	20	5	'8		
Pepper — *Capsicum* spp.	T, TB, RB	20—30	6	14	Photos 2492, 2498—2500	Light and KNO₃
Pumpkin — *Cucurbita pepo*	B, T, S	20—30	4	7	Keep substratum on dry side [par. (a)(3)]	
Radish — *Raphanus sativus*	B, T	20	4	6	Photos 2554, 19555, 19556	
Rhubarb — *Rheum rhaponticum*	TB, TS	20—30	7	21	Light	
Rutabaga — *Brassica napus* var. *napobrassica*	B, T	20—30	3	14		
Salsify — *Tragopogon porrifolius*	B, T	15	5	10		Prechill at 10°C for 3 days
Sorrel — *Rumex acetosa*	P, TB, TS	20—30	3	14	Light	Test at 15°C
Soybean — *Glycine max*	B, T, S	20—30; 25	5	'8	Photos 2371, 2372, 2378	
Spinach — *Spinacia oleracea*	TB, T	15; 10	7	21	Keep substratum on dry side [par. (a)(3)]	
Spinach, New Zealand — *Tetragonia expansa*	TS	10—30	5	28	Do	
Alternate method	B, T	15	5	21	Remove pulp from "seeds"	
Squash — *Cucurbita moschata* and *C. maxima*	B, T, S	20—30	4	7	Keep substratum on dry side [par. (a)(3)]; photos 19537, 19538	
Tomato — *Lycopersicon esculentum*	B, P, RB, T	20—30	5	14	Photo 2513	Light; KNO₃

Tomato, husk — *Physalis pubescens*	P, TB	20—30	7	28	Light; KNO$_3$
Turnip — *Brassica rapa*	B, T	20—30	3	7	
Watermelon — *Citrullus vulgaris*	B, T, S	20—30; 25	4	14	Keep substratum on dry side [par. (a)(3)] Test at 30°C

[a] The symbols used for substrate are: B, between blotters; TB, top of blotters; T, paper toweling; S, sand; TS, top of sand or soil; P, covered Petri dishes; C, creped cellulose paper; and RB, blotters with raised covers. Conditions as specified in Rules and Regulations, p. 24. [5]

[b] Hard seeds often present (1) or firm ungerminated seeds frequently present (2).

[c] Specific requirements given in Rules and Regulations. [5] Information concerning the availability of photographs of seedlings may be obtained from the U.S. Department of Agriculture, Agricultural Marketing Service, Livestock, Poultry, Grain and Seed Div., Seed Standardization Branch, Building 306, Beltsville Agric. Res. Center — East, Beltsville, Md., 20705. Specific photos or paragraphs referred to in this column may be found in the source listed for this table.

[d] Rhizomatous derivatives of a johnsongrass × sorghum cross or a johnsongrass × sudangrass cross.

From Agricultural Marketing Service, Livestock, Poultry, Grain and Seed Division, Rules and Regulations under the Federal Seed Act, U.S. Department of Agriculture, Washington, D.C., 1975.

Table 94
SPECIFIC REQUIREMENTS FOR THE CERTIFICATION OF PLANT MATERIALS UNDER THE AOSCA SYSTEM

Minimum Land, Isolation, Field, and Seed Standards

Crop kind	Foundation				Registered				Certified			
	Land[a]	Isolation[b]	Field[c]	Seed[d]	Land[a]	Isolation[b]	Field[c]	Seed[d]	Land[a]	Isolation[b]	Field[c]	Seed[d]
Alfalfa	4[1]	600[44,48]	1,000	0.1	3[1]	300[3,44,48]	400	0.25	1[1,2]	165[44,49]	100	1.0
Alfalfa hybrid	4[1]	1,320[3]	1,000[42]	0.1	—	—	—	—	1[1,2]	165[3,43,44]	100[42]	1.0
Barley	1[7]	0[23]	3,000	0.05	1[7]	0[23]	2,000	0.1	1[7]	0[23]	1,000	0.2
Barley hybrid	1[30]	660[21,32]	3,000	0.05	1[30]	660[21,32]	2,000	0.1	1[30]	330[21,32]	1,000	0.2
Birdsfoot trefoil	5[1]	600[5,44]	1,000	0.1	3[1]	300[5,44]	400	0.25	2[1]	165[6,44]	100	1.0
Clover (all kinds)	5[1,9]	600[5,18,44]	1,000	0.1	3[1,9]	300[5,18,44]	400	0.25	2[1,9]	165[18,44]	100	1.0
Corn												
Inbred lines	0	660[10,11]	1,000[13,46]	0.1[15]	—	—	—	—	—	—	—	—
Foundation												
Single cross	0	660[10,11]	1,000[13,46]	0.1[15]	—	—	—	—	—	—	—	—
Backcross	0	660[10,11]	1,000[13,46]	0.1[15]	—	—	—	—	—	—	—	—
Hybrid	—	—	—	—	—	—	—	—	0	660[11,12]	—	0.5
Open-pollinated	—	—	—	—	—	—	—	—	0	660[11,12]	200	0.5
Sweet	—	—	—	—	—	—	—	—	0	660[11,14]	—	0.5
Cotton	0	0[19]	10,000	0.03	0	0[19]	5,000	0.05	0	0[19]	1,000	0.1
Cowpeas	1[7]	0[23]	2,000	0.1	1[7]	0[23]	1,000	0.2	1[7]	0[23]	500	0.5
Crambe	1[7]	660	2,000	0.05	1[7]	660[24]	1,000	0.1	1[7]	660[24]	500	0.25
Crownvetch	5[1]	600[5,44]	1,000	0.1	3[1]	300[5,44]	400	0.25	2[1]	165[6,44]	100	1.0
Fava beans	1[7]	0[23]	2,000	0.05	1[7]	0[23]	1,000	0.1	1[7]	0[23]	500	0.2
Field and garden beans	1[7]	0[23]	2,000	0.05	1[7]	0[23]	1,000	0.1	1[7]	0[23]	500	0.2
Field peas	1[7]	0[23]	2,000	0.05	1[7]	0[23]	1,000	0.1	1[7]	0[23]	500	0.2
Flat peas	4[1]	600[5,44]	1,000	0.1	3[1]	300[3,5,44]	400	0.25	1[1,2]	165[3,44]	100	1.0
Flax	1[7]	0[23]	5,000	0.05	1[7]	0[23]	2,000	0.1	1[7]	0[23]	1,000	0.2
Grasses-cross pollinated	5	900[4,18,20]	1,000	0.1	1[8]	300[4,18,20]	100	1.0	1[8]	165[4,18,20]	50	2.0[47,50]
Grass strains at least 80% apomictic and highly self-fertile species	5	60[4,18,20]	1,000	0.1	1[8]	30[4,18,20]	100	1.0	1[8]	15[4,18,20]	50	2.0[16]

	I a	I b	I c	I d	II a	II b	II c	II d	III a	III b	III c	III d
Lespedeza	5[1]	10[4]	1,000	0.1	3[1]	10[4]	400	0.25	2[1]	10[4]	100	1.0
Millet-cross pollinated	1[8]	1,320[40]	20,000[27]	0.005	1[8]	1,320[40]	10,000[27]	0.01	1[8]	660[40]	5,000[27]	0.02
Millet-self pollinated	1[8]	0[23]	3,000	0.05	1[8]	0[23]	2,000	0.10	1[8]	0[23]	1,000	0.2
Milkvetch	5[1]	600[5,44]	2,000	0.05	3[1]	300[5,44]	1,000	0.1	2[1]	165[44]	200	0.5
Mung beans	1[7]	0[23]	2,000	0.1	1[7]	0[23]	1,000	0.2	1[7]	0[23]	500	0.5
Mustard	4	1,320	2,000	0.05	—	—	—	—	2	660[24]	500	0.25
Oats	1[7]	0[23]	3,000	0.2	1[7]	0[23]	2,000	0.3	1[7]	0[23]	1,000	0.5
Okra	1[7]	1,320	0[27]	0.0	1[7]	1,320	2,500[27]	0.5	1[7]	825	1,250[27]	1.0
Onion	1[7]	5,280	200[22]	0.0	1[7]	2,640	200[22]	0.5[22]	1[7]	1,320	200[22]	1.0[22]
Peanuts	1[7]	0[23]	1,000	0.1	1[7]	0[23]	500	0.2	1[7]	0[23]	200	0.5
Pepper	1[7]	200[25]	0	0.0	1[7]	100[25]	300	0.5	1[7]	30[25]	150	1.0
Rape—self pollinated	4	660[24]	2,000	0.05	—	—	—	—	2	330[24]	500	0.25
Rape—cross pollinated	4	1,320[24]	2,000	0.05	—	—	—	—	2	330[24]	500	0.25
Rice	1[7]	10[39]	10,000	0.05	1[7]	10[39]	5,000	0.1	1[7]	10[39]	1,000	0.2
Rye	1[7]	660[18]	3,000	0.05	1[7]	660[18]	2,000	0.1	1[7]	660[18]	1,000	0.2
Safflower	2[7]	1,320	10,000	0.01	2[7]	1,320	2,000	0.05	2[7]	1,320	1,000	0.1
Sainfoin	5[1]	600[5,44]	1,000	0.1	3[1]	300[5,44]	400	0.25	2[1]	165[6,44]	100	1.0
Sorghum	1[7]	990	50,000[27]	0.005	1[7]	990	35,000[27]	0.01	1[7]	660[29]	20,000[27]	0.05
Hybrid seedstock	—	990	50,000[27]	0.005	—	—	—	—	—	—	—	—
Com. hybrid	1[33]	0[23]	—	—	1[33]	0[23]	—	—	1[33]	660[21,29,31]	20,000[27]	0.1
Soybeans	1	2,640[41,45]	1,000	0.1	1	—	500	0.2	1	0[23]	200	0.5
Sunflower	1	2,640[41,45]	200	0.02	—	2,640[41,45]	200	0.02	1	2,640[41,45]	200	0.1[34]
Sunflower hybrid	1	—	250[35]	0.02	—	—	—	—	1	2,640[41,45]	250[35]	0.1[34]
Tomato	1[7]	200[25]	0	0.0	1[7]	100[25]	300	0.5	1[7]	30[25]	150	1.0
Tobacco	0[36]	150[37]	0	0.01	0[36]	150[37]	0	0.01	0[36]	150[37]	0	0.01
Tobacco hybrid	—	—	—	—	—	—	—	—	0[36]	150[38]	0	0.01
Triticale	1[7]	0[23]	3,000	0.05	1[7]	0[23]	2,000	0.1	1[7]	0[23]	1,000	0.2
Vetch	5[1,7]	10[17,44]	1,000	0.1	3[1,7]	10[17,44]	400	0.25	2[1,7]	10[17,44]	100	1.0
Watermelon	1[7]	2,640[26]	0[28]	0.0	1[7]	2,640[26]	0[28]	0.5	1[7]	1,320[26]	500[28]	1.0
Wheat	1[7]	0[23]	3,000	0.05	1[7]	0[23]	2,000	0.1	1[7]	0[23]	1,000	0.2
Wheat hybrid	1[30]	660[21,32]	3,000	0.05	1[30]	660[21,32]	2,000	0.1	1[30]	330[21,32]	1,000	0.2

a   Number of years that must elapse between destruction of a stand of a kind and establishment of a stand of a specific class of a variety of the same crop kind.

b   Distance in feet from any contaminating sources.

c   Minimum number of plants or heads in which one plant or head of another variety or off-type is permitted.

d   Maximum percentage of seed of other varieties or off-types permitted.

Notes to body of Table 94

1. The land must be free of volunteer plants of the crop kind during the year immediately prior to establishment; and no manure or other contaminating material shall be applied the year previous to seeding or during the establishment and productive life of the stand.

2. Two years are required for the production of Certified seed of varieties adapted to the northern and central regions, following varieties adapted to the southern region.

3. Isolation distance for Certified seed production of varieties adapted to the northern and central regions shall be 500 ft from varieties adapted to the southern region.

4. Isolation between classes of the same variety may be reduced to 25% of the distance otherwise required.

5. This distance applies when fields are 5 acres or larger in area. For smaller fields, the distances are 900 and 450 for the Foundation and Registered classes, respectively.

6. Fields of less than 5 acres require 330 ft.

7. Requirement is waived if the previous crop was grown from Certified seed of the same variety.

8. Requirement is waived if the previous crop was of the same variety and of a Certified class equal or superior to that of the crop seeded.

9. Reseeding varieties of crimson clover may be allowed to volunteer back year after year on the same ground. When a new variety is being planted where another variety once grew, the field history requirements pertain.

10. No isolation is required for the production of hand-pollinated seed.

11. When the contaminant is of the same color and texture, the isolation distance may be modified by (1) adequate natural barriers or (2) differential maturity dates, provided there are no receptive silks in the seed parent at the time the contaminant is shedding pollen. In addition, for open-pollinated corn, no isolation is required between a seed field of "dent sterile" popcorn and dent field corn.

12. Where the contaminating source is corn of the same color and texture as that of the field inspected or white endosperm corn optically sorted, the isolation distance is 410 ft and may be modified by the planting of pollen parent border rows according to the following table.

Minimum distance from contaminant	Minimum number of border rows required	
	Field size up to 20 acres	Field size 20 acres or more
410	0	0
370	2	1
330	4	2
290	6	3
245	8	4
205	10	5
165	12	6
125	14	7
85	16	8
0	Not permitted	10

13. Refers to off-type plants in the pollen parent that have shed pollen or to the off-type plants in the seed parent at the time of the last inspection.

14. The required minimum isolation distance for sweet corn is 660 ft from the contaminating source, plus four border rows when the field to be inspected is 10 acres or less in size. This distance may be decreased by 15 ft for each increment of 4 acres in the size of the field to a maximum of 40 acres, and further decreased 40 ft for each additional border row to a maximum of 16 rows. These border rows are for pollen-shedding purposes only.

15. Refers to off-type ears. Ears with off-colored or different textured kernels limited to 0.5% or a total of 25 off-colored seeds or different textured kernels per 1000 ears.

16. The variety of bluegrass, Merion, allowed is 3%.

17. All cross-pollinating varieties must be 400 ft from any contaminating source.

18. Isolation between diploids and tetraploids shall at least be 15 ft.

19. Minimum isolation shall be 100 ft if the cotton plants in the contaminating source differ by easily observed morphological characteristics from the field to be inspected. Isolation distance between upland and Egyptian types is 1320, 1320, and 660 ft for Foundation, Registered, and Certified, respectively.

20. Border removal applies only to fields of 5 acres or more. These distances apply when there is no border removal. Removal of a 9-ft border (after flowering) decreases the required distance to 600, 225, and 100 ft for cross-pollinated species and to 30, 15, and 15 ft for apomictic and self-pollinated species. Removal of a 15-ft border allows a further decrease to 450, 150, and 75 ft for cross-pollinated species.

21. Isolation distances between two fields of the same kind may be reduced to a distance adequate to prevent mechanical mixture if the sum of the percentages of plants in bloom in both fields does not exceed 5% at a time when more than 1% of the plants in either field are in bloom.

22. Refers to bulbs.

23. Distance adequate to prevent mechanical mixture is necessary.

24. Required isolation between classes of the same variety is 10 ft.

25. The minimum distance may be reduced by 50% if different classes of the same variety are involved.

26. The minimum distance may be reduced by 50% if the field is adequately protected by natural or artificial barriers.

27. These ratios are for definite other varieties. The ratios for doubtful other varieties are

	Foundation	Registered	Certified
Millet	1/10,000	1/5,000	1/2,500
Sorghum	1/20,000	1/10,000	1/1,000
Hybrid sorghum	1/20,000	Not applicable	1/1,000
Okra	None	1/750	1/500

28. Whiteheart fruits may not exceed 1 per 100, 40, and 20 for Foundation, Registered, and Certified classes, respectively. Citron or hard rind is not permitted in Foundation or Registered classes and may not exceed 1 per 1,000 in the Certified class.

29. This distance applies if the contaminating source does not genetically differ in height from the pollinator parent or has a different chromosome number. If the contaminating source does (genetically) differ and has the same chromosome number, the distance shall be 990 ft. The minimum isolation from grass sorghum or broomcorn with the same chromosome number shall be 1,320 ft.

30. Requirement is waived for the production of pollinator lines if the previous crop was grown from a Certified seed of the same variety. Sterile lines and crossing blocks must be on land free of contaminating plants.

31. If the contaminating source is similar to the hybrid in all important characteristics, the isolation may be reduced by 66 ft for each pair of border rows of the pollinator parent, down to a minumum of 330 ft. These rows must be located directly opposite or diagonally to the contaminating source. The pollinator border rows must be shedding pollen during the entire time 5% or more of the seed parent flowers are receptive.

32. An unplanted strip at least 2 ft in width shall separate male steriles and pollinators in interplanted blocks.

33. Unless the preceding crop was another kind, the preceding soybean crop was planted with a class of certified seed of the same variety, or the preceding soybean crop and the variety being planted are of a contrasting pubescence or hilum color, in which case, no time need elapse.

34. May include not more than 0.04% purple or white seeds.

35. Standards apply equally to seed and pollen parents, which may not exceed 1:1000 each of wild-type branching, purple plants, or white seeded.

36. A new plant bed must be used each year unless the bed is properly treated with a soil sterilant prior to seeding.

37. This distance is applied between varieties of the same type and may be waived if four border rows of each variety are allowed to bloom and set seed between the two varieties, but are not harvested for seed. Isolation between varieties of different types shall be 1,320 ft, unless protected by bagging or by topping all plants in the contaminating source before bloom.

38. This distance is applied between plants of the same type, except that when male sterile and male fertile varieties are adjacent in a field, four border rows of male sterile shall be allowed to bloom and set seed. Seed set on these rows shall not be harvested for seed, except when the male fertile plants are used as the pollen parent. When plants are of different types, the distance shall be 1320 ft, except if protected by bagging or topping all plants in the contaminating source before bloom.

39. Isolation between varieties shall be 50 ft if ground broadcast and 100 ft if aerial seeded.

40. Isolation between millets of different genus shall be 6 ft.

41. Does not apply to *Helianthus similes, H. ludens,* or *H. agrestis.*

42. The ratio of male sterile (A) strains and pollen (B or C) strains shall not exceed 2:1.

43. Parent lines (A and B) in a crossing block or seed and pollen lines in a hybrid production field shall be separated by at least 6 ft and shall be managed and harvested in a manner to prevent mixing.

44. Distance between fields of Certified classes of the same variety may be reduced to 10 ft, regardless of class or size of field.

45. An isolation distance of 5,280 ft is required between oil and non-oil sunflower types and between either type and other volunteer or wild types.

46. Detasseling, cutting, or pulling of cytoplasmic male sterile seed parent is permitted.

47. All varieties of perennial ryegrass seed are allowed 3.0%.

48. This distance applies for fields over 5 acres. For alfalfa fields of 5 acres or less that produce the Foundation and Registered seed classes, the minimum distance from a different variety or a field of the same variety that does not meet the varietal purity requirements for certification shall be 900 and 450 ft, respectively.

49. There must be at least 10 ft or a distance adequate to prevent mechanical mixture between a field of another variety (or noncertified area within the same field) and the area being certified. The 165-ft isolation requirement is waived if the area of the "isolation zone" is less than 10% of the field eligible for the Certified class. The isolation zone is that area calculated by multiplying the length of the common border(s) with other varieties of alfalfa by the average width of the field (being certified) falling within the 165-ft isolation. Areas within the isolation zone nearest the contamination source shall not be certified.

50. Seed of Critana thickspike wheatgrass (*Agropyron dasystachum*) may contain up to 30% slender wheatgrass (*Agropyron trachycaulum*) types.

From AOSCA Certification Handbook, Publ. No. 23, Association of Official Seed Certifying Agencies, Raleigh, N.C., 1982, 10. With permission.

Table 95
ALFALFA SEED STANDARDS UNDER THE AOSCA SEED CERTIFICATION SYSTEM

Factor	Standards for each class (%)		
	Foundation	Registered	Certified
Pure seed (minimum)	99.0	99.00	99.00
Inert matter (maximum)	1.00	1.00	1.00
Weed seeds (maximum)	0.10	0.20	0.50
Objectionable or noxious weeds seed (maximum)[a]	None	None	None
Total other crop seeds (maximum)	0.20	0.35	1.00
Other varieties (maximum)	0.10	0.25	1.00
Other kinds (maximum)[b]	0.10	0.10	0.50
Germination and hard seed (minimum)	80.00	80.00	80.00

[a]   Objectionable and/or noxious weed seeds shall include the following: bindweed (*Convolvulus arvensis*), Canada thistle (*Cirsium arvense*), dodder (*Cuscuta* spp.), dogbane (*Apocynum cannabinum*), Johnson grass (*Sorghum halepense*), leafy spurge (*Euphorbia esula*), perennial sow thistle (*Sonchus arvensis*), quackgrass (*Agropyron repens*), Russian knapweed (*Centaurea repens*), and white top (*Lepidium draba, L. repens, Hymenophysa pubescens*).

[b]   Sweet clover seed shall not exceed 9 per lb for Foundation seed, 90 per lb for Registered seed, and 180 per lb for Certified seed.

From *AOSCA Certification Handbook,* Pub. No. 23, Association of Official Seed Certifying Agencies, Raleigh, N.C., 1982, 20. With permission.

Table 96
EXAMPLE OF CERTIFICATION STANDARDS — SOYBEAN

I.   Application of Genetic Certification Standards — The Genetic Certification Standards (pages 1 through 17 of the Handbook) are basic.

II. Land Requirements — Soybeans shall be grown on land on which the previous crop was of another kind, or planted with a class of certified seed of the same variety or with a variety of a contrasting pubescence or hilum color.

III. Field Standards
    A.   General
        Isolation
        Fields of soybeans shall be separated from any other variety or uncertified seed of the same variety by a distance adequate to prevent mechanical mixture.
    B.   Specific

	Maximum permitted — ratio of plants		
	Foundation	**Registered**	**Certified**
Other varieties	1:1000	1:500	1:200

IV. Seed Standards

Factor	Standards for each class (%)		
	Foundation	**Registered**	**Certified**
Pure seed (minimum)	NS	98.00	98.00
Inert matter (maximum)	NS	2.00	2.00
Weed seeds (maximum)[a]	0.05	0.05	0.05
Objectionable weed seed (maximum)[b]	None	None	None
Total other crop seeds (maximum)	0.20	0.30	0.60
Other varieties (maximum)[c]	0.10	0.20	0.50
Other kinds[d]	0.10	0.10	0.10
Germination and hard seed (minimum)[e]	NS	80.00	80.00

[a]   Total weed seed shall not exceed 10 per lb.

[b]   Objectionable weeds, as designated by the certifying agency.

[c]   Off-colored beans due to environmental factors shall not be considered other varieties. Other varieties shall be considered to include off-type seeds that can be differentiated from the variety that is being analyzed.

[d]   Not to exceed three per pound in any class. Corn or sunflower seed maximum; Foundation — NS, Registered — None; and Certified — one per pound. NS = no standards.

[e]   Germination for edible or large-seeded varieties may be lowered to 70.00%. Each agency will identify those varieties considered edible or large seeded.

From *AOSCA Certification Handbook,* Pub. No. 23, Association of Official Seed Certifying Agencies, Raleigh, N.C., 1982, 81. With permission.

SEEDING RATES

Table 97
SEEDING RATE AND PLANT CHARACTERISTICS FOR SELECTED CROPS

Common name	Seeding rate per acre (lb) Close drills	Rows	Seeds per Pound (thousands)	Gram (no.)	Weight per bushel (lb)	Germination time (d)	Temp. type[a]
Field crops							
Barley	72—96		13	30	48	7	C
Bean							
Broad (horse)	120—200		1—2	3	47	10	C
Green (field)		40—75	1—2	4	60	8	W
(dryland)		5—20					
Lima		60—120	0.4	2	56	9	W
Mung		10—15	11	24	—	7	W
Tepary		10—12	2	5	—	—	W
Beet							
Mangel		6—8	22	48	—	—	C
Sugar		2—8	22	48	15	—	C
Buckwheat	36—60		20	45	48	6	W
Castorbean		7—10	1	2	50		W
Chickpea	60		1	2		7	C
Chicory		1—2	420	940		14	C
Corn							
Grain		6—18	1.2	3	56	7	W
Silage		8—18					
Pop		3—6	3	7	56	7	W
Sweet		12—18	2	4	50	7	W
Cotton (upland)			4	8	28—33	12	W
Humid		24—40					
Mississippi Valley		24—48					
Dry areas		16—32					
Irrigated		25—30					
Cotton (Pima or Sea Island)		35—40	4	9		12	W
Cowpea	75—120	30—45	2—6	8	60	8	W
Crambe	8—5		86	190	27		C
Field Pea							
Large seeded	120—180		4	8	60	8	C
Small seeded	90—120						
Flax							
For seed	28—42		82	180	56	7	C
For fiber	84		113	250			
Guar	40—60	10	5	11	60		W
Guayule			600	1,300			W
Lentil		5—8	9	20	60		W
Millet							
Browntop	10—20	4—10	140	300	14		W
Italian (foxtail)	20—30		220	470	50	10	W
Pearl[b]	16—20	4—6	85	190		7	W
Proso	15—35		80	180	56	7	W
Mustard		3—4			58		
Black			570	1,250		7	C
White	4		73	160		5	C
Oats	48—128		14	30	32	10	C
Pea (see Field pea)							

Table 97 (continued)
SEEDING RATE AND PLANT CHARACTERISTICS FOR SELECTED CROPS

Common name	Seeding rate per acre (lb) Close drills	Rows	Seeds per Pound (thousands)	Gram (no.)	Weight per bushel (lb)	Germination time (d)	Temp. type[a]
Peanut		20—40	1	1—3	20—30	10	W
Pepper							
Tabasco		0.25	75	165		14	W
Potato (tubers)		1,000			60		C
Pumpkin (hills)		3—4	2	4		7	W
Rape	3—4		160	345		7	C
Rice	67—160		15	65	45	14	W
Rutabaga		1—2	200	430	60	14	C
Rye	28—112		18	40	56	7	C
Safflower	20—100		8—13	22	45		W
Sesame		5	100	220	46	10	W
Sorghum							
Grain (milo)		2—5	15	33	56	10	W
Sorgo	15—75	4—8	28	50	50	10	W
Soybean							
Small seed	60	15—20	8	18	60	8	W
Medium seed	90	20—30	2—3	6—13	60		
Large seed	120	30—45	1	2	60		
Sunflower		3—7	3—9	13	24	7	W
Silage		5—10					
Tobacco		1/250	5,000	11,000		14	W
Turnip	4—5	1—2	154	535	55	7	C
Wheat							
Bread	30—120		12—20	35	60	7	C
Club	60—90		20—24	48	60	7	C
Durum	60—90		8—16	26	60	10	C

Grasses: Forage/Grazing/Turf

Bahiagrass	10—12		150	336	—	21	W
Bentgrass							
Colonial	40—60		8,000	19,000	20—40	28	C
Creeping	40—60		7,700	17,000		28	C
Velvet	40—60		11,000	24,000		21	C
Bermudagrass	6—8		1,800	3,900	40 (14)	21	W
(unhulled)	10—15		1,300	2,900			
Bluegrass							
Kentucky	15—25		2,200	4,800	14	28	C
Roughstalk	15—25		2,500	5,600		21	C
Bluestem							
Big	15—20		150	340		28	W
Little	12—20		260	560		28	W
Sand	10—15		100	230		28	W
Brome							
Smooth	15—20		137	300	14	14	C
Buffalograss (burs)	15—20		50	110		28	W
Caryopses	5		330	738			
Canarygrass	20—25		68	150	50		
Reed	8—12		550	1,200	44—48	21	C
Carpetgrass	8—12		1,350	2,500	18—36	21	W
Dallisgrass	8—25		340	485	12—15	21	W

Table 97 (continued)
SEEDING RATE AND PLANT CHARACTERISTICS FOR SELECTED CROPS

Common name	Seeding rate per acre (lb)		Seeds per		Weight per bushel (lb)	Germination time (d)	Temp. type[a]
	Close drills	Rows	Pound (thousands)	Gram (no.)			
Fescue							
Meadow	10—25		230	500	14—24	14	C
Red	15—40		400	900		21	C
Sheep	25		530	1,167	10—30	21	C
Tall	10—25		227	500			C
Grama							
Blue	8—12		900	1,980		28	W
Side oats	15—20		200	442		28	W
(caryopses)			730	1,600			
Guineagrass	Veg.		1,000	2,200		28	W
Harding grass	25—30		340	750		28	
Indiangrass			170	365		21	W
Johnsongrass	20—30		130	290	28	35	W
Lovegrass							
Lehmann	1—3		4,245	9,400			W
Weeping	0.25		1,500	3,300		14	W
Molassesgrass			6,800	15,000		21	W
Orchardgrass	20—25		590	1,440	14	18	C
Rhodesgrass	10—12		1,700	4,700	8—12	14	W
Ricegrass							
Indian		8—10	140	310		42	C
Ryegrass							
Annual	25—30		227	500	24	14	C
Perennial	25—30		330	500	24	14	C
Smilograss			990	2,000		42	
Sudangrass	20—35	4—6	55	120	40	10	W
Switchgrass			370	815		28	W
Timothy	4—12		1,230	2,500	45	10	C
Wheatgrass							
Crested	12—20	4—6	190	425	20—24	14	C
Fairway	12—20	4—6	320	714		14	C
Western	12—20	3—7	110	235		35	C
Wildrye, Canada	10—12		120	261		21	C

Legumes: Forage/Grazing/Green Manure

Common name	Close drills	Rows	Pound (thousands)	Gram (no.)	Weight per bushel (lb)	Germination time (d)	Temp. type[a]
Alfalfa			220	500	60	7	C
Humid areas	10—20						
Irrigation	10—15						
Semiarid	8—10						
Burclover (out of burs), California	15—20		209	303	50	14	C
Button clover	15—20		150				C
Clover							
Alsike	2—8		680	1,500	60	7	C
Alyce	10—20		275	660	60	21	W
Crimson	15—25		150	330	60	7	C
(unhulled)	40—60						
Red	8—12		260	600	60	7	C
White, Ladino	5—7		860	1,900	60	10	C
Crotalaria, Showy	15—25	4—6	30	80	60	10	W
Crownvetch	5—10		140	300			C

Table 97 (continued)
SEEDING RATE AND PLANT CHARACTERISTICS FOR SELECTED CROPS

Common name	Seeding rate per acre (lb) Close drills	Rows	Seeds per Pound (thousands)	Gram (no.)	Weight per bushel (lb)	Germination time (d)	Temp. type[a]
Indigo, Hairy	3—10		200		55		W
Kudzu	Veg.	6—10	37	81	54	14	W
Lespedeza							
Common	25—30		343	750	25	14	W
Korean	20—25		240	525	45	14	W
Sericea	30—40		372	820	35	28	W
(scarified)	15—20		335	820	60		
Lupine							
Blue	70—90		3	7	60	10	C
Yellow	45—60		4	9	60	21	C
White	120—160		3	7	60	7	C
Pigeonpea		8—10	8	18	60		
Roughpea	20		14	40	55	14	C
Sainfoin	30—35		23	50	55	10	C
Serradella	15—20				36		C
Sweetclover							
White (hulled)	12—15		250	570	60	7	C
Yellow (unhulled)	30—45		250	570	60	7	C
Trefoil							
Big	4—6		1,000	1,900	60	7	C
Birdsfoot	8—12		375	800	60	7	C
Velvet bean		30—45	1	2	60	14	W
Vetch							
Common	40—80		7	19	60	10	C
Purple	40—70		9	20	60	10	C
Winter (hairy)	20—40		21	36	60	14	C

[a] Temperature type: C = cool-weather growth; W = warm-weather growth.
[b] For forage grazing in U.S.

Adapted from various sources.

Table 98
VEGETABLE SEEDING GUIDE

Common name	Seeds per 100-ft row (oz)	Seed per acre		Depth of planting (in.)	Distance within rows (in.)	Ready to use from seeding (d)
		Drilled (lb)	Transplanted (oz)			
Asparagus, root	66 rts		6000 rts	2	18	2 years
Beans, dwarf	16	60		1	2	42—75
Beans, pole	8	30		1	3—4 ft	65—90
Beet and swiss chard	1	10		1	2	45—60
Beet, mangel, and sugar	1	5		1	4	90—120
Broccoli	0.25	2	4	0.5	18—24	90—100
Brussels sprouts	0.25	2	4	0.5	12—16	100—120
Cabbage, Chinese	0.25	2	4	0.5	10—12	75
Cabbage, early	0.25	2	4	0.5	12—18	90—100
Cabbage, late	0.25	2	4	0.5	16—24	110—120
Cantaloupe, muskmelon	0.5	3		1	4/4 ft	85—150
Carrot	0.5	3		0.5	2	55—80
Cauliflower	0.25	2	4	0.5	14—18	95—110
Celery	0.25	1	4	0.5	2	120—150
Collards	0.25	2	4	0.5	14—18	100—120
Corn, sweet	4	10		1	4/3 ft	55—90
Cucumber	0.5	3		1	4/3 ft	50—70
Eggplant	0.13	2	4	0.5	18	125—140
Endive	1	4		0.5	12	100
Kale	0.25	3	16	0.5	18	55—60
Kohlrabi	0.25	4	16	0.5	6	50—70
Leek	0.5	4		0.5	4	120—150
Lettuce	0.25	3	16	0.5	6	70—90
Mustard	0.25	4		0.25	4—5	60—90
Okra	2	8		1	24	90—140
Onion, seed	1	4		0.5	2	125—150
Onion, sets	1 qt	12 bu		1	2	100
Parsley	0.25	3		0.13	3	65—90
Parsnip	0.5	3		0.5	2	130
Peas	16	200		1—2	1	45—75
Pepper	0.13	2	4	0.5	15	130—150
Pumpkin	0.5	4		1	4/6 ft	75—90
Radish	1	15		0.5	1	20—75
Rhubarb, root	40 rts		3500 rts	3—4	30	2—3 yrs
Rutabaga	0.25	2		0.5—1	6	90
Soybeans, vegetable	8	30		1	2	90—120
Spinach	1	15		0.5	2	45+
Squash, summer	0.5	4		1	4/4 ft	65—70
Squash, winter	0.5	2		1	4/6 ft	125
Tomato	0.13	2	2	0.5	3—4 ft	125—150
Tomato, pelleted	1	6	12 lb	0.5	3—4 ft	125—150
Turnip	0.5	2		0.25	2	45—90
Turnip, greens	1	5		0.25	—	45—90
Watermelon	1	3		1	4/6 ft	100—130

Adapted from Ware, G. W. and McCollum, J. P., *Producing Vegetable Crops,* 3rd ed., Interstate Printers and Publishers, Danville, Ill., 1980. With permission.

PLANTS AND ENVIRONMENT

Table 99
RESISTANCE OF CROPS TO FROST IN DIFFERENT DEVELOPMENTAL PHASES

	Temperature (°C) harmful to plant in the phases of		
	Germination	Flowering	Fruiting
Highest resistance to frost			
Spring wheat	−9,−10	−1,−2	−2,−4
Oats	−8,−9	−1,−2	−2,−4
Barley	−7,−8	−1,−2	−2,−4
Peas	−7,−8	−2,−3	−3,−4
Lentils	−7,−8	−2,−3	−2,−4
Vetchling	−7,−8	−2,−3	−2,−4
Coriander	−8,−10	−2,−3	−3,−4
Poppies	−7,−10	−2,−3	−2,−3
Kok-saghyz	−8,−10	−3,−4	−3,−4
Resistance to frost			
Lupine	−6,−8	−3,−4	−3,−4
Spring vetch	−6,−7	−3,−4	−2,−4
Beans	−5,−6	−2,−3	−3,−4
Sunflower	−5,−6	−2,−3	−2,−3
Safflower	−4,−6	−2,−3	−3,−4
White mustard	−4,−6	−2,−3	−3,−4
Flax	−5,−7	−2,−3	−2,−4
Hemp	−5,−7	−2,−3	−2,−4
Sugar beets	−6,−7	−2,−3	—
Fodder beets	−6,−7	—	—
Carrot	−6,−7	—	—
Turnip	−6,−7	—	—
Medium resistance to frost			
Cabbage	−5,−7	−2,−3	−6,−9
Soy beans	−3,−4	−2,−3	−2,−3
Italian millet	−3,−4	−1,−2	−2,−3
European yellow lupine	−4,−5	−2,−3	—
Low resistance to frost			
Corn	−2,−3	−1,−2	−2,−3
Millet	−2,−3	−1,−2	−2,−3
Sudan grass	−2,−3	−1,−2	−2,−3
Sorghum	−2,−3	−1,−2	−2,−3
Potatoes	−2,−3	−1,−2	−1,−2
Rustic tobacco	−2,−3	—	−2,−3
No resistance to frost			
Buckwheat	−1,−2	−1,−2	−0.5,−2
Castor plant	−1,−1.5	−0.5,−1	−2
Cotton	−1,−2	−1,−2	−2,−3
Melons	−0.5,−1	−0.5,−1	−1
Rice	−0.5,−1	−0.5,−1	−0.5,−1
Sesame	−0.5,−1	−0.5,−1	—
Hemp mallow	−0.5,−1	—	—
Peanuts	−0.5,−1	—	—
Cucumbers	−0.5,−1	—	—
Tomatoes	0,−1	0,−1	0,−1
Tobacco	0,−1	0,−1	0,−1

Source: Chang, J.-H., *Climate and Agriculture*, Aldine Publishing, Chicago, 1968.

Table 100
HEAT KILLING TEMPERATURE FOR DIFFERENT PLANTS AND PLANT PARTS

Plant	Heat-killing temperature (°C)	Exposure time
Zea mays	49—51	10 min
Shoots of iris	55	—
Potato leaves	42.5	1 hr
Pine and spruce seedlings	54—55	5 min
Cortical cells of trees	57—59	30 min
Barley grains (soaked 1 hr)	65	6—8 min
Wheat grains (9% H_2O)	90.8	8 min
Wheat (soaked for 24 hr)	60	45—75 sec
Trifolium pratense seeds	70	short time
Grapes (ripe)	63	—
Tomatoes	45	—
Apples	49—52	—

From Levitt, J., *Responses of Plants to Environmental Stress,* Academic Press, New York, 1972. With permission.

Table 101
EFFECTS OF HUMIDITY ON PLANT GROWTH AND YIELD

Crop	Average VPD mbar	Fresh g/plant	Dry g/plant	Economic yield g/plant
Barley	3		40	6.9
	14		37	8.1
Bean	4	1083	121	359
	16	838	99	284
Beet	3	212	18	99
	13	155	14	65
Corn	4	2197	545	410
	15	1739	470	352
	23	1724	429	254
Cotton	7		839	10.6
	16		799	164
	31		774	48
	39		615	10.1
Onion	3			649
	14			658
Pepper	6	3560	2101	2230
	29	2508	1121	1000
Radish	2	28.9	1.26	18.4
	13	24.3	1.08	16.1
Wheat	2		22.8	10.5
	14		20.9	8.7

Note: VPD = vapor pressure deficit of the atmosphere.

Source: Hoffman, G. J., Humidity effects on yield and water relations of nine crops, *Trans. ASAE,* 16, 164—167, 1973.

Table 102

EFFECTS OF HUMIDITY ON THE DEVELOPMENT OF COTTON

Average VPD mbar	Dry Weights (g per plant)				Shoot-root ratio (g/g)	Leaf area (dm²/plant)	Height (cm)
	Root	Stem	Leaf	Seed cotton			
7	93	463	273	11	8.8	468	208
16	78	379	178	164	9.2	273	157
31	142	373	211	48	4.5	310	127
39	153	258	194	10	3.0	292	129

Note: VPD, vapor pressure deficit.

From Hoffman, G. J., Rawlins, S. L., Garber, M. J., and Cullen, E. M., Water relations and growth of cotton as influenced by salinity and relative humidity, reprinted from *Agronomy Journal,* Volume 63, page 826, 1971 by permission of the American Society of Agronomy.

WATER STRESS AND IRRIGATION

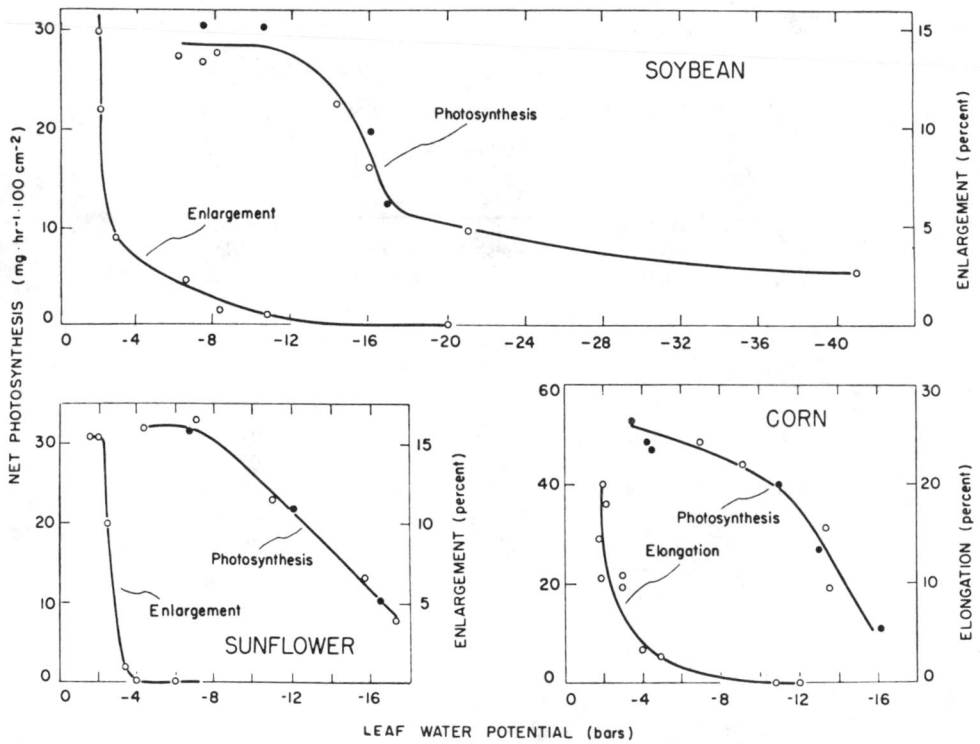

FIGURE 12.   Relationship between leaf water stress, leaf enlargement, and net photosynthesis in soybean, sunflower, and corn. Note that both photosynthesis and leaf enlargement are reduced at lower water deficits in sunflower than in the other two species. (From Boyer, J. S., *Plant Physiol.,* 46, 234, 1970. With permission of the American Society of Plant Physiology.)

Table 103

EFFECTS OF WATER STRESS APPLIED AT VARIOUS
STAGES OF GROWTH ON DRY WEIGHT OF
VEGETATIVE PARTS AND SEED YIELD OF SOYBEAN

Stage when stressed	Days after emergence	Vegetative dry weight per plant[a]	Seed weight[b]
		(g)	(g)
At flower induction	30	65.6	46.9
During flowering	44	76.1	49.8
Start of pod formation	57	80.4	36.5
During pod filling	65	84.6	32.5
Unstressed control	—	89.9	58.5

Note: Plants were stressed to a leaf water potential of -23 bars, then rewatered.

[a] At maturity.
[b] Air-dry weight of seeds per plant.

Modified from Sionit, N. and Kramer, P. J., Effects of water stress during different stages of growth of soybean, *Agron. J.,* 69, 274—278, 1977.

Table 104

AVERAGE YIELD OF ALFALFA SEED
FROM PLANTS SUBJECTED TO VARIOUS
DEGREES OF SOIL WATER STRESS

Soil water stress (mbar)	100	500	1000	2000
Seed yield (lb/acre)	718	1267	870	288

Modified from Hageman, R. W., Willardson, Lt. S., Marsh, A. W., and Ehlig, C. F., Irrigating for maximum alfalfa seed yield, *Calif. Agric.,* 29(11), 14—15, 1975.

Table 105

**EFFECT OF DELAY OF
IRRIGATION AFTER CUTTING
ALFALFA**

Days between cutting and irrigation	Average yield (kg/ha)	Average yield (%)
1	2980	100.0
2	2660	89.3
3	2570	86.2
4	2550	85.6
6	2080	69.8

Source: Guggenheim, A., Alfalfa Growing (transl.), Israel Ministry of Agriculture, Extension Service Bull. 130/105, 1979.

Table 106
YIELD AND WATER USE OF CORN UNDER SIX MOISTURE TREATMENTS AT PROSSER, WASHINGTON

Treatment and number	Yield at 15.5% moisture (kg/ha)	Water use (mm)
1 — Irrigated at tassel, no subsequent irrigation	6046	323
2 — Irrigated at tassel, two subsequent irrigations	8458	569
3 — Irrigated at tassel, three subsequent irrigations	8801	574
4 — Wilted 6—8 days at tassel, one subsequent irrigation	4245	404
5 — Wilted 6—8 days at tassel, two subsequent irrigations	5034	495
6 — Approached wilting percentage at tassel, irrigated at tassel, three subsequent irrigations	8719	554

Reproduced from Robins, J. S. and Domingo, C. E., Some effects of severe moisture deficits at specific growth stages in corn, *Agron. J.,* 45(12), 618, 1953, by permission of the American Society of Agronomy, Inc.

Table 107
YIELD OF CORN IN NEBRASKA IRRIGATION TRIALS

Treatment number	Rainfall (mm)	Water stored in soil (mm)	Irrigation (mm)	Total moisture applied (mm)	Number of irrigations	Yield (kg/ha)
1	63.5	119.38	360.6	535	6	9,735
2	63.5	167.7	223.5	455	4	8,781
3	63.5	182.9	180.3	427	3	7,508
4	63.5	177.8	124.5	366	2	6,936
5	63.5	195.6	91.4	351	1	6,172
6	63.5	200.7	228.6	493	3	8,717
7	63.5	182.9	142.2	389	2	7,000
8	63.5	154.9	177.8	396	3	9,163
9	63.5	182.9	127.0	373	2	7,127
10	63.5	162.6	121.9	348	2	7,636
11	63.5	170.2	251.5	485	4	8,017
12	63.5	172.7	91.4	328	1	6,427
13	63.5	198.1	—	262	0	4,391

From Howe, O. W. and Rhoades, H. F., Irrigation practice for corn production in relation to stage of plant development, *Proc. Soil Sci. Am.,* 19(1), 94, 1955. With permission.

Table 108
WATER- USE EFFICIENCY IN
NEBRASKA IRRIGATION TRIALS

Treatment number	Yield of maize in kg/ha/mm of water
No Irrigation	
13	16.8
One Irrigation	
5	17.5
12	19.6
Two Irrigations	
7	18.0
4	19.0
9	19.0
10	22.0
Three Irrigations	
3	17.5
6	17.8
8	23.0
Four Irrigations	
114	16.5
2	19.2
Six Irrigations	
1	17.8

From Howe, O. W. and Rhoades, H. F., Irrigation practice for corn production in relation to stage of plant development, *Proc. Soil Sci. Am.*, 19(1), 94, 1955. With permission.

Table 109
IRRIGATION WATER USED AND CROP PRODUCTIVITY

Crop	Days irrigated (number)	Water used (t/ha)	Dry matter produced (t/ha)	$\dfrac{\text{Water}}{\text{Dry matter}}$
Alfalfa	300	20,000	13.3	1,500
Barley	150	6,250	3.4	1,800
Cotton	240	10,175	3.2	3,170
Potatoes	120	7,000	7.8	900
Sorghum	90	6,275	4.5	1,395
Sugar beet	300	10,570	13.3	720
Wheat	150	5,655	4.1	1,380
Cantaloupe	210	4,715	1.7	2,770
Lettuce	105	2,100	2.1	1,000
Carrots	210	4,100	2.1	1,950

REFERENCES
1. **Erie, L. J., French, O. F., and Harris, K.,** Consumptive use of water by crops in Arizona, *Ariz. Agric. Exp. Stn. Bull.,* No. 169, 1948.
2. **Troughton, J. H.** Photosynthetic mechanisms in higher plants, in *Photosynthesis and Productivity in Different Environments,* Cooper, J. P., Ed., Cambridge University Press, London, 1975, 351—391.

Table 110
EXTRATERRESTRIAL RADIATION (R) EXPRESSED IN EQUIVALENT EVAPORATION IN mm/DAY

Northern Hemisphere												Lat.	Southern Hemisphere											
Jan.	Feb.	Mar.	Apr.	May	June	July	Aug.	Sept.	Oct.	Nov.	Dec.	(degrees)	Jan.	Feb.	Mar.	Apr.	May	June	July	Aug.	Sept.	Oct.	Nov.	Dec.
3.8	6.1	9.4	12.7	15.8	17.1	16.4	14.1	10.9	7.4	4.5	3.2	50	17.5	14.7	10.9	7.0	4.2	3.1	3.5	5.5	8.9	12.9	16.5	18.2
4.3	6.6	9.8	13.0	15.9	17.2	16.5	14.3	11.2	7.8	5.0	3.7	48	17.6	14.9	11.2	7.5	4.7	3.5	4.0	6.0	9.3	13.2	16.6	18.2
4.9	7.1	10.2	13.3	16.0	17.2	16.6	14.5	11.5	8.3	5.5	4.3	46	17.7	15.1	11.5	7.9	5.2	4.0	4.4	6.5	9.7	13.4	16.7	18.3
5.3	7.6	10.6	13.7	16.1	17.2	16.6	14.7	11.9	8.7	6.0	4.7	44	17.8	15.3	11.9	8.4	5.7	4.4	4.9	6.9	10.2	13.7	16.7	18.3
5.9	8.1	11.0	14.0	16.2	17.3	16.7	15.0	12.2	9.1	6.5	5.2	42	17.8	15.5	12.2	8.8	6.1	4.9	5.4	7.4	10.6	14.0	16.8	18.3
6.4	8.6	11.4	14.3	16.4	17.3	16.7	15.2	12.5	9.6	7.0	5.7	40	17.9	15.7	12.5	9.2	6.6	5.3	5.9	7.9	11.0	14.2	16.0	18.3
6.9	9.0	11.8	14.5	16.4	17.2	16.7	15.3	12.8	10.0	7.5	6.1	38	17.9	15.8	12.8	9.6	7.1	5.8	6.3	8.3	11.4	14.4	17.0	18.3
7.4	9.4	12.1	14.7	16.4	17.2	16.7	15.4	13.1	10.6	8.0	6.6	36	17.9	16.0	13.2	10.1	7.5	6.3	6.8	8.8	11.7	14.6	17.0	18.2
7.9	9.8	12.4	14.8	16.5	17.1	16.8	15.5	13.4	10.8	8.5	7.2	34	17.8	16.1	13.5	10.5	8.0	6.8	7.2	9.2	12.0	14.9	17.1	18.2
8.3	10.2	12.8	15.0	16.5	17.0	16.8	15.6	13.6	11.2	9.0	7.8	32	17.8	16.2	13.8	10.9	8.5	7.3	7.7	9.6	12.4	15.1	17.2	18.1
8.8	10.7	13.1	15.2	16.5	17.0	16.8	15.7	13.9	11.6	9.5	8.3	30	17.8	16.4	14.0	11.3	8.9	7.8	8.1	10.1	12.7	15.3	17.3	18.1
9.3	11.1	13.4	15.3	16.5	16.8	16.7	15.7	14.1	12.0	9.9	8.8	28	17.7	16.4	14.3	11.6	9.3	8.2	8.6	10.4	13.0	15.4	17.2	17.9
9.8	11.5	13.7	15.3	16.4	16.7	16.6	15.7	14.3	12.3	10.3	9.3	26	17.6	16.4	14.4	12.0	9.7	8.7	9.1	10.9	13.2	15.5	17.2	17.8
10.2	11.9	13.9	15.4	16.4	16.6	16.5	15.8	14.5	12.6	10.7	9.7	24	17.5	16.5	14.6	12.3	10.2	9.1	9.5	11.2	13.4	15.6	17.1	17.7
10.7	12.3	14.2	15.5	16.3	16.4	16.4	15.8	14.6	13.0	11.1	10.2	22	17.4	16.5	14.8	12.6	10.6	9.6	10.0	11.6	13.7	15.7	17.0	17.5
11.2	12.7	14.4	15.6	16.3	16.4	16.3	15.9	14.8	13.3	11.6	10.7	20	17.3	16.5	15.0	13.0	11.0	10.0	10.4	12.0	13.9	15.8	17.0	17.4
11.6	13.0	14.6	15.6	16.1	16.1	16.1	15.8	14.9	13.6	12.0	11.1	18	17.1	16.5	15.1	13.2	11.4	10.4	10.8	12.3	14.1	15.8	16.8	17.1
12.0	13.3	14.7	15.6	16.0	15.9	15.9	15.7	15.0	13.9	12.4	11.6	16	16.9	16.4	15.2	13.5	11.7	10.8	11.2	12.6	14.3	15.8	16.7	16.8
12.4	13.6	14.9	15.7	15.8	15.7	15.7	15.7	15.1	14.1	12.8	12.0	14	16.7	16.4	15.3	13.7	12.1	11.2	11.6	12.9	14.5	15.8	16.5	16.6
12.8	13.9	15.1	15.7	15.7	15.5	15.6	15.6	15.2	14.4	13.3	12.5	12	16.6	16.3	15.4	14.0	12.5	11.6	12.0	13.2	14.7	15.8	16.4	16.5
13.2	14.2	15.3	15.7	15.5	15.3	15.3	15.5	15.3	14.7	13.6	12.9	10	16.4	16.3	15.5	14.2	12.8	12.0	12.4	13.5	14.8	15.9	16.2	16.2
13.6	14.5	15.3	15.6	15.3	15.0	15.1	15.4	15.3	14.8	13.9	13.3	8	16.1	16.1	15.5	14.4	13.1	12.4	12.7	13.7	14.9	15.8	16.0	16.0
13.9	14.8	15.4	15.4	15.1	14.7	14.9	15.2	15.3	15.0	14.2	13.7	6	15.8	16.0	15.6	14.7	13.4	12.8	13.1	14.0	15.0	15.7	15.8	15.7
14.3	15.0	15.5	15.5	14.9	14.4	14.6	15.1	15.3	15.1	14.5	14.1	4	15.5	15.8	15.6	14.9	13.8	13.2	13.4	14.3	15.1	15.6	15.5	15.4
14.7	15.3	15.6	15.3	14.6	14.2	14.3	14.9	15.3	15.3	14.8	14.4	2	15.3	15.7	15.7	15.1	14.1	13.5	13.7	14.5	15.2	15.5	15.3	15.1
15.0	15.5	15.7	15.3	14.4	13.9	14.1	14.8	15.3	15.4	15.1	14.8	0	15.0	15.5	15.7	15.3	14.4	13.9	14.1	14.8	15.3	15.4	15.1	14.8

Table 111
SPECIFIC REPRODUCTIVE PLANT RESPONSES TO EVAPORATIVE DEMAND STRESS

Process or parameter affected	Description of plant response
Flowering and fruit set	Stress just prior to internode elongation just before flowering, ear emergence from the leaf sheath, and flower opening adversely affects yields. Stress generally increases flower and early fruit abscission and decreases seed and fruit set; however, in some cases, mild stress prior to blooming actually increases or initiates bloom set, particularly in tropical plants.
Fruit enlargement	The rate of fruit expansion decreases with mild stress in fruit tissue. A diurnal cycle in fruit size has been frequently observed in response to the diurnal cycle of evapotranspiration.
Fruit ripening	Evaporative demand stress hastens ripening (maturity) in some species and delays ripening (maturity) in others. When delayed, ripening tends to be less uniform than with nonstressed controls.
Total yield	Evaporative demand stress generally reduces yields. Yield reduction depends on the timing and duration of the stress. Plants such as corn (maize) store photosynthate in reserves during early growth for use later in producing kernels. Water stress during flowering has the most dramatic effect.

Note: References included in original.

Table 112
GENERALIZED SENSITIVITY TO WATER STRESS OF PLANT PROCESSES OR PARAMETERS

Sensitivity to stress[a]

Very sensitive → Relatively insensitive

Reduction in tissue Ψ Required to Affect Process[b]

Process or parameter affected	0 bar	10 bars	20 bars	Remarks
Cell growth	——— — —			
Wall synthesis	————			Fast-growing tissue
Protein synthesis	————			Fast-growing tissue
Protochlorophyll formation	———			Etiolated leaves
Nitrate reductase level	———			
ABA accumulation	— — —			
Cytokinin level	————			
Stomatal opening	— — ———————— — —			Depends on species
CO₂ assimilation	— — ———————— — —			Depends on species
Respiration	— — ———			
Proline accumulation	— — ———			
Sugar accumulation	————			

[a] Length of the horizontal lines represents the range of stress levels within which a process becomes first affected. Dashed lines signify deductions based on more tenuous data.

[b] With Ψ of well-watered plants under mild evaporative demand as the reference point.

From Hsiao, T. C., *Annu. Rev. Plant Physiol.*, 24, 519—570, 1973. Reproduced with permission from the *Annual Review of Plant Physiology*, Volume 24. © 1973 by Annual Reviews, Inc.

Table 113
THE WATER REQUIREMENTS OF CROPS AND WEEDS, AKRON, COLORADO[13]

Crop	Water requirement[a]	Weed	Water requirement[a]
Alfalfa	844	Annual sunflower	577
Barley	518	Cocklebur	415
Bromegrass	977	Gumweed	585
Buckwheat	540	Knotweed	678
Corn	349	Lambsquarter	658
Flax	783	Nightshade	487
Millet	285	Pigweed, prostrate	260
Oats	583	Pigweed, rough	305
Potatoes	575	Purslane	281
Red clover	759	Ragweed	912
Rye	634	Russian thistle	314
Sorghum	305		
Soybeans	646		
Sugar beets	377		
Sweet clover	731		
Wheat	545		

[a] Pounds of water per pound of dry matter produced.

Source: Shantz, H. L. and Piemeisel, L. N., The water requirement of plants at Akron, Colorado, *J. Agric Res.*, 34, 1093, 1927.

Table 114
TRANSPIRATION RATIO CONSTANTS ADJUSTED FOR RELATIVE HUMIDITY OF THE ATMOSPHERE

Plant species	n	k	SD	SE
Proso (*Panicum miliaceum*)	3	0.178	0.009	0.0016
Millet (*Chaetochloa italica*)				
Kursk S.P.I. 30029	1	0.219		
Turkestan S.P.I. 20694	1	0.129		
Other	10	0.179	0.018	0.0056
Sorghum (*Andropogon sorghum*)				
Sorgo	10	0.183	0.007	0.0022
Grain sorghum	7	0.1665	0.014	0.0053
Sudan grass (var. *Aethiopicus*)	5	0.129	0.007	0.0031
Dwarf milo	2	0.148	0.012	0.008
Corn (*Zea mays*) (includes 8 early hybrids)	33	0.1437	0.0137	0.0016
Teosinte (*Euchalena mexicana*)	2	0.1515	0.0035	0.0025
Sugar beet (*Beta vulgaris*)	2	0.128	0.0141	0.0116
Tumbleweed (*Amaranthus graicizans*)	2	0.190	0.0014	0.0010
Pigweed (*A. retroflexus*)	6	0.165	0.018	0.007
Russian thistle (*Salsola pestifer*)	1	0.158		
Lambs quarter (*Chenopodium album*)	2	0.0842	0.0167	0.0118
Polygonum aviculare	1	0.0738		
Barley (*Hordeum vulgare*)	12	0.093	0.0059	0.0017
Buckwheat (*Fagopyrum vulgare*)	1	0.090		
Wheat				
Emmer (*Triticum dicoccum*)	2	0.093	0.0056	0.0040
Durum (*T. durum*)	26	0.095	0.0105	0.0021
Common (*T. aestivum*)	37	0.0874	0.0121	0.0016
Cotton (*Gossypium hirsutum*)	6	0.089	0.1095	0.0045

Table 114 (continued)
TRANSPIRATION RATIO CONSTANTS ADJUSTED FOR
RELATIVE HUMIDITY OF THE ATMOSPHERE

Plant species	n	k̄	SD	SE
Potato (*Solanum tuberosum*)	6	0.0955	0.0130	0.0053
Oats (*Avena sativa*)	18	0.0835	0.007	0.0016
Cabbage (*Brassica oleracea capitata*)	1	0.108		
Turnip (*B. rapa*)	1	0.091		
Rape (*B. napus*)	1	0.078		
Rye (*Secale cereale*)	6	0.077	0.0053	0.0022
Grama grass (*Bouteloua gracelis*)	4	0.156	0.0361	0.0180
Watermelon (*Citrullus vulgaris*)	1	0.097		
Cantaloupe (*Cucumis melo*)	1	0.093		
Cucumber (*Cucumis sativus*)	1	0.081		
Squash, Hubbard (*Cucurbita maxima*)	1	0.078		
Pumpkin (*C. pepe*)	1	0.070		
Rice	2	0.0755	0.0035	0.0025
Legumes				
Guar (*Caymopsis*)	1	0.096		
Cowpea (*Vigna sinensis*)	5	0.090	0.0123	0.0055
Chickpea (*Cicer arietinum*)	1	0.0745		
Clover, crimson (*Trifolium inarnatum*)	2	0.0865	0.0205	0.0145
Clover, red (*T. pratense*)	1	0.074		
Bean, navy (*Phaseolus vulgaris*)	1	0.085		
Bean, Mexican (*P. vulgaris*)	1	0.075		
Vetch (*Vicia ervillea, V. villosa*)	3	0.090	0.007	0.0041
Bean, horse (*Vicia faba*)	2	0.0745	0.0007	0.0004
Vetch (*Vicia atropurpurea*)	1	0.062		
Bean, soy (cultivated) (*Soja max*)	1	0.086		
Bean, soy (wild) (*S. max*)	1	0.071		
Clover, sweet (*Melilotus alba*)	2	0.0665	0.0092	0.0065
Pea, Canada field (*Pisum sativum*)	2	0.070	0.007	0.005
Lupine (*Lupinus alba*)	1	0.060		
Alfalfa (*Medicago falcata*)	1	0.069		
Alfalfa, Peruvian (*M. Sativa*)	1	0.089		
Alfalfa, Grimm (*M. Sativa*)	13	0.0553	0.0070	0.0004
Flax (*Linum usitatissimum*)	10	0.060	0.0118	0.0037

Note: Results calculated from data of Shanz and Piemeisel (*J. Agric. Res.*, 34, 1093, 1927) and based on the equation:

$$Y = k \, Tr/(100 - H)$$

where Y = yield of dry matter harvested in grams, k = plant species or variety constant where soil fertility is not limiting, Tr = water transpired during the growing season, and H = percent of relative humidity in the atmosphere.

Table 115
MEAN ANNUAL AND MAY TO OCTOBER CLASS A PAN EVAPORATION

Location	Annual (in.)	May—Oct (in.)	Location	Annual (in.)	May—Oct (in.)
Ala.			Mont.		
Birmingham	56	40	Billings	53	41
Ariz.			Great Falls	50	40
Phoenix	105	72	Neb.		
Yuma	127	86	Lincoln	62	47
Ark.			North Platte	67	50
Little Rock	58	42	Nev.		
Calif.			Las Vegas	120	85
Fresno	90	72	Winnemucca	69	54
Los Angeles	59	41	N.J.		
Sacramento	70	53	Seabrook	42	31
Colo.			N.H.		
Denver	60	43	Concord	33	25
Grand Junction	52	40	N.M.		
Conn.			Albuquerque	90	65
Hartford	37	28	N.Y.		
D.C.			Albany	35	27
Washington	47	33	N.C.		
Fla.			Raleigh	54	37
Jacksonville	59	37	N.D.		
Miami	66	38	Bismarck	47	39
Tampa	65	40	Fargo	39	33
Ga.			Ohio		
Atlanta	55	38	Columbia	44	34
Idaho			Okla.		
Pocatello	51	41	Oklahoma City	85	60
Ill.			Ore.		
Chicago	38	30	Burns	56	45
Springfield	44	35	Portland	33	25
Ind.			Pa.		
Indianapolis	43	34	Harrisburg	44	32
Iowa			Pittsburgh	39	29
Des Moines	50	40	S.C.		
Kan.			Columbia	56	38
Dodge City	91	65	S.D.		
Ky.			Huron	49	40
Lexington	47	35	Tenn.		
La.			Nashville	50	37
New Orleans	66	43	Texas		
Shreveport	65	45	Abilene	97	65
Maine			Amarillo	96	65
Caribou	24	20	Brownsville	80	51
Portland	31	24	El Paso	105	69
Mass.			Fort Worth	80	55
Boston	35	27	Houston	73	47
Mich.			Utah		
Lansing	40	32	Milford	67	53
Minn.			Salt Lake City	55	44
Duluth	32	26	Va.		
Minneapolis	41	33	Richmond	52	36
Mo.			Wash.		
Kansas City	60	45	Seattle	30	23
St. Louis	48	37	Spokane	55	45

Table 115 (continued)
MEAN ANNUAL AND MAY TO OCTOBER CLASS A PAN EVAPORATION

Location	Annual (in.)	May—Oct (in.)	Location	Annual (in.)	May—Oct (in.)
Wis.			Wyo.		
Green Bay	35	28	Cheyenne	56	41
Madison	40	32			

Source: U.S. Department of Commerce, Climatic Atlas of the United States, Environmental Science Services Administration, Environmental Data Service, Washington, D.C., 1968.

Table 116
PAN COEFFICIENT Kp FOR CLASS A PAN FOR DIFFERENT GROUND COVER AND LEVELS OF MEAN RELATIVE HUMIDITY AND 24 HR WIND

| | | Case A — Pan surrounded by short green crop | | | | Case B — Pan surrounded by dry-fallow land | | |
| | | RH mean % | | | | RH mean % | | |
Wind km/day	Upwind distance of green crop (m)	Low <40	Medium 40—70	High >70	Upwind distance of dry fallow (m)	Low <40	Medium 40—70	High >70
Light	0	0.55	0.65	0.75	0	0.7	0.8	0.85
<175	10	0.65	0.75	0.85	10	0.6	0.7	0.8
	100	0.7	0.8	0.85	100	0.55	0.65	0.75
	1000	0.75	0.85	0.85	1000	0.5	0.6	0.7
Moderate	0	0.5	0.6	0.65	0	0.65	0.75	0.8
175—425	10	0.6	0.7	0.75	10	0.55	0.65	0.7
	100	0.65	0.75	0.8	100	0.5	0.6	0.65
	1000	0.7	0.8	0.8	1000	0.45	0.55	0.6
Strong	0	0.45	0.5	0.60	0	0.6	0.65	0.7
425—700	10	0.55	0.6	0.65	10	0.5	0.55	0.65
	100	0.6	0.65	0.7	100	0.45	0.5	0.6
	1000	0.65	0.7	0.75	1000	0.4	0.45	0.55
Very strong	0	0.4	0.45	0.5	0	0.5	0.6	0.65
>700	10	0.45	0.55	0.6	10	0.45	0.5	0.55
	100	0.5	0.6	0.65	100	0.4	0.45	0.5
	1000	0.55	0.6	0.65	1000	0.35	0.4	0.45

[a] For extensive areas of bare-fallow soils and no agricultural development, reduce Kp values by 20% under hot windy conditions, and by 5 to 10% for moderate wind, temperature, and humidity conditions. (Empirical constants Kp have been developed to convert pan evaporation data Ep to PET-potential evapotranspiration, according to the relationship PET = Ep × Kp).

From Doorenbos, J. and Pruitt, W. O., Crop Water Requirements, Irrigation and Drainage Paper No. 24, Food and Agriculture Organization, Rome, 1975. With permission of the Food and Agriculture Organization of the United Nations.

Table 117
PAN COEFFICIENT Kp FOR COLORADO SUNKEN PAN FOR DIFFERENT GROUND COVER AND LEVELS OF MEAN RELATIVE HUMIDITY AND 24 HR WIND[a]

Wind km/day		Case A Pan surrounded by short green crop				Case B[b] Pan surrounded by dry-fallow land			
	Upwind distance of green crop (m)	RH mean %			Upwind distance of dry fallow (m)	RH mean %		
		Low <40	Medium 40—70	High >70		Low <40	Medium 40—70	High >70
Light	0	0.75	0.75	0.8	0	1.1	1.1	1.1
<175	10	1.0	1.0	1.0	10	0.85	0.85	0.85
	≥100	1.1	1.1	1.1	100	0.75	0.75	0.8
					1000	0.7	0.7	0.75
Moderate	0	0.65	0.7	0.7	0	0.95	0.95	0.95
175—425	10	0.85	0.85	0.9	10	0.75	0.75	0.75
	≥100	0.95	0.95	0.95	100	0.65	0.65	0.70
					1000	0.6	0.6	0.65
Strong	0	0.55	0.6	0.65	0	0.8	0.8	0.8
425—700	10	0.75	0.75	0.75	10	0.65	0.65	0.65
	≥100	0.8	0.8	0.8	100	0.55	0.6	0.65
					1000	0.5	0.55	0.6
Very strong	0	0.5	0.55	0.6	0	0.7	0.75	0.75
>700	10	0.65	0.7	0.7	10	0.55	0.6	0.65
	≥100	0.7	0.75	0.75	100	0.5	0.55	0.6
					1000	0.45	0.5	0.55

[a]   See footnotes, Table 116.
[b]   For extensive areas of bare-fallow soils and no agricultural development, reduce Kp by 20% under hot, windy conditions; by 50 to 10% for moderate wind, temperature, and humidity conditions.

From Dorrenbos, J. and Pruitt, W. O., Crop Water Requirements, Irrigation and Drainage Paper No. 24, Food and Agriculture Organization, Rome, 1975. With permission of the Food and Agriculture Organization of the United Nations.

Table 118
APPROXIMATE RANGE OF CROP FACTORS
k_c (%)

ETC (crop) seasonal	k_c (%)	ETC (crop) seasonal	k_c (%)
Alfalfa	90—105	Onions	25— 40
Avocado	65— 75	Orange	60— 75
Bananas	90—105	Potatoes	25— 40
Beans	20— 25	Rice	45— 65
Cocoa	95—110	Sisal	65— 75
Coffee	95—110	Sorghum	30— 45
Cotton	50— 65	Soybeans	30— 45
Dates	85—110	Sugarbeets	50— 65
Deciduous trees	60— 70	Sugarcane	105—120
Flax	55— 70	Sweet potatoes	30— 45
Grains (small)	25— 30	Tobacco	30— 35
Grapefruit	70— 85	Tomatoes	30— 45
Maize	30— 45	Vegetables	15— 30
Oil seeds	25— 40	Vineyards	30— 55
		Walnuts	65— 75

[a]   Estimates of PET (potential evapotranspiration) for various regions may
be converted to estimates of the evapotranspiration rates of specific
crops by means of a specific constant for each crop, Kc (crop factor).
The Kc will vary with stage of growth and other factors.

From Doorenbos, J. and Pruitt, W. O., Crop Water Requirements,
Irrigation and Drainage Paper No. 24, Food and Agriculture Organi-
zation, Rome, 1975. With permission of the Food and Agriculture
Organization of the United Nations.

Table 119
CLASSIFICATION OF IRRIGATION WATER BASED ON
BORON CONCENTRATION

Boron in-dex	Concentration	Boron toxicity hazard
1	<0.5	Generally safe for sensitive crops
2	0.5—1.0	Sensitive crops will generally show slight to moder-ate injury
3	1.0—2.0	Semitolerant crops will generally show slight to moderate injury
4	2.0—4.0	Tolerant crops will generally show slight to moder-ate injury
5	>4.0	Hazardous for nearly all crops

Source: Richards, L. A., Ed., Diagnosis and Improvement of Saline and Alkali Soils,
U.S. Dep. Agric. Agric. Handb., No. 60, 1954.

Table 120
TENTATIVE CLASSIFICATION OF IRRIGATION WATER BASED ON CHLORIDE CONTENT

Chloride index	Concentration (meq/l)	Chloride hazard
1	<2	Generally safe even with sensitive plants
2	2—4	Sensitive plants (low tolerance) will generally show slight to moderate injury
3	4—8	Medium tolerant plants will generally show slight to moderate injury
4	>8	Slight to moderate injury for some tolerant plants

Table 121
MAXIMUM PERMISSIBLE CHLORIDE CONCENTRATIONS IN SOIL SOLUTION FOR VARIOUS FRUIT-CROP VARIETIES AND ROOTSTOCKS

Crop	Rootstock or variety	Limit of tolerance to chloride in soil solution (meq/l)
Rootstock		
Citrus	Rangpur lime, Cleopatra mandarin	50
	Rough lemon, tangelo, sour orange	30
	Sweet orange, citrange	20
Stone fruit	Marianna	50
	Lovell, Shalil	20
	Yunnan	14
Avocado	West Indian	16
	Mexican	10
Variety		
Grape	Thompson seedless, Perlette	50
	Cardinal, Black Rose	20
Cane berries	Boysenberry	20
	Olallie blackberry	20
	Indian Summer raspberry	10
Strawberry	Lassen	16
	Shasta	10

From Bernstein, L., *Water Quality Criteria,* ASTM STP 416, American Society for Testing and Materials, Philadelphia, Pa., 1967, 51. Copyright ASTM, 1916 Race Street, Philadelphia, Pa. 19103. Reprinted with permission.

Table 122
TYPICAL SEEPAGE RATES IN UNLINED CANALS

Soil type	Losses in m^3/m^2 of wetted perimeter in 24 hr
Impervious clay loam	0.075—0.10
Medium clay over hardpan	0.10 —0.15
Clay loam-silt loam	0.15 —0.23
Gravelly clay loam, sandy loam, sand and clay	0.23 —0.30
Sandy loam	0.30 —0.45
Loose sandy soil	0.45 —0.53
Gravelly sandy soil	0.60 —0.76
Very gravelly soils	0.90 —1.80

From Kraatz, D. B., Irrigation Canal Lining, Irrigation and Drainage Series No. 2, Food and Agriculture Organization, Rome, 1971. With permission of the Food and Agriculture Organization of the United Nations.

Table 123
MEAN VELOCITY VALUES AT INCIPIENT SCOUR

	Permissible mean velocity V, (m/sec)			
	Clean Water		Water containing colloids	
Type of soil	m/sec	ft/sec	m/sec	ft/sec
Very fine sand	0.45	1.48	0.75	2.46
Sandy loam	0.55	1.80	0.75	2.46
Silty loam	0.60	1.97	0.90	2.95
Alluvial silt, without colloids	0.60	1.97	1.00	3.28
Dense clay	0.75	2.46	1.00	3.28
Hard clay, colloidal	1.10	3.61	1.50	4.92
Very hard clay	1.80	5.90	1.80	5.90
Fine gravel	0.75	2.46	1.50	4.92
Medium and coarse gravel	1.20	3.94	1.80	5.90
Stones	1.50	4.92	1.80	5.90

Table 124
MAXIMUM NONEROSIVE VELOCITIES IN CANALS

Soil type	Maximum nonerosive velocity	
	m/sec	ft/sec
Fine sand (quicksand)	0.2 —0.3	0.75—1.00
Sand	0.3 —0.75	1.00—2.50
Sand loam	0.75—0.90	2.50—3.00
Loam to clay loam	0.85—1.1	2.80—3.60
Heavy clay	1.1 —1.5	3.60—5.00
Concrete	1.5 —2.5	5.00—8.00

From FAO, Irrigation Canal Lining, Irrigation and Drainage Paper No. 2, Food and Agriculture Organization, Rome, 1971. With permission of the Food and Agriculture Organization of the United Nations.

Table 125
SUGGESTED LENGTHS FOR IRRIGATION BORDERS

Irrigation stream delivered to each border	Soil textures					
	Sandy		Silt loam		Clay	
	Width[a]	Length[a]	Width[a]	Length[a]	Width[a]	Length[a]
FPS system (ft³/sec)						
1	20	200—300	25	330—550	30	550—770
1—2	30	330—440	35	550—770	40	660—880
2—4	35	440—550	40	550—770	50	660—1,000
4—6	40	550—660	50	660—880	60	880—1,320
6—8	66	660—770	66	880—1,320	66	1,320—1,500
Metric system (m³/hr/width)						
100	6.0	60—90	7.5	100—170	9.0	170—235
100—200	9.0	100—135	11.0	170—235	12.0	200—270
200—400	11.0	135—170	12.0	170—235	15.0	200—300
400—600	12.0	170—200	15.0	200—270	18.0	270—400
600—800	20.0	200—235	20.0	270—400	20.0	400—460

[a] Units are feet in FPS system, meters in metric system.

Adapted from Irrigation Advisers' Guide, Water and Power Resources Service, U.S. Department of the Interior, Washington, D.C., 1951, Table 18.

Table 126
RECOMMENDATIONS FOR SPRINKLER SYSTEMS EFFICIENCIES

| | | System efficiency | |
	Irrigation interval	Av wind — up to 6 k/hr (%)	Av wind — 6 to 12 k/hr (%)
Net water requirement 50 mm or 2 in.	10 days	70	68
	7—10 days	68	65
	5—6 days	65	62
Net water requirement 100 mm or 4 in.	20 days	75	70
	12—20 days	70	68
	10 days	68	65

Table 127
RECOMMENDED MAXIMUM ALLOWABLE SPRINKLING RATES

| | Maximum allowable sprinkling rate (in./hr) | | | | | | | |
| | 0—5% | | 5—8% | | 8—12% | | Over 12% | |
Description of soil and profile conditions	With cover	Bare	With cover	Bare	With cover	Bare	With cover	Bare
Sandy soil; homogeneous profile to depth of 1.8 m	2.00	2.00	2.00	1.50	1.50	1.00	1.00	0.50
Sandy soil over heavier soil	1.75	1.50	1.25	1.00	1.00	0.75	0.75	0.40
Light sandy-loam soil; homogeneous profile to 1.8 m	1.75	1.00	1.25	0.80	1.00	0.60	0.75	0.40
Sandy-loam soil over heavier soil	1.25	0.75	1.00	0.50	0.75	0.40	0.50	0.30
Silty-loam soil; hom. profile to 1.8 m	1.00	0.50	0.80	0.40	0.60	0.30	0.40	0.20
Silty-loam soil over heavier soil	0.60	0.30	0.50	0.25	0.40	0.15	0.30	0.10
Clay soil; silty clay loam soil	0.20	0.15	0.15	0.10	0.12	0.08	0.10	0.06

Source: Handbook of Engineering Practices for Region 7, Planning Sprinkler Irrigation Systems, Pacific Region, Soil Conservation Service, U.S. Department of Agriculture, Washington, D.C., 1949, p. VI-11 (1-6).

Table 128
ECONOMIC LIVES OF SELECTED IRRIGATION SYSTEM ELEMENTS

	In years
Dams	50—150
Wells	25—50
Ponds and reservoirs	20—50
Pumps	20—25
Motors	10—20
Canals and ditches	25—50
Pipes	
Steel	20—40
Concrete and asbestos-cement	20—40
Aluminium (portable)	15
Plastic (underground)	20—40
Hydrants	20—40
Water meters	20
Sprinklers	5—10
Mechanically moving sprinkling systems	10—15
Fittings of portable pipes	15

Table 129
ENERGY REQUIRED TO PUMP 1 ha-cm OF WATER[a]

Total head m	Direct energy (pump efficiency = .70) MJ	Electricity (Motor efficiency = .88) kWh	Diesel[b] (Engine efficiency = 0.30) ℓ	Natural gas[b] engine efficiency = 0.20 m³
10	14.0	4.4	1.21	1.92
50	70.0	22.1	6.07	9.59
100	140.1	44.2	12.14	19.18
200	280.2	88.4	24.28	38.37
300	420.3	132.6	36.42	57.55

[a] 1 hectare - cm = 100m³ = .9729 acre·in ≈ 1 acre·in.

[b] Heating value of diesel = 38.4 MJ/I; heating value of Natural Gas = 36.5 MJ/m³. Direct energy calculated as:

100 m³ × 1000 kg/m³ × 9.806 65 N/kg × 1MJ/10⁶N·m × total head (m)/Pump efficiency

Conversion may be made as follows: 4186 J = 1 kcal; 1 kwh = 859 kcal.

Table 130
PRACTICAL GUIDE FOR JUDGING MOISTURE AVAILABLE FOR CROPS

Available soil moisture remaining (%)	Appearance or feel of different soils			
	Sand	Loamy sand to sandy loam	Fine sandy loam to silt loam	Clay loam to clay
0	Dry, loose, single grained	Dry, loose, flows through fingers	Powdery dry and readily broken down to powdery state	Hard, baked, cracked, possibly crumbs on surface
50 or less	Dry, will not form ball with pressure	Appears dry, will not form ball	Somewhat crumbly, but will form ball	Somewhat pliable, will form ball
50 to 75	Appears dry, will not form ball	Tends to ball, but does not hold shape	Forms ball, will slick slightly with pressure	Forms ball, ribbons out between thumb and forefinger
75 to 100	Sticks together slightly, may form very weak ball	Forms weak ball that breaks easily, will not slick	Forms a ball, very pliable, slicks readily if high in clay	Ribbons easily between fingers, very slick feeling
At field capacity (100)	Upon squeezing, no free water appears on soil, but wet outline of ball is left on hand. Color is darkened.			

Adapted from Francis, C. J. and Turelle, J. W., Irrigating Corn, *U.S. Dep. Agric. Farmers' Bull.,* No. 2059, October, 1953.

Table 131
APPROXIMATE SOIL-WATER CHARACTERISTICS FOR TYPICAL SOIL CLASSES[a]

	Sandy soil	Loamy soil	Clayey soil
Dry weight per cubic foot	90 lb	80 lb	75 lb
Field capacity — percent of dry weight	10%	20%	35%
Permanent wilting percentage	5%	10%	19%
Percent available water	5%	10%	16%
Water available to plants:			
Pounds per cubic foot	4 lb	8 lb	12 lb
Inches per foot depth	$3/4$ in	$1^1/_2$ in	$2^1/_4$ in
Gallons per cubic foot	$^1/_2$ gal	1 gal	$1^1/_2$ gal
Water not usable to plants:			
Gallons per cubic foot	$^1/_2$ gal	1 gal	$1^1/_2$ gal
Approximate depth of soil that will be wetted by each 1 inch of water applied if half the available water has been used	24 in	16 in	11 in
Approximate number of second-feet to efficiently irrigate 1 acre	20 cfs	5 cfs	1.5 cfs
Suggested lengths of irrigation runs	330 ft	660 ft	1320 ft

[a]   After Booher, L. J., Department of Irrigation, University of California at Davis.

From Knott, J. E., *Handbook for Vegetable Growers,* John Wiley & Sons, New York, 1966, 81. Copyright by John Wiley & Sons, 1966. With permission.

Table 132

RELATIVE TOLERANCE OF CROPS TO WATERLOGGING

	To O_2 deficit (lab experiment)	To excess CO_2 (lab experiment)	To high ground-water levels (field tests)	To waterlogging (practical experience)
Highly tolerant	Rice Willow Sugar cane Several grasses	Citrus	Sugar cane Potato Broad beans	Plum Strawberry Several grasses
Medium tolerant	Oats Barley Onion Cotton Citrus Soya Apple	Apple Tomato Sunflower	Sugar beet Wheat Barley Oats Peas Cotton	Citrus Banana Apple Pear Blackberry
Sensitive	Maize Peas Beans Tobacco	Tobacco	Maize	Peach Cherry Raspberry Date palm Olive

The Limiting Values for the Different Groups

	Concentrations affecting root growth (%)		Groundwater depth of 0—50 m
	O_2 deficit	Excess CO_2	
Highly tolerant	0—1	>20	80—100 of normal yield
Medium tolerant	2—5	10—20	60—80 Of normal yield
Sensitive	About 10	<10	<60 Of normal yield

From Irrigation, Drainage and Salinity, An International Sourcebook. © Unesco/FAO 1973. Reproduced by permission of Unesco and FAO.

Table 133

MAXIMUM DESIGN VELOCITIES IN DRAIN TILE

	Velocity	
Type of soil	m/s	ft/s
Sand and sandy loam	1.0	3.5
Silt and silt loam	1.5	5.0
Silty clay loam	1.8	6.0
Clay and clay loam	2.0	7.0
Coarse sand or gravel	2.7	9.0

PLANT PRODUCTIVITY

Table 134
MAXIMUM PHOTOSYNTHETIC PRODUCTIVITY AND MEASURED MAXIMUM YIELDS IN SELECTED PLANTS

	g/m² day	Tons/acre year	Metric tonnes/ha year
Theoretical maximum			
U.S. average annual	61	100	223
U.S. Southwest average annual	71	117	262
U.S. Southwest, summer	106	172	385
Maximum measured			
C-4 plants			
Sugar cane	38	(62)	(138)
Napier grass	39	(64)	(139)
Sudan grass (sorghum)	51	(83)	(186)
Corn (*Zea mays*)	52	(85)	(190)
Non-C-4 plants			
Sugar beet	31	(51)	(113)
Alfalfa	23	(37)	(84)
Chlorella	28	(46)	(102)
Annual Yield			
C-4 plants			
Sugar cane	31	50	112
Sudan grass	10	16	36
Corn (*Z. mays*)	4	6	13
Non-C-4 plants			
Alfalfa	8	13	29
Eucalyptus	15	24	54
Sugar beet	9	15	33
Algae	24	39	87

Source: Proc. Conf. Capturing the Sun through Bioconversion, The Bio-Energy Council, Washington, D.C.

Table 135
RATE OF PHOTOSYNTHESIS AT AIR LEVEL AND ELEVATED LEVELS OF CO_2 (MILLIGRAMS CO_2/dm²/hr)

Plant	Air	Elevated CO_2
Corn, grain, sorghum, sugar cane	60—75	100
Rice	40—75	135
Sunflower	50—65	130
Soybean, sugar beet	30—40	56
Cotton	40—50	100

Table 136

RECORDED RATES OF MAXIMUM NET CO$_2$ UPTAKE (P$_{max}$) OF MAJOR CROPS (AT APPROXIMATELY 300 ppm CO$_2$)

Crop	P$_{max}$			Light (visible)		Temp. (°C)	Notes
	Individual leaf	Isolated plant	Whole crop (mg/dm² ground area h⁻¹)	Illumination (lx)	Energy (J/ cm²/min)		
	(mg/dm²/leaf area/h)						
Cereal grains:							
Wheat (C$_3$)							
Triticum spp.	30—45	—	—	—	1.79	25	¹⁴C In the field
	20	—	—	—	3.58	—	¹⁴C In the field
	20—22	—	—	—	4.12	—	
	40	—	—	—	1.09	30	
	58—72	—	—	—	1.00	21	Field chamber
	—	—	38—74	—	—	—	Field chamber; P efficiency 2.6%
	—	—	61	—	3.70	—	
	—	—	46	—	—	—	Field chamber
	—	—	45—76	—	2.32	—	Energy budget method; LAI 3.2
	—	—	48	33,000	—	20	Field chamber
Barley (C$_3$)							
Hordeum vulgare	15—18	—	—	light saturation		—	
	—	—	60—70	—	5.64	—	Field chamber 1 week after anthesis
	18	—	—	—	0.21	—	Field chamber 1 week after anthesis
	—	—	38	—	4.18	14	Energy budget method; P efficiency 2.1%

Table 136 (continued)
RECORDED RATES OF MAXIMUM NET CO$_2$ UPTAKE (P$_{max}$) OF MAJOR CROPS (AT APPROXIMATELY 300 ppm CO$_2$)

Crop	P$_{max}$			Light (visible)		Temp. (°C)	Notes
	Individual leaf (mg/dm²/leaf area/h)	Isolated plant	Whole crop (mg/dm² ground area h')	Illumination (lx)	Energy (J/cm²/min)		
Rice (C$_3$) *Oryza sativa*	40	—	—	—	1.25	30	12 varieties
	34—47	—	—	80,000	—	—	Erect leaf type
	—	—	68	—	5.00	—	Lax leaf type
	—	—	40	near light saturation	5.00	—	Stand at 'boot' stage
	—	—	55				
Maize (C$_4$) *Zea mays*	58	—	—	—	8.40	28	
	50	—	—	50,000	—	33	
	58	—	—	—	3.76	32	
	86	—	—	90,000	4.51	—	
	50—67	—	—	—	—	—	
	—	—	100	—	5.57	—	Field chamber; LAI 4.4
	—	—	94—144	—	2.24	—	Energy budget method; 10 min av values; LAI 4.2
	—	—	104—121	—	2.90	—	Momentum budget method; 10 min av. values; LAI 3.0

Species							Notes
Sorghum (C₄) *Sorghum* spp.	90	—	—	—	1.92	35	—
	79—91	—	—	—	1.83	28	—
	70	—	—	—	2.51	—	—
	70	—	—	—	2.98	30	—
	70	—	—	—	4.80	42	—
	—	100	—	—	—	—	Small field plots; July in Kansas, U.S.
Forage crops Ryegrass (C₃) *Lolium* spp.	31	—	—	>Light saturation	1.46	15	—
	32	—	—		0.71	20	—
	30	13	—	—	1.54	21	—
	—	8	—	—	1.7—2.1	—	Mature plant; LAI 7.0
	—	47	—	—	1.7—2.1	—	Young seedling (equivalent to leaf)
	—	—	40	—	1.7—2.1	—	Field chamber; LAI 30
	—	—	30	—	1.55	23	Simulated swards
	—	—	39	—	1.67	30	Simulated swards
	—	—	84	—	5.0	—	Field chamber; plots 35 days old
Cocksfoot (C₃) *Dactylis glomerata*	22	—	—	>Light saturation		25	—
	36	—	—		1.4	25	—
	30	—	—		—	20	—
	—	14	—	32,000	0.76	20	—
	—	—	48	>Light saturation		12—20	Field chamber; LAI 8.0
Tall fescue (C₃) *Festuca arundinacea*	33	—	—	>Light saturation		29	

Table 136 (continued)
RECORDED RATES OF MAXIMUM NET CO₂ UPTAKE (P*max*) OF MAJOR CROPS
(AT APPROXIMATELY 300 ppm CO₂)

Crop	P$_{max}$			Light (visible)		Temp. (°C)	Notes
	Individual leaf (mg/dm²/leaf area/h)	Isolated plant	Whole crop (mg/dm² ground area h¹)	Illumination (lx)	Energy (J/cm²/min)		
Red clover (C₃)							
Trifolium pratense	22	—	—	35,000	—	25	Excised leaves
	28	—	—	—	8.4	—	
White clover (C₃)							
Trifolium repens	—	17	—	—	1.54	21	Diffuse light; LAI 8.0
	—	20	—	—	0.59	20	
	—	—	33	80,000	—	26	Field chamber; LAI 3.0
Subterranean clover (C₃)							
Trifolium subterraneum	—	—	10	21,500	—	20	Simulated swards; LAI 2.5
	—	—	31	33,000	—	25	LAI 5.0
Lucerne (C₃)							
Medicago sativa	20	—	—	60,000	—	—	
	47	—	—	—	2.38	26	
	40	—	—	>Light saturation	—	20	After removal of other leaves
	86	—	—		2.22	—	
	—	—	80	80,000	—	26	Field chamber; LAI 5.5
	—	—	45	33,000	—	20	Field chamber

Species						Photosynthetic efficiency	Remarks
Rhodes grass (C₄) *Chloris gayana*	48	—	—	—	2.93	—	Field plots; Queensland 2.1% Photosynthetic efficiency
	—	—	58	—	—	—	
	—	—	61	—	2.39	—	
Pangola grass (C₄) *Digitaria decumbens*	45	—	—	43,000	—	30	
Bulrush millet (C₄) *Pennisetum typhoides*	100	—	—	—	3.59	35	
Kikuyu grass (C₄) *Pennisetum clandestinum*	—	—	50	—	1.63	25	Av for 14-hr light period
Buffalo grass (C₄) *Cenchrus ciliari*	49	—	—	—	1.20	35	
	106	—	—	100,000	—	40	
Green panic (C₄) *Panicum maximum*	72	—	—	75,000	—	30	
Columbus grass (C₄) *Sorghum almum*	85	—	—	100,000	—	30	
Bermuda grass (C₄) *Cynodon dactylon*	—	16	—	—	1.17	30	
Calopo (C₃, Tropical legume) *Calopogonium muconoides*	38	—	—	100,000	—	30	
Siratro (C₃, tropical legume)	37	—	—	75,000	—	30	
Phaseolus atro-purpureus *Glycine javanica* (C₃, tropical legume)	30	—	—	100,000	—	30	

Table 136 (continued)
RECORDED RATES OF MAXIMUM NET CO$_2$, UPTAKE (P$_{max}$) OF MAJOR CROPS (AT APPROXIMATELY 300 ppm CO$_2$)

Crop	P$_{max}$			Light (visible)		Temp. (°C)	Notes
	Individual leaf (mg/dm²/leaf area/h)	Isolated plant	Whole crop (mg/dm² ground area h⁻¹)	Illumination (lx)	Energy (J/cm²/min)		
Sugar crops							
Sugar cane (C$_4$)							
Saccharum spp.	51—86	—	—	>Light saturation			
	100	—	—	—	2.96	25	
	52	—	—	—	0.84	20	
	53	—	—	—	0.60	22	
	—	55	—	—	5.0	—	3-mo-old plants
	—	37	—	—	5.0	—	18-mo-old plants
Sugar beet (C$_3$)							
Beta vulgaris	29	—	—	—	2.96	25	
	25	—	—	—	1.45	20	
	24	—	—	42,000	—	24	
Legumes and oil seeds							
Soybean (C$_3$)							
Glycine max	29—43	—	—	—	2.00	29	
	37	—	—	90,000	—	—	
	—	—	72	64,000	—	30	Maximum daily rate
	—	—	94	64,000	—	30	Maximum hourly rate; crop at initial flowering
	34	—	—	50,000	—	24	
	56	—	—	81,000	—	25	
	50	—	—	150,000	—	30	
	—	—	55	—	2.50	30	Field enclosure

Species						
Field bean (C₃)						
Vicia faba	25	—	—	>Light saturation	29	
Sunflower (C₃)						
Helianthus annuus	50	—	—	8.40	—	
	45	—	—	>Light saturation	—	
	28—38	—	—	1.46	25	
	42	—	—	5.40	25	
	30	—	35,000	—	25	
	61	—	—	2.51	—	
Peanut (C₃)						
Arachis hypogaea	—	10	—	1.17	30	20-day-old plant
Pepper (C₃)						
Capsicum frutescens	—	6	—	1.17	30	94-day-old plant
Fiber crop						
Cotton (C₃)						
Gossypium spp.	60	—	—	2.51	—	
	84	—	—	0.59	30	
	32	—	—	>Light saturation	28	
	66	—	—	1.49	—	
	38	12—20	—	>Light saturation	25	120- and 80-day old plants
	—	16	—	1.09	30	123-day-old plants
	—	—	72	—	30	Field enclosure
	—	—	37—55	>Light saturation	25—35	Field enclosure
	—	—	32	33,000	20	Field enclosure
Fruit						
Tomato (C₃)						
Lycopersicon esculentum	32	—	—	1.49	20	
Citrus spp. (C₃)	11—13	18	—	—	21	
	12	—	—	1.54	22	>Light saturation
	—	—	—	>Light saturation	20	>Light saturation

Table 136 (continued)
RECORDED RATES OF MAXIMUM NET CO₂ UPTAKE (P _max_) OF MAJOR CROPS (AT APPROXIMATELY 300 ppm CO₂)

Crop	P$_{max}$			Light (visible)		Temp. (°C)	Notes
	Individual leaf	Isolated plant	Whole crop (mg/dm² ground area h¹)	Illumination (lx)	Energy (J/ cm²/min)		
	(mg/dm²/leaf area/h)						
Grape (C₃) *Vitis vinifera*	11	—	—	>Light saturation		25	
Other crops Tobacco (C₃)							
Nicotiana tabaccum	21	—	—	>Light saturation			
	23	—	—	25,000	—	25	
Softwood trees Sitka spruce (C₃)	18[a]	—	—	—	1.07	20	
Pinus radiata (C₃)	—	—	126	—	1.49	—	
Typha latifolia (C₃)	69	—	—	—	1.19	25	Energy budget method; 1 hr av values; LAI 4

Note: References included in the original.

[a] Whole shoot.

Table 137
ANNUAL PRODUCTIVITY AND MEAN GROWTH RATES OVER THE PERIOD OF GROWTH

Crop	Location	Latitude	Product	Growing period (days)	Annual production (t/ha)	Mean growth rate (g/m²/d)ᵃ
Forage crops						
Napier grass (C₄)						
Pennisetum purpureum	El Salvador	14°N	Forage	365	85.2	23.2
	Puerto Rico	18°N	Forage	365	84.7	23.2
		18°N	Forage	365	66.5	18.2
	Trinidad	10°N	Forage	365	41.7	11.4
Kikuyu grass (C₄)						
Pennisetum clandestinum	Hawaii	21°N	Forage	365	35.3	9.7
	Queensland, Australia	26°S	Forage	365	30.0	8.2
		26°S	Forage	365	29.2	8.0
Bulrush millet (C₄)						
Pennisetum typhoides	Northern Territory, Australia	14°S	Forage	112	21.7	19.4
Pennisetum glaucum (C₄)	Illinois, U.S.	40°N	Forage	110	16.1	14.6
Sugar cane (C₄)						
Saccharum sp.	Hawaii	21°N	Total DM	365	64.1	17.6
			Sugar	365	22.4	
Pangola grass (C₄)						
Digitaria decumbens	Guadeloupe, F.W.I.	16°N	Forage	365	54.6	15.0
	Colombia	3°N	Forage	365	50.6	13.9
	Cuba	23°N	Forage	365	39.4	10.8
	Queensland, Australia	27°S	Forage	365	24.2	6.6
	Hawaii	21°N	Forage	365	18.0	4.9
Sorghum (C₄)						
Sorghum bicolor	California, U.S.	33°N	Forage	210	46.6	22.2
	Illinois, U.S.	40°N	Forage	110	16.9	15.4

Table 137 (continued)
ANNUAL PRODUCTIVITY AND MEAN GROWTH RATES OVER THE PERIOD OF GROWTH

Crop	Location	Latitude	Product	Growing period (days)	Annual production (t/ha)	Mean growth rate (gm²d⁻¹)ᵃ
Sudan grass (C₄)						
Sorghum vulgare	California, U.S.	38°N	Forage	160	30.0	18.8
Guinea grass (C₄)						
Panicum maximum	Puerto Rico	18°N	Forage	365	48.8	13.4
	Puerto Rico	18°N	Forage	365	48.6	13.3
	El Salvador	14°N	Forage	365	29.9	8.2
	Nigeria	7°N	Forage	328	23.4	7.1
Panicum barbinode (C₄)	Trinidad	10°N	Forage	365	43.6	11.9
Panicum purpurascens (C₄)	Puerto Rico	18°N	Forage	365	40.6	11.1
Maize (C₄)						
Zea mays	Italy	45°N		140	40.0	28.6
	Nigeria	7°N		75	29.5	39.3
	Japan	36°N		128	26.5	20.7
	Kentucky, U.S.	38°N		129	21.9	17.0
	Ottawa, Canada	45°N	Forage	140	18.8	13.4
	Great Britain	51°N	Forage	160	17.1	10.7
	Netherlands	52°N	Forage	88ᵇ	15.0	17.0
	Kentucky, U.S.	38°N		127	8.6	6.8
Tripsacum laxum (C₄)	Trinidad	10°N	Forage	365	36.2	9.9
Para grass (C₄)						
Brachiaria mutica	Cuba	22°N	Forage	365	33.8	9.3
Paspalum plicatulum (C₄)	Queensland, Australia	27°S	Forage	365	31.9	8.7
Setaria (C₄)						
Setaria sphacelata	Queensland, Australia	27°S	Forage	365	31.8	8.8
Jaragua grass (C₄)						
Hyparrhenia rufa	El Salvador	14°N	Forage	365	31.5	8.6
Alfalfa (C₃)						
Medicago sativa	California, U.S.	38°N	Forage	250	32.5	13.0

Bermuda grass (C₄)					
Cynodon dactylon	Texas, U.S.	33°N	Forage	365	29.6	8.1
	Georgia, U.S.	31°N	Forage	365	27.2	7.5
	California, U.S.	38°N	Forage	365	27.2	7.5
	Georgia, U.S.	31°N	Forage	365	26.0	7.1
	Alabama, U.S.	33°N	Forage	365	24.6	6.7
	Tennessee, U.S.	36°N	Forage	365	23.1	6.3
Stylosanthes guyanesis (C₃)	Ghana	7°N	Forage	365	21.1	5.8
Perennial ryegrass (C₃)					
Lolium perenne	Great Britain	52°N	Forage	365	29.0	7.9
	New Zealand	40°S	Forage	365	26.6	7.3
	Netherlands	52°N	Forage	365	22.3	6.1
	Great Britain	52°N	Forage	365	19.5	5.3
Lolium perenne× multiflorum (C₃)	Great Britain	52°N	Forage	365	24.8	6.8
Tall fescue (C₃)					
Festuca arundinacea	Great Britain	52°N	Forage	365	29.0	7.9
Red Clover (C₃)					
Trifolium pratense	New Zealand	40°S	Forage	365	26.4	7.2
Cocksfoot (C₃)					
Dactylis glomerata	New Zealand	40°S	Forage	365	22.0	6.0
Kale (C₃)					
Brassica oleracea	Great Britain	54°N	Forage	140	21.0	15.0
	Great Britain	53°N	Forage	180	8.3	4.6
			Forage	420	24.4	5.8
Roots and tubers					
Cassava (C₃)					
Manihot esculenta	Tanzania	7°S	Total DM	330	40.8	12.4
			Tubers	330	23.2	—
	Sierra Leone	8°N	Total DM	230	29.3	12.7
Tapioca (C₃)					
Manihot utilissima	Malaysia	3°N	Total DM	270	37.5	13.9
			Tubers	270	22.4	—
Sugar beet (C₃)					
Beta vulgaris	California, U.S.	38°N	Total DM	240	33.1	13.8
	California, U.S.	38°N	Total DM	230	30.3	13.2
			Roots	230	15.2	—
			Sugar	230	10.2	—

Table 137 (continued)
ANNUAL PRODUCTIVITY AND MEAN GROWTH RATES OVER THE PERIOD OF GROWTH

Crop	Location	Latitude	Product	Growing period (days)	Annual production (t/ha)	Mean growth rate (gm²d¹)ᵃ
Potato (C₃)						
Solanum tuberosum	Japan	43°N	Total DM	175	22.9	13.1
	Netherlands	52°N	Total DM	—	21.0	—
	Great Britain	52°N	Roots	217	18.0	—
		52°N	Sugar	153ᵇ	8.0	—
	Netherlands	52°N	Total DM	162	22.0	13.6
		52°N	Tubers	162	18.2	—
	Great Britain	52°N	Tubers	164	11.0	—
			Tubers	122ᵇ	12.0	—
Sweet potato (C₃)						
Ipomoea batatas	Japan	36°N	Tubers	125	6.3	—
			Total DM	125	12.1	9.7
Grain crops						
Maize (C₄)						
Zea mays	Egypt	30°N	Grain	—	11.6	—
	Peru	12°N	Grain		10.3	—
	Iowa, U.S.	42°N	Grain	141	9.0	—
	Kentucky, U.S.	38°N	Grain	127	8.6	—
	Thailand	15°N	Grain	103	7.1	—
	Great Britain	51°N	Grain	160	5.5	—
Sorghum (C₄)						
Sorghum sp.	Philippines	15°N	Grain	80	6.6	—
Rice (C₃)						
Oryza sativa	New South Wales, Australia	32°S	Grain	190	13.8	—
	Peru	7°S	Grain	205	11.7	—
			Total DM	205	22.0	10.7
	Northern Territory, Australia	15°S	Grain	125	10.7	—
	Philippines (dry season)	15°N	Grain	122	10.1	—

Soybean (C₃) *Glycine max*	Philippines (wet season)	15°N	Grain	115	7.1	—
	Japan	36°N	Grain	120	4.9	—
			Total DM	120	12.5	10.4
			Total DM	161	19.7	12.3
Wheat (C₃) *Triticum vulgare*	Japan	39°N	Total DM	113	9.4	8.3
	Syria	33°N	Grain	—	8.0	—
	California, U.S.	38°N	Grain	—	7.0	—
	Sudan	17°N	Grain	—	7.0	—
	Netherlands	52°N	Total DM	69[b]	12.0	—
	Great Britain	52°N	Grain	130	4.8	—
			Grain	290	5.7	—
	Victoria, Australia	37°S	Grain	40[b]	4.6	—
Barley (C₃) *Hordeum vulgare*	Great Britain	52°N	Grain	61	6.0	—
Tree crops Oil palm (C₃) *Elaeis guineensis*	Malaysia	3°N	Total DM	365	29.8	8.2
			Oil	365	5.0	—
	Nigeria	3°N	Total DM	365	28.2	7.7
		7°N	Total DM	365	19.5	5.3
Scots pine (C₃) *Pinus sylvestris*	Great Britain	53°N	Total DM	365	19.0	5.2
Birch (C₃) *Betula* sp.	Great Britain	53°N	Total DM	365	7.9	2.2

Note: References included in the original.

[a] g/m²d = Grams dry weight/meter² ground/day.

[b] Period of bulking or grain filling only.

Table 138

MAXIMUM CROP GROWTH RATES OVER SHORT PERIODS AND ESTIMATES OF EFFICIENCY OF UTILIZATION OF RADIATION

Crop	Location	Latitude	Length of growth period (days)	Maximum growth rates (g/m²/d)	Average total radiation J/cm²/day	Utilization of visible radiation (%)
Bulrush millet (C₄) *Pennisetum typhoides*	Northern Territory, Australia	14°S	14	54.0	2132	9.5
	Australia		7	44.0	2575	6.6
Sudan grass (C₄) *Sorghum vulgare*	California, U.S.	39°N	35	51.0	2884	6.7
Maize (C₄) *Zea mays*	Nigeria	7°N	7	78.0		
	California, U.S.	39°N	12	52.0		6.4
	New York, U.S.	41°N		52.0ᵃ		
	Japan	36°N	21	51.6	2031	10.3
	Kentucky, U.S.	38°N	14	40.0		
	Nigeria	7°N	75	39.3		
	California, U.S.	39°N	10	38.0	3072	5.1
	Turin, Italy	45°N		32.8	1756	6.9
	Great Britain	51°N	21	24.0		
		52°N		13.6	1559	3.2
Sugar cane (C₄) *Saccharum* sp.	Hawaii	20°N	90	37.0	1676	8.4
Bermuda grass (C₄) *Cynodon dactylon*	Georgia, U.S.	31°N	40 (cut to cut)	21.2		
	Alabama, U.S.	32°N	32 (cut to cut)	13.4		
Carrot (C₃) *Daucus carota*	California, U.S.	33°N	48	90.6		
Sunflower (C₃) *Helianthus annuus*	Japan	36°N	7	76.0		
			28	68.0		
			7	30.9		
	Czechoslovakia	50°N	82	14.1		
Corncockle (C₃) *Agrostemma githago*	Great Britain	52°N	4	62.7		

Rice (C₃) Oryza sativa	Japan	36°N	8	55.0		6.3
		33°N	21	35.8	2032	
		36°S	14	33.0		
Typha angustifolia (C₃)	South Australia	32°S	40	23.0	2720	3.0
	Czechoslovakia	49°N	130	47.5		
				39.0ᵇ		
Tall fescue (C₃)	Great Britain	52°N	6	43.6	2201	7.8
Festuca arundinacea	Great Britain	52°N	13 (cut to cut)	16.4		
Phragmites communis (C₃)	Czechoslovakia	49°N	18	41.6		
			130	56.9ᵇ		
Sugar beet (C₃) Beta vulgaris	Czechoslovakia	50°N		40.0		
		52°N	12	31.0ᶜ	1229	9.5
Cocksfoot (C₃) Dactylis glomerata	Japan	43°N	21	27.8	1726	6.1
	Netherlands	52°N		20.6		
	Great Britain	52°N	6	40.5	2201	7.3
Timothy (C₃) Phleum pratense	Great Britain	52°N	6	36.4	2201	6.5
Ryegrass (C₃)	Great Britain	52°N	30 (cut to cut)	28.0	1983	5.6
Lolium perenne	Netherlands	52°N	10 (cut to cut)	20.0		
	New Zealand	40°S	32 (cut to cut)	18.9		
Cotton (C₃) Gossypium hirsutum	Great Britain	52°N	13 (cut to cut)	14.1		
	Georgia, U.S.	32°N		27.0		
Soybean (C₃) Glycine max	Japan	39°N	21	26.7	1212	9.7
	Iowa, U.S.	42°N	28	13.1		
Lucerne (C₃) Medicago sativa	California, U.S.	39°N		23.0		
Barley (C₃) Hordeum vulgare	Great Britain	52°N	42	23.0	2023	4.3
	Netherlands	52°N		17.7		
Subterranean clover (C₃) Trifolium subterraneum	Australia	35°C	14	22.9	2801	2.9

Table 138 (continued)
MAXIMUM CROP GROWTH RATES OVER SHORT PERIODS AND ESTIMATES OF EFFICIENCY OF UTILIZATION OF RADIATION

Crop	Location	Latitude	Length of growth period (days)	Maximum growth rates (g/m²/d)	Average total radiation J/cm²/day	Utilization of visible radiation (%)
Potato (C₃) *Solanum tuberosa*	Netherlands	52°N	84	22.8		
	New South Wales, Australia	34°S		22.8		
Tall oat grass *Arrhenatherum elatius*	Great Britain	52°N	13	21.9	1024	
Red clover (C₃) *Trifolium pratense*	New Zealand	40°S	18	21.1		
Kale (C₃) *Brassica oleracea*	Great Britain	52°N	14	21.0	1597	4.9
	New South Wales, Australia	34°S	10	18.5		
	Rome, Italy	42°N		15.9	2358	2.5
	Great Britain	53°N		14.9	1551	3.6
Sweet potato (C₃) *Ipomoea batatas*	Japan	36°N	23	20.0		
Pangola grass (C₄) *Digitaria decumbens*	Queensland, Australia	28°S		19.3		
French beans (C₃) *Phaseolus vulgaris*	New South Wales, Australia	34°S	10	18.3		
Tapioca (C₃) *Manihot utilissima*	Malaysia	3°N		18.0		
Cassava (C₃) *Manihot esculenta*	Tanzania	7°S		17.2		
	Sierra Leone	8°N		15.8		
Wheat (C₃) *Triticum vulgare*	Netherlands	52°N	69	17.5		
Oats (C₃) *Avena sp.*	Netherlands	52°N	53	16.9		
White clover (C₃) *Trifolium repens*	New Zealand	40°S	32	13.6		

Townsville lucerne (C$_3$)	Northern Territory,	14°S	14	13.4
Stylosanthes humilis	Australia	14°S		28.0
Oil palm (C$_3$)	Malaysia	3°N	365	11.0
Elaeis guineensis				

Note: References included in the original.

a Derived from aerodynamic gas exchange.
b Growth in hydroponic culture.
c Includes roots.

Table 139
EXAMPLES OF PLANT POPULATION RESPONSE CURVES FOR DIFFERENT TYPES OF CROP YIELD

Reproductive Yield

Wheat[2]

Plants/m²	1.4	7	35	184	1078
Grain yield	46	173	247	234	185

Barley[3]

Plants/m²	114	253	527	831
Grain yield (g/m²)	407	458	443	398

Maize (mean of nine hybrids)[4]

Plants/m²	1.0	2.0	3.0	4.0	7.9	15.8
Grain yield — low N (g/m²)	308	472	501	425	390	322
Grain yield — medium N (g/m²)	320	531	605	622	602	507
Grain yield — high N (g/m²)	367	604	721	783	791	702

Peas[5]

20-cm rows

Seeds/m² and rectangularity	67 (2.7:1)	100 (4:1)	200 (8:1)
Seed yield (g/m²)	278	286	235

40-cm rows

Seeds/m² and rectangularity	33 (5.3:1)	50 (8:1)	100 (16:1)
Seed yield (g/m²)	239	245	212

60-cm rows

Seeds/m² and rectangularity	22 (8:1)	33 (12:1)	67 (24:1)
Seed yield (g/m²)	225	229	217

Soybean[6]

Seeds/m²	10.7	13.3	17.8	26.7	53.3
Seed yield (g/m²)	369	390	412	412	393

Cotton (mean of 3 years)[7]

Plants/m²	1.0	2.0	4.0	7.9	15.8
Seed cotton yield (g/m²)	61.0	71.1	73.6	75.8	63.7

Vegetative Yield

Kale[1]

Plants/m²	2.8	5.1	16.3	37.1	93.5
Dry matter yield (g/m²)	417	437	480	505	512

Potatoes[1]

Plants/m²	1.5	1.9	2.3	3.0	4.7
Total tuber yield (t/ha)	27.5	27.1	27.8	29.1	32.8

Carrots[8]

Plants/m²	11	108	323	538
Root yield, early harvest (t/ha)	12.6	40.2	60.3	75.4
Root yield, late harvest (t/ha)	30.1	82.9	113.0	133.0

Perennial ryegrass[9]

Tillers/m²	2249	3056	4551	5724	7026	7209
Dry matter yield (g/m²)	14.3	37.6	99.2	184.6	232.9	268.3

Graded Yield

Parsnips[10]

Plants/m²	6.5	8.6	11.8	16.1	22.6	32.3	44.1	61.3	83.9
Total roots (t/ha)	19.8	24.1	29.9	33.2	37.2	39.7	41.4	41.7	41.7
Roots 3.8—6.4 cm (t/ha)	0.5	1.8	3.5	11.3	23.1	30.4	36.2	33.7	27.9
Roots >3.8 cm (t/ha)	16.3	22.9	26.6	31.9	35.9	37.9	37.4	34.2	27.9
Roots >5.1 cm (t/ha)	16.1	22.9	26.1	30.9	30.9	27.4	17.3	8.3	3.8

Table 140
TYPICAL RANGES OF SEED RATES, SPACINGS, AND PLANT POPULATIONS FOR A NUMBER OF REPRESENTATIVE CROPS

Crop	Planting material	Seed rate (kg/ha)	Spacing	Plant population
Cereals				
Wheat	Seed	50—200	10—30 cm rows	100—400/m²
Rice	Seed	75—125	15—30 cm rows, broadcast	125—300/m²
	Seedlings	25—40	10—25 cm between "hills;" 1—5 plants/hill	75—150/m²
Maize	Seed	10—40	60—100 cm rows; 10—30 cm in rows (or wider) with 2 to 3 plants per hill	2.5—10/m²
Sorghum	Seed	3—12	50—100 cm rows	6—24/m²
Grain legumes				
Peas	Seed	125—275	15—45 cm rows	50—150/m²
Field beans (*Vicia*)	Seed	175—325	15—60 cm rows	30—100/m²
Soy	Seed	35—75	25—90 cm rows; 3—15 cm in rows	40—100/m²
Groundnut	Shelled nuts	40—100	30—60 cm rows; 10—30 cm in rows	7.5—22.5/m²
Kidney bean	Seed	20—100	30—90 cm rows; 7.5—30 cm in rows	10—45/m²
Roots and vegetables				
Potatoes	"Seed" tubers	1.5—3.5 (t/ha)	60—100 cm rows; 25—50 cm in rows	3—6/m²
Sweet potatoes	Tubers, stem cuttings, leaf cuttings	—	90—150 cm rows or ridges; 30—50 cm in rows; "mounds" 90—150 cm apart with several plants/mound	2—5/m²
Cassava	Stem cuttings	—	90—150 cm between plants (often interplanted)	0.4—1/m²
Carrots	Seed	1—5	20—40 cm rows; Multiple rows or narrow beds with rows 5—10 cm apart; 3—5 cm in rows	50—400/m²
Onions	Seed (small bulbs on noncommercial scale)	5—20	25—50 cm rows; multiple rows or bands with rows 5—10 cm apart; 2—4 cm in rows	50—400/m²
Cabbage	Seed, transplants	0.5—1.5	35—60 cm rows; 25—35 cm in rows	3—10/m²
Fodder and forage				
Temperate grasses	Seed	20—35	Broadcast 15—20 cm rows	—
Kale	Seed	4—6	40—60 cm rows	25—40/m²
Beverages				
Tea	Cuttings, sometimes seedlings	—	1—1.5 m between bushes	8—10,000/ha
Coffee	Seedlings, sometimes cuttings	—	2—3.5 m between bushes	900—1,500/ha
Cocoa	Seeds, seedlings, cuttings	—	2½—4 m between trees (often closely spaced in early years and then selectively thinned)	600—1,000/ha

Table 140 (continued)
TYPICAL RANGES OF SEED RATES, SPACINGS, AND PLANT POPULATIONS FOR A NUMBER OF REPRESENTATIVE CROPS

Crop	Planting material	Seed rate (kg/ha)	Spacing	Plant population
Oils and oilseeds				
Sunflower	Seed	5—10	60—90 cm rows; 20—30 cm in row	4—8/m²
Oil palm	Seedlings	—	7—8 m between palms	160—200/ha
Coconut	Seedlings	—	7.5—9 m between palms	120—180/ha
Fruit				
Apples	Budded root stocks	—	1 m × 2.5 m to 7 m × 7 m	200—4,000/ha
Citrus	Budding	—	4—7 m between trees	200—600/ha
Bananas	Suckers, rhizomes	—	2.5—5 m between plants	150—400/ha
Pineapples	Crowns, slips, suckers	—	About 1.5 m between double rows 50—60 cm apart; 20—35 cm in rows	4—6/m²
Fibers				
Cotton	Seed	35—40, natural seed; 20—30, "defuzzed" seed	60—100 cm rows 20—40 cm in rows (often two plants/hill)	4—8/m²
Miscellaneous				
Sugar beet	Seed (actually fruit containing several seeds) often "rubbed" to give single plant/seed	10—20	40—60 cm rows; 20—30 cm in rows	6—10/m²
Sugar cane	Stem cuttings, each with 2 to 3 nodes	—	1.2—1.8 m rows; cuttings end to end	1—1.5 "stools"/m²
Rubber	Buddings, seedlings	—	4—6 m between trees (often thinned out further in later years)	350—500/ha initially — may be reduced to 250—300/ha
Tobacco	Seedlings	—	90—120 cm rows; 60—90 cm in rows	0.75—1.5/m²

ROOTS

Table 141
ROOTING DEPTH AND LATERAL SPREAD OF ROOTS FOR SELECTED PLANT SPECIES

Scientific name[a]	Maximum rooting depth[b] (cm)	Working rooting depth[c] (cm)	Lateral spread[d] (cm)	Site characteristics[e]	Remarks
Agropyron repens (L.) Beauv.	245	180	—	C	Many rhizomes 60 to 90 cm long
A. smithii Rydb.	245	210	—	B	Many rhizomes; dense sod
Agrostis canina L.	—	68	60	E	
(*A. vulgaris* With.) = *A. tenuis* Sibth.	100	60	22	E	
Allium cepa L.	100	80	30	H	20 to 25 roots attached to base of bulb
A. sativum L.	100	100	55	J	25 roots attached to base of stalk
Amaranthus retroflexus L.	240	200	180	K	Plot pre-irrigated to 1.5-m depth
(*Andropogon furcatus* Muhl.) = *A. gerardii* Vitm.	270	200	35	C	Numerous and densely branched roots to 200 cm
Asparagus officinalis L.	320	135	180	H	Dense rooting in surface 30 cm of soil
Avena sativa L.	200	150	25	C	Profuse rooting to 60 cm of soil
Beta vulgaris L. [sugar beet]	180	120	45	J	
B. vulgaris L. [table beet]	300	180	120	J	
(*Brassica caulorapa* [DC.] Pasq.) = *B. oleracea* var. *gongylodes* L.	260	210	80	H	
(*B. napobrassica* [L.] Mill.) = *B. napus* var. *napobrassica* (L.) Reichenb.	190	150	45	H	
B. oleracea var. *botrytis* L.	137	120	75	J	
B. oleracea var. *capitata* L.	235	150	100	J	
B. rapa L.	168	150	75	J	
Bromus inermis Leyss.	195	85	Rhizomatous	H	
Buchloe dactyloides (Nutt.) Engelm.	215	120	50	S	
Convolvulus arvensis L.	480	430	—	H	Rhizomatous
Cucumis sativus L.	120	110	210	H	

Table 141 (continued)
ROOTING DEPTH AND LATERAL SPREAD OF ROOTS FOR SELECTED PLANT SPECIES

Scientific name[a]	Maximum rooting depth[b] (cm)	Working rooting depth[c] (cm)	Lateral spread[d] (cm)	Site characteristics[e]	Remarks
Cucurbita maxima Duch.	210	60	500	H	
C. pepo L.	180	120	500	H	
Cynodon dactylon (L.) Pers.	245	200	—	S	
Dactylis glomerata L.	130	100	—	B	
Daucus carota subsp. sativus (Hoffm.) Arcang.	220	150	75	H	
Digitaria decumbens Stent	245	200	—	S	
Elymus canadensis L.	90	90	60	L	Roots penetrate diagonally downward
(Festuca elatior L.) = F. pratensis Huds.	120	90	15	B	
F. ovina L.	70	30	23	E	
Fragaria chiloensis (L.) Duch.	94	60	30	L	
F. virginiana Duch.	35	20	30	I	
Glycine max (L.) Merr.	225	200	50	Z	Wayne variety
Helianthus annuus L.	270	150	150	H	
Hordeum vulgare L.	195	135	30	H	
Kochia scoparia (L.) Schrad.	220	180	330	K	
Lactuca sativa var. crispa L.	225	150	60	H	
Lespedeza capitata Michx.	240	150	90	C	
Lycopersicon esculentum Mill.	188	188	60	BB	Rhizotron data
Medicago sativa L.	610	300	15	H	
Melilotus alba Medik.	207	150	60	H	
Panicum antidotale Retz.	365	240	75	FF	
Paspalum dilatatum Poir.	245	200	—	R	
P. notatum Fluegge	245	200	—	R	
Phalaris arundinacea L.	80	65	40	E	
Phleum pratense L.	85	45	30	E	
Pisum sativum L.	90	90	60	H	Depth greatly depends upon variety

				Much surface absorption
Poa pratensis L.	210	150	30	L
Raphanus sativus L.	220	120	100	J
Secale cereale L.	230	150	25	T
Sesamum indicum L.	120	—	—	GG
Solanum tuberosum L.	150	90	40	C
(*Sorghum vulgare* Pers.) = *S. bicolor* (L.) Moench	180	180	60	H
Trifolium pratense L.	300	120	45	H
T. repens L.	240	120	45	H
Triticum aestivum L.	200	150	15	H
T. durum Desf.	225	180	35	C
Zea mays L.	188	180	100	H

a   Scientific names are as originally listed in references cited in original table; however, if we considered the listed name a synonym in modern usage, we put it in parentheses, followed by the correct name. We provided all authorities (authors; describers) appended to names.

b   The maximum depth is that at which a root was found at the particular site.

c   Average depth reached by many roots or root branches and to which depth a considerable absorption must take place.

d   Radial distance to which roots penetrate from the base of the plant.

e   B, Moist site near Lincoln, Neb. C, Loess soil near Peru, Neb. E, Rich heavy clay underlain at 15 to 25 cm with tenacious gravelly clay at Geneva, N.Y. H, Deep, dark brown fertile Carrington silt loam soil at Lincoln, Neb. I. Forested site on soil consisting of half decomposed granite and half rich brown soil at 2400-m near Pike's Peak, Colo. J, Fine sandy loam surface soil at Norman, Okla, with excellent tilth becoming compact sandy clay at 45 to 60 cm. K, Pullman silty clay loam surface soil becoming clay from 22 to 45 cm, then clay loam to 130 cm at which layer high in $CaCo_3$ is encountered. L, Tall grass prairie site near Lincoln, Neb. Soil is silt loam surface soil over silt loam subsoil to several meters depth. R, Hardlands near Limon, Colo. S, Fertilized sand to 270-cm depth at Tifton, Ga. T, Sandhills near Central City, Neb. Z, Ida silt loam, a loess soil near Castana, Ia. BB, Loamy sand to 188 cm in the Auburn, Ala., rhizotron. FF, 8 km northwest of Sterling, Colo. Soil is silt loam surface soil and a hard compact subsoil grading into a hardpan at 60-cm depth. GG, Sand sheet soil, Kordofan Province, Sudan.

Adapted from Taylor, H. M. and Terrell, E. E. in *CRC Handbook of Agricultural Productivity*, Vol. 1, Rechigl, Miloslav, Jr., Ed., CRC Press, Boca Raton, Fla., 1982, 186. References are included in original.

Table 142

VALUES OF ROOT LENGTH UNDER UNIT AREA OF GROUND SURFACE (L$_A$) FOR PLANTS GROWING UNDER FIELD OR RHIZOTRON CONDITIONS

Scientific name	Soil	Depth sampled (cm)	L$_A$ (cm/cm^2)[a]
Allium cepa L.	Alluvial fine sandy loam	61	26—52
Avena fatua L.	Loam	Whole plant	1170
A. fatua L.	Loam	165	1700
A. sativa L.	Unknown	15	43
A. sativa L.	Loam		
(*Brassica arvensis* [L.] Rabenh.) = *B. kaber* (DC.) Wheeler	Loam	Whole plant	150
Camellia sinensis (L.) Ktze.	Alluvial sandy loam	70	100—130
Conringia orientalis (L.) Dumort.	Loam	Whole plant	220
Glycine max (L.) Merr.	Unknown	15	64
G. max (L.) Merr.	Ida silt loam	180	55
G. max (L.) Merr.	Nicollet sandy clay loam	120	11
Gossypium hirsutum L.	Fuquay loamy sand	188	334
G. hirsutum L.	Hanford sandy loam	183	120
(*Hordeum distichum* L.) = *H. vulgare* L.	Loam	Whole plant	950
(*H. distichum* L.) = *H. vulgare* L.	Loam	165	1700
Lactuca sativa L.	Dunnington Heath sandy loam	60	47
Lycopersicon esculentum Mill.	Fuquay loamy sand	188	700
Medicago sativa L.	Loam	10	210
Pinus sylvestris L.	Ca. 1 m acid sand over chalk	183	126
Poa pratensis L.	Unknown	15	360
Prunus persica (L.) Batsch	Sand to fine sandy loam	91 or 122	17—68
(*Pseudotsuga taxifolia* [Lamb] Britt.) = *P. menziesii* (Mirbel) Franco	Coarse sand or sandy loam	107	77
Pyrus communis L.	Sand to fine sandy loam	91 or 122	26—69
Salsola kali L.	Loam	Whole plant	100
Secale cereale L.	Unknown	15	60
Trifolium repens L.	Clay loam	30	310
(*Triticum vulgare* [Vill.] = *T. aestivum* L.	Loam	Whole plant	200
(*Triticum vulgare* [Vill.] = *T. aestivum* L.	Loam	165	1130
Zea mays L.	Chalmers silt loam	75	153
Zea mays L.	Fuquay loamy sand	188	2100
Zea mays L.	Raub silt loam	60	71—92
Zea mays L.	Panoche clay loam	183	410
Zea mays L.	Hanford sandy loam	183	306

Note: References included in the original.

[a] Old roots are less effective than are new roots in absorbing either water or minerals. This must be considered when comparing plants at equal rooting densities (L$_A$).

TILLAGE AND ROTATIONS

Table 2

A COMPARISON OF CHARACTERISTICS OF SEVERAL TILLAGE SYSTEMS

Quality	Moldboard plow (conventional)	Plow-plant	Disk-plant	Field cultivate or chisel plow	Till-plant	No-tillage
Time and labor demand	Most	Moderate	Moderate	Moderate	Low	Least
Fuel consumption	Greatest	Intermediate	Moderate	Moderate	Low	Least
Dependence on herbicides	Least	Slight	Intermediate	Intermediate	Intermediate	Greatest
Control erosion	Poorest	Fair	Fair	Intermediate	Good	Greatest
Opportunity to double crop	Poorest	Poor	Fair	Fair	Fair	Greatest
Requires learning new system	No	Least	Intermediate	Intermediate	High	Greatest
Support of traffic during wet harvest season	Poor	Poor	Good	Poor	Good	Greatest
Adaptable to poorly drained soils	Best	Fair	Fair	Intermediate	Good	Least
Minimize insect and disease problems	Best	Good	Good	Good	Intermediate	Least
Special planter required	No	No	No	No	Yes	Yes
Special problems with soil fertility	No	No	No	No	No	No

Table 144
INFLUENCE OF CROP ROTATION TILLAGE AND SOIL TYPE ON CORN GRAIN YIELD

Soil type	Rotation	Years 2—6			Remaining years		
		No-till	Plow	Prob[a]	No-till	Plow	Prob[a]
		kg/ha					
Wooster:	Continuous corn	7060	6370	0.001	9400	8420	<0.001
silt loam	Corn-soybeans	6470	5900	0.015	9480	8720	0.03
	Corn-oats-hay	6880	6850	ND	10450	9720	0.01
Hoytville:	Continuous corn	5720	6490	<0.001	6820	8000	<0.001
silty clay loam	Corn-soybeans	6720	7000	ND	7920	8260	ND
	Corn-oats-hay	6850	6800	ND	8180	8390	ND

[a] Probability level at which yield differences between years of tillage treatments is equal to zero. ND means no difference (*P*>0.02).

Table 145
EFFECT OF AGE OF GRASS ON SOIL AGGREGATION AND ON THE FLUID CONDUCTIVITY IN DISTURBED SOIL COLUMNS

Plots		Geometric mean diameter (μ) of aggregates		Water-stable aggregates >295 μ in diameter (%)	IAP[a](μ²)	HC[b] (cm/hr)
Time in grass (year)	No. of	Air-dry	Water-stable			
0	8	561 [c]	56.4 [c]	4.0 [c]	3.7[c]	2.2[c]
2	32	573	55.1	3.7	3.9	2.7
4	32	463	57.8	5.6	4.3	3.0
6	32	425	70.9	10.0	4.6	3.6
8	32	504	85.3	14.8	4.9	3.1[d]
20	6	524	140.9	27.1	5.3	4.6

[a] Intrinsic air permeability.
[b] Hydraulic conductivity.
[c] Where values are in brackets the differences between and among the treatments are not statistically significant at 1% (Duncan's multiple range test [4]); values not in brackets are significantly different from each other.
[d] Not significantly different from value 3.0.

From Mazurak, A. P. and Ramig, R. E., Aggregation and air-water permeabilities in a chernozem soil cropped to perennial grasses and fallow-grain, *Soil Science*, 94, 151—157, 1962, © by Williams & Wilkins 1962. With permission.

Table 146
EARTHWORMS AND SOIL AGGREGATION IN VARIOUS
ONE-YEAR CROPPING SYSTEMS

Annual Cropping Plan	Earthworms per acre number (thousands)	Weight (lb)	Soil aggregation (%)
Corn	51	59	25
Drilled soybeans for hay	98	70	31
Drilled soybeans for grain; residue turned in for winter cover	151	148	30
Drilled soybeans for grain; residue left on surface	332	337	34
Winter grain, idle during summer	238	135	48
Winter grain, legume during summer	394	258	60
Legume sod	342	306	69
Grass sod	423	226	70

From Hopp, H. and Hopkins, H. T., The effect of cropping systems on the winter population of earthworms, *J. Soil Water Conserv.*, 1, 85—88, 98, 1946. With permission.

Table 147

**SOME EXAMPLES OF IMPORTANT PLANT DISEASES
CONTROLLED BY FARM MANAGEMENT AND CULTURAL
PRACTICES**

Crop plant	Disease	Cultural control practice
Banana	"Moko"	Crop rotation
Bean	Bacterial blights	Crop rotation
	Anthracnose	Crop rotation
	White mold	Wide spacing
Cabbage	Club root	Lime soil (pH of 7.2)
	Black rot	Thermotherapy (seed)
	Mosaic	Weed control
Cacao	Pod rot	Sanitation
	Viral diseases	Sanitation
Celery	Early blight	Wide spacing
Citrus	Canker	Sanitation
Corn	Leaf blights	Sanitation
	Stalk rot	Sanitation
	Seedling blight	Plant in warm soil
Cotton	Angular leafspot	Crop rotation
	Charcoal rot	Irrigation ("out of season")
	Texas root rot	Acidify soil
Elm	Dutch elm disease	Sanitation
Flax	Rust	Sanitation
Oat	Halo blight	Crop rotation
Onion	Downy mildew	Crop rotation
	White rot	Irrigation ("out of season")
Pea	Root rot	Crop rotation
Pepper	Mosaic	Barrier crops (corn)
Pine	Blister rusts	Isolate from primary hosts
	Root diseases	Ensure infection by mycorrhizae
	Brown spot	Controlled burning (for long-leaved pine)
Potato	Scab	Crop rotation; acidify soil (pH of 5.5)
	Ring rot	Sanitation
	Soft rot	Enhance suberization of harvested tubers
	Golden nematode	Crop rotation
Rice	Blast	Sanitation
Rubber	Root rots	Soil amendments
Rye	Ergot	Crop rotation, sanitation
Sugarbeet	Leafspot	Wide spacing, crop rotation
Sugarcane	Ratoon slanting	Thermotherapy (of setts)
Sweet potato	Soft rot	Enhance suberization of harvested roots
Tobacco	Mosaic	Sanitation
	Black shank	Crop rotation flooding
	Black root rot	Crop rotation
	Bacterial wilt	Crop rotation
Vegetables (various species)	Cottony rot	Flooding
Wheat	Eyespot	Crop rotation
	Leaf blotch	Crop rotation, sanitation
	Take-all	Crop rotation
	Seedling blight	Plant in cool soil

Note: References included in the original.

Table 148
AVERAGE ANNUAL RAINFALL, RUNOFF, AND SOIL LOSS BY CROPPING TREATMENT

Treatment no.[a]	Cropping treatment	Slope (%)	Row direction[b]	Rainfall (in.)	Runoff (in.)	Soil loss (t/acre)
1	Fallow	7	(C)	55.71	18.35	60.43
2A	Cotton continuously	7	A	55.71	6.33	12.28
2B		7	W	55.71	9.84	21.86
3A	Corn continuously	7	A	55.71	4.54	4.02
3B		3	W	55.71	6.54	2.40
3B		7	W	55.71	7.03	12.11
3B		11	W	55.71	2.40	20.90
4A	3-year rotation of					
	1st coastal bermudagrass	7	(C)	54.78	2.42	1.99
	2d coastal bermudagrass	7	(C)	54.78	1.78	0.16
	Corn	7	W	54.78	4.42	4.43
	Rotation average	7	W	54.78	2.87	2.19
4B	3-year rotation of					
	1st coastal bermudagrass	7	(C)	54.78	2.25	2.26
	2d coastal bermudagrass	7	(C)	54.78	2.74	0.19
	Corn	7	A	54.78	2.99	1.08
	Rotation average	7	A	54.78	2.66	1.17
5	3-year rotation of					
	1st tall fescue	7	(C)	56.73	5.42	0.88
	2d tall fescue	7	(C)	56.73	2.80	0.09
	Corn	7	A	56.73	4.56	1.97
	Rotation average	7	A	56.73	4.26	0.98
6	4-year rotation of					
	1st tall fescue	7	(C)	57.56	11.04	4.05
	2d tall fescue	7	(C)	57.56	5.07	0.05
	Corn	7	A	57.56	6.04	2.93
	Cotton	7	A	57.56	7.23	6.25
	Rotation average	7	A	57.56	7.34	3.32

[a] Years of record: treatment numbers 1 to 3, 1961—67; treatment number 4, 1962—67; treatment numbers 5 and 6, 1961—65.
[b] A = rows on contour across slope; W = rows uphill and downhill with the slope.
[c] No rows.

From Carreker, J. R., Wilkinson, S. R., Barnett, A. P., and Box, J. F., Soil and water management systems for sloping land, *U.S. Dep. Agric. Agric. Res. Serv.,* Rep. 160, 1—76, 1977.

Table 149
AVERAGE ANNUAL RAINFALL, RUNOFF, AND SOIL LOSS FROM CROPS

	Years of record	Rainfall (in.)	Runoff (in.)	Soil loss (t/acre)
Continuous fallow	1960—1966	48.53	11.84	5.14
Continuous peanuts	1951—1958	42.92	2.88	1.23
Continuous corn	1955—1966	47.05	1.24	0.77
Meadow (bermudagrass and clover)	1951—1954	43.66	0.35	0.34
Three-year rotation	1951—1966	46.20	1.83	0.98
Oats	1951—1966	46.20	1.20	0.51
Peanuts	1951—1966	46.20	1.83	1.32
Corn	1951—1966	46.20	2.47	1.12
Four-year rotation	1952—1966	46.09	1.96	0.71
1st year meadow	1952—1966	46.09	3.11	0.92
2nd year meadow	1953—1966	46.33	0.31	0.46
Corn	1952—1966	46.09	0.80	0.56
Peanuts	1952—1966	46.09	1.65	0.91

From Thomas, A. W., Carreker, J. R., and Carter, R. L., Water, soil, and nutrient losses on Tifton loamy sand, *Ga. Agric. Exp. Stn. Res. Bull.,* 64, 1—33, 1969.

Table 150

GENERALIZED VALUES OF THE COVER AND MANAGEMENT FACTOR, C, IN THE 37 STATES EAST OF THE ROCKY MOUNTAINS[a]

		Productivity level[b]	
		High	Mod.
Line no.	Crop, rotation, and management[c]	C value	
	Base value: continuous fallow, tilled up and down slope	1.00	1.00
Corn			
1	C, RdR, fall TP, conv (1)	0.54	0.62
2	C, RdR, spring TP, conv (1)	0.50	0.59
3	C, RdL, fall TP, conv (1)	0.42	0.52
4	C, RdR, wc seeding, spring TP, conv (1)	0.40	0.49
5	C, RdL, standing, spring TP, conv (1)	0.38	0.48
6	C, fall shred stalks, spring TP, conv (1)	0.35	0.44
7	C(silage)-W(RdL, fall TP) (2)	0.31	0.35
8	C, RdL, fall chisel, spring disk, 40—30% rc (1)	0.24	0.30
9	C(silage), W wc seeding, no-till pl in c-k W (1)	0.20	0.24
10	C(RdL)-W(RdL, spring TP) (2)	0.20	0.28
11	C, fall shred stalks, chisel pl, 40—30% rc (1)	0.19	0.26
12	C-C-C-W-M, RdL, TP for C, disk for W (5)	0.17	0.23
13	C, RdL, strip till row zones, 55—40% rc (1)	0.16	0.24
14	C-C-C-W-M-M, RdL, TP for C, disk for W (6)	0.14	0.20
15	C-C-W-M, RdL, TP for C, disk for W (4)	0.12	0.17
16	C, fall shred, no-till pl, 70—50% rc (1)	0.11	0.18
17	C-C-W-M-M, RdL, TP for C, disk for W (5)	0.087	0.14
18	C-C-W-M, RdL, no-till pl 2d & 3rd C (5)	0.076	0.13
19	C-C-W-M, RdL, no-till pl 2d C (4)	0.068	0.11
20	C, no-till pl in c-k wheat, 90—70% rc (1)	0.062	0.14
21	C-C-C-W-M-M, no-till pl 2d & 3rd C (6)	0.061	0.11
22	C-W-M, RdL, TP for C, disk for W (3)	0.055	0.095
23	C-C-W-M-M, RdL, no-till pl 2d C (5)	0.051	0.094
24	C-W-M-M, RdL, TP for C, disk for W (4)	0.039	0.074
25	C-W-M-M-M, RdL, TP for C, disk for W (5)	0.032	0.061
26	C, no-till pl in c-k sod, 95—80% rc (1)	0.017	0.053
Cotton[d]			
27	Cot, conv (Western Plains) (1)	0.42	0.49
28	Cot, conv (South) (1)	0.34	0.40
Meadow			
29	Grass and legume mix	0.004	0.01
30	Alfalfa, lespedeza, or Sericia	0.020	
31	Sweet clover	0.025	
Sorghum, grain (Western Plains)[d]			
32	RdL, spring TP, conv (1)	0.43	0.53
33	No-till pl in shredded 70—50% rc	0.11	0.18
Soybeans[d]			
34	B, RdL, spring TP, conv (1)	0.48	0.54
35	C-B, TP annually, conv (2)	0.43	0.51
36	B, no-till pl	0.22	0.28
37	C-B, no-till pl, fall shred C stalks (2)	0.18	0.22

Table 150 (continued)
GENERALIZED VALUES OF THE COVER AND MANAGEMENT
FACTOR, C, IN THE 37 STATES EAST OF THE ROCKY
MOUNTAINS[a]

Line no.	Crop, rotation, and management[c]	Productivity level[b] High C value	Mod.
	Base value: continuous fallow, tilled up and down slope	1.00	1.00
Wheat			
38	W-F, fall TP after W (2)	0.38	
39	W-F, stubble mulch, 500 lbs rc (2)	0.32	
40	W-F, stubble mulch, 1000 lbs rc (2)	0.21	
41	Spring W, RdL, Sept TP, conv (N. and S. Dak.) (1)	0.23	
42	Winter W, RdL, Aug TP, conv (Kans.) (1)	0.19	
43	Spring W, stubble mulch, 750 lbs rc (1)	0.15	
44	Spring W, stubble mulch, 1250 lbs rc (1)	0.12	
45	Winter W, stubble mulch, 750 lbs rc (1)	0.11	
46	Winter W, stubble mulch, 1250 lbs rc (1)	0.10	
47	W-M, conv (2)	0.054	
48	W-M-M, conv (3)	0.026	
49	W-M-M-M, conv (4)	0.021	

[a] This table is for illustrative purposes only and is not a complete list of cropping systems or potential practices. Values of C differ with rainfall pattern and planting dates. These generalized values show approximately the relative erosion-reducing effectiveness of various crop systems, but locationally derived C values should be used for conservation planning at the field level. Tables of local values are available from the Soil Conservation Service.

[b] High level is exemplified by long-term yield averages greater than 75 bu corn or 3 tons grass-and-legume hay; or cotton management that regularly provides good stands and growth.

[c] Numbers in parentheses indicate number of years in the rotation cycle. No. (1) designates a continuous one-crop system.

[d] Grain sorghum, soybeans, or cotton may be substituted for corn in lines 12, 14, 15, 17-19, 21-25 to estimate C values for sod-based rotations.

Abbreviations defined:

B	— soybeans	F	— fallow
C	— corn	M	— grass and legume hay
c-k	— chemically killed	pl	— plant
conv	— conventional	W	— wheat
cot	— cotton	wc	— winter cover

lbs rc	— pounds of crop residue per acre remaining on surface after new crop seeding
% rc	— percentage of soil surface covered by residue mulch after new crop seeding
70-50% rc	— 70% cover for C values in first column; 50% for second column
RdR	— residues (corn stover, straw, etc.) removed or burned
RdL	— all residues left on field (on surface or incorporated)
TP	— turn plowed (upper 5 or more inches of soil inverted, covering residues)

From Control of Water Pollution from Cropland, Vol. 1, A Manual for Guideline Development, Agricultural Research Service, U.S. Department of Agriculture, Washington, D.C., 1975.

PEST AND OTHER LOSSES

LOSSES IN PLANTS

Table 151
PERCENTAGE YIELD LOSS FROM DEAD SOYBEAN PLANTS
(ALL VEGETATIVE STAGES)

Orig. Stand Plants/Acre (1000)	REMAINING PLANTS PER ACRE (1000)																						
	120	115	110	105	100	95	90	85	80	75	70	65	60	55	50	45	40	35	30	25	20	15	10
125+	1	2	3	4	6	8	10	12	14	16	18	21	24	27	30	33	36	40	44	49	54	59	65
120	0	1	2	3	5	7	9	11	13	15	17	20	23	26	29	32	35	39	43	48	53	58	64
115			0	1	2	4	6	8	10	12	14	16	19	22	25	28	31	34	38	42	47	52	57	63
110				0	1	3	5	7	9	11	13	15	18	21	24	27	30	33	37	41	46	51	56	62
105					0	1	3	5	7	9	11	13	16	19	22	25	28	31	35	39	44	49	54	61
100						0	1	3	5	7	9	11	14	17	20	23	26	29	33	37	42	47	52	59
95							0	1	3	5	7	9	12	15	18	21	24	27	31	35	40	45	50	57
90								0	1	3	5	7	10	13	16	19	22	25	29	33	38	43	48	55
85									0	2	4	6	9	12	15	18	21	24	28	32	37	42	47	54
80										0	2	4	7	10	13	16	19	22	26	30	35	40	45	52
75											0	2	5	8	11	14	17	20	24	28	33	38	43	50
70												0	3	6	9	12	15	18	21	25	30	35	41	48
65													0	3	6	9	12	16	19	23	27	32	39	46
60														0	3	7	10	13	16	20	25	30	37	45
55															0	3	6	10	13	17	22	27	34	43
50																0	4	8	12	16	20	25	32	41
45																	0	4	8	13	18	24	31	40
40																		0	5	11	17	23	30	39

From Soybean Loss Instructions, NClA 6302, revised, National Crop Insurance Association, Colorado Springs, Col., 1987. With permission.

Table 152
GROSS PERCENTAGE YIELD LOSS FROM SOYBEAN LEAF AREA
DESTROYED (DETERMINATE SOYBEANS)

DEFOLIATION %

Stages	10	15	20	25	30	35	40	45	50	55	60	65	70	75	80	85	90	95	100
V9-V12	0	0	0	0	0	0	0	3	4	4	5	6	7	8	8	8	9	9	10
V13-V^n	0	0	0	0	0	0	3	4	8	9	9	10	11	12	14	16	19	22	25
R1-R2	0	0	0	0	0	3	6	8	11	12	13	14	15	17	20	26	32	36	40
R2.5	0	0	0	0	3	5	6	8	11	12	13	15	16	18	22	30	36	40	45
R3	0	0	0	3	5	6	7	9	12	13	14	16	17	20	25	35	40	45	50
R3.5	0	0	3	5	6	7	8	10	12	13	15	17	18	21	28	36	41	47	63
R4	0	3	5	6	7	8	9	11	12	14	16	18	19	22	30	37	43	49	76
R4.5	3	4	5	6	7	8	10	12	13	15	17	19	22	24	34	40	46	58	80
R5	3	4	5	7	8	9	11	13	15	16	18	20	23	26	35	44	50	66	84
R5.5	3	4	5	7	8	9	11	13	15	16	18	20	23	26	35	44	50	66	84
R6	2	3	4	5	6	7	8	9	11	12	13	15	17	19	25	32	36	49	62

Note: V = vegetative stage; R = reproductive stage.

Table 153

SOYBEAN PLANTS PER ACRE BASED ON NUMBER OF PLANTS IN 10 FEET OF ROW (CONVERSION TABLE)

		ROW WIDTH (inches)																	
		40	38	36	34	32	30	28	26	24	22	20	18	16	14	12	10	8	7	6
P	125	96	91	86	81	77	72	67	62	57	53	48	43	38	33	29	24	19	17	14
L	120	92	87	83	78	73	69	64	60	55	51	46	41	37	32	28	23	18	16	14
A	115	88	84	79	75	70	66	62	57	53	48	44	40	35	31	26	22	18	15	13
N	110	84	80	76	72	67	63	59	55	51	46	42	38	34	29	25	21	17	15	13
T	105	80	76	72	68	64	60	56	52	48	44	40	36	32	28	24	20	16	14	12
S	100	77	73	69	65	61	57	54	50	46	42	38	34	31	27	23	19	15	13	11
	95	73	69	65	62	58	55	51	47	44	40	36	33	29	25	22	18	15	13	11
	90	69	65	62	59	55	52	48	45	41	38	34	31	28	24	21	17	14	12	10
	85	65	62	59	55	52	49	46	42	39	36	33	29	26	23	20	16	13	11	10
P	80	61	58	55	52	49	46	43	40	37	34	31	28	24	21	18	15	12	11	9
E	75	57	55	52	49	46	43	40	37	34	32	29	26	23	20	17	14	11	10	9
R	70	54	51	48	46	43	40	37	35	32	29	27	24	21	19	16	13	11	9	8
	65	50	47	45	42	40	37	35	32	30	27	25	22	20	17	15	12	10	9	7
	60	46	44	41	39	37	34	32	30	28	25	23	21	18	16	14	11	9	8	7
	55	42	40	38	36	34	32	29	27	25	23	21	19	17	15	13	11	8	7	6
A	50	38	36	34	33	31	29	27	25	23	21	19	17	15	13	11	10	8	7	6
C	45	34	33	31	29	28	26	24	22	21	19	17	15	14	12	10	9	7	6	5
R	40	30	29	27	26	24	23	21	19	18	17	15	14	13	11	9	8	6	5	4
E	35	27	25	24	23	21	20	19	17	16	15	13	12	11	9	8	7	5	5	4
	30	23	22	21	20	18	17	16	15	14	13	11	10	9	8	7	6	5	4	3
	25	19	18	17	16	15	14	13	12	11	11	10	9	8	7	6	5	4	3	3
	20	15	15	14	13	12	11	11	10	9	8	8	7	6	5	5	4	3	3	2
	15	11	11	10	10	9	9	8	7	7	6	6	5	5	4	3	3	2	2	2
(1000)	10	8	7	7	7	6	6	5	5	5	4	4	3	3	3	2	2	2	1	1

Instructions: Count the number of plants in a representative 10 ft row. Find the number nearest that total in the appropriate row width column. Then go across to the far left column to find the number of plants per acre. If the number of plants counted is greater than the top number in the row width column, divide the number of plants by 2, proceed as above, and multiply the plants per acre found in the left column by 2 to arrive at the actual number of plants per acre.

Example 1:	**Example 2:**
Row width = 30″	Row width = 22″
50 original plants in 10′ of row	68 plants in 10′ of row
49 is nearest in 30″ column	68 ÷ 2 = 34
Original plants per acre = 85,000	80,000 × 2 = 160,000 plants per acre

From National Crop Insurance Association Soybean Loss Instructions, NC1A 6302 (revised), Colorado Springs, 1987. With permission.

Table 154

SUMMARY OF DATA ON THE EFFECT OF INSECT DENSITY AND OF CONTROL MEASURES ON YIELDS OF VARIOUS COMMODITIES

Crop/pest	No. of data sets	Calculated losses (%) With best control	Calculated losses (%) Without control (untreated check)	Yield index (%) with no pest infestation Y_0	Slope k	Coefficient of determination r^2
Alfalfa						
Alfalfa weevil	3	0.53 ± 0.24	27.0 ± 22.1	151.0 ± 32.7	−0.82 ± 0.43	0.32 ± 0.15
Aphids	4	1.25 ± 0.75	48.75 ± 27.0	119.76 ± 28.0	−1.25 ± 0.79	0.55 ± 0.21
Clover root curculio	2	2.5 ± 2.5	6.0 ± 6.0	108.0 ± 1.0	−0.05 ± 0.09	0.19 ± 0.18
Leafhoppers	4	2.59 ± 1.86	26.3 ± 15.5	155.8 ± 48.2	−0.36 ± 0.25	0.10 ± 0.01
Meadow spittle-bug	1	1.0	8.0	111	−0.09	0.03
Mirid bugs	2	4.5 ± 0.5	19.5 ± 12.5	192.5 ± 72.5	−0.28 ± 0.10	0.12 ± 0.11
Alfalfa (seed)						
Leafhoppers	5	6.80 ± 4.27	14.00 ± 8.58	127.60 ± 17.69	−0.20 ± 0.16	0.23 ± 0.10
Leafminers	5	1.45 ± 1.39	4.2 ± 2.35	112 ± 4.51	0.06 ± 0.03	0.05 ± 0.03
Lygus bugs	2	3.5 ± 0.5	43.5 ± 8.50	323.5 ± 204.5	−1.57 ± 1.16	0.48 ± 0.10
Mirids	1	0	0	186	.00026	0.000
Thrips	1	12	38	135	−0.51	0.77
Apples						
Codling moth	1	16	100	134	−5.32	0.79
Red-banded leaf-roller	2	0	0	84.5 ± 10.5	1.06	0.38 ± 0.01
Barley						
Greenbug	1	7	84	1741.63	−14.55	0.04
Beans						
Bean thrips	2	0	0	85.5 ± 8.5	0.24 ± 0.16	0.24 ± 0.12
Corn earworm	3	6.0 ± 3.5	37.0 ± 31.7	446.3 ± 300.9	−6.01 ± 6.29	0.36 ± 0.19
European corn borer	1	0	5	102	−0.05	0.03
Leafhoppers	2	1 ± 0	3.0 ± 1.0	103 ± 1.0	−0.04 ± 0.01	0.03 ± 0.01

Table 154 (continued)
SUMMARY OF DATA ON THE EFFECT OF INSECT DENSITY AND OF CONTROL MEASURES ON YIELDS OF VARIOUS COMMODITIES

Crop/pest	No. of data sets	Calculated losses (%) With best control	Calculated losses (%) Without control (untreated check)	Yield index (%) with no pest infestation Y_o	Slope k	Coefficient of determination r^2
Lygus bugs	4	0 ± 0	0 ± 0	86.8 ± 3.86	0.33 ± 0.12	0.29 ± 0.14
Mexican bean beetle (lima)	2	5.0 ± 3.0	97.5 ± 2.50	164.50 ± 64.50	-2.13 ± 0.06	0.27 ± 0.16
Mexican bean beetle (snap)	8	8.75 ± 5.86	29.00 ± 15.98	171.13 ± 26.02	-0.41 ± 0.85	0.32 ± 0.12
Potato leafhopper	2	0 ± 0	41.0 ± 33.0	100.5 ± 9.5	-0.38 ± 0.29	0.53 ± 0.15
Spider mites	4	1.51 ± 0.95	23.5 ± 8.81	199.8 ± 63.0	-0.04 ± 0.77	0.29 ± 0.20
Tarnished plant bugs	1	0	0	239	7.51	0.32
Wireworms	2	3 ± 0	47.0 ± 39.0	115.0 ± 8.0	-0.57 ± 0.49	0.48 ± 0.37
Cabbage						
Cabbage looper	3	0.75 ± 0.75	9.50 ± 5.72	234.0 ± 50.38	0.12 ± 0.51	0.09 ± 0.03
Imported cabbage worm	2	0 ± 0	41.0 ± 41.0	297.50 ± 29.50	-0.96 ± 1.23	0.14 ± 0.04
Carrot						
Carrot rust fly	1	2	97	2761	-26.71	0.9
Celery						
Leafminers	1	0	0	84	0.17	0.07
Cole crops						
Aphids	2	1.5 ± 0.5	56.0 ± 23.0	288.5 ± 96.5	-1.84 ± 1.21	0.58 ± 0.29
Cabbage looper	1	0	0	141	0.46	0.44
Caterpillars	7	11.5 ± 11.5	32.0 ± 19.0	383.0 ± 97.0	-1.41 ± 1.04	0.12 ± 0.10
Diamond back moth	2	0.5 ± 0.5	0.5 ± 0.5	139.0 ± 16.0	-0.01 ± 0.01	0.04 ± 0.04
Looper and cabbageworm	1	3	25	132	-0.34	0.08

Collards					
Cabbage looper	1	0	49	664	-3.29	0.22
Cabbage web-worm	1	0	21	650	-1.39	0.5
Corn, field						
Ant	1	3	20	138	-0.28	0.61
Corn rootworms	12	4.75 ± 1.23	16.17 ± 4.18	0.42 ± 10.58	-0.27 ± 0	0.25 ± 0.08
Garden symphylid	1	7	14	133	-0.19	0.31
Southwestern corn borer	11	9.91 ± 2.71	34.45 ± 10.40	146.64 ± 17.91	-0.48 ± 0.49	0.60 ± 0.10
Sugarcane borer	1	0	0	109	0.35	0.42
White grubs	1	11	43	149	-0.65	0.86
Wireworms	1	18.36	48.31	296.0	-1.43	0.34
Leafhopper on silage corn	3	38.33 ± 14.81	74.67 ± 15.72	308.67 ± 100.65	-2.47 ± 1.00	0.69 ± 0.17
Corn, sweet						
Cutworms	1	7	22	132	-0.29	0.92
European corn-borer	3	0.33 ± 0.33	4.33 ± 4.33	107.33 ± 3.18	0.14 ± 0.19	0.42 ± 0.30
Fall armyworm	2	27.0 ± 1.0	67.5 ± 1.5	1130.0 ± 898.0	-7.81 ± 6.28	0.62 ± 0.06
Rootworms	2	0 ± 0	45.5 ± 45.5	145.5 ± 26.5	-0.66 ± 0.92	0.21 ± 0.15
Spider mites	1	0	0	95	0.10	0.03
Wireworms	2	4.0 ± 2.0	29.0 ± 19.0	137.0 ± 17.0	-0.36 ± 0.21	0.51 ± 0.50
Cotton						
Aphids	3	0.33 ± 0.33	22.3 ± 22.3	133.7 ± 22.2	0.07 ± 0.68	0.36 ± 0.25
Boll weevil	17	19.00 ± 6.69	30.94 ± 8.63	176.42 ± 50.68	-0.40 ± 1.01	0.37 ± 0.08
Cabbage looper	1	20	35	152	-0.53	0.63
Cotton aphid	4	0.0005 ± 0.0005	4.75 ± 4.75	121.0 ± 59.36	0.98 ± 0.80	0.25 ± 0.20
Cotton fleahopper	4	12.5 ± 3.88	39.00 ± 11.57	185.75 ± 30.1	-0.81 ± 0.37	0.59 ± 0.18
Cotton leaf perforator	1	0	0	122	0.1	0.08

Table 154 (continued)

SUMMARY OF DATA ON THE EFFECT OF INSECT DENSITY AND OF CONTROL MEASURES ON YIELDS OF VARIOUS COMMODITIES

Crop/pest	No. of data sets	Calculated losses (%)		Yield index (%) with no pest infestation Y_o	Slope k	Coefficient of determination r^2
		With best control	Without control (untreated check)			
Heliothis	29	12.07 ± 2.17	90.82 ± 14.21	315.76 ± 55.15	-2.00 ± 0.58	0.59 ± 0.06
Pink bollworm	2	10.0 ± 6.0	35.5 ± 13.5	169.0 ± 30.0	-0.64 ± 0.34	0.50 ± 0.25
Spider mites	2	0 ± 0	0 ± 0	84.0 ± 3.0	4.70 ± 4.10	0.60 ± 0.30
Thrips	4	18.01 ± 6.08	67.75 ± 22.40	167.25 ± 23.14	-1.29 ± 0.50	0.46 ± 0.21
Grapes						
Grape phylloxera	2	1.50 ± 0.50	6.50 ± 2.50	108.50 ± 4.50	-0.07 ± 0.02	0.04 ± 0.02
Lettuce						
Cabbage looper	1	2	81	107.56	-87.56	0.83
Oats						
Cereal leaf beetle	2	0.65 ± 0.35	26.00 ± 21.00	163.00 ± 35.00	-0.50 ± 0.43	0.52 ± 0.48
Onion						
Onion maggot	1	15	31	164	-0.51	0.56
Onion thrips	6	7.50 ± 3.59	54.67 ± 11.76	822.83 ± 659.80	-19.43 ± 18.66	0.63 ± 0.14
Pasture and range						
Capsids on blue-grass	1	3	26	126	-0.33	0.70
Fall armyworm	1	0	0	102	0.19	0.25
Leafhoppers	1	33	100	170	-1.93	0.90
Peanuts						
Cornstalk borer	1	2	7	117	-0.08	0.1
Peanut worm	1	0	0	74	0.23	0.17
Southern corn rootworm	1	1.0 ± 0	15.50 ± 1.50	119.0 ± 3.0	-0.18 ± 0.01	0.28 ± 0.05
Thrips	6	0.35 ± 0.21	7.83 ± 6.31	100.17 ± 4.56	0.01 ± 0.09	0.24 ± 0.10

Peas						
Aphids	3	4.0 ± 2.31	48.67 ± 22.33	123.67 ± 10.90	−0.45 ± 0.25	0.12 ± 0.06
Pea moth on field peas	2	2.5 ± 2.5	9.0 ± 9.0	107.50 ± 1.50	−0.13 ± 0.06	0.06 ± 0.01
Peas (southern)						
Beet armyworm	1	18	54	207	−1.13	0.52
Thrips	1	20	58	273	−1.58	0.76
Peppers						
Aphids	1	0	24	151	−0.36	0.11
Corn earworm	2	0.5 ± 0.5	48.50 ± 48.50	228.0 ± 100.0	−0.14 ± 0.18	0.38 ± 0.38
European corn borer	6	0.0 ± 0.0	54.50 ± 19.89	171.33 ± 36.12	−3.27 ± 2.18	0.25 ± 0.15
Fall armyworm	2	0 ± 0	0 ± 0	124.50 ± 19.50	0.16 ± 0.08	0.02 ± 0.02
Pepper weevil	1	4	37	145	−0.54	0.2
Potatoes						
Aphids	13	0.54 ± 0.37	15.00 ± 7.45	124.00 ± 6.44	−0.17 ± 0.13	0.14 ± 0.05
Armyworms	2	1.28 ± 0.98	59.0 ± 17.0	745.0 ± 197	−4.74 ± 2.42	0.45 ± 0.31
Colorado potato beetle	13	1.04 ± 0.58	46.62 ± 10.38	450.85 ± 101.19	−3.05 ± 1.13	0.48 ± 0.08
European corn borer	5	1.50 ± 0.65	54.25 ± 26.64	598.20 ± 251.35	−3.41 ± 1.69	0.34 ± 0.08
Flea beetles	7	0.74 ± 0.35	43.29 ± 10.30	175.29 ± 46.10	−0.81 ± 0.52	0.44 ± 0.08
Green peach aphid	3	0 ± 0	3.67 ± 3.67	120.00 ± 4.04	−0.08 ± 0.04	0.16 ± 0.04
Leafhoppers	3	0 ± 0	36.00 ± 26.63	139.00 ± 9.50	−10.19 ± 9.56	0.44 ± 0.22
Potato aphid	1	0.001	5	118	−0.06	0.03
Potato leafhopper	6	0.37 ± 0.20	43.17 ± 18.62	162.50 ± 17.98	−0.85 ± 0.43	0.39 ± 0.13
Potato psyllid	3	1.0 ± 1.0	32.0 ± 17.2	271.7 ± 124.6	−0.52 ± 0.17	0.12 ± 0.08
Potato tuberworm	1	1	91	560	−5.1	0.39
Spider mites	1	4	5	84	−0.04	0.09
Wireworms	14	0.22 ± 0.21	4.86 ± 3.47	111.07 ± 4.70	−0.08 ± 0.06	0.11 ± 0.03
Rape (seed)						
Cabbage aphid	1	0	100	146	−46	0.75
Rice						
Leafhopper	1	9	42	133	−0.56	1.0
Rice water weevil	8	1.51 ± 1.50	5.25 ± 4.55	107.50 ± 3.97	−0.01 ± 0.08	0.18 ± 0.10

Table 154 (continued)
SUMMARY OF DATA ON THE EFFECT OF INSECT DENSITY AND OF CONTROL MEASURES ON YIELDS OF VARIOUS COMMODITIES

Crop/pest	No. of data sets	Calculated losses (%)		Yield index (%) with no pest infestation Y_0	Slope k	Coefficient of determination r^2
		With best control	Without control (untreated check)			
Safflower						
Lygus	1	3	21	109	−0.23	0.47
Thrips	2	3.5 ± 3.5	16 ± 16	122 ± 17.0	−0.06 ± 0.39	0.62 ± 0.29
Sorghum						
Chinch bug	1	0	0	139	+1.29	1.0
Cornleaf aphid	2	0 ± 0	0 ± 0	126.50 ± 9.50	0.18 ± 0.17	0.21 ± 0.03
Greenbug	18	2.80 ± 1.55	19.39 ± 7.16	119.94 ± 4.51	−0.24 ± 0.17	0.35 ± 0.07
Mites	3	0 ± 0	0 ± 0.	97.33 ± 4.26	0.22 ± 0.12	0.54 ± 0.02
Sorghum midge	2	0.2 ± 0.2	5.0 ± 5.0	136.00 ± 31.00	+0.11 ± 0.27	0.16 ± 0.14
Sorghum web-worm	1	0	3	5184	−1.44	0.8
Southwestern corn borer	1	4	24	186	−0.45	0.60
White grubs	3	9.33 ± 4.70	39.00 ± 26.27	592.33 ± 446.64	−4.54 ± 4.37	0.59 ± 0.20
Soybeans						
Caterpillar com-plex	1	0	46	198	−0.91	0.98
Corn earworm	2	0.65 ± 0.35	1.50 ± 0.50	100.00 ± 8.00	−0.02 ± 0.01	0.42 ± 0.41
Green cloverworm	4	3.00 ± 2.98	8.75 ± 6.55	117.75 ± 9.98	−0.12 ± 0.10	0.26 ± 0.21
Looper	3	10.50 ± 3.50	25.50 ± 5.50	130.67 ± 11.62	−0.24 ± 0.13	0.34 ± 0.17
Mexican bean bee-tle	7	0.41 ± 0.28	26.00 ± 13.33	121.57 ± 16.59	−0.77 ± 0.73	0.44 ± 0.13
Mites	1	0	89	221	−1.97	0.93
Soybean looper	6	4.83 ± 1.92	15.67 ± 7.06	138.67 ± 21.45	−0.28 ± 0.15	0.5 ± 0.14
Stink bugs	2	8.5 ± 7.5	15.00 ± 2.00	120.50 ± 0.50	−0.19 ± 0.03	0.12 ± 0.04
Velvet bean cater-pillar	10	2.35 ± 0.94	16.64 ± 3.73	143.64 ± 13.23	−0.30 ± 0.09	0.51 ± 0.09

Squash					
Squash vineborer	1	0	0	94	0.02	0.0005
Strawberry					
Spider mites	2	2.0 ± 1.0	23.00 ± 8.00	191.00 ± 63.00	−0.49 ± 0.30	0.34 ± 0.27
Spittlebugs	1	0	0	113	0.11	0.06
Sugar beets					
Root aphids	1	0	0	105	0.4	0.03
Root maggot	6	8.17 ± 2.89	22.67 ± 6.82	173.80 ± 22.29	−0.44 ± 0.17	0.37 ± 0.09
Webworms	1	0	0	101	1.0	0.02
Sugarcane					
Sugarcane borer	5	5.74 ± 2.51	28.6 ± 12.36	127.80 ± 5.77	−0.45 ± 0.16	0.39 ± 0.17
Sunflowers					
Sunflower moth	15	7.13 ± 2.07	40.27 ± 8.66	222.40 ± 30.96	−1.83 ± 0.76	0.58 ± 0.08
Sweet potatoes					
Flea beetle	1	0	0	56	0.46	0.02
Sweet potato weevil	3	1.67 ± 1.20	29.67 ± 3.71	149.00 ± 21.66	−0.46 ± 0.13	0.34 ± 0.05
Wireworms	1	16	38	130	−0.50	0.08
Tobacco					
Aphids	3	0.33 ± 0.33	0.20 ± 0.20	112.00 ± 7.64	14.40 ± 14.30	0.67 ± 0.21
Budworm	3	1.33 ± 0.88	13.00 ± 6.66	128.00 ± 15.04	−0.11 ± 0.18	0.61 ± 0.09
Flea beetle	1	0	0.24	137	−0.32	0.23
Green peach aphid	6	0.04 ± 0.03	22.17 ± 13.45	184.67 ± 48.13	0.56 ± 1.39	0.43 ± 0.18
Loopers	1	4	10	143	−0.15	0.12
Tomato					
Aphids	2	0 ± 0	6.50 ± 4.50	105.00 ± 13.00	−0.08 ± 0.06	0.09 ± 0.01
Armyworms	1	0	0	85	0.08	0.01
Cabbage looper	1	0	15	125	−0.19	0.36
Colorado potato beetle	1	0	93	805	−7.48	0.81
Cutworms	1	0.3	16	123	−0.2	0.36
Leafminers	1	10	38	124	−0.47	0.59
Tomato pinworm	2	5.0 ± 5.0	50 ± 50	201.50 ± 118.50	−0.96 ± 0.97	0.42 ± 0.39
Whiteflies	1	12	38	143	−0.55	0.99
Worms (horn-worms and fruit-worms)	1	0	0	89	0.06	0.02

Table 154 (continued)

SUMMARY OF DATA ON THE EFFECT OF INSECT DENSITY AND OF CONTROL MEASURES ON YIELDS OF VARIOUS COMMODITIES

Crop/pest	No. of data sets	Calculated losses (%) With best control	Calculated losses (%) Without control (untreated check)	Yield index (%) with no pest infestation Y_0	Slope k	Coefficient of determination r^2
Wheat						
Banks grass mite	1	18	61	120	−0.73	0.02
Brown wheat mite	1	21	100	193	−6.47	0.33
Cutworms	3	7.67 ± 3.18	54.67 ± 12.44	164.00 ± 48.00	−0.73	0.52 ± 0.11
Greenbug	2	0 ± 0	0 ± 0	266.50 ± 147.50	1.51 ± 1.47	0.01 ± 0.004
Hessian fly	3	0 ± 0	0 ± 0	121.00 ± 7.02	0.13 ± 0.07	0.04 ± 0.02

Note: References included in original.

Note: Regression equations were calculated to establish the mathematical relationship among yield, pest density, and losses. The regression equations have the form $Y = Y_0 + kP$, where Y = yield index, P = pest population index, Y_0 and k are constants, and r^2 is the coefficient of determination.

Source: *J. Econ. Entomol.* 1942—1978 and *Insecticide and Acaricide Tests* 1976—1978.

Table 155

EFFECT OF CROP PLANTING TIME ON INSECT ABUNDANCE AND/OR POTENTIAL DAMAGE[a]

Corn[b]

Insect	Early	Late
Seed-corn beetle[c]	+	-
Seed-corn maggot	+	-
Wireworm	+	-
White grub	?	-
Corn root aphid	-	+
Crape colaspis	+	-
N. corn rootworm[c]	-	+
W. corn rootworm[c]	-	+
S. corn rootworm	-	+
Cutworms	0	0
Billbug	0	0
Thrips	-	+
Mite	-	+
Eur. corn borer	+	-
S. W. corn borer	-	+
Corn earworm	-	+
Fall armyworm	-	+
Armyworm	+	-
Chinch bug	-	+
Corn leaf aphid	-	+
Grasshoppers	?	+

Sorghum

Insect	Early	Late
Wireworm	+	-
White grub	+	-
Flea beetles	-	+
Chinch bug	-	+
Thrips	+	-
Cutworms	-	+
S. corn rootworm	-	+
Greenbug	+	-
Corn leaf aphid	-	+
Fall armyworm	-	+
Beet armyworm	-	+
Corn earworm	-	+
Sorghum midge	-	+
Sorghum webworm	-	+
False chinch bug	?	?
Leaf-footed plant bug	-	+
Stink bugs	-	+
S. W. corn borer	-	+
Sugarcane borer	-	+
Sugarcane rootstock weevil	0	0
Spider mites	+	1
Yellow sugarcane aphid	+	

Cotton

Insect	Early	Late
Thrips	+	-
Cotton aphid	+	-
Cotton fleahopper	+	-
Boll weevil	-	+
Bollworm	-	+
Tobacco budworm	-	+
Pink bollworm	-	+
Cotton leafworm	-	+
Cabbage looper	-	+
Lygus bug	-	+
Stink bug	-	+
Spider mites	-	+
Cotton square borer	+	-
Salt marsh cat.	0	0
Garden webworm	-	+
Cutworms	+	-
Leaf miners	-	+

Wheat

Insect	Early	Late
Wireworm	+	-
White grub	+	-
Grasshoppers	+	-
Hessian fly	+	-
Greenbug	+	-
Corn leaf aphid	+	-
Oat bird-cherry aphid	+	-
Cutworms	+	-
Armyworms	+	-
Chinch bug	+	-
Wheat strawworm	0	0
Wheat stem maggot	0	0
Common stalk borer	+	0
Lesser cornstalk borer	0	0
Flea beetles	+	-
Leafhoppers	+	-
Leaf-feeding sawflies	?	?
Thrips	+	-
Wheat curl mite	+	-
Wheat stem sawfly	0	0
Wheat jointworm	0	0

Note: + = Increase in population density or damage, − = reduction in population density or damage, 0 = no effect, ? = effect unknown.

[a] Compiled by the author in consultation with entomologists familiar with the subject. Most of the information presented is from published research results, but some may represent the views of entomologists.

[b] North Central States.

[c] Influence is different in Southeast.

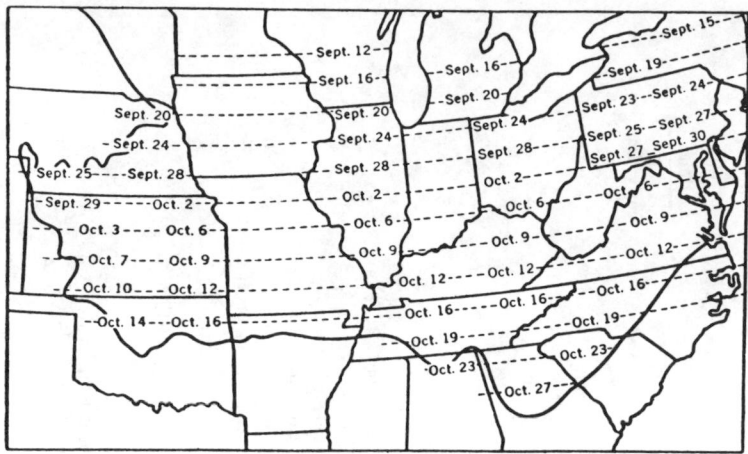

FIGURE 13.   Fly-free dates for sowing winter wheat to avoid Hessian fly damage, illustrating effectiveness of delayed planting. (From Walton, W. R. and Packard, C. M., The Hessian fly and how losses from it can be avoided, *U.S. Dep. Agric. Farmers' Bull.*, 1627, 1936.)

Table 156
AVERAGE YIELDS OF WHEAT AND PERCENTAGE OF INFESTATION BY THE HESSIAN FLY FOR 8 YEARS IN FIELDS PLANTED BEFORE THE SAFE-SEEDING DATE AS CONTRASTED TO THOSE PLANTED AFTER THE SAFE-SEEDING DATE[2]

Location of field	Average yield		Average % of infestation	
	From wheat sown before the safe-seeding date, (bu)	From wheat sown after the safe-seeding date, (bu)	In wheat sown before the safe-seeding date (%)	In wheat sown after the safe-seeding date (%)
Rockford, Ill.	21.8	28.1	24.5	1.7
Bureau, Ill.	27.4	32.9	45.5	5.2
La Harpe, Ill.	30.8	36.5	38.0	1.8
Urbana, Ill.	29.5	37.1	32.6	5.4
Virden, Ill.	23.6	28.4	48.0	6.3
Centralia, Ill.	14.5	21.9	81.0	8.0
Carbondale, Ill.	21.5	23.9	16.0	1.0
Grand Chain, Ill.	15.5	21.4	32.3	1.0
Average	23.1	28.8	39.7	3.8

From Metcalf, C. L. and Flint, W. P., *Destructive and Useful Insects,* McGraw-Hill, New York, 1951. With permission.

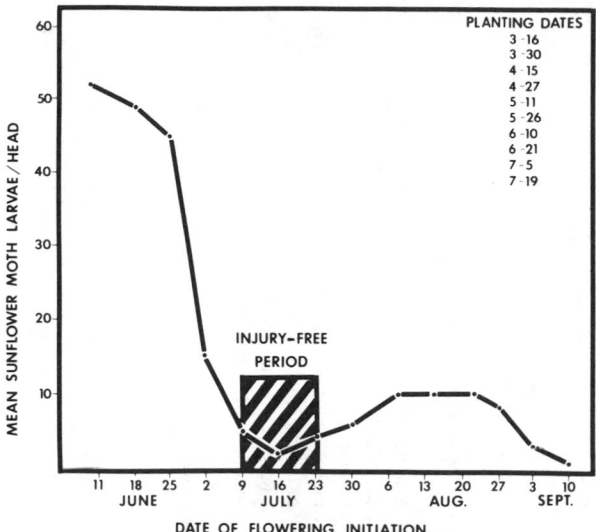

FIGURE 14.   Seasonal sunflower moth larval density, illustrating
injury free period to avoid damage. (**Source:** Mitchell, T. L., Ward,
C. R., Teetes, G. L., Schaefer, C. A., Bynum, E. D., and Brigham,
R. D., Sunflower pest population levels in relation to date of planting
on the Texas High Plains, *Southwest Entomology,* 3, 279, 1978.)

Table 157
ECONOMIC THRESHOLDS FOR KEY AND OCCASIONAL PESTS OF SORGHUM

Common and scientific name	Geographical distribution[a]	Pest status[b]	Type of damage	Economic threshold
Greenbug, *Schizaphis graminum*	C	Key	Suck plant sap and inject toxins which may kill plant, virus vector	Seedling — visible damage with colonies Preboot — before any entire leaves are killed Headed — cause of death for two normal-sized leaves
Yellow sugarcane aphid, *Sipha flava*	Nw	Occ	Suck plant sap and inject toxins which may kill young plants	At first sign of damage to young plants — prior to stand loss
Corn leaf aphid, *Rhopalosiphum maidis*	C	Occ	Suck plant sap, virus vector	Seedling — visible damage prior to stand loss Near-harvest — prior to accumulation of honeydew in heads
Shoot fly, *Atherigona soccata*[c]	Af, As	Key	Injure seedling growing point causing dead heart and tillering with small, late-maturing heads	Unknown
Sorghum borer, *Chilo partellus*[d]	Af, As	Key	Some leaf feeding, larvae tunnel stalks causing lodging, dropped heads, and greatly reduced yields	Unknown
Fall armyworm, *Spodoptera frugiperda*[c]	Nw	Occ	Leaf feeder in whorl and destruction of seed head	At heading — two larvae per head
Chinch bug, *Blissus leucopterus*	Nw	Occ	Suck sap from lower leaves and stem	Seedling — 1—2 adults per five plants; 15—30 cm tall plants — 75% of plants infested

Table 157 (continued)
ECONOMIC THRESHOLDS FOR KEY AND OCCASIONAL PESTS OF SORGHUM

Common and scientific name	Geographical distribution[a]	Pest status[b]	Type of damage	Economic threshold
Sorghum midge, *Contarinia sorghicola*[f]	C	Key	Destroy developing seeds	When 25—30% of heads have begun to bloom — mean of one adult per head
Sorghum webworm, *Celama sorghiella*	Nw	Occ	Destruction of seeds in head	Five larvae per head
Corn earworm *Heliothis zea*[e]	Nw	Occ	Destruction of seeds in head	Two larvae per head
Jowar earhead bug, *Calocoris angustatus*	As	Key	Feed on developing seed causing smaller, lighter distorted seed	Unknown
Banks grass mite, *Oligonychus pratensis*	Nw	Sec	Suck plant sap, causing discoloration and death of leaves	After heading — rapid population buildup away from leaf midrib
White grub, *Phyllophaga crinita*	Nw	Occ	Seedling death, stunting and/or lodging	Prior to planting: one grub per 0.0283 m^3
Wireworms, species of *Eleodes, Conoderus, Aeolus*, etc.	Nw	Occ	Destroy planted seedling, feed on base of plant	2 Or more larvae per 30 cm

a   Distribution key: C = Cosmopolitan, Af = Africa, As = Asia, and Nw = New World.

b   Key — key pest: Occ = occasional pest and sec = secondary pest.

c   32 species of flies attacking sorghum in Africa (reported in 1972).

d   Other important stem borers are *Diatraea grandiosella, Diatraea saccharalis, Eldana saccharina, Busseola fusca, Sesamia cretica*.

e   Other important armyworms are *Pseudaletia separata, Spodoptera exempta, Spodoptera exigua, Pseudaletia convecta*.

f   11 species of gall midges attacking sorghum (reported in 1956).

g   Other important head caterpillars are *Stenachroia elongella, Eublemma* spp., *Dichocrosis punctiferalis, Cryptoblabes adoceta, Heliothis armigera*.

Table 158
LOSS ESTIMATES FROM PLANT PATHOGENS WITH CURRENT FUNGICIDE USE AND AN ESTIMATE OF LOSSES IF NO FUNGICIDES WERE USED[a]

Crop	Acres treated (%)	Current crop pest less, (%)	Additional crop pest pest loss without fungicide control (%)
Field Crops			
Corn	1	10	0
Cotton	4	12	1
Wheat	0	14	0
Soybeans	2	14	1
Rice	0	7	0
Tobacco	7	11	3
Peanuts	85	28	45
Sorghum	0	9	0
Sugar beets	13	16	3
Other grain	0	20	0
Alfalfa	0	24	0
Other hay	0	15	0
Other field crops	1	13	0
Pasture	0	5	0
Total			
Vegetable Crop			
Lettuce	21	12	5
Cole	15	9	5
Carrots	25	8	6
Potatoes	49	20	20
Tomatoes	98	23	44
Sweet corn	9	8	1
Onions	70	21	28
Cucumbers	42	15	17
Beans	24	25	6

Crop	Acres treated (%)	Current crop pest less, (%)	Additional crop pest pest loss without fungicide control (%)
Cantaloupe	53	21	9
Peas	2	23	0
Peppers	48	14	10
Sweet potatoes	1	18	0
Watermelons	40	10	2
Asparagus	7	9	1
Other vegetables	10	10	2
Total			
Fruit and nut crops			
Apples	61	8	60
Cherries	51	24	35
Peaches	65	21	30
Pears	34	17	35
Prunes and plums	35	10	35
Grapes	40	27	15
Oranges	48	16	25
Grapefruit	52	2	45
Lemons	47	29	20
Other citrus	28	18	20
Strawberries	42	26	10
Other fruits	26	20	30
Pecans	46	21	45

| Other nuts | 46 | 12 | 10 |
| Total | | | |

a   Estimates obtained from Pimentel, D. et al., A Cost-Benefit Analysis of Pesticide Use in U.S. Food Production, New York College of Agriculture and Life Sciences, Cornell University, Ithaca, N.Y., 1978.

Modified from Palm, E. W., in *Handbook of Pest Management in Agriculture*, Vol. 1, Pimentel, D., Ed., CRC Press, Boca Raton, Fla., 1981.

Table 159
ESTIMATION OF PERCENTAGE LOSSES DUE TO DISEASES OF SELECTED MAJOR FIELD CROPS IN THE U.S.[a]

Crop/disease	% Loss	Crop/disease	% Loss
Corn		**Cotton**	
Stalk rots (fungus)	3.0	Seedling diseases	2.6
Helminthosporium leaf blights	2.3	Verticillium wilt	2.3
Seedling blights	1.6	Boll rots	2.2
Root rots	1.3	Bacterial blight	1.7
Ear rots	1.2	Fusarium wilt	1.1
Smut	.7	Phymatotrichum root rot	.9
Bacterial leaf blight	.5	Ascochyta blight	.2
Physoderma brown spot	.2	Others	1.0
Others	1.2	**Total**	12.0
Total	12.0		
		Tobacco	
Wheat		Leaf decay during curing	2.7
Stem rust	4.0	Tobacco mosaic	1.4
Leaf rust	2.5	Wildfire	1.3
Root rots	1.0	Miscellaneous stalk and root diseases	1.1
Septoria leaf and glume blotch	1.0	Black shank	1.0
Wheat streak mosaic	1.0	Black root rot	.9
Loose smut	.8	Brown spot	.8
Cercosporella foot rot	.6	Blue mold	.6
Scab	.5	Blackfire	.3
Soilborne mosaics	.5	Miscellaneous leaf diseases	.3
Common bunt	.4	Frogeye	.2
Powdery mildew	.4	Fusarium wilt	.2
Take-all	.2	Granville wilt	.2
Bacterial diseases	.1	**Total**	11.0
Dwarf bunt	.1		
Miscellaneous leaf and head blights	.1	**Oats**	
Miscellaneous virus diseases	.1	Yellow dwarf	3.8
Fusarium and Typhula snow molds	.1	Crown rust	3.7
Stripe rust	.1	Blast (sterility)	2.5
Others	.5	Root necrosis	2.4
Total	14.0	Stem rust	2.3
		Septoria foliage blight	1.8
Soybeans		Scab	.9
Phytophthora	2.3	Helminthosporium leaf blotch	.7
Bacterial blight	2.2	Smuts	.6
Downy mildew	1.6	Soilborne mosaic	.5
Bacterial pustule	1.4	Bacterial stripe blight	.4
Pod and stem blight	1.2	Victoria blight	.4
Stem canker	1.1	Blue dwarf	.3
Bud blight	.8	Halo blight	.3
Brown spot	.8	Helminthosporium culm rot	.2
Brown stem rot	.8	Physiologic leaf spot	.1
Purple stain	.7	Others	.1
Fusarium root rot	.6	**Total**	21.0
Frogeye leaf spot	.1		
Others	.4		
Total	14.0		

From Losses in Agriculture, *U.S. Dep. Agric. Agric. Handb.*, No. 291, 1965 (estimates for period 1951—1960).

Table 160

ESTIMATION OF PERCENTAGE LOSSES DUE TO DISEASES OF SELECTED
MAJOR FRUIT AND VEGETABLE CROPS IN THE U.S.

Crop/disease	% Loss	Crop/disease	% Loss
Potatoes		Grapes	
Late blight	4.0	Leaf roll	8.4
Leaf roll	3.0	Black measles	5.0
Verticillium wilt	2.7	Fanleaf, yellow mosaic, and vein	5.0
Scab	1.5	banding	
Early blight	1.0	Summer bunch rot	2.2
Latent mosaic	1.0	Dead arm	2.0
Mild mosaic	1.0	Powdery mildew	1.8
Rhizoctonia black scurf	1.0	Fruit rots	1.0
Rugose mosaic	1.0	Pierce's disease	.5
Blackleg	.5	Armillaria root rot	.2
Fusarium wilt	.5	Black rot	.2
Ring rot	.5	Yellow vein	.2
Spindle tuber	.2	Anthracnose	.1
Bacterial brown rot	.1	Downy mildew	.1
Others	1.0	Others	.3
Total	19.0	**Total**	27.0
Tomatoes		Oranges	
Tobacco mosaic	5.0	Root and foot rot	3.2
Gray leaf spot	3.0	Psorosis	2.0
Verticillium wilt	3.0	Exocortis	1.9
Bacterial spot	1.0	Tristeza	1.9
Blossom-end rot	1.0	Fruit and leaf spots	1.3
Curly top	1.0	Wood and heart rot	1.3
Early blight (alternaria leaf spot)	1.0	Blossom blight	.1
Fusarium wilt	1.0	"Stubborn" disease	.1
Bacterial wilt	.5	Twig blight	.1
Late blight	.5	Xyloporosis	.1
Leaf mold	.5	**Total**	12.0
Septoria leaf spot	.5		
Others	3.0		
Total	21.0		

From Losses in Agriculture, *U.S. Dep. Agric. Agric. Handb.*, No. 291, 1965 (estimates for period 1951—1960).

Table 161
ESTIMATED LOSSES DUE TO DISEASES AFFECTING FIELD CROPS IN U.S.

Crop	Loss from potential production %
Corn	12
Wheat	14
Cotton	12
Oats	21
Soybeans	14
Tobacco	11
Barley	14
Peanuts	28
Sorghum	9
Sugarbeets	16
Beans, dry	17
Rice	7
Sugar cane	23
Flax (seed)	10
Others	3—14

From Losses in Agriculture, *U.S. Dep. Agric. Agric. Handb.*, No. 291, 1965 (estimates for period 1951—1960).

Table 162
ESTIMATED LOSSES DUE TO DISEASES AFFECTING FORAGE/ PASTURE CROPS IN U.S.

Crop	Loss from potential production (%)
Alfalfa (hay)	24
All other hay plants	15
Forage plants for seed	4—52
Cropland pastures	9
Forestland pastures	3
Grassland	5
Total	

From Losses in Agriculture, *U.S. Dep. Agric. Agric. Handb.*, No. 291, 1965 (estimates for period 1951—1960).

Table 163
ESTIMATED LOSSES DUE TO DISEASES AFFECTING FRUIT/VEGETABLE/ ORNAMENTAL CROPS

Crop	Loss from potential production (%)
Potatoes	19
Tomatoes	21
Grapes	27
Oranges	12
Strawberries	26
Beans (green)	20
Peaches	14
Apples	8
Lettuce	12
Onions	20
Ornamental plants and shade trees	1—22
Cherries	24
Peas	23
Celery	17
Lemons	25
Pears	17
Other vegetables	2—21
Other fruits and nuts	2—38
Total	

From Losses in Agriculture, *U.S. Dep. Agric. Agric. Handb.*, No. 291, 1965 (estimates for period 1951—1960).

Table 164

CROPS AND DISEASES IN WHICH ROTATION HAS HAD A MAJOR EFFECT IN DISEASE CONTROL

Crop	Disease	Pathogen	Crop interval (year)	Commentary
Alfalfa	Root rot	*Cylindrocladium ehrenbergii*	1—2	Rotate with corn, small grains, or forage grasses
	Stem nematode	*Ditylenchus dipsaci*	2	Use any cultivated crop
	Violet root rot	*Helicobasidium purpureum*	1—2	Rotate with cereal crops
Barley	Leaf blotch	*Septoria passerinii*	2	Rotate when barley fields not contiguous or close
Bean, dry	Anthracnose	*Colletotrichum lindemuthianum*	2	Use any crop but beans
	Ashy stem blight	*Macrophomina phaseoli*	2—3	Avoid beans or sweet potato
	Common blight	*Xanthomonas phaseoli*	2	Avoid bean crops
	Dry root rot	*Fusarium solani*	5	Avoid bean crops
	Halo blight	*Pseudomonas phaseolicola*	2	Avoid bean crops
	Leaf spot	*Cercospora cruenta*	5+	Rotate with nonlegumes
	Root rot	*Pythium butleri, Rhizoctonia solani, Sclerotium rolfsii*	4—5	Rotate with cereals, clover, or alfalfa
	Watery soft rot	*Sclerotinia sclerotiorum*	2	Use small grains, corn, or hay crop
	Wilt	*Corynebacterium flaccumfaciens*	2	Avoid bean crops
Bean, lima	Pod blight	*Diaporthe phaseolorum*	1	Avoid beans in rotation
	Scab	*Elsinoë phaseoli*	1	Avoid beans in rotation
Bean, red kidney	Root rot	*Fusarium solani*	3	Wheat before bean
Bean, snap	Hypocotyl rot	*Rhizoctonia solani*	1	After corn; plow down residues
Beets, sugar	Black root	*Aphanomyces cochlioides*	2—5	Beets after corn, soybeans, and cereals, but not legumes Montana: beets, barley, alfalfa (3 years); oats, beans, alfalfa (2 years); corn, beets; beets, fallow, potato Iowa: 3—4 years between beets; >6 with 3—4 years alfalfa where disease severe Ohio: corn better than oats before beets. Legumes effective if fall but not spring plowed
	Cyst nematode	*Heterodera schactii*	2—5	If spotty, 2 years legumes, grains, or solanaceous crops between beet crops; if >25% of beet crop lost, plant 3—5 years between beet crops

Table 164 (continued)
CROPS AND DISEASES IN WHICH CROP ROTATION HAS HAD A MAJOR EFFECT IN DISEASE CONTROL

Crop	Disease	Pathogen	Crop interval (year)	Commentary
Beet, table	Leaf spot	Cercospora beticola	4—5	Rotate with corn, potato, other crops
		Xanthomonas beticola	2+	Rotate with corn, potato, other crops
		Phoma betae	5+	Rotate with corn, potato, other crops
Bentgrass	Nematode	Anguina agrostis	1	Seed fields: permit no Agrostis spp. to go to seed for 1 year
Broccoli	Anthracnose	Gloeosporium concentricum	1	Rotate with noncrucifers
Cabbage	Black leg	Phoma lingam	3+	Rotate with noncrucifers
Carrot	Root knot	Meloidogyne hapla	1	Plant corn or weed-free sod (Poa pratensis) 1 year between root crops
	Root scab and blight	Xanthomonas carotae	2	Long rotations preferable
Cantaloupe	Root knot	Meloidogyne incognita	2	Rotate with soybean or soybean and crotalaria
Cauliflower	Anthracnose	Gloeosporium concentricum	1	Fungus on cabbage, broccoli, and Brussels sprouts
	Peppery leaf spot	Pseudomonas maculicola	1	Rotation with noncrucifers and destruction of residues
Celery	Brown spot	Cephalosporium apii	1	Rotate with lettuce, onions or potato (N.Y.)
	Pin nematode	Paratylenchus hamatus	2	Lettuce or spinach between celery crops
Cereals and grasses	Cyst nematode	Heterodera avenae	1—2	Use legumes in rotation
	Root knot	Meloidogyne naasi	1	Use root crops in rotation
Corn	Downy mildew	Sclerospora sorghi	1	Rotate with nongrass crop
	Leaf blight	Helminthosporium carbonum, H. maydis, H. turcicum	1—3	Rotate with nongrass crop
	Meadow nematode	Pratylenchus leiocephalus	1—2	Alternate with peanut crop
Cotton	Stalk rot	Fusarium moniliforme, F. roseum	1	After soybeans
	Root rot	Macrophomina phaseoli	1	After alfalfa or barley
	Seedling blight	Phymatotrichum omnivorum	4	After cereals or grasses; plow residues deeply
		Ascochyta gossypii	1	After any other crop; plow residues after harvest
Crucifers	Leaf spot	Alternaria brassicae	1	Rotate with noncrucifers
Cucumber	Anthracnose	Colletotrichum lagenarium	2	Any crop but muskmelon, watermelon, gourd

Crop	Disease	Pathogen		Remarks
Hops	Wilt	*Verticillium albo-atrum*	1	Rotate with cereals, root crops, vegetables, not potato or raspberry
Lettuce	Leaf spot	*Cercospora lactucae*	4+	
		Septoria lactucae	4+	
Onion	Bloat	*Ditylenchus dipsaci*	2	Rotate with beets, carrots, crucifers, lettuce, spinach (N.Y.)
	Nematode	*Balonolaimus gracilis*	5—10	Use watermelon or tobacco and eradicate weeds
	White rot	*Sclerotinia cepivorum*	8—10	
Parsnip	Canker and leaf spot	*Itersonilia perplexans*	5—10	Rotate with non-host crops
Pea	Anthracnose	*Colletotrichum pisi*	2+	
	Bacterial blight	*Pseudomonas pisi*	2+	
	Leaf spot	*Septoria pisi*	4+	
	Root rot	*Aphanomyces euteiches*	6—10	Use more than half crop sequence in corn, grains, vegetables; not forage
Pepper	Seedling blight	*Rhizoctonia solani*	1+	Rotate with crops such as cereals or corn
	Anthracnose	*Colletotrichum piperatum, C. capsici*	1	Avoid solanaceous crops in rotation
	Bacterial spot	*Xanthomonas vesicatoria*	1	Control solanaceous weeds
Peppermint	Wilt	*Verticillium albo-atrum*	5	Peppermint for 2 years; corn or corn and onions 5 years (midwest U.S.)
Potato	Corky ring spot (Spraing)	*Tobacco rattle virus*	1	Barley better than corn, potato, or sugar beet in rotation
	Golden nematode	*Heterodera rostochiensis*	8	Exclude potato or tomato in rotation; in short rotations, use beans, corn, red clover, rye-grass
	Root knot	*Meloidogyne hapla*	1	Use corn or weed-free sod (*Poa pratensis*) 1 year between crops
	Root rot	*Rhizoctonia solani*	3—6	Use alfalfa before potato (Neb. irrigated field)
	Scab	*Streptomyces scabies*	1+	Soybean as green manure before potato (Calif.)
	Wilt	*Fusarium oxysporum*	3—6	Alfalfa before potato (Neb. irrigated fields)
		Verticillium albo-atrum	2—3	Grain-clover (1—2 years) potato (in some locations, red clover is a symptomless carrier and increases wilt of potato in a rotation)

Table 164 (continued)

CROPS AND DISEASES IN WHICH CROP ROTATION HAS HAD A MAJOR EFFECT IN DISEASE CONTROL

Crop	Disease	Pathogen	Crop interval (year)	Commentary
Sorghum	Root knot	*Meloidogyne naasi*	1	Rotate with root crops
	Stalk rot	*Fusarium moniliforme*	2	Winter wheat-grain sorghum-fallow-with no tillage
Soybean	Brown spot	*Septoria glycinea*	1	Use any nonsusceptible crop
	Brown stem rot	*Cephalosporium gregatum*	5	5 years corn before soybean or corn-soybean-oats-clover
	Downy mildew	*Peronospora manshurica*	1	Use nonhost crops
	Cyst nematode	*Heterodera glycines*	1—5	Use nonhost crops
	Leaf spot	*Cercospora sojina*	1	Use nonhost crops
		Phyllosticta sojaeicola	1	Rotation and plow residues
		Pseudomonas glycinea	1	Rotate and plow residues
Spinach	Downy mildew	*Peronospora spinaciae*	3	
Squash	Root and stem rot	*Fusarium solani*	2—3	
Sweet potato	Black rot	*Ceratocystis fimbriata*	2+	Use any other crop
Tobacco	Black shank	*Phytophthora parasitica*	4	Rotate with cotton, lespedeza, peanut, soybean, oats, rye, or wheat, but not legumes before tobacco
	Black root rot	*Thielaviopsis basicola*	5	Many crops in rotation but use small grain just before tobacco
	Granville wilt	*Pseudomonas solanacearum*	3—5	Rotate with corn, cotton, cowpea, soybean, velvet bean, red top, or small grains
	Root knot	*Heterodera marioni*	2—3	Corn-oats-tobacco; cotton-peanut-tobacco; peanut-oats-tobacco
Tomato	Anthracnose	*Colletotrichum phomoides*	2	Rotate only if susceptible crops not nearby
	Bacterial canker	*Corynebacterium michiganense*	1	Rotate with other crops and control solanaceous weeds
	Bacterial speck	*Pseudomonas tomato*	1	
	Bacterial spot	*Xanthomonas vesicatoria*	1	
	Blight and fruit rot	*Helminthosporium lycopersici*	1	Rotate and plow diseased residues

	Disease	Pathogen		Control
Vegetables	Septoria blight	*Septoria lycopersici*	4	Rotate other crops and control horse nettle
	Soil rot	*Rhizoctonia solani*	1	Rotate with pangola grass (Florida)
	Nematodes	*Pratylenchus* spp.	1	Vegetable crops after oats and peanuts then after corn or lupines
Wheat, spring	Common root rot	*Helminthosporium sativum* *Fusarium roseum*	4	Use noncereal crops (Sask.)
	Eyespot	*Cercosporella herpotrichoides*	2—3	Rotate other crops but not barley or wheat (England)
	Flag smut	*Urocystis tritici*	1	Any crop but susceptible wheat
	Foot rot	*Helminthosporium sativum*	1	Use nongrass crops (e.g., legumes)
	Root gall nematode	*Subanguina radicicola*	1	Use legumes and root crops
	Seed gall	*Anguina tritici*	1—2	Rotate with noncereals
	Scab	*Fusarium roseum*	1	Where scab severe, 1 year between cereals and plow in residues
	Septoria leaf and glume blotch	*Septoria avenae, S. nodorum, S. tritici*	2	Rotate with noncereals
Wheat, winter	Cephalosporium stripe	*Cephalosporium gramineum*	1—2	Rotate with corn or legumes
	Snow mold	*Fusarium nivale, Sclerotinia borealis, Typhula* spp.	1	Rotate with spring cereals or legumes
	Take-all	*Gaeumannomyces graminis*	2—3	Rotate with noncereals (Kansas)

Note: References included in original.

Table 165

CROPS AND DISEASES IN WHICH CROP RESIDUE MANAGEMENT HAS HAD A MAJOR EFFECT IN DISEASE CONTROL

Crop	Disease	Pathogen	Recommendation
Banana	Wilt	*Fusarium oxysporum*	Plant prior crop sugarcane and plow in residues
Bean, snap	Root rot	*Thielaviopsis basicola*	Plow in alfalfa and cabbage residues, or castor, peanut, safflower, or soybean meals (5000 ppm)
Beet, table	Powdery mildew	*Erysiphe polygoni*	Plow under bean refuse
	Bacterial blights and leaf spots	*Xanthomonas beticola* and other bacteria	Plow in or destroy beet refuse
	Leaf spot	*Cercospora beticola*	Plow in refuse
Bentgrass	Nematode	*Anguina agrostis*	Plow under bentgrass sod at least 15 cm with complete sod turnover
Cotton	Root rot	*Phymatotrichum omnivorum*	Green manure before cotton; spring: cereals; winter: cowpea, guar, sesbania, sweet clover with other control methods (e.g., 4-year rotation)
Corn	Seedling blight	*Ascochyta gossypii*	Plow down residues after harvest
	Anthracnose	*Colletotrichum graminicola*	Plow down corn refuse to prevent early infection
	Gray leaf spot	*Cercospora zeae-maydis*	Plow under infected leaves to delay early onset of disease
	Southern corn leaf blight	*Helminthosporium maydis*	Inoculum on crop refuse reduced by chopping residues, harrow and plow
	Stalk rot	*Fusarium moniliforme, F. roseum*	Leave corn residues on soil surface
	Yellow leaf blight	*Phyllosticta maydis*	Clean plow to bury debris to delay early onset of disease
Peanut	Stem rot	*Sclerotium rolfsii*	Deep plow crop refuse with "nondirting" procedure
Sorghum	Downy mildew	*Sclerospora sorghi*	Bury surface soil and debris to 25—30 cm
Soybean	Anthracnose	*Colletotrichum dematium*	Plow down residues after harvest
		Glomerella glycines	
Wheat	Seedling blight	*Rhizoctonia solani*	If wheat after corn, bury corn residue by plowing
Wheat, winter	Cephalosporium stripe	*Cephalosporium graminoeum*	Deep plow stubble, bury below 8 cm
	Take-all	*Gaeumannomyces graminis*	Drill wheat directly in stubble and leave residues above ground; early rotavation between successive wheat crops

Note: References included in original.

Table 166
STRATEGIES APPLIED AT VARIOUS STAGES OF CROP PRODUCTION TO MANAGE DISEASES OF ALFALFA

Stage	Strategies
Harvesting	Insect control; timely harvests; foliar disease monitoring; crop sanitation; weed control; fertility maintenance
Planting	Seeding rate; firm seedbed; viable seed; planting depth; weed control; clean seed
Preplant	Crop rotation; resistant variety; species combination; site selection; weed control; pH and fertility

Table 167
PERCENTAGE OF U.S. CORN PRODUCTION AFFECTED BY VARIOUS CORN DISEASES AND ESTIMATED DEGREE OF CONTROL ACHIEVED BY HOST RESISTANCE, CULTURAL PRACTICES, OR CHEMICALS

Disease	Pathogen	U.S. production affected (%)	In disease-affected areas, estimated % of control achieved by		
			Resistance	Cultural practices	Chemical
Seed rot	*Pythium* sp.	100	10	40	50
Seedling blight	Various fungi	1	65	30	5
Stewart's bacterial wilt and leaf blight	*Erwinia stewartii*	47	99	0	1
Northern leaf blight	*Helminthosporium turcicum*	80	99	0	1
Southern leaf blight	*H. maydis*	13	99	0	1
Helminthosporium leaf spot	*H. carbonum*	58	100	0	0
Yellow leaf blight	*Phyllosticta maydis*	13	60	40	0
Eyespot	*Kabatiella zeae*	12	10	90	0
Sorghum downy mildew	*Sclerospora sorghi*	2	50	50	0
Common rust	*Puccinia sorghi*	81	100	0	0
Southern rust	*P. polysora*	3	100	0	0
Brown spot	*Physoderma maydis*	3	100	0	0
Other leaf blights	Various fungi and bacteria	2	98	2	0
Maize dwarf mosaic	MDM virus	2	100	0	0
Maize chlorotic dwarf	MCD virus	2	100	0	0
Diplodia stalk rot	*Diplodia maydis*	39	75	25	0
Gibberella stalk rot	*Gibberella zeae*	78	75	25	0
Fusarium stalk rot	*Fusarium moniliforme*	90	75	25	0
Charcoal rot	*Macrophomina phaseolina*	5	30	70	0
Anthracnose	*Colletotrichum graminicola*	30	75	25	0
Common smut	*Ustilago maydis*	70	100	0	0
Head smut	*Spacelotheca reiliana*	<1	70	20	10
Ear rots	Various fungi	100	100	0	0
Nematodes	Various sp.	3	80	10	10
Storage rots	Various fungi	86	0	100	0

Table 168
MINIMUM POPULATION DENSITIES
OF VARIOUS PHYTOPARASITIC
NEMATODES REQUIRED TO CAUSE
SOYBEAN YIELD LOSS

Nematode	Threshold density (no. per 100 cm³ soil)
Belonolaimus longicaudatus	5
Helicotylenchus dihystera	>200
Meloidogyne hapla	100—200
M. incognita	20
M. javanica	35 juveniles and eggs
Pratylenchus penetrans	<200
Trichodorus porous	20—40
Tylenchorhynchus claytoni	>400

Source: After Barker, K. R. and Olthof, T. H. A., Relationships between nematode population densities and crop responses, *Annu. Rev. Phytopathol.*, 327, 1976.

Table 169
GROWTH OF SOYBEAN-INFECTING NEMATODES ON SELECTED CROPS[a]

Nematode	Alfalfa	Barley	Corn	Cotton	Cowpeas	Crotalaria	Fescue	Grain sorghum	Lespedeza	Lupines	Millet	Oats	Peanuts	Rye	Soybeans	Sugarcane	Tobacco	Vetch	Wheat
Tylenchorhynchus spp.	+	−	−	−	−	−	−	+	+		+	+	−	+	+	+	−	+	−
Heterodera glycines	−	−	−	−	−	−	−	−	+	−	−	−	−	−	+	−	−	+	−
Meloidogyne spp.																			
M. incognita	+	+	−	+	+	−	−	−	+	+	+	+	−	−	+	+	+	+	+
M. javanica	+	+	+	−	+	−	−	−	+	+	+	+	−	−	+	+	+	+	+
M. arenaria	+	−	−	−	+	−	−	−	+	−	−	−	+	−	+	+	+	+	−
M. hapla	+	−	−	−	−	−	−	−	+	−	−	−	+	−	+	−	+	+	−
Hoplolaimus spp.	+	+	+	+	+	−	−	+	+	+	+	+	−	+	+	+	−	+	+
Pratylenchus spp.	+	+	+	−	−	−	−	+	+	−	+	+	−	−	+	−	+	+	+
Rotylenchulus reniformis		−	−	+	+	+	−	+	−	+	+	−	−	−	+	+	+	+	−
Belonolaimus spp.	+	+	+	+	−	−	−	+	+	+	+	+	−	+	+	+	−	+	+
Criconemoides spp.	+	−	−	−	−	−	−	−	+	−	−	−	+	−	+	−	−	−	−
Xiphinema spp.	+	+	+	−	+	−	−	+	+	−	+	+	−	+	+	+	−	+	+
Trichodorus spp.	+	+	+	−	−	−	−	+	+	−	+	−	−	+	+	+	−	+	+

Note: + = susceptible host and − = nonhost or intermediately suitable; blank = information not available.

[a] Not including resistant varieties.

Modified after Good, J. M., Management of plant parasitic nematode populations, *Tall Timbers Conf. Ecological Animal Control Habitat Managem.*, 1972, 109.

Table 170
ESTIMATED ANNUAL AVERAGE PERCENTAGE LOSS FROM
WEEDS (1972—76)

Commodity	Range in loss from potential production (%)[a]
Row crops	
Corn, cotton, soybeans, etc.	8—17
Drill crops	
Barley, rice, wheat, etc.	10—17
Vegetables	
Cole crops, root crops, vegetable legumes, etc.	3—15
Fruits	
Citrus, pome, and stone fruits	3—6
Small fruits, strawberries, etc.	10—25
Deciduous tree nuts	
Almonds, pecans, and walnuts	5—9
Forage seed crops	
Grass seed crops	13—17
Legume seed crops	17—22
Pasture and range: 31 eastern states[b]	20
Rangelands: 17 western states[b]	13
Hay in 48 states	10

[a] Estimate based on full production with the cause eliminated. Adapted from *U.S. Dep. Agric. Agric. Handb.,* No. 291, 1965.
[b] Excludes cropland used for pasture.

Adapted from Chandler, J. M., in *Handbook of Pest Management in Agriculture,* Vol. 1, Pimentel, D., Ed., CRC Press, Boca Raton, Fla.

Table 171
SELECTED WEED DENSITY STUDIES AND
RESULTANT SOYBEAN YIELD LOSSES
FOLLOWING FULL-SEASON COMPETITION

Pest species	Weed density[a]	Reduction in soybean yield (%)
Cocklebur, common	1/305 cm	10
	1/152 cm	28
	1/76 cm	43
	1/38 cm	52
Cocklebur, common	1/480 cm	15
	1/240 cm	20
	1/120 cm	39
	1/60 cm	63
	1/30 cm	87
	1/15 cm	100
	1/7.5 cm	100
Foxtail, giant	1/60 cm	3
	1/30 cm	4
	1/10 cm	5
	1/5 cm	9
	1/2.5 cm	17
Pigweed, smooth	1/100 cm	18
	1/50 cm	27
	1/25 cm	32
	1/12.5 cm	45
	1/2.5 cm	51
Pigweed (spp. complex)	1/360 cm	2.7
	1/270 cm	6.6
	1/180 cm	14.5
	1/90 cm	18.3
	1/30 cm	44.0
Morning glory	1/60 cm	12
	1/30 cm	20
	1/15 cm	26
	1/7.5 cm	27
	1/3.75 cm	44
Sunflower, common	1/360 cm	13.7
	1/270 cm	29.8
	1/180 cm	34.7
	1/90 cm	59.0
	1/30 cm	78.9
	1/15 cm	82.3
Velvetleaf	1/360 cm	2.7
	1/270 cm	6.6
	1/180 cm	14.5
	1/90 cm	18.3
	1/30 cm	44.0

[a]   1/10 cm = one weed per 10 cm of row.

Note: References included in the original.

Format adapted from Pareja, M. R., M.S. thesis, Iowa State University Library, Ames, 1977.

Table 172
CRITICAL PERIODS OF WEED COMPETITION WITH SOYBEANS[a]

Pest species	Critical duration of weed competition[b]	Critical weed-free requirement[c]
Barnyardgrass	4—7	3—4
Cocklebur, common	4—6	4
Foxtail, giant	—	<3
Foxtail, yellow	5	—
	5—11	—
	6—8	—
Mallow, venice	—	<3
	7	—
Morning glory, tall	6—12	—
Morning glory, tall/ morning glory, ivyleaf	6—8	—
Pigweed	5	—
Sesbania, hemp	4	>8
Sicklepod	4—8	2—4
Sida, prickly	—	<3
Smartweed, Pennsylvania	5—11	—
Velvetleaf	—	<3
Mixed weed species	4	—

Note: The chances of obtaining acceptable suppression are reduced the longer implementation of controls is delayed.

[a] Most applicable to a vigorous, well-managed crop under normal growing conditions.
[b] Weeks before weeds emerging with soybeans cause a significant yield loss.
[c] Weeks weeds must be suppressed following emergence before crop has a permanent competitive advantage.

Note: References included in the original.

Table 173
POTENTIAL SEED PRODUCTION OF INDIVIDUAL PLANTS AND SEED WEIGHT RELATIONSHIPS

Pest species	Seeds per plant[a]	Seed no./lb
Barnyardgrass	7,160	324,286
Bindweed, field	50	14,934
Corn, volunteer	600	1,200
Foxtail, giant	4,030	238,947
Johnsongrass	80,000	130,000
Lambsquarters, common	72,450	648,570
Mallow, venice	58,600	129,714
Milkweed, common	770	108,095
Morning glory, annuals	≤15,200	—
Nightshade, black	8,000	—
Nutsedge, yellow	2,420	2,389,474
Pigweed, redroot	117,400	1,194,737
Ragweed, common	3,380	114,937
Ragweed, giant	1,650	26,092
Sandbur	1,110	67,259
Smartweed, Pennsylvania	3,140	126,111
Sunflower, common	7,200	69,050
Thistle, Canada	680	288,254
Velvetleaf	4,300	51,886

Note: References included in the original.

[a] Except for johnsongrass (plant size unknown), the yield of perennials was calculated from one main stem.

Table 174
INFLUENCE OF CLOSE OR WIDE SPACING ON CROP YIELD REDUCTIONS AND WEED YIELDS WITH FULL-SEASON COMPETITION

Crop	Location	Weeds	Relative crop yield w/o weeds[a]		Relative crop yield w/weeds		Relative weed yield w/crops[b]	
			Close	Wide	Close	Wide	Close	Wide
Corn	Iowa	Foxtail	100	63	90	55	100	209
Sweet corn	Oregon	Redroot pigweed	100	59	81	38	ND[c]	ND
Dry beans	New York	Redroot pigweed	100	55	69	45	100	179
Soybeans	Nebraska	Mixed grass; broad leaved	100	73	63	40	100	199
	Missouri	Mixed grass; broad leaved	100	102	52	22	100	133
Snap beans	Oregon	Redroot pigweed	100	53	87	18	ND	ND
Tomato	New York[d]	Redroot pigweed	100	69	92	25	100	1238
Cotton	Alabama	Mixed grass; broad leaved	100	98	3	<1	ND	ND
Flax	North Dakota	Wild buckwheat	100	102	90	74	ND	ND
Sugarbeets[e]	Washington	PW, BYG, nightshade	100	91	77	46	100	361
Sorghum	Texas	Mixed grass; broad leaved	100	101	87	80	100	200
	Nebraska	Mixed grass; broad leaved	100	97	50	27	100	147
		Mixed grass, broad leaved	100	97	32	22	100	111

Note: References included in the original.

[a]   Close spacing without weeds = 100.
[b]   Weed yield with close crop spacing = 100.
[c]   No data.
[d]   Equidistant spacing.
[e]   Sugarbeets maintained weed free for first 12 weeks.

Table 175

INFLUENCE OF VARIETY ON CROP YIELD REDUCTIONS AND WEED YIELDS WITH FULL-SEASON COMPETITION

Crop	Location	Weeds	No. of varieties tested	Row width (cm)	Relative crop yield w/o weeds[a]		Relative crop yield w/weeds		Relative weed yield w/crops[b]	
					Most compet. variety	Least compet. variety	Most compet. variety	Least compet. variety	Most compet. variety	Least compet. variety
Sorghum	Nebraska	Mixed broad leaves, grasses	10	76	100	102	79	54	100	159
		Pigweed, foxtail, tall waterhemp	12	75	100	44	56	6	100	141
Soybean	Mississippi	Johnson grass	6	100	100	127	73	75	ND[a]	ND
		Cocklebur	6	100	100	107	37	27	ND	ND
	Nebraska	Green foxtail, tall waterhemp	10	76	100	106	59	51	100	112
	Mississippi	Cocklebur	5	33	100	95	68	45	ND	ND
Carrots	Brazil	Purple nutsedge	2	33	ND	ND	61	50	ND	ND
Potatoes	New York	Lambsquarter, pigweed	4	86	100	121	94	71	100	570
		Yellow nutsedge	5	91	ND	ND	ND	ND	100	564
Dry beans		Redroot pigweed	2	20[d]	100	100	76	60	100	140
Transplanted tomato		Redroot pigweed	2	Mixed[d]	100	131	78	76	100	152
Rice	Arkansas	Barnyard grass	3	15	100	93	62	33	ND	ND

Note: **References included in the original.**

[a] Most competitive variety without weeds = 100.

[b] Weed yield with most competitive variety = 100.

[c] No data.

[d] Equidistant spacing.

LOSSES IN ANIMALS

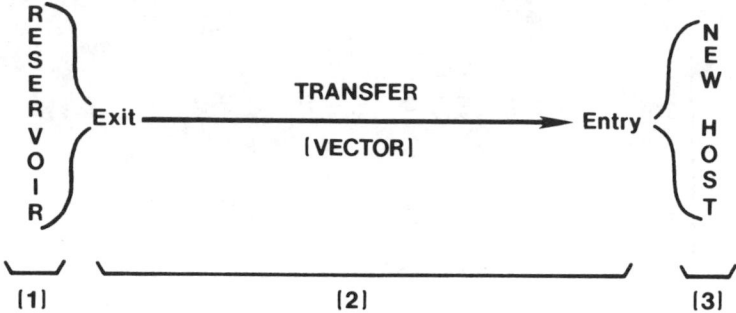

FIGURE 15.   Diagrammatic representation of infectious livestock disease transmission. Phases of control: (1) reservoir neutralization, (2) transfer inhibition, and (3) host resistance modification.

Table 176
ANIMAL RESERVOIRS AND VECTORS OF SELECTED LIVESTOCK INFECTIOUS DISEASES

Disease	Etiologic agent	Principal livestock host	Vector type	Reservoir	Carrier state (duration)
African horse sickness[a]	Virus	Horse	Biological (*Culicoides* spp.)	Horse	Chronic (months)
African swine fever	Virus	Swine	Biological (*Ornithodorus* spp.)	Swine, tick	Chronic (lifetime)
Bluetongue[a]	Virus	Sheep, cattle	Biological (*Culicoides* spp.)	Sheep, cattle	Convalescent (4 months); chronic (4 years)
Encephalitides Western equine[a] Eastern equine[a] Venezuelan[a]	Virus	Horse	Biological (Mosquito spp.)	Bird	Convalescent (weeks)
Epizootic bovine abortion	Virus	Cattle	Biological (*Ornithodorus* spp.)	Cattle, tick	
Equine infectious anemia	Virus	Horse	Mechanical (stable fly—tabanid)	Horse	Chronic (lifetime)
Vesicular stomatitis	Virus	Cattle	Biological (sand fly)	Cattle, wildlife	Convalescent (months)
Anaplasmosis[a]	Rickettsia	Cattle	Mechanical (biting Diptera—tick)	Cattle	Chronic (lifetime)
Eperythrozoonoses	Rickettsia	Swine	Mechanical (sucking lice)	Swine	Chronic (lifetime)
Pink eye	Bacteria	Cattle	Mechanical (face fly)	Cattle	Chronic (1 year)
Babesiosis:	Hemo-protozoa		Biological	Tick,	Transovarian
Bovine		Cattle,	(tick ssp.)	cattle	Chronic (2 years)
Equine		horse,		horse	Chronic (lifetime)
Porcine		swine		swine	Chronic (lifetime)
Habronemiasis	Nematode	Horse	Biological (housefly—stable fly)	Horse	Chronic (lifetime)

Table 177
CONTROL PROGRAMS FOR INFECTIOUS LIVESTOCK
DISEASES BY RESERVOIR-DISEASE CYCLE

Control program	Number of reservoir host species	Type of vector transmission	Proficient disease control	
			Reduction[a]	Elimination
A	Single	Mechanical	Vector abatement	Reservoir test and slaughter
B	Single	Biological	Vector abatement	Vector eradication
C	Multiple	Mechanical	Vector abatement	Reservoir immunization
D	Multiple	Biological	Vector abatement	Reservoir immunization, vector eradication

[a] Immunization may also be necessary to reduce the incidence of disease.

Table 178
AVERAGE NUMBER OF ADULT FEMALE
TICKS PER MONTH ON CATTLE
GRAZED CONTINUOUSLY IN THE SAME
PASTURE AND ON CATTLE
ALTERNATELY GRAZED IN TWO
PASTURES

	Average number adult female ticks per one side of animal[a]	
Months	Continuously grazed on same pastures	Alternately grazed on two pastures
	1955	
January	18	17
February	62	35
March	8[b]	33[b]
April	101[b]	70[b]
May	109	57[c]
June	124	0
July	135[c]	0[c]
August	42	0
September	18	0[e]
October	18	0
November	27[d]	0
December	26	7
	1956	
January	74	0[e]
February	52	0
March	85	2
April	84[d]	2
May	16	5[e]
June	88[d]	0
July	17	0
August	21	0
September	3	0[e]
October	0	0
November	9	0
December	23[d]	0

[a]   Approximate average number ticks per animal per month.
[b]   Animals sprayed to point of run-off with 0.16% arsenical preparation for tick control.
[c]   Animals dusted with 1.5 oz. of 0.6% gamma BHC for tick control.
[d]   Animals sprayed with 1% DDT emulsion for tick control.
[e]   Pasture rotation dates.

Source: Wilkinson, P. R., Pasture spelling as a control measure for cattle ticks in Southern Queensland, *Aust. J. Agric. Res.,* 15, 822, 1964.

Table 179
RELATIVE IMPORTANCE OF LOSSES IN LIVESTOCK DUE TO ARTHROPOD PESTS[a]

Arthropod	Rank	Arthropod	Rank
Cattle		Sheep	
Horn fly	1	Keds	1
Cattle grub	2	Bots	2
Stable fly	3	Swine	
Ticks	4	Lice	1
Lice	5	Mange mites	2
Mosquitoes	6	Poultry	
Face fly	7	Lice	1
Scabies mites	8	Mites	2
Mange mites	9	Others (ticks, chiggers)	3

[a] Ranked on the basis of Losses in Agriculture, *U.S. Dep. Agric. Agric. Handb.*, No. 291, 1965.

Table 180
RELATIVE IMPORTANCE OF LOSSES IN LIVESTOCK ATTRIBUTABLE TO DISEASES, NUTRITIONAL DISORDERS, AND PARASITES

Cause	Rank	Cause	Rank
Cattle		Swine	
Diseases		Diseases	
Respiratory		Respiratory	2
Feedlots	2	Enteric	2
Dairy	4	Reproductive	4
Enteric		Other	5
Calf scours	3	Nutritional disorders	
Feedlots	4	Metritis-mastitis-agalactia	3
Dairy	5	Iron deficiency	5
Reproductive	4	Hypoglycemia	5
Bacterial		Parasites	1
Pinkeye	4	Poultry	
Mastitis	1	Diseases	
Mycotic	5	Bacterial	1
Nutritional disorders		Salmonellosis	7
Milk fever	7	Viral	
Ketosis	8	Newcastle	3
Grass tetany	9	Infectious bronchitis	5
Parasites		Neoplasms	6
Helminths	3	Mycotic	8
Coccidiosis	6	Nutritional disorders	4
Anaplasmosis	4	Parasites	2
Sheep			
Diseases			
Bacterial	1		
Viral			
Bluetongue	5		
Chronic progressive pneumonia	3		
Mycotic	4		
Nutritional disorders	6		
Parasites	2		

Source: Losses in Agriculture, *U.S. Dep. Agric. Agric. Handb.*, No. 291, 1965.

Table 181
RELATIVE IMPORTANCE OF LOSSES IN LIVESTOCK PRODUCTION FROM SPECIFIC TOXIC CONDITIONS CAUSED BY ANY OF SEVERAL PLANTS WITH SIMILAR ACTIONS

Condition	Source	Rank
Nitrate poisoning	Many plants accumulate nitrate	1
Photosensitization	*Ammi* spp.; many plants cause poisoning and/or secondary photosensitization	2
Mycotoxicosis	*Fusarium* spp.	3
Wheat pasture poisoning		3
Bermudagrass staggers		3
Ergotism	*Claviceps* spp.	4
Prussic acid poisoning	Johnsongrass, sorghums	5
Oxalate poisoning	*Halogeton* spp.	5
Pulmonary edema		6
Pulmonary adenomatosis		6
Grass tetany		6

Source: *U.S. Dept. Agric. Agric. Handb.,* No. 291, 1965.

Table 182
SOME PLANT TOXINS IN PASTURE SPECIES

Species	Source	Effects
Tall fescue	Alkaloids and/or endophytic fungus *Acremonium coernophialium, Epichloe typhina*	Reduced intake, production; increased body temperature, respiration, fescue toxicosis
Reed canarygrass	Indole alkaloids	Reduced intake, palatability, animal performance
Phalaris spp.	Unidentified toxin in *Phalaris* staggers	*Phalaris* staggers
Ryegrass	Tremorgens, unidentified toxins	Ryegrass staggers, neuromuscular disease, mainly in sheep
Brachiaria and temperate grasses	Heptatoxin produced by *Pithomyces chartarum* growing on plant litter	Facial eczema, liver damage, and photosensitization
Sorghum and *Cynodon* spp.	Cyanoglycosides	Cyanide toxicity; thiocyanates — goiter. Sulfur deficiency in animals grazing sorghums
Setaria, Cenchrus, Pennisetum	Oxalates metabolized in rumen, but absorbed at high concentrations	Hypocalcemia and uremia, osteodystrophia fibrosa in horses
Crownvetch, *Astragalus, Indigofera*		Growth depression, toxicity in monogastrics on crown vetch
Lathyrus spp.		Neurotoxic
Crotalaria spp.		Hepatotoxic

Modified from Heath, M. E., Barnes, R. F., and Metcalfe, D. S., *Forages, The Science of Grassland Agriculture,* 4th ed., Iowa State University Press, Ames, 1985.

Table 183
RELATIVE IMPORTANCE OF LOSSES IN
LIVESTOCK PRODUCTION ATTRIBUTABLE TO
SOME OF THE MORE PREVALENT POISONOUS
PLANTS

Plant	Common name	Rank
Quercus spp.	Oaks	1
Astragalus spp.	Locoweed	2
Gutierrezia microcephala	Broomweed, snakeweed	2
Senecio spp.	Groundsels	2
Hymenoxys odorata	Bitter rubberweed	3
Baileya multiradiata	Desert baileya	4
Brassica spp.	Mustards	4
Delphinium spp.	Larkspur	4
Helenium spp.	Sneezeweeds	4
Prosopis glandulosa	Honey mesquite	4
Solanum spp.	Nightshades	4
Crotalaria spp.	Rattlebox	5
Euphorbia spp.	Spurge	5
Kallstroemia spp.	Carpetweed	5
Lupinus spp.	Lupines	5
Pteridium aquilinum	Bracken fern	5
Acacia berlandieri	Guajillo	6
Agave lecheguilla	Lechuguilla	6
Amaranthus spp.	Pigweed	6
Asclepias spp.	Milkweed	6
Cassia occidentalis	Coffee senna	6
Festuca spp.	Fescue	6
Nerium oleander	Oleander	6
Nolina texana	Texas sacahuista	6
Psilostrophe spp.	Paperflower	6
Ricinus communis	Castorbean	6
Sesbania spp.	Sesbania	6
Veratrum spp.	False hellebore	6
Xanthium spp.	Cocklebur	6
Flourensia cernua	Tarbrush, blackbrush	7
Centaurium spp.	Texas star	8
Isocoma wrightii	Rayless goldenrod	8
Lobelia inflata	Indian tobacco	8
Panicum spp.	Panicgrass	8
Peganum harmala	African-rue	8
Phyllanthus abnormis	Abnormal leafflower	8

Source: *U.S. Dep. Agric. Agric. Handb.*, No. 291, 1965.

Table 184
ESTIMATED LOSSES OF SHEEP AND LAMBS TO PREDATORS IN 1977

Region	Percentage loss of inventory plus lamb crop
California, Oregon, and Washington	5.98
Arizona, Colorado, Idaho, Montana, Nevada, New Mexico, Utah, and Wyoming	7.30
Texas	6.66
Iowa, Kansas, Minnesota, Missouri, Nebraska, North Dakota, and South Dakota	6.91
Illinois, Indiana, Michigan, and Wisconsin	3.13
New York and West Virginia	4.70
Arkansas, Kentucky, Louisiana, and Oklahoma	2.12

Modified from Drummond, R. L., Lambert, G., Smalley, H. E., Jr., and Terrill, C. E., Estimated losses of livestock to pests, in *CRC Handbook of Pest Management in Agriculture,* Vol. 1, Pimentel, D., Ed., CRC Press, Boca Raton, Fla., 1981, 111.

HORTICULTURAL CROPS

STORAGE

Table 185
STORAGE REQUIREMENTS AND PROPERTIES OF FRUITS

Commodity	Temperature (°F)	Relative humidity (%)	Approx. storage life	Highest freezing point (°F)	Water content (%)	Specific heat (Btu/lb, °F)[a]
Apples	30—40	90—95	1—12 months	29.3	84.1	0.87
Apricots	31—32	90—95	1—3 weeks	30.1	85.4	0.88
Avocados[b]	40—55	85—90	2—8 weeks	31.5	76.0	0.81
Bananas, green	56—58	90—95	—[c]	30.6	75.7	0.81
Berries						
Blackberries	31—32	90—95	2—3 d	30.5	84.8	0.88
Blueberries	31—32	90—95	2 weeks	29.7	83.2	0.86
Cranberries	36—40	90—95	2—4 months	30.4	87.4	0.90
Currants	31—32	90—95	1—4 weeks	30.2	84.7	0.88
Gooseberries	31—32	90—95	3—4 weeks	30.0	88.9	0.91
Loganberries	31—32	90—95	2—3 d	29.7	83.0	0.86
Raspberries	31—32	90—95	2—3 d	30.0	82.5	0.86
Strawberries	32	90—95	5—7 d	30.6	89.9	0.92
Cherries, sour	32	90—95	3—7 d	29.0	83.7	0.87
Cherries, sweet	30—31	90—95	2—3 weeks	28.8	80.4	0.84
Coconuts	32—35	80—85	1—2 months	30.4	46.9	0.58
Dates[d]	0 or 32	75	6—12 months	3.7	22.5	0.38
Figs, fresh	31—32	85—90	7—10 weeks	27.6	78.0	0.82
Grapefruit, Calif. and Ariz.	58—60	85—90	6—8 weeks	—	87.5	0.90
Grapefruit, Fla. and Tex.	50—60	85—90	6—8 weeks	30.0	89.1	0.91
Grapes, vinifera[e]	30—31	90—95	1—6 months	28.1	81.6	0.85
Grapes, American	31—32	85	2—8 weeks	29.7	81.9	0.86
Kiwifruit	31—32	90—95	3—5 months	29.0	82.0	0.86
Lemons	[f]	85—90	1—6 months	29.4	87.4	0.90
Limes	48—50	85—90	6—8 weeks	29.1	89.3	0.91
Lychees	35	90—95	3—5 weeks	—	81.9	0.86
Mangoes[g]	55	85—90	2—3 weeks	30.3	81.7	0.85
Nectarines	31—32	90—95	2—4 weeks	30.4	81.8	0.85
Olives, fresh[h]	41—50	85—90	4—6 weeks	29.4	80.0	0.84
Oranges, Calif. and Ariz.	38—48	85—90	3—8 weeks	29.7	85.5	0.88
Oranges, Fla. and Tex.	32—34	85—90	8—12 weeks	30.6	86.4	0.89
Papayas	45	85—90	1—3 weeks	30.4	88.7	0.91
Passion fruit	45—50	85—90	3—5 weeks	—	75.1	0.80
Peaches	31—32	90—95	2—4 weeks	30.3	89.1	0.91
Pears[i]	29—31	90—95	2—7 months	29.2	83.2	0.87
Persimmons, Japanese	30	90	3—4 months	28.1	78.2	0.83
Pineapples[j]	45—55	85—90	2—4 weeks	30.0	85.3	0.88
Plums and prunes[k]	31—32	90—95	2—5 weeks	30.5	86.6	0.89
Pomegranates	41	90—95	2—3 months	26.6	82.3	0.86
Quinces	31—32	90	2—3 months	28.4	83.8	0.87
Tangerines, mandarins, and related citrus	40	90—95	2—4 weeks	30.1	87.3	0.90

[a] Specific heat: $S = 0.008 \times$ (percent water in food) $+ 0.20$.

[b] Optimum storage temperature differs with cultivar because they differ in sensitivity to chilling injury.

Table 185 (continued)
STORAGE REQUIREMENTS AND PROPERTIES OF FRUITS

c   Combined use of reduced level of oxygen and increased level of carbon dioxide delays ripening. Best holding temperature is 56°F for 'Gros Michel' and 58°F for 'Valery': ordinarily held for about 2 to 4 d.
d   For best retention of flavor, texture, color, and aroma, store 'Deglet Noor' and similar cultivars at 32°F or lower.
e   For maximum storage life, precool after harvest by forced-air or tunnel cooling.
f   Conditioned at about 58°F and 85 to 90% relative humidity. In ventilated storage, should keep 1 to 4 months and sometimes 6 months.
g   Should keep 2 or 3 weeks at 55°F.
h   Above 50°F, olives ripen and shrivel.
i   Very sensitive to temperature; precise control is needed.
j   Hawaiian pineapples harvested at half-ripe stage can be held for 2 weeks at 45 to 55°F and still have one week's shelf life.
k   Not adapted to long cold storage; storage beyond 5 weeks often results in flesh browning and abnormal flavors.

Data from Hardenburg, R. E., Watada, A. E., and Wang, C. Y., The Commercial Storage of Fruits, Vegetables, and Florist and Nursery Stocks, *U.S. Dep. Agric. Agric. Handb.,* No. 66, rev. September 1986.

Table 186
STORAGE REQUIREMENTS AND PROPERTIES OF VEGETABLES

Commodity	Temperature (°F)	Relative humidity (%)	Approx. storage life	Highest freezing point (°F)	Water content (%)	Specific heat (Btu/lb, °F)[a]
Artichoke, globe	32	95—100	2—3 weeks	29.9	83.7	0.87
Artichoke, Jerusalem	31—32	90—95	4—5 months	28.0	79.8	0.84
Asparagus	32—35	95—100	2—3 weeks	30.9	93.0	0.94
Beans, dry	40—50	40—50	6—10 months	—	15.0	0.32
Beans, green[b]	40—45	95	7—10 d	30.7	88.9	0.91
Beans, lima[c]	37—41	95	5—7 d	31.0	66.5	0.73
Bean sprouts	32	95—100	7—9 d	—	88.8	0.91
Beets, bunched	32	98—100	10—14 d	31.3	—	—
Beets, topped	32	98—100	4—6 months	30.3	87.6	0.90
Broccoli	32	95—100	10—14 d	30.9	89.9	0.92
Brussels sprouts	32	95—100	3—5 weeks	30.5	84.9	0.88
Cabbage, late	32	98—100	5—6 months	30.4	92.4	0.94
Carrots, bunched	32	95—100	2 weeks	—	—	—
Carrots, mature	32	98—100	7—9 months	29.5	88.2	0.91
Carrots, immature	32	98—100	4—6 weeks	29.5	88.2	0.91
Cauliflower	32	95—98	3—4 weeks	30.6	91.7	0.93
Celery	32	98—100	2—3 months	31.1	93.7	0.95
Chard	32	95—100	10—14 d	—	91.1	0.93
Collards	32	95—100	10—14 d	30.6	86.9	0.90
Corn, sweet	32	95—98	5—8 d	30.9	73.9	0.79
Cucumbers	50—55	95	10—14 d	31.1	96.1	0.97
Eggplants	46—54	90—95	1 week	30.6	92.7	0.94
Endive	32	95—100	2—3 weeks	31.9	93.1	0.95
Greens, leafy	32	95—100	10—14 d	—	—	—
Horseradish	30—32	98—100	10—12 months	28.7	74.6	0.80
Kale	32	95—100	2—3 weeks	31.1	86.6	0.89
Kohlrabi	32	98—100	2—3 months	30.2	90.3	0.92
Leeks	32	95—100	2—3 months	30.7	85.4	0.88
Lettuce	32	98—100	2—3 weeks	31.7	94.8	0.96
Melons						
Cantaloupe[d]	36—41	95	15 d	29.9	92.0	0.94

Table 186 (continued)
STORAGE REQUIREMENTS AND PROPERTIES OF VEGETABLES

Commodity	Temperature (°F)	Relative humidity (%)	Approx. storage life	Highest freezing point (°F)	Water content (%)	Specific heat (Btu/lb, °F)[a]
Crenshaw	45	90—95	2 weeks	30.1	92.7	0.94
Honeydew	45	90—95	3 weeks	30.3	92.6	0.94
Watermelons	50—60	90	2—3 weeks	31.3	92.6	0.94
Mushrooms	32	95	3—4 d	30.4	91.1	0.93
Okra	45—50	90—95	7—10 d	28.7	89.8	0.92
Onion, green	32	95—100	3—4 weeks	30.4	89.4	0.91
Onion, dry[e]	32	65—70	1—8 months	30.6	87.5	0.90
Onion sets	32	65—70	6—8 months	30.6	87.5	0.90
Parsley	32	95—100	2—2.5 months	30.0	85.1	0.88
Parsnips	32	98—100	4—6 months	30.4	78.6	0.83
Peas, green	32	95—98	1—2 weeks	30.9	74.3	0.79
Peas, southern	40—41	95	6—8 d	—	66.8	0.73
Peppers, dry	32—50	60—70	6 months	—	12.0	0.30
Peppers, sweet	45—55	90—95	2—3 weeks	30.7	92.4	0.94
Potatoes, early	f	90—95	f	30.9	81.2	0.85
Potatoes, late	g	90—95	5—10 months	30.9	77.8	0.82
Pumpkins	50—55	50—70	2—3 months	30.5	90.5	0.92
Radishes, spring	32	95—100	3—4 weeks	30.7	94.5	0.96
Radishes, winter	32	95—100	2—4 months	—	—	—
Rhubarb	32	95—100	2—4 weeks	30.3	94.9	0.96
Rutabagas	32	98—100	4—6 months	30.0	89.1	0.91
Salsify	32	95—98	2—4 months	30.0	79.1	0.83
Spinach	32	95—100	10—14 d	31.5	92.7	0.94
Squash, summer	41—45	95	1—2 weeks	31.1	94.0	0.95
Squash, winter	50	50—70	—[h]	30.5	85.1	0.88
Sweetpotatoes	55—60[i]	85—90	4—7 months	29.7	68.5	0.75
Tomatoes, mature-green	55—70[j]	95—95	1—3 weeks	31.0	93.0	0.94
Tomatoes, firm-ripe	46—50	90—95	4—7 d	31.1	94.1	0.95
Turnips	32	95	4—5 months	30.1	91.5	0.93
Turnip greens	32	95—100	10—14 d	31.7	90.3	0.92
Watercress	32	95—100	2—3 weeks	31.4	93.3	0.95
Yams	61	70—80	6—7 months	—	73.5	0.79

[a] Specific heat: $S = 0.008 \times$ (percent water in food) $+ 0.20$. In metric units of kJ/kg/°C, $S = 0.0335 \times$ (percent water in food) $+ 0.8374$.

[b] Highly perishable — cool rapidly after harvest.

[c] Precool, preferably by hydrocooling, immediately after harvest.

[d] Three-quarter slip.

[e] Adequately cure before storage; storage quality is influenced by cultivar and conditions under which they are grown.

[f] Not stored, except briefly. If free from bruising and decay, can be held 4 to 5 months at 40°F when cured 4 or 5 d at about 65°F.

[g] Cure by holding at 50 to 60°F and high humidity for 10 to 14 d. Storage at 45°F and 95% relative humidity is recommended for Russet and Burbank potatoes. Most potatoes remain dormant for 5 to 8 months at 40°F.

[h] Acorn-type squash should keep 5 to 8 weeks at 50°F, while Butternut squash should keep 2 to 3 months at 50°F.

[i] Cure by holding at 85°F and 90 to 95% relative humidity for 4 to 7 d.

[j] Ripening of tomatoes is initiated by the ethylene they produce. In commercial practice, mature-green tomatoes are treated with supplemental ethylene; they cannot be held at temperatures that delay ripening.

Data from Hardenburg, R. E., Watanada, A. E., and Wang, C. Y., The Commercial Storage of Fruits, Vegetables, and Florist and Nursery Stocks, *U.S. Dep. Agric. Agric. Handb.*, No. 66, rev. September, 1986.

Table 187
STORAGE CONDITIONS FOR CUT FLOWERS AND NURSERY STOCK

Commodity	Storage temperature (°C)	Approximate storage period	Highest freezing point (°C)
Cut flowers[a]			
Calla lily	4	1 week	—
Camellia[b]	7	3—6 d	−0.7
Carnation[c]	−0.5 to 0	3—4 weeks	−0.7
Chrysanthemum	−0.5 to 0	3—4 weeks	−0.8
Daffodil (narcissus)[d]	0—0.5	1—3 weeks	−0.1
Dahlia	4	3—5 d	—
Gardenia[b]	0—1	2 weeks	−0.6
Gladiolus[e]	2—5	5—8 d	−0.3
Iris, tight buds	−0.5 to 0	1—2 weeks	−0.8
Lily, Easter	0—1	2—3 weeks	−0.5
Lily-of-the-valley[f]	−0.5 to 0	2—3 weeks	—
Orchid, cattleya[b,g]	7—10	2 weeks	−0.3
Peony, tight buds	0—1	2—6 weeks	−1.1
Rose (dry pack)[h]	−0.5 to 0	2 weeks	−0.5
Snapdragon	4	1—2 weeks	−0.9
Tulips[i]	−0.5 to 0	2—3 weeks	—
Florist greens[a,j]			
Asparagus (plumosa)[k]	2—4	2—3 weeks	−3.3
Fern, dagger, and wood[k]	0	2—3 months	−1.7
Holly[k]	0—4	3—5 weeks	−2.8
Huckleberry[k]	0	1—4 weeks	−3.0
Laurel	0	2—4 weeks	−2.5
Magnolia	2—4	2—4 weeks	−2.8
Rhododendron	0	2—4 weeks	−2.5
Salal[k]	0	2—3 weeks	−2.9
Bulbs, corms[l]			
Begonia, tuberous	2—7	3—5 months	−0.5
Caladium	21	—	−1.3
Crocus	17	2—3 months	—
Dahlia	4—9	5 months	−1.8
Gladiolus[m]	7—10	5—8 months	−2.1
Hyacinth[n]	17—20	2—5 months	−1.5
Iris, Dutch[o]	20—25	4—12 months	—
Lilium	−0.5 to 0.5	1—10 months	−1.7
Lily, gloriosa	10—17	3—4 months	—
Narcissus[p]	13—17	2—4 months	−1.3
Peony	0.2	5 months	—
Tulip[q]	17	2—6 months	−2.4
Nursery stock			
Rose budwood	−2 to 0.5	1—2 years	—
Rose bushes	−0.5 to 2	4—5 months	—
Strawberry plants	−1 to 0	8—10 months	—
Trees and shrubs	0—2	5—6 months	—
Herbaceous perennials	−2.8 to −2.2	4—8 months	—
Christmas trees[r]	−5 to 0	6—7 weeks	—

[a] High relative humidity of 90 to 95% is recommended in refrigerated storage rooms for cut flowers and florist greens. Some flowers for which a temperature of 4°C is recommended probably could be stored longer and safely at lower temperatures.
[b] Not placed in water for handling or storage, but may be misted.
[c] Dry-pack storage is better than storage in water and 0° is better than 4°.
[d] Will keep 1 to 3 weeks at 0 to 0.5°C with dry packaging.
[e] Cut when one or two lowest florets show color, but still in bud stage.

Table 187 (continued)
STORAGE CONDITIONS FOR CUT FLOWERS AND NURSERY STOCK

[f]   Cut when top floret is yellow-green; store with dry packaging.
[g]   Stems of orchids should be placed in vials of water (some may be stored by dry-pack methods).
[h]   Precool and hydrate after harvest; dry pack in moisture-proof containers.
[i]   Keeps well in dry storage.
[j]   Most stored with stems in water, for 1 to 2 weeks at 4°C at the retail level.
[k]   Usually held in moisture-retentive shipping cases.
[l]   Desirable relative humidity for most bulbs is 70 to 90%, with adequate ventilation.
[m]   Cure for 10 d at 27 to 29°C prior to dry storage.
[n]   Cure at 25 to 27°C for several weeks; store in wire-bottom trays.
[o]   Cure for 10 to 15 d at 32°C, then at 18 to 24°C for 2 to 3 weeks.
[p]   Cure at 30°C for 4 d.
[q]   Cure at 26°C for one week.
[r]   Overwrapping trees with polyethylene prevents drying.

Data from Hardenburg, R. E., Watada, A. E., and Wang, C. Y., The Commercial Storage of Fruits, Vegetables, and Florist and Nursery Stocks, *U.S. Dep. Agric. Agric. Handb.,* No. 66, revised September, 1986.

Table 188
SENSITIVITY OF SOME FRUITS AND VEGETABLES TO FREEZING[12]

Most sensitive[b]	Moderately sensitive[c]	Least sensitive[d]
Apricots	Apples	Beets[a]
Asparagus	Broccoli, sprouting	Brussels sprouts
Avocados	Cabbage, new	Cabbage, old
Bananas	Carrots[a]	Dates
Beans, snap	Cauliflower	Kale
Berries (except cranberries)	Celery	Kohlrabi
Cucumbers	Cranberries	Parsnips
Eggplant	Grapefruit	Rutabagas
Lemons	Grapes	Salsify
Lettuce	Onions, dry	Turnips[a]
Limes	Oranges	
Okra	Parsley	
Peaches	Pears	
Peppers, sweet	Peas	
Plums	Radishes[a]	
Potatoes	Spinach	
Squash, summer	Squash, winter	
Sweetpotatoes		
Tomatoes		

[a]   Without tops.
[b]   Commodities likely to suffer injury by one light freezing.
[c]   Commodities able to recover from one or two light freezings.
[d]   Commodities that can be lightly frozen several times without sustaining serious damage.

Source: Hardenburg, R. E., Watada, A. E., and Wang, C. Y., The commercial storage of fruits, vegetables, and florist and nursery stocks, *U.S. Dep. Agric. Agric. Handb.,* No. 66, revised September, 1986.

Table 189
FRUITS AND VEGETABLES SUSCEPTIBLE TO CHILLING
INJURY WHEN STORED AT MODERATELY LOW BUT
NONFREEZING TEMPERATURES

Commodity	Approximate lowest safe temperature °C	Character of injury when stored betwwen 0°C and safe temperature[a]
Apples—certain varieties[b]	2—3	Internal browning, brown core, soggy breakdown, soft scald
Avocados[b]	4.5—13	Grayish brown discoloration of flesh
Bananas[b]	11.5—13	Dull color when ripened
Beans, snap	7	Pitting and russeting
Cranberries	2	Rubbery texture, red flesh
Cucumbers	7	Pitting, water-soaked spots, decay
Eggplants	7	Surface scald, alternaria rot
Grapefruit[b]	10	Scald, pitting, watery breakdown
Lemons[c]	11—13	Pitting, membranous stain, red blotch
Limes	7—9	Pitting
Mangos	10—13	Grayish scaldlike discoloration of skin, uneven ripening
Melons		
Cantaloups[d]	2—5	Pitting, surface decay
Honey dew	7—10	Pitting, surface decay, failure to ripen
Casaba, Crenshaw and Persian	7—10	Ditto
Watermelons	4.5	Pitting, objectionable flavor
Okra	7	Discoloration, water-soaked areas, pitting
Olives, fresh	7	Internal browning
Oranges, California and Arizona	3	Pitting, brown stain
Papayas	7	Pitting, failure to ripen, off flavor, decay
Peppers, sweet	7	Sheet pitting, alternaria rot of pods and calyxes
Pineapples	7—10	Dull green when ripened
Potatoes	3	Mahogany browning, sweet flavor
Pumpkins and hard-shell squashes	10	Decay, especially alternaria rot
Sweet potatoes	13	Decay, pitting, internal discoloration
Tomatoes		
Ripe	7—10	Watersoaking, decay
Mature-green	13	Alternaria rot, poor color when ripe

[a]   Often these symptoms are apparent only after removal to warm temperatures, as in marketing.
[b]   Depends on the variety.
[c]   Depends on previous history of lemons. Short periods (2 to 3 weeks) at the market at 0°C usually not harmful.
[d]   Depends on maturity and length of storage. Fullslip cantaloups may be held 1 to 2 weeks at 0 to 1°C.

Adapted from Hardenburg, R. E., Watanda, A. E., and Wang, C. Y., The commercial storage of fruits, vegetables, and florist and nursery stocks, *U.S. Dep. Agric. Agric. Handb.*, No. 66, rev., September, 1986.

Please send us some
of your mailing labels.

Thank you,
UNCA

Table 190

VARIATIONS IN THE NUTRIENT COMPOSITION OF CANNED SAMPLES OF A SINGLE SWEET POTATO CULTIVAR, JEWEL, GROWN IN THE SAME YEAR FROM A SINGLE SEED PIECE SOURCE BUT IN SEVEN DIFFERENT GEOGRAPHICAL LOCATIONS

Nutrient	State							Average	Standard deviation
	Ala.	N.C.	Ark.	Tenn.	Miss.	La.	Va.		
Solids (%)	32.7	28.1	26.5	27.6	27.9	30.0	27.5	28.6	2.1
Fat (%)	0.3	0.3	0.2	0.3	0.2	0.3	0.2	0.3	0.1
Protein (%)	1.4	1.2	1.0	1.1	1.1	1.4	1.0	1.2	0.2
Ash (%)	0.4	0.5	0.4	0.4	0.8	0.5	0.5	0.5	0.1
Carbohydrate (%)	30.6	26.1	24.9	25.8	25.8	27.8	25.8	26.7	1.9
Calories	131	112	105	110	109	120	109	114	9
Calcium (mg%)	9	12	20	5	11	10	12	11	5
Iron (mg%)	0.7	1.3	0.9	0.9	0.5	1.2	0.9	0.9	0.3
Thiamine (mg%)	0.01	0.03	0.03	0.02	0.02	0.03	0.01	0.02	0.01
Riboflavin (mg%)	0.03	0.03	0.03	0.04	0.03	0.04	0.03	0.03	0
Niacin (mg%)	0.3	0.3	0.1	0.2	0.3	0.3	0.4	0.3	0.1
Vitamin A (IU)	6500	6585	4415	5250	4985	5025	4650	5345	862
Vitamin C (mg%)	9.0	7.0	8.0	9.0	10.0	13.0	12.0	10	2

From S-101 Tech. Comm. on Sweet Potato Quality, 1979, Louisiana State Agric. Exp. Sta., Baton Rouge, La.

SELECTED NUTRIENTS

Table 191
AVERAGE AMOUNT OF SELECTED NUTRIENTS IN 100 g OF FRUIT (EDIBLE PORTION)

Fruit[a]	Protein (g)	Vitamin A (IU)	Thiamin (mg)	Riboflavin (mg)	Ascorbic acid (mg)	Niacin (mg)	Iron (mg)	Phosphorus (mg)	Calcium (mg)
Apples[b]	0.19	53	0.017	0.014	5.7	0.077	0.18	7	7
Apricots	1.40	2612	0.030	0.040	10.0	0.600	0.54	19	14
Avocado	1.98	612	0.108	0.122	7.9	1.921	1.02	41	11
Banana	1.03	81	0.045	0.100	9.1	0.540	0.31	20	6
Blackberries	0.72	165	0.030	0.040	21.0	0.400	0.57	21	32
Blueberries	0.67	100	0.048	0.050	13.0	0.359	0.17	10	6
Cantaloupe	0.88	3224	0.036	0.021	42.2	0.574	0.21	17	11
Cherries, sweet	1.20	214	0.050	0.060	7.0	0.400	0.39	19	15
Cranberries	0.39	46	0.030	0.020	13.5	0.100	0.20	9	7
Dates[c]	1.97	50	0.090	0.100	0.0	2.200	1.15	40	32
Figs	0.75	142	0.060	0.050	2.0	0.400	0.37	14	35
Grapefruit, white	0.69	10	0.037	0.020	33.3	0.269	0.06	8	12
Grapes[d]	0.66	73	0.092	0.057	10.8	0.300	0.26	13	11
Lemon	1.10	29	0.040	0.020	53.0	0.100	0.60	16	26
Limes	0.70	10	0.030	0.020	29.1	0.200	0.60	18	33
Nectarines	0.94	736	0.017	0.041	5.4	0.990	0.15	16	5
Oranges[e]	0.94	205	0.087	0.040	53.2	0.282	0.10	14	40
Peaches	0.70	535	0.017	0.041	6.6	0.990	0.11	12	5
Pears	0.39	20	0.020	0.040	4.0	0.100	0.25	11	11
Plums	0.79	323	0.043	0.096	9.5	0.500	0.10	10	4
Raspberries	0.91	130	0.030	0.090	25.0	0.900	0.57	12	22
Strawberries	0.61	27	0.020	0.066	56.7	0.230	0.38	19	14
Tangerines	0.63	920	0.105	0.022	30.8	0.160	0.10	10	14
Watermelon	0.62	366	0.080	0.020	9.6	0.200	0.17	9	8

a   Raw.
b   With skin.
c   Domestic, natural dry.
d   European type, adherent skin.
e   All commercial varieties.

From Composition of Foods: Fruits and Fruit Juices, *U.S. Dep. Agric. Agric. Handb.*, No. 8—9, rev. August 1982.

Table 192
AVERAGE AMOUNT OF SELECTED NUTRIENTS IN 100 g OF VEGETABLES (EDIBLE PORTION)

Vegetables[a]	Protein (g)	Vitamin A (IU)	Thiamin (mg)	Riboflavin (mg)	Ascorbic acid (mg)	Niacin (mg)	Iron (mg)	Phosphorus (mg)	Calcium (mg)
Asparagus[b]	2.59	829	0.099	0.121	27.1	1.052	0.66	61	24
Beans, lima	6.81	370	0.140	0.096	10.1	1.040	2.45	130	32
Beans, snap	1.89	666[c]	0.074	0.097	9.7	0.614	1.28	39	46
Beets	1.06	13	0.031	0.014	5.5	0.273	0.62	31	11
Broccoli	2.97	1,409	0.082	0.207	62.8	0.755	1.15	48	114
Brussels sprouts	2.55	719	0.107	0.080	62.0	0.607	1.20	56	36
Cabbage	0.96	86	0.057	0.055	24.3	0.230	0.39	25	33
Carrots	1.09	24,554	0.034	0.056	2.3	0.506	0.62	30	31
Cauliflower	1.87	14	0.063	0.052	55.4	0.552	0.42	35	27
Celery[d]	0.66	127	0.030	0.030	6.3	0.300	0.48	26	36
Corn, sweet	3.32	217[e]	0.215	0.072	6.2	1.614	0.61	103	2
Cucumber[d]	0.54	45	0.030	0.020	4.7	0.300	0.28	17	14
Eggplant	0.83	64	0.076	0.020	1.3	0.600	0.35	22	6
Lettuce[d,f]	1.01	330	0.046	0.030	3.9	0.187	0.50	20	19
Mustard greens	2.26	3,031	0.041	0.063	25.3	0.433	0.70	41	74
Onions	0.90	0	0.042	0.008	5.7	0.080	0.20	23	27
Peas, green	5.36	597	0.259	0.149	14.2	2.021	1.54	117	27
Peppers, sweet	0.62	338[g]	0.053	0.035	111.4	0.363	0.88	15	4
Potatoes[h]	1.71	—	0.098	0.019	7.4	1.312	0.31	40	8
Radishes[d]	0.60	8	0.005	0.045	22.8	0.300	0.29	18	21
Spinach	2.97	8,190	0.095	0.236	9.8	0.490	3.57	56	136
Sweet potatoes[h]	1.65	17,054	0.053	0.140	17.1	0.640	0.56	27	21
Tomatoes, red[d]	0.89	1,133	0.060	0.050	17.6	0.600	0.48	23	7
Turnips	0.71	0	0.027	0.023	11.6	0.299	0.22	19	22

a Cooked, boiled, and drained, except as noted.
b Includes cuts and spears.
c Green varieties; yellow varieties, 81 IU.
d Raw.
e Yellow varieties; only trace in white varieties.
f Iceberg.
g Green varieties; red varieties, 3,760 IU.
h Boiled, cooked without skin, flesh.

From Composition of Foods: Vegetables and Vegetable Products, *U.S. Dep. Agric. Agric. Handb.*, No. 8—11, rev. August 1984.

Table 193
AVERAGE AMOUNT OF PROTEIN AND AMINO ACIDS IN 100 g OF
SELECTED NUTS (EDIBLE PORTION)[a]

Nutrients (g)	Almonds[b]	Brazil nuts[b]	Cashew[c]	Filberts[b]	Peanuts	Pecans	Pistachio	Walnuts[d]
Protein	19.95	14.34	15.31	13.04	25.67	7.75	20.58	14.29
Amino Acids								
Tryptophan	0.358	0.260	0.237	0.216	0.310	0.199	0.283	0.189
Threonine	0.739	0.460	0.592	0.448	0.743	0.253	0.722	0.448
Isoleucine	0.866	0.601	0.731	0.568	0.997	0.322	0.975	0.566
Leucine	1.552	1.187	1.285	1.100	1.928	0.520	1.677	0.992
Lysine	0.666	0.541	0.817	0.399	0.992	0.292	1.278	0.388
Methionine	0.227	1.014	0.274	0.162	0.263	0.186	0.381	0.280
Cystine	0.358	0.349	0.283	0.229	0.329	0.209	0.513	0.345
Phenylalanine	1.113	0.746	0.791	0.686	1.467	0.409	1.184	0.628
Tyrosine	0.705	0.457	0.491	0.453	1.232	0.284	0.714	0.439
Valine	1.028	0.911	1.040	0.662	1.161	0.386	1.410	0.723
Arginine	2.495	2.390	1.741	2.155	3.456	1.105	2.186	2.103
Histidine	0.558	0.402	0.399	0.327	0.748	0.227	0.536	0.359
Alanine	0.943	0.570	0.702	0.708	1.133	0.338	0.994	0.609
Aspartic acid	2.349	1.355	1.505	1.604	3.451	0.708	2.116	1.475
Glutamic acid	5.934	3.151	3.624	3.537	6.094	1.545	4.916	2.809
Glycine	1.236	0.657	0.803	0.704	1.773	0.377	1.095	0.755
Proline	1.255	0.762	0.690	0.509	1.246	0.360	0.947	0.553
Serine	0.901	0.746	0.849	0.669	1.434	0.376	1.351	0.782

[a] Dried.
[b] Dried, unblanched.
[c] Dry roasted.
[d] English.

From Composition of Food: Nut and Seed Products, *U.S. Dep. Agric. Agric. Handb.*, No. 8—12, rev. September 1984.

Table 194

TOXIC COMPOUNDS OF PLANT ORIGIN FOUND IN HORTICULTURAL CROPS

Class and specific examples	Plant source	Effect
Nitrites and nitrates	Green leafy vegetables	Methemoglobinemia Nitrosamine formation
Trace elements		
Arsenic	Green vegetables	Trivelent, form most toxic, Enzyme inhibition
Copper	Nuts, leafy plants, grapes	Liver malfunction, counteracted by high zinc and iron
Cobalt	Low levels in many vegetables	Anemia, heart failure
Selenium	Vegetables, esp. cabbage	Enzyme inhibition. Counteracted by sulfate and protein in diet
Cadmium	Traces in vegetables and fruit	Hypertension, reproductive organs. Some interaction with Zn, Cu, Fe
Chromium	Fruits and vegetables	Liver, kidney damage at high levels
Hemagglutinins	Legumes, esp. beans	Clumping of red-blood cells Goitrogenic effects
Allergens	Beans, tomatoes, strawberries and nuts	
Toxic peptides	Mushrooms	Liver damage
Aminonitriles and unusual amino acids	*Lathyrus sativus* beans	Paralysis
Amines		
Vaso-active compounds	Tomatoes, potatoes, bananas	Heart, hypertension effects
Hallucinogens	Mushrooms, spices	Hallucinations
Glucosinolates (thioglucosides)	Crucifers, Brassica species and onion, garlic, chives	Goiter. Form goitrins from isothioscyanates
Antivitamins		
Thiamin antagonists	Beans, berries, cabbage	Deficiency systems counteracted by vitamins
Niacin antagonists	Corn	
Pantothenic acid antagonists	Pea seedlings	
Vitamin D antagonists	Green vegetables	
Enzyme Inhibitors		
Trypsin and chymotripsin inhibitors	Beans, potatoes, peas	Growth inhibition
Amylase inhibitors	Beans, unripe bananas	
Catalase inhibitors	Green bananas	
Cholinesterose inhibitors	Potatoes, tomatoes, eggplant	Gastrointestinal and nervous disorders
Cyanogenetic glycosides	Almond, apricot, cherry, etc., beans, cassava	Cyanide poisoning
Plant phenolics	Fruits, vegetables, tree nuts	Carcinogenic, estrogenic
Gossypol		Affects liver and spleen
Tannine		function
Flavonoids		
Oxalates	Rhubarb, spinach	Kidney stones, gastric hemorrhage, low calcium absorption

Source: Toxicants occuring naturally in foods, National Academy of Science, Washington, D.C., 1973.

OILSEEDS — COMPOSITION AND FEEDING

Table 195
GRADES, GRADE REQUIREMENTS, AND GRADE DESIGNATIONS FOR SOYBEANS

Item	U.S. Grade[a]			
	No. 1	No. 2	No. 3[b]	No. 4[c]
Test weight, lb/bushel, min.	56.0	54.0	52.0	49.0
Moisture, %, max.	13.0	14.0	16.0	18.0
Splits, %, max.	10.0	20.0	30.0	40.0
Damaged kernels, % max.				
Total damaged	2.0	3.0	5.0	8.0
Heat damaged	0.2	0.5	1.0	3.0
Foreign material, % max.	1.0	2.0	3.0	5.0
Brown, black, and/or bicolored beans in yellow or green soybeans, %, max.	1.0	2.0	5.0	10.0

[a] U.S. sample grade shall be soybeans that do not meet the requirements for any of the grades from U.S. No. 1 to U.S. No. 4 inclusive, or that are musty, sour, or heating, or that have any commercially objectionable foreign odor, or that contain stones, or that are otherwise of distinctly low quality.

[b] Soybeans that are purple-mottled or stained shall be graded not higher than U.S. No. 3.

[c] Soybeans that are materially weathered shall be graded not higher than U.S. No. 4.

Source: Official United States Standards for Grain, Grain Division, Agricultural Marketing Service, U.S. Department of Agriculture, Washington, D.C. 1975, 8.1—8.6.

Table 196
QUALITY OF OILSEED PROTEINS BY VARIOUS MEASURES

	BV[a]	NPU[b]	PER[c]
Soybeans	73	61	2.32
Cottonseed	67	53	2.25
Peanuts	55	43	1.05
Sunflower	70	58	2.10

[a] BV = biological value.
[b] NPU = net protein utilization.
[c] PER = protein efficiency ratio.

From Smith, K. J., Soybean meal: production, composition and utilization, *Feedstuffs,* 49(3), 22, 1977. With permission.

Table 197

AVERAGE COMPOSITION OF SOYBEAN PRODUCTS COMMONLY USED IN ANIMAL FEEDS

	Soybeans, full-fat cooked	Soybean meal, dehulled, solvent	Soybean meal, expeller	Soybean meal, solvent	Soybean mill feed	Soybean hulls
International feed number	5-04-597	5-04-612	5-04-600	5-04-604	5-04-594	1-04-560
Dry matter (%)	90	90	90	89	89	91
Crude protein (%)	37.0	48.5	42.6	44.0	13.3	12.5
Ether extract (%)	18.0	1.0	4.0	0.8	1.6	
Crude fiber (%)	5.5	3.9	6.2	7.3	33.0	35.5
Energy,[a] kcal/kg						
Swine						
DE	4056	3860	3483	3350		
ME	3540	3485	2990	3090	1870	
Poultry						
ME_n	3300	2440	2430	2230	720	
MEpro	2170	1730	1720	1570	440	
Beef						
ME	3060		2760	2610		2550
NE_m	2170		1850	1720		1560
NE_g	1380		1230	1150		990
TDN (%)	84.6		76.5	72.1		70.5
Dairy						
ME	3370	2800	2780	2800		2750
NE_m	2170	1680	1870	1680		1630
NE_g	1380	1120	1240	1120		1080
NE_{lact}	1920	1660	216	1660		1630
TDN (%)	84.6	72.1	76.5	72.1		71.0

[a] DE = digestible energy; ME = metabolizable energy; ME_n = metabolizable energy, nitrogen corrected; ProE = productive energy; NE_m = net energy, maintenance; NE_g = net energy, gain; TDN = total digestible nutrients; NE_{lact} = net energy, lactation.

Source: National Research Council, *Nutrient Requirements of Domestic Animals,* National Academy of Sciences, Washington, D.C.

Table 198
AVERAGE AMINO ACID COMPOSITION OF SOYBEAN PRODUCTS COMMONLY USED IN ANIMAL FEEDS (%)

	Soybeans, full-fat, cooked	Soybean meal, dehulled, solvent	Soybean meal, expeller	Soybean meal, solvent	Soybean meal, feed
International feed number	5-04-597	5-04-612	5-04-600	5-04-604	5-04-594
Crude protein	37.0	48.5	42.6	44.0	13.3
Arginine	2.8	3.7	3.0	3.3	0.94
Histidine	0.9	1.3	1.1	1.2	0.18
Isoleucine	2.0	2.6	2.8	2.4	0.40
Leucine	2.8	3.8	3.6	3.5	0.57
Lysine	2.40	3.18	2.78	2.93	0.48
Methionine	0.5	0.7	0.7	0.7	0.10
Cystine	0.6	0.7	0.6	0.7	0.21
Phenylalanine	1.8	2.1	2.1	2.3	0.37
Tyrosine	1.2	2.0	1.4	1.3	0.23
Threnonine	1.50	1.91	1.71	1.81	0.30
Tryptophan	0.55	0.67	0.61	0.62	0.10
Valine	1.8	2.7	2.2	2.3	0.37
Glycine	2.0	2.3	2.4	2.3	0.40
Serine	—	2.9	2.0	2.4	—

Source: National Research Council, *Nutrient Requirements of Domestic Animals,* National Academy of Sciences, Washington, D.C.

Table 199
AVERAGE MINERAL COMPOSITION OF SOYBEAN PRODUCTS COMMONLY USED IN ANIMAL FEEDS

	Soybeans, full-fat, cooked	Soybean meal, dehulled solvent	Soybean meal, expeller	Soybean meal, solvent	Soybean mill feed	Soybean hulls
International feed number	5-04-597	5-04-612	5-04-600	5-04-604	5-04-594	1-04-560
Calcium (%)	0.25	0.27	0.27	0.29	0.37	0.54
Phosphorus (%)	0.58	0.62	0.61	0.65	0.19	0.16
Potassium (%)	1.61	2.02	1.83	2.00	1.50	—
Chlorine (%)	0.03	0.05	0.07	0.05	—	—
Magnesium (%)	0.21	0.27	0.26	0.27	0.12	—
Sodium (%)	0.28	0.34	0.27	0.34	—	—
Sulfur (%)	0.22	0.43	0.33	0.43	0.06	—
Copper (mg/kg)	15.8	36.3	18.0	36.3	—	—
Iron (mg/kg)	80	120	140	120	—	—
Manganese (mg/kg)	29.8	27.5	30.7	29.3	28.5	12.7
Selenium (mg/kg)	0.11	0.10	0.10	0.10	—	—
Zinc (mg/kg)	16	45	60	27	—	—

Source: National Research Council, *Nutrient Requirements of Domestic Animals,* National Academy of Sciences, Washington, D.C.

Table 200
NUTRITIONAL QUALITY GUIDE FOR SOYBEAN MEAL

Degree of cooking	Increase in pH[a]	Color test	Response to duPont test	PDI[b]	Compatibility with urea under normal conditions	Compatibility with urea under extreme conditions[c]
Raw	1.90—2.10	Red	Active	80—90	Unsafe	Unsafe
Undercooked	0.80—1.50	Pink	Active	45—80	Borderline	Unsafe
Properly cooked						
a	0.05—0.30	Pink	Active	20—30	Safe	Unsafe
b	0.05—0.10	Light pink	Inactive	12—30	Safe	Safe
Overcooked	0.02—0.05	Amber	Active	8—12	Safe	Safe

[a]　Using an approved Caskey-Knapp urease activity test such as AOCS Tentative Method BA 9-58, inquire of Oil Chemists' Society, 508 So. 6th St., Champaign, Ill. 61820.

[b]　This refers to Protein Dispensibility Index A.O.C.S. Method Ba 10-65, Revised 1969.

[c]　Five percent or more of urea, 20% or more of soybean meal together with 20% or more of molasses and/or hot, humid storage conditions.

From Hayward, J. W., Precision processing of soybean meal, in *Feedstuffs*, 47(16), 62, 1975. With permission.

Table 201
GROSS COMPOSITION OF PEANUTS[a]

	Blanched, full fat raw cotyledons		Shells (%)	Testa (%)	Germ (%)	Blanched defatted cotyledons (%)
	Range (%)	Avg. (%)				
Moisture	[a]	[a]	[a]	9.01	—	2.7
Protein	25.4—33.8	27.6	4.8—7.2	11.0—13.4	26.5—27.8	43.2
Lipid (oil)	44.5—56.3	52.1	1.2—2.8	0.5—1.9	39.4—43.0	16.6
Total carbohydrates	6.0—24.9	13.3	10.6—21.2	48.3—52.2	—	31.2
Reducing sugars	0.1—0.4	0.2	0.3—1.8	1.0—1.2	7.9	—
Disaccharides (as sucrose)	2.9—6.4	4.46	1.7—2.5	—	12.0	—
Pentosans	2.2—2.7	2.5	16.1—17.8	—	—	—
Starch	0.9—5.3	4.0	0.7	—	—	—
Hemicellulose	—	3.0	10.1	—	—	—
Crude fiber	1.6—1.9	—	65.7—79.3	21.4—34.9	1.6—1.8	—
Ash	1.8—2.9	2.44	1.9—4.6	2.1	2.9—3.2	6.3
Calories	—	564/100g	—	—	—	415.8/100g

[a]　Varies with curing and storage techniques; usually 5—8%.

From Cobb, W. Y. and Johnson, B. R., *Peanuts — Culture and Uses,* American Peanut Research and Education Society, Stone Printing, Roanoke, Va., 1973, chap. 6. With permission.

Table 202

VITAMIN CONTENT OF PEANUTS (UNITS/100 g DRY WEIGHT)

Fat-soluble	Cotyledons	Defatted flour
Vitamin A	26 I.U.	—
Carotene (provitamin A)	Trace (<1 μg)	—
Vitamin D	([a])	—
Vitamin E[b]	26.3—59.4 mg (avg. 41.6)	—
α — tocopherol	11.9—25.3 mg (avg. 17.1)	—
γ — tocopherol	10.4—34.2 mg (avg. 22.9)	—
δ — tocopherol	0.58—2.50 mg (avg. 1.62)	—
Vitamin K	([a])	—
Water-soluble		
B-complex		
Vitamin B_1 — thiamine	0.99 mg	0.75 mg
Vitamin B_2 — riboflavin	0.13 mg	0.35 mg
Vitamin B_6 — pyridoxine	0.30 mg	—
Vitamin B_{12} — cyanocobalamine	([a])	—
Niacin — nicotinic acid	12.8—16.7 mg	2.5 mg
Choline	165—174 mg	252 mg
Folic acid	0.28 mg	—
Inositol	180 mg	—
Biotin	0.034 mg	—
Pantothenic acid	2.715 mg	—
Vitamin C	5.8 mg	—

[a]　No evidence for presence.
[b]　Results expressed as mg/100 g oil.

From Cobb, W. Y. and Johnson, B. R., *Peanuts — Culture and Uses,* American Peanut Research and Education Society, Stone Printing, Roanoke, Va., 1973, chap. 6. With permission.

Table 203
COMPOSITION AND NUTRITIONAL VALUE OF SUNFLOWER AND SOYBEAN MEAL

Component	Sunflower meal with hull	Sunflower meal without hull	Soybean meal
Ash (%)	6.7	7.3	6.6
Crude fiber (%)	26.5	12.7	5.9
Ether extract (%)	2.1	2.6	1.3
N-free extract (%)	28.7	27.6	33.8
Protein (N×6.25) (%)	36.0	49.8	52.4
Energy			
DE cattle (Mcal/kg)	2.21	2.85	3.63
DE swine (kcal/kg)	2888	3205	3748
ME cattle (Mcal/kg)	1.81	2.34	2.98
ME swine (kcal/kg)	2562	2755	3201
TDN cattle (%)	50.1	64.7	82.4
TDN swine (%)	65.5	72.7	85.0
Calcium (%)	0.43	0.57	0.33
Magnesium (%)	0.75	—	0.30
Phosphorus (%)	1.08	0.54	0.73
Manganese (mg/kg)	5.50	—	30.70
Choline (mg/kg)	4712	—	3139
Niacin (mg/kg)	318.7	—	30.1
Panothenic acid (mg/kg)	44.8	16.1	17.7
Riboflavin (mg/kg)	3.6	—	3.5
Thiamin (mg/kg)	37.8	—	7.4

Source: National Academy of Science, *Atlas of Nutritional Data on United States and Canadian Feeds,* National Academy of Sciences, Washington, D.C., 1971.

Table 204
AMINO ACID COMPOSITION OF SUNFLOWER AND SOYBEAN OILSEED MEAL (g amino acid/ 100 g N)

Amino acid	Sunflower	Soybean
Essential		
Arginine	56.2	42.2
Histidine	14.4	17.0
Isoleucine	29.2	29.6
Leucine	38.4	44.0
Lysine	19.5	38.3
Methionine	11.4	8.1
Cysteine + cystine	13.4	12.9
Tryptophan	10.0	11.1
Phenylalanine	29.4	28.8
Valine	34.9	33.3
Threonine	22.3	22.1
Nonessential		
Alanine	24.5	26.1
Aspartic acid	54.9	70.8
Glutamic	143.0	113.0
Glycine	33.9	26.3
Proline	31.1	32.1
Serine	26.3	29.0
Tyrosine	14.2	19.0
Total	608.0	603.7
N recovered (%)	97.3	91.3

From Tkachuk, et al., *Cereal Chem.,* 46, 206—218, 1969. With permission.

Table 205
COMPOSITION OF COTTONSEED

Source	Moisture (%)	Oil (%)	Crude protein[a] (%)	Linters (%)	Kernels in seed (%)	Weight of 100 g of seed (g)
United States (American Upland)	9.0—12.8	18.2—19.5	18.7—21.3	10.4—11.4	50.2—63.3	11.1
India	6.7—10.7	13.9—19.4	16.8—20.9	9.2—12.8	44.1—50.8	4.8—8.6
Indian varieties	6.8—10.6	15.2—18.0	—	3.6—5.4	—	5.2—6.5
American varieties	9.0—12.1	16.4—18.3	—	12.3—13.2	—	9.0—10.8
Egypt	7.8—9.5	20.0—21.0	17.1—19.8	—	—	9.1—9.6
Brazil (tree type)	6—7	20—22	19.1—21.2	1—2	55—56	—
Sudan	5	20.5	17.2	39.6	60.7	—

[a] Nitrogen × 6.25.

From Altschul, A. M., Lyman, C. M., and Thurber, F. H., *Processed Plant Protein Foodstuffs*, Altschul, A. M., Ed., Academic Press, New York, 1958, 469. With permission.

Table 206

ANTEMORTEM AND POSTMORTEM FINDINGS IN SOME
NONRUMINANTS ATTRIBUTED TO CHRONIC TOXICITY OF GOSSYPOL
OR FREE GOSSYPOL PIGMENTS IN COTTONSEED MEALS

Animal species	Antemortem symptoms	Postmortem findings
Rats	Loss of appetite; growth rate depression; diarrhea; anorexia; hair loss; anemia; infertile males	Intestinal dilation and impaction; hemorrhagic congestion of stomach and intestine; congestion in lungs and kidneys; nuculear vacuolation and swelling of spermatids; decreased sperm count; azoospermia
Cats	Spastic paralysis, usually of hind legs; rapid pulse; dyspnea; cardiac irregularity	Edema of lungs and heart, heart enlargement; degeneration of the sciatic nerve
Dogs	Posterior incoordination; stupor; lethargy; diarrhea; anorexia; weight loss; vomiting	Lung edema; hypertrophy and edema of the heart; congestion and hemorrhages of the liver, small intestine and stomach; fibrosis of the spleen and gall bladder; congestion of the splanchnic organs
Rabbits	Stupor; lethargy; loss of appetite; diarrhea; hypoprothrombinemia; spastic paralysis; decrease in litter weights	Hemorrhages in small intestine, lungs, brain, and leg bones; enlarged gallbladder; edema and impaction of the large intestine
Poultry	Loss of weight; decreased appetite; leg weakness; lowering of hemoglobin and red blood cell count; lowering of protein and albumin/globulin ratio of serum; decreased egg size; egg yolk discoloration; decreased egg hatchability	Fluid in body cavities; enlargement of gallbladder and pancreas; liver discoloration; many vacuoles and foamy spaces in the liver; ceroid-like pigment deposition in the liver, spleen, and intestinal mucosa
Swine	"Thumps" or labored breathing; dyspnea; weakness; emaciation; increase in glutamic oxaloacetic transaminase	Widespread congestion and edema of many organs; fluid in body cavities; edematous bladder and thyroid gland; flabby, dilated heart with microscopic lesions; renal lipidosis; atrophied spleens; myocardial injury

Adapted from Berardi, L. C. and Goldblatt, L. A., *Toxic Constituents of Plant Foodstuffs*, Liener, I. E., Ed., Academic Press, New York, 1980, 183. With permission.

Table 207

INDIVIDUAL TOCOPHEROLS IN
VEGETABLE OILS

Oilseed	Tocopherols (mg/kg)			
	α	β	γ	δ
Sunflower (R)[a]	608	17	11	—
Cottonseed, crude	402	1.5	572	7.5
Maize, germ (R)	134	18	412	39
Peanut	169	5.4	144	13
Rapeseed (R)	70	16	178	7.4
Safflower	223	7	33	3.9
Sesame	12	5.6	244	32
Soybean (R)	116	34	737	275

[a] Refined.

From Muller-Mulot, W., Rapid method for the quantitative determination of individual tocopherols in oils and fats, *J. Am. Oil Chem. Soc.*, 53, 732—736, 1976. With permission.

CEREAL GRAINS

COMPOSITION AND FEEDING

GENERAL

Table 208
SOME PHYSICAL PROPERTIES OF CEREAL GRAINS

Name	Length (mm)	Width (mm)	Grain wt (mg)	Bulk density (kg/m³)	Unit density (kg/m³)
Rye	4.5 — 10	1.5 — 3.5	21	695	—
Sorghum	3 — 5	2.5 — 4.5	23	1360	—
Paddy rice	5 — 10	1.5 — 5	27	575—600	1370 — 1400
Oats	6 — 13	1 — 4.5	32	356 — 520	1360 — 1390
Wheat	5 — 8	2.5 — 4.5	37	790 — 825	1400 — 1435
Barley	8 — 14	1 — 4.5	37	580 — 660	1390 — 1400
Maize	8 — 17	5 — 15	285	745	1310
Bullrush millet	3 — 4	2 — 3	11	760	1322
Wild rice	8 — 20	0.5 — 2	22	388 — 775	—
T'of	1 — 1.5	0.5 — 1	0.3	880	—
Findi	1	1 — 1.5	0.4	790	—
Finger millet	1.5	1.5	—	—	—

From Muller, H. G., in *Industrial Uses of Cereals*, Pomeranz, Y., Ed., American Association of Cereal Chemists, St. Paul, Minn., 1973, 24.

Table 209
APPROXIMATE GRAIN SIZE AND PROPORTIONS OF THE PRINCIPAL PARTS COMPRISING THE MATURE KERNEL OF DIFFERENT CEREALS

Cereal	Grain weight (mg)	Embryo (%)	Scutellum (%)	Pericarp (%)	Aleurone (%)	Endosperm (%)
Barley	36—45 (41)	1.85	1.53	18.3	79.0	79.0
Bread wheat	30—45 (40)	1.2	1.54	7.9	6.7—7.0	81—84
Durum wheat	34—46 (41)	1.6	1.6	12.0	86.4	86.4
Maize	150—600 (350)	1.15	7.25	5.5	5.5	82
Oats	15—23 (18)	1.6	2.13	28.7—41.4	28.7—41.4	55.8—68.3
Rice	23—27 (26)	2—3	1.5	1.5	4—6	89—94
Rye	15—40 (30)	1.8	1.73	12.0	85.1	85.1
Sorghum	8—50 (30)	7.8—12.1	7.8—12.1	7.3—9.3	7.3—9.3	80—85
Triticale	38—53 (48)	3.7	3.7	14.4	81.9	81.9

From Simmonds, D. H., in *Cereals '78: Better Nutrition for the World's Millions*, Pomeranz, Y., Ed., American Association of Cereal Chemists, St. Paul, Minn., 1978, 109.

Table 210
PROXIMATE ANALYSIS OF IMPORTANT CEREAL GRAINS (% DRY WEIGHT)

Cereal	Nitrogen	Protein[a]	Fat	Fiber	Ash	N.F.E.[b]
Barley						
Grain	1.2—2.2	11	2.1	6.0	3.1	—
Kernel	1.2—2.5	9	2.1	2.1	2.3	78.8
Maize	1.4—1.9	10	4.7	2.4	1.5	72.2
Millet	1.7—2.0	11	3.3	8.1	3.4	72.9
Oats						
Grain	1.5—2.5	14	5.5	11.8	3.7	—
Kernel	1.7—3.9	16	7.7	1.6	2.0	68.2
Rice						
Brown	1.4—1.7	8	2.4	1.8	1.5	77.4
Milled			0.8	0.4	0.8	—
Rye	1.2—2.4	10	1.8	2.6	2.1	73.4
Sorghum	1.5—2.3	10	3.6	2.2	1.6	73.0
Triticale	2.0—2.8	14	1.5	3.1	2.0	71.0
Wheat (bread)	1.4—2.6	12	1.9	2.5	1.4	71.7
Wheat (durum)	2.1—2.4	13			1.5	70.0
Wild rice	2.3—2.5	14	0.7	1.5	1.2	74.4

[a] Typical or average figure.
[b] N.F.E. = Nitrogen-free extract. This is an approximate measure of the starch content.

Modified from Simmonds, D. H., in *Cereals '78: Better Nutrition for the World's Millions,* Pomeranz, V., Ed., *American Association of Cereal Chemists,* St. Paul, 1978, 107. With permission. (References included in the original.)

Table 211

AVERAGE COMPOSITION OF MAJOR CEREAL FEED
GRAINS, PERCENT DRY BASIS

	Corn	Barley	Wheat Hard winter	Wheat Soft white	Oats	Sorghum
Crude protein	10.9	13.9	14.4	11.2	13.6	11.9
Ether extract	4.5	2.2	1.8	2.0	5.6	3.2
Crude fiber	2.4	5.6	2.8	2.6	12.2	2.6
Ash	1.4	2.6	1.9	1.9	3.4	1.9
N-free extract	80.8	75.7	79.1	82.3	65.2	80.4
Starch	72.2	64.6	63.4	67.2	41.2	70.8
Sugars	1.9	2.5	2.9	4.1	1.5	1.5

Essential amino acids

	Corn	Barley	Wheat Hard winter	Wheat Soft white	Oats	Sorghum
Arginine	0.46	0.60	0.69	0.52	0.73	0.44
Histidine	0.30	0.29	0.31	0.24	0.21	0.26
Isoleucine	0.41	0.54	0.58	0.46	0.48	0.50
Leucine	1.44	0.91	1.00	0.73	0.87	1.62
Lysine	0.30	0.48	0.41	0.35	0.43	0.28
Phenylalanine	0.56	0.69	0.71	0.51	0.57	0.63
Threonine	0.40	0.43	0.43	0.35	0.40	0.41
Tryptophan	0.10	0.18	0.20	0.14	0.17	0.16
Valine	0.50	0.68	0.67	0.50	0.61	0.59
Methionine	0.17	0.18	0.24	0.18	0.16	0.14
Cysteine	0.27	0.26	0.35	0.30	0.21	0.23

Note: References included in the original.

Table 212
CARBOHYDRATE CONTENTS OF CEREAL GRAINS AND THEIR PRODUCTS

	Carbohydrate			Carbohydrate	
	Total (g/100 g)	Fiber (g/100 g)		Total (g/100 g)	Fiber (g/100 g)
Barley, pearled	78.8	0.5	Rice		
Buckwheat, whole grain	72.9	9.9	Brown	77.4	0.9
Buckwheat flour			White	80.4	0.3
Dark	72.0	1.6	Rice bran	50.8	11.5
Light	79.5	0.5	Rice polish	57.7	2.4
Bulgur			Rye	73.4	2.0
Club wheat	79.5	1.7	Rye flour		
Hard red winter wheat	75.7	1.7	Light	77.9	0.4
			Medium	74.8	1.0
White wheat	78.1	1.3	Dark	68.1	2.4
Maize			Sorghum grain	73.0	1.7
Field corn	72.2	2.0	Wheat		
Sweet corn, raw	22.1	0.7	Hard red spring	69.1	2.3
Maize flour (corn flour)	76.8	0.7	Hard red winter	71.7	2.3
			Soft red winter	72.1	2.3
Malt, dry	77.4	5.7	White	75.4	1.9
Malt extract, dried	89.2	trace	Durum	70.1	1.8
Millet, proso	72.9	3.2	Wheat flour		
Oatmeal, dry	68.2	1.2	80% extraction	74.1	0.5
Popcorn			Straight, hard wheat	74.5	0.4
Unpopped	72.1	2.1	Straight, soft wheat	76.9	0.4
Popped, plain	76.7	2.2	Patent, all purpose	76.1	0.3
			Wheat bran	61.9	9.1
			Wheat germ	46.7	2.5

REFERENCES
1. **Lockhart, H. B. and Nesheim, R. O.,** Nutritional quality of cereal grains, in *Cereals '78: Better Nutrition for the World's Millions,* Pomeranz, Y., Ed., American Association of Cereal Chemists, St. Paul, 1978.
2. **Watt, B. K. and Merrill, A. L.,** Composition of Foods, Raw, Processed, and Prepared, *U.S. Dept. Agric. Agric. Handb.,* No. 8, 1963.

Table 213

AMINO ACID COMPOSITION OF CEREALS (% BY WEIGHT)

Amino acid	Rice (brown)	Wheat (HRS)	Maize (field)	Sorghum	Pearl millet	Barley	Oats	Rye	Triticale	WHO requirement (1973)[a]
Tryptophan	1.08	1.24	0.61	1.12	2.18	1.25	1.29	1.13	1.08	1.0
Threonine	3.92	2.88	3.98	3.58	4.00	3.38	3.31	3.70	3.11	4.0
Isoleucine	4.69	4.34	4.62	5.44	5.57	4.26	5.16	4.26	3.71	4.0
Leucine	8.61	6.71	12.96	16.06	15.32	6.95	7.50	6.72	6.87	7.0
Lysine	3.95	2.82	2.88	2.72	3.36	3.38	3.67	4.08	2.77	5.5
Methionine	1.80	1.29	1.86	1.73	2.37	1.44	1.47	1.58	1.44	3.5
Cystine	1.36	2.19	1.30	1.66	1.33	2.01	2.18	1.99	1.55	
Phenylalanine	5.03	4.94	4.54	4.97	4.44	5.16	5.34	4.72	5.26	6.0
Tyrosine	4.57	3.74	6.11	2.75	—	3.64	3.69	3.22	2.14	—
Valine	6.99	4.63	5.10	5.71	5.98	5.02	5.95	5.21	4.39	—
Arginine	5.76	4.79	3.52	3.79	4.60	5.15	6.58	4.88	4.99	—
Histidine	1.68	2.04	2.06	1.92	2.11	1.87	1.84	2.28	2.48	—
Alanine	3.56	3.50	9.95	—	—	4.60	6.11	5.13	3.53	—
Aspartic acid	4.72	5.46	12.42	—	—	5.56	4.13	7.16	5.00	—
Glutamic acid	13.69	31.25	17.65	21.92	—	22.35	20.14	21.26	31.80	—
Glycine	6.84	6.11	3.39	—	—	4.55	4.55	4.79	4.05	—
Proline	4.84	10.44	8.35	—	—	9.02	5.70	5.20	12.06	—
Serine	5.08	4.61	5.65	5.05	—	4.65	4.00	4.13	4.70	—
Total protein (%)	7.5	14.0	10.0	11.0	11.4	12.8	14.2		17.3	—

[a] Nutritionally essential amino acids.

Source: Simmonds, D. H., in *Cereals '78: Better Nutrition for the World's Millions*, Pomeranz, Y., Ed., American Association of Cereal Chemists, St. Paul, Minn., 1978, 120.

Table 214
PROTEIN QUALITY OF CEREAL GRAINS (PER)

	Actual	Estimate[a]		Actual	Estimate[a]
Barley	—	1.6	Sorghum	1.8	0.9
Buckwheat	—	1.8	Wheat	1.5	1.3
Foxtail Millet	—	1.0	Wheat germ	˙2.5	2.5
Job's Tears	—	1.0	Wheat gluten	—	0.7
Maize					
Normal	1.2	1.2	Wheat flour	—	1.1
Opaque-2	2.3	1.9	80—90% extraction		
Pearl millet	1.8	1.6	Wheat flour	—	1.0
Proso millet	—	1.4	70—80% extraction		
Rogi (finger) millet	0.8	—	Wheat flour	—	0.8
			60—70% extraction		
Rice					
Brown	1.9	1.8			
Polished	1.7	1.7			
Rye	1.6	1.6	Bulgur wheat	—	1.2
Oats	1.9	1.7	Triticale	1.6	1.4

Note: References included in the original.

[a] Estimated from amino acid content assuming availability of amino acids the same as the amino acids in casein.

Table 215

MINERAL CONTENTS OF CEREAL GRAINS AND CEREAL PRODUCTS

	Ca (mg/100 g)	Fe (mg/100 g)	Mg (mg/100 g)	P (mg/100 g)	K (mg/100 g)	Na (mg/100 g)	Cu (mg/100 g)	Mn (mg/100 g)	Zn (mg/100 g)
Barley	80	10	120	420	560	3	0.76	1.63	1.53
Buckwheat	110	4	390	330	450	—	0.95	3.37	0.87
Maize									
Grain	30	2	120	270	280	1	0.21	0.51	1.69
Bran	30	—	260	190	730	—	—	1.61	—
Germ	90	90	280	560	130	—	1.10	0.90	
Millet (proso)	50	10	160	280	430	—	2.16	2.91	1.39
Oats	100	10	170	350	370	2	0.59	3.82	3.4
Rice									
Grain	40	3	60	230	150	9	0.33	1.76	1.8
White	30	1	20	120	130	5	0.29	1.09	1.3
Rye	60	10	120	340	460	1	0.78	6.69	3.05
Sorghum	40	4	170	310	340	—	0.96	1.45	1.37
Wheat									
Grain	50	10	160	360	520	3	0.72	4.88	3.4
Bran	140	70	550	1170	1240	9	1.23	11.57	9.8
Triticale	20	4	—	—	385	—	0.52	4.26	0.02

Note: References included in the original.

Table 216

VITAMIN CONTENTS OF CEREAL GRAINS AND CEREAL PRODUCTS

	Thiamin (mg/100 g)	Riboflavin (mg/100 g)	Niacin (mg/100 g)	Vitamin B₆ (mg/100 g)	Folic acid (mcg/100 g)	Pantothenic acid (mg/100 g)	Biotin (mcg/100 g)	Vitamin E (IU/100 g)
Barley (pearled)	0.23	0.13	4.52	0.26	67	0	6	1
Buckwheat	0.60							
Maize	0.37	0.12	2.2	0.47	26	1	21	2
Oats	0.67	0.11	0.8	0.21	104	1	13	3
Millet	0.73	0.38	2.3			1		1
Rice								
Brown	0.34	0.05	4.7	0.62	20	2	12	2
Polished	0.07	0.03	1.6	0.04	16	1	5	1
Rye	0.44	0.18	1.5	0.33	34	1	—	2
Sorghum	0.38	0.15	3.9					
Wheat	0.57	0.12	7.4	0.35	78	1	6	1
Wheat germ	2.01	0.68	4.2	0.92	328	2		
Wheat bran	0.72	0.35	21.0	1.38	258	3	14	—
Wheat flour (patent)	0.13	0.04	2.1	0.05	25	1	1	—

Note: References included in the original.

Table 217

FATTY ACID COMPOSITION OF CEREALS AND RELATED PRODUCTS

Fatty acid (gm/100 g food, edible portion)[a,b]

Food	Water	Total lipid	Saturated Sum	14:0	16:0	18:0	20:0	Unsaturated Sum	16:1	18:1	18:2	18:3
Barley (*Hordeum vulgare*)												
Whole grain	14	2.8	0.48	0.01	0.45	0.02	0	1.52	0.01	0.24	1.14	0.13
Buckwheat, domestic (*Fagopyrum* spp.)												
Whole grain	11	2.4	0.46	0.01	0.36	0.05	0.04	1.66	0	0.80	0.74	0.12
Corn (*Zea mays*)												
Whole grain, raw	13.8	4.1	0.47	0	0.40	0.06	0.01	3.07	0.01	0.91	2.12	0.03
Flour	12	2.6	0.30	0	0.25	0.04	0.01	2.00	—	0.64	1.34	0.02
Germ	0	30.8	3.93	0	3.30	0.54	0.09	25.57	0.04	7.58	17.7	0.25
Grits, degermed, enriched or unenriched dry form	12	0.8	0.09	0	0.08	0.01	—	0.60	—	0.18	0.41	0.01
Corn oil (commercial)	0	100	12.73	0	10.7	1.74	0.29	82.96	0.14	24.6	57.4	0.82
Cornmeal, white or yellow												
Bolted (nearly whole grain)	12	3.4	0.39	0	0.33	0.05	0.01	2.54	—	0.75	1.76	0.03
Degermed, enriched, or unenriched dry form	12	1.2	0.14	0	0.12	0.02	—	0.90	—	0.27	0.62	0.01
Farina, enriched, regular												
Dry form	10.3	1.5	0.24	—	0.22	0.01	0.01	0.83	—	0.11	0.69	0.03
Millet, pearl millet (*Pennisteum glaucum*)												
Whole grain	11.8	4.1	0.86	0	0.68	0.16	0.02	2.67	0.02	0.83	1.69	0.13
Oatmeal or rolled oats (*Avena sativa*)												
Dry form	8.3	7.4	1.37	0.02	1.21	0.10	0.04	5.65	0.02	2.60	2.87	0.16
Rice (*Oryza sativa*)												
Brown dry form	12	2.3	0.62	0.03	0.54	0.04	0.01	1.36	0.01	0.54	0.78	0.03
White, fully milled or polished, enriched dry form	12	0.8	0.21	0.01	0.19	—	—	0.47	—	0.19	0.27	0.01

Table 217 (continued)
FATTY ACID COMPOSITION OF CEREALS AND RELATED PRODUCTS

Food	Water	Total lipid	Fatty acid (gm/100 g food, edible portion)[a,b]									
			Saturated					Unsaturated				
			Sum	14:0	16:0	18:0	20:0	Sum	16:1	18:1	18:2	18:3
Bran oil	0	100	19.50	0.69	16.7	1.58	0.53	74.56	0.26	39.2	33.5	1.60
Rye (*Secale cereale*)												
Whole grain	12.1	2.2	0.27	—	0.25	0.02	0	1.30	0.01	0.22	0.95	0.12
Sorghum (*Sorghum vulgare*)												
Whole grain	11	3.3	0.48	0.01	0.44	0.03	0	2.74	0.04	1.15	1.46	0.09
Triticale (*Triticum × Secale*)												
Whole grain flour	14	3.4	0.49	0.02	0.45	0.02	0	1.91	0	0.28	1.48	0.15
Wheat (*Triticum aestivum*)												
Whole grain												
Hard red spring	14	2.7	0.37	—	0.36	0.01	—	1.56	0.01	0.25	1.20	0.10
Hard red winter	14	2.5	0.35	—	0.33	0.02	—	1.47	0.01	0.28	1.08	0.10
Soft red winter	14	2.4	0.35	—	0.33	0.02	0	1.40	0.01	0.25	1.07	0.07
White	14	2.0	0.30	—	0.28	0.02	0	1.14	0.01	0.18	0.88	0.07
Flours												
Hard red spring	14	1.5	0.23	—	0.21	0.01	0.01	0.84	—	0.12	0.69	0.03
Hard red winter	14	1.5	0.20	—	0.19	0.01	—	0.74	—	0.10	0.64	0.03
Soft red winter	14	1.4	0.22	—	0.20	0.01	0.01	0.76	—	0.08	0.65	0.03
All purpose	14	1.4	0.23	—	0.22	0.01	—	0.72	—	0.11	0.58	0.03
Bran	14	4.6	0.74	0.01	0.69	0.04	—	3.09	0.02	0.71	2.20	0.16
Germ	14	10.9	1.88	—	0.01	1.81	0.06	8.18	0.04	1.54	5.86	0.74
Wheat, durum (*Triticum durum*)												
Whole grain	14	3.3	0.54	—	0.51	0.03	0	1.88	0.01	0.40	1.36	0.11
Semolina	14	1.8	0.33	—	0.31	0.02	0	0.90	0.01	0.17	0.68	0.04

Note: References included in the original.

[a]   14:0 = myristic; 16:0 = palmitic; 18:0 = stearic; 20:0 = arachidic; 16:1 = palmitoleic; 18:1 = oleic; 18:2 = linoleic; 18:3 = linolenic.

[b]   — = <0.005 g.

Table 218
DIGESTION COEFFICIENTS OF CEREALS
FOR CATTLE AND SWINE

Digestion Coefficient[a]

Cereal	Organic matter	Protein	Fiber	Nitrogen-free extract
Cattle				
Barley	81	70	6	88
Corn	84	63	13	88
Milo	78	57	40	83
Oats	68	74	13	76
Swine				
Barley	81	71	3	89
Corn	86	76	3	90
Milo	86	76	66	90
Oats	67	78	14	74

[a] The difference between the nutrient consumed and the nutrient excreted, expressed as percentage. Values are obtained by difference when each grain was fed as part of a complete diet.

Adapted from Schneider, B. H., *Feeds of the World,* West Virginia University Agricultural Experiment Station, Morgantown, 1947.

Table 219
RELATIVE FEEDING VALUES OF CEREALS FOR
CATTLE, SWINE, AND CHICKENS

Grain	Total digestible nutrients		Metabolizable energy	Net energy milk Gain	
	Cattle	Swine	Chickens	Cattle	
Corn, #2 dent	100	100	100	100	100
Barley	91	88	77	93	95
Milo	88	96	103	81	83
Oats	84	79	74	76	77
Wheat, hard red winter	97	99	90	95	96

[a] Corn is arbitrarily set at 100. All values, dry basis. National Research Council data.

Adapted from Church, D. C., *Livestock Feeds and Feeding,* Portland, Oregon, 1977, 89.

WHEAT

Table 220
U.S. WHEAT GRADES

| | Minimum test weight per bushel (pounds) | | Percent maximum limits of— | | | | | | |
| | | | | | | | | Wheat of other classes[3] | |
Grade	Hard Red Spring wheat or White Club wheat	All other classes and sub-classes	Heat damaged kernels	Damaged kernels (total)[1]	Foreign material	Shrunken and broken kernels	Defects (total)[2]	Con-strasting classes	Wheat of other classes (total)[4]
U.S. No. 1 ..	58.0	60.0	0.2	2.0	0.5	3.0	3.0	1.0	3.0
U.S. No. 2 ..	57.0	58.0	0.2	4.0	1.0	5.0	5.0	2.0	5.0
U.S. No. 3 ..	55.0	56.0	0.5	7.0	2.0	8.0	8.0	3.0	10.0
U.S. No. 4 ..	53.0	54.0	1.0	10.0	3.0	12.0	12.0	10.0	10.0
U.S. No. 5 ..	50.0	51.0	3.0	15.0	5.0	20.0	20.0	10.0	10.0

U.S. Sample
grade ..... U.S. sample grade shall be wheat which:
   (1) Does not meet the requirements for the grades U.S. Nos. 1, 2, 3, 4, or 5; or
   (2) Contains a quantity of smut so great that 1 or more of the grade requirements cannot be determined accurately; or
   (3) Contains 8 or more stones, 2 or more pieces of glass, 3 or more crotalaria seeds (*Crotalaria* spp.), 3 or more castor beans (*Ricinus communis*), 4 or more particles of an unknown foreign substance(s) or a commonly recognized harmful or toxic substance(s), or 2 or more rodent pellets, bird droppings, or an equivalent quantity of other animal filth per 1,000 g of wheat; or
   (4) Has a musty, sour, or commercially objectionable foreign odor (except smut or garlic odor); or
   (5) Is heating or otherwise of distinctly low quality.

[1]Includes heat-damaged kernels.
[2]Defects (total) include damaged kernels (total), foreign material, and shrunken and broken kernels. The sum of these 3 factors may not exceed the limit for defects.
[3]Unclassed wheat of any grade may contain not more than 10 percent of wheat of other classes.
[4]Includes contrasting classes.

From the Official U.S. Standards for Grain, U.S. Department of Agriculture.

Table 221
COMPOSITION OF ENDOSPERM, GERM, AND BRAN IN COMMERCIAL WHEAT SAMPLES

	Endosperm (%)	Germ (%)	Bran (%)
Moisture	14.0	11.7	13.2
Protein	9.6	28.5	14.4
Fat	1.4	10.4	4.7
Mineral	0.7	4.5	6.3
Starch	71.0	14.0	8.6
Hemicellulose	1.8	6.8	26.2
Sugars	1.1	16.2	4.6
Cellulose	0.2	7.5	21.4
Carbohydrate (total)	74.1	44.5	60.8

From Kent-Jones, and Amos, *Modern Cereal Chemistry,* Food Trade Press, London, 1967. With permission.

Table 222
WHOLE WHEAT VALUES IN ANIMAL FEEDS

Nutrient value	Hard wheat	Soft wheat
Dry matter (%)	88.0	86.0
Crude protein (%)	13.5	10.8
Crude fat (%)	1.9	1.7
Crude fiber (%)	3.0	2.8
Calcium (%)	0.05	0.05
Total phos. (%)	0.41	0.3
Ash (%)	2.0	2.0
Ruminant T.D.N. (%)	76.0	79.0
Ruminant dic. prot. (%)	10.9	8.5
Poultry P.E. (cal/kg)	2200	2200
Poultry M.E. (cal/kg)	3086	3086
Swine M.E. (cal/kg)	3220	3416
Swine T.D.N. (%)	79.0	83.0
Amino acids		
Methionine (%)	0.25	0.14
Cystine (%)	0.30	0.20
Lysine (%)	0.40	0.30
Tryptophane (%)	0.18	0.12
Threonine (%)	0.35	0.28
Isoleucine (%)	0.69	0.43
Histidine (%)	0.17	0.20
Valine (%)	0.69	0.48
Leucine (%)	1.00	0.60
Arginine (%)	0.60	0.40
Phenylalanine (%)	0.78	0.49
Glycine (%)	0.60	0.50
Poultry amino acid available (%)	92.0	95.0
Vitamins		
A(IU/g)	—	—
E(mg/kg)	15.5	15.5
Thiamine (mg/kg)	5.2	4.8
Riboflavin (mg/kg)	1.1	1.2
Pantothenic acid (mg/kg)	13.5	12.8
Biotin (Mcg/kg)	100.0	100.0
Folic acid (Mcg/kg)	426.0	300.0
Choline (mg/kg)	778.0	778.0
B12 (Mcg/kg)	—	—
Niacin (mg/kg)	56.1	48.4
Minerals		
Sodium (%)	0.06	0.06
Potassium (%)	0.50	0.40
Magnesium (%)	0.11	0.10
Sulfur (%)	N.A.	N.A.
Manganese P.P.M.	62.2	51.3
Iron P.P.M.	50.0	43.0
Copper P.P.M.	10.6	9.7
Zinc P.P.M.	14.0	14.0
Selenium P.P.M.	0.06—0.07	0.06—0.07

From Feedstuffs Ingredient Analysis Table, *Feedstuffs,* 51(29), 33, 1979. With permission.

Table 223
WHEAT MILL FEED VALUES IN ANIMAL FEEDING

Nutrient value	Bran	Germ meal	Middlings	Shorts
Dry matter (%)	89.0	88.0	89.0	89.0
Crude protein (%)	14.8	25.0	17.7	16.8
Crude fat (%)	4.0	7.0	3.6	4.2
Crude fiber (%)	10.0	3.5	7.0	8.2
Calcium (%)	0.14	0.01	0.15	0.11
Total phos. (%)	1.17	1.00	0.91	0.76
Ash (%)	6.40	5.30	5.50	8.20
Ruminant T.D.N. (%)	62.0	84.0	81.0	77.0
Ruminant dig. prot. (%)	11.5	19.5	12.2	12.0
Poultry P.E. (cal/kg)	968.0	1600.0	1600.0	1584.0
Poultry M.E. (cal/kg)	1256.0	3000.0	2460.0	2640.0
Swine M.E. (cal/kg)	2320.0	2400.0	2000.0	2910.0
Swine T.D.N. (%)	57.0	80.0	73.0	72.0
Amino acids				
Methionine (%)	0.20	0.42	0.12	0.18
Cystine (%)	0.30	0.46	0.19	0.25
Lysine (%)	0.60	1.37	0.60	0.70
Tryptophane (%)	0.30	0.30	0.20	0.23
Threonine (%)	0.48	0.94	0.50	0.50
Isoleucine (%)	0.60	0.79	0.70	0.70
Histodine (%)	0.30	0.62	0.40	0.32
Valine (%)	0.70	1.12	0.80	0.77
Leucine (%)	0.90	1.10	1.10	1.00
Arginine (%)	1.07	1.83	1.00	0.95
Phenylalanine (%)	0.57	0.93	0.50	0.70
Glycine (%)	0.90	1.38	0.35	0.98
Poultry am. acid available (%)	75.0	85.0	80.0	75.0
Vitamins				
A(IU/g)	—	—	5.2	5.2
E(mg/kg)	10.8	133.0	57.6	29.9
Thiamine (mg/kg)	6.0	21.9	18.9	19.9
Riboflavin (mg/kg)	3.1	6.1	1.5	2.0
Pantothenic acid (mg/kg)	29.0	23.2	13.6	17.6
Biotin (Mcg/kg)	110.0	N.A.	110.0	N.A.
Folic acid (Mcg/kg)	1800.0	2800.0	560.0	1420.0
Choline (mg./kg)	980.0	3175.0	1100.0	930.0
B12 (Mcg/kg)	—	—	—	—
Niacin (Mg/kg)	321.0	—	52.6	100.0
Minerals				
Sodium (%)	0.06	0.02	0.60	0.07
Potassium (%)	1.20	0.90	0.60	0.88
Magnesium (%)	0.55	0.22	0.29	0.26
Sulfur (%)	0.22	0.31	0.16	0.23
Manganese P.P.M.	100.0	128.0	43.0	115.0
Iron P.P.M.	170.0	41.0	60.0	100.0
Copper P.P.M.	10.3	8.6	4.4	12.1
Zinc P.P.M.	95.0	115.0	62.0	105.5
Selenium P.P.M.	0.50—1.00	0.50—1.00	0.50—1.00	0.50—1.00

From Feedstuffs Ingredient Analysis Table, *Feedstuffs,* 51(29), 33, 1979. With permission.

TRITICALE

Table 224
PROTEIN, LYSINE, AND THREONINE CONTENTS OF WINTER WHEATS AND TRITICALES

Cultivar	Colorado location	Grain			Flour		
		Protein (%)	Lysine (g/16 g of N)	Threonine (g/16 g of N)	Protein (%)	Lysine (g/16 g of N)	Threonine (g/16 g of N)
Lancer	Akron	12.4	2.60	2.76	11.4	2.19	2.70
	Burlington	10.7	2.65	3.05	9.3	1.90	2.76
	Springfield	16.3	2.15	2.73	13.2	1.85	2.79
	Fort Collins	12.7	1.93	3.03	10.6	1.63	2.56
	Julesburg	14.0		2.64	11.3	1.61	1.94
	Nunn	13.0	2.31	3.88	10.6	1.42	3.50
Triticale							
TR 131	Akron	18.8	2.74	3.19	14.3	2.04	2.63
	Burlington	15.0	2.68	3.42	11.9	1.35	3.31
	Springfield	19.7	2.07	3.46	15.8		
	Fort Collins	17.6	2.04	2.76	13.5	1.92	2.22
	Julesburg	17.3	2.70	3.12	14.1	2.32	2.73
	Nunn	18.3	2.32	2.70	13.8	2.14	2.50

Reprinted with permission from Lorenz, K., Maga, J., Sizer, C., and Welsh, J., Variability in the limiting amino acid and fatty acid composition of winter wheats and triticales, *J. Agric. Food Chem.*, 50, 215, 1973. Copyright 1973 American Chemical Society.

Table 225
AMINO ACID CONTENT OF TRITICALE

Amino acid	Percent total grain	Amino acid/mg nitrogen (μmol)	g/16 g Nitrogen
Lysine	0.439	1.00	3.3
Histidine	0.296	0.79	2.3
Arginine	0.710	1.43	5.6
Aspartic acid	0.879	2.23	6.5
Threonine	0.421	1.34	2.9
Serine	0.621	2.17	4.7
Glutamic acid	4.851	11.7	32.0
Proline	1.196	6.44	9.5
Glycine	0.673	2.70	4.3
Alanine	0.465	2.08	3.9
Cystine	0.163	0.70	1.1
Valine	0.578	2.20	4.8
Methionine	0.204	0.79	1.2
Isoleucine	0.461	1.74	3.6
Leucine	0.910	—	7.3
Tyrosine	0.417	0.81	2.8
Phenylalanine	0.633	1.58	4.6

Note: References included in the original.

Table 226
VITAMIN CONTENTS OF WHEAT
AND TRITICALE AND THEIR
MILLING FRACTIONS (μg/g, DRY
BASIS)

	Wheat chris	Winter triticales	
		TR385	TR386
Grain			
Thiamine	9.9	9.8	8.7
Riboflavin	3.1	2.5	4.1
Niacin	48.3	17.9	16.3
Biotin	0.056	0.067	0.062
Folacin	0.56	0.59	0.61
Pantothenic acid	9.1	6.5	6.8
Vitamin B_6	4.7	4.2	5.0
Flour			
Thiamine	0.7	0.2	0.4
Riboflavin	1.5	1.4	1.4
Niacin	9.5	6.5	6.5
Biotin	0.013	0.012	0.010
Folacin	0.09	0.09	0.10
Pantothenic acid	2.5	2.4	2.4
Vitamin B_6	0.48	0.43	0.44
Shorts			
Thiamine	10.1	5.2	4.7
Riboflavin	1.8	1.8	1.6
Niacin	23.5	14.4	13.7
Biotin	0.055	0.046	0.045
Folacin	0.59	0.33	0.25
Pantothenic acid	7.0	5.0	4.6
Vitamin B_6	5.3	3.4	2.5
Bran			
Thiamine	13.2	12.5	11.8
Riboflavin	5.5	5.1	5.5
Niacin	171.4	58.2	58.0
Biotin	0.162	0.123	0.120
Folacin	1.59	1.32	1.20
Pantothenic acid	31.7	19.9	18.5
Vitamin B_6	13.0	8.5	9.0

From Michela, P. and Lorenz, K., The vitamins of triticale, wheat and rye, *Cereal Chem.*, 53, 853, 1976. With permission.

Table 227
FORAGE YIELDS AND QUALITY CHARACTERISTICS OF CEREAL GRAINS HARVESTED AT THE SOFT DOUGH STAGE[a]

Crop	Forage yield (tons dry matter/acre)	Dry matter (%)	Protein (%)	Fiber (%)	Fat (%)
Triticale	6.54	35.6	7.3	32.1	2.4
Wheat	5.87	36.6	7.9	30.9	2.2
Rye	5.62	36.0	7.2	34.2	2.4
Barley	4.67	26.6	8.7	29.7	2.3
Oats	5.55	29.1	7.5	31.3	3.0

[a] Means over four varieties and three planting dates for each drop at Five Points, California, 1972.

From Lorenz, K., Maga, J., Sizer, C., and Welsh, J., Variability in the limiting amino acid and fatty acid composition of winter wheats and triticales, *J. Agric. Food Chem.*, 50, 215, 1973. With permission.

Table 228
EFFECT OF TYPE OF SILAGE ON INTAKE OF DRY MATTER, YIELD, AND COMPOSITION OF MILK

	Intake (kg/100 kg body weight)	Milk composition (%)			Yield of milk (kg/day)
		Fat	Protein	Lactose	
Corn silage	9.41	3.86	3.34	4.91	25.05
Triticale silage	7.49	4.03	3.11	4.82	21.43

Source: Fisher, L. J., Evaluation of triticale silage for lactating cows, *Can. J. Anim. Sci.*, 52, 373, 1972.

Table 229
FEEDLOT PERFORMANCE AND CARCASS TRAITS OF STEERS FED TRITICALE AND SORGHUM GRAIN RATIONS

	Triticale	Sorghum
Initial wt. (kg)	284.50	284.70
Final wt. (kg)	422.00	449.20
Total gain (kg)	137.50	164.50
Daily gain (kg)	0.94	1.13
Daily feed intake (kg)	7.35	10.06
Feed/gain (kg)	7.82	8.90
Warm carcass wt. (kg)	267.20	285.40
Daily carcass gain (kg)	0.72	0.84
Feed/carcass gain (kg)	10.121	11.98
Dressing (%)	63.30	63.60
Marbling score	4.60	4.60
Carcass grade	11.10	11.30
Backfat thickness (cm)	1.20	1.20

Source: Sherrod, L. B., Triticale as a feed grain for ruminants, in Int. Triticale Symp., ICASALS Pub. No. 76-1, Yang, S. P., Ed., Lubbock, Texas, 1976, 131.

Table 230

GROWTH RATE AND FEED INTAKE OF CROSSBRED STEERS

Grain in ration	Initial wt (lb)	Daily feed intake (lb)	Daily gain (lb)	Feed per pound gain (lb)
Barley	716	20.9	3.38	6.19
Sorghum	712	20.3	3.12	6.51
Triticale	743	20.5	3.30	6.20
Barley + triticale	739	21.0	3.40	6.18
Sorghum + triticale	696	19.8	3.21	6.17

Source: Lofgreen, G. P., Triticale, Milo and Barley Comparison, *Calif. Agric. Exp. Stn. Feeders Day Report,* 1969, 67.

Table 231
DIGESTIBILITY OF TRITICALE AND SORGHUM GRAIN IN RATIONS FOR CATTLE AND SHEEP

Ingredient and Chemical Composition of Digestibility Trial Rations

Ration	Triticale	Sorghum grain
Ingredient composition (%)		
Triticale	79.0	—
Sorghum grain	—	71.6
Cottonseed hulls	20.0	20.0
Cottonseed meal	—	6.3
Urea	—	1.0
Calcium carbonate	0.5	0.5
Trace mineral salt	0.5	0.5
Elemental sulfur		0.1
Chemical analyses, DM basis (%)		
Organic matter	97.2	97.4
Ash	2.8	2.6
Crude protein	14.5	14.4
Calcium	0.28	0.30
Phosphorus	0.29	0.30
Gross energy, kcal/g DM	4.41	4.32

Digestibility of Triticale and Sorghum Grain Rations and of the Grain Portions by Cattle and Sheep

	Triticale		Sorghum grain	
	Cattle	Sheep	Cattle	Sheep
Daily DM intake (kg)	5.25	1.08	5.22	1.07
Ration digestibility (%)				
Dry matter	72.80	72.90	69.60	72.40
Organic matter	73.50	74.00	70.30	73.30
Crude protein, apparent[a]	68.80	70.50	52.30	55.60
Crude protein, true[a]	88.50	89.90	72.80	75.10
Gross energy	70.80	72.30	66.60	70.20
Daily N retention, g	—	3.29	—	3.59
N retained, % intake	—	13.20	—	14.50
Digestible energy, kcal/g DM	3.12	3.19	2.87	3.03
TDN, DM basis (%)	70.90	72.40	65.20	68.70

Source: Sherrod, L. B., Triticale as a feed grain for ruminants, in Int. Triticale Symp., ICASALS Pub. No. 76-1, Yang, S. P., Ed., Lubbock, Texas, 1976, 131.

OATS AND WILD-RICE

Table 232
U.S. OAT GRAIN STANDARDS GRADES AND GRADE REQUIREMENTS

	Minimum limits		Maximum limits		
Grade	Test weight per bushel (lb)	Sound oats (%)	Heat-damaged kernels (%)	Foreign material (%)	Wild oats (%)
U.S. No. 1	36.0	97.0	0.1	2.0	2.0
U.S. No. 2	33.0	94.0	0.3	3.0	3.0
U.S. No. 3[a]	30.0	90.0	1.0	4.0	5.0
U.S. No. 4[b]	27.0	80.0	3.0	5.0	10.0

U.S. Sample grade

U.S. Sample grade shall be oats which:

(a)   Do not meet the requirements for the grades U.S. No. 1, 2, 3, or 4.
(b)   Contain more than seven stones which have an aggregate weight in excess of 0.2% of the sample weight or more than 2 crotalaria seeds (*Crotalaria* spp.) per 1000 g of oats or more than 16% of moisture.
(c)   Have a musty, sour, or commercially objectionable foreign odor (except smut or garlic odor), or
(d)   Are heating or otherwise of distinctly low quality.

[a]   Oats that are slightly weathered shall be graded not higher than U.S. No. 3.
[b]   Oats that are badly stained or materially weathered shall be graded not higher than U.S. No. 4.

From *Federal Register*, 39 F.R. 32124, Sept. 5, 1974.

Table 233
U.S. OAT GRAIN STANDARDS, SPECIAL GRADES, AND SPECIAL GRADE REQUIREMENTS

Grades	Requirement
Bleached oats	Oats which, in whole or in part, have been treated with sulfurous acid or any other bleaching agent
Bright oats	Oats, except bleached oats, that are of good natural color
Ergoty oats	Oats which contain ergot in excess of 0.10%
Extra-heavy oats	Oats which have a test weight per bushel of 40 lb or more
Garlicky oats	Oats which contain four or more green garlic bulblets or an equivalent quantity of dry or partly dry bulblets in 500 g of oats
Heavy oats	Oats which have a test weight per bushel of 38 lb or more, but less than 40 lb
Smutty oats	Oats which have kernels covered with smut spores or which contain smut masses and smut balls in excess of 0.2%
Thin oats	Oats which contain more than 20.0% of oats and other matter, except "fine seeds," which may be removed from a test portion of the original sample by approved devices in accordance with the procedures prescribed in the Grain Inspection Manual. For the purpose of this paragraph "approved devices" shall be the 0.064 × 3/8 oblong-hole sieve and the 5/64 triangular-hole sieve
Tough oats	Oats which contain more than 14.0% but not more than 16.0% of moisture
Weevily oats	Oats which are infested with live weevils or other insects injurious to stored grain. As applied to oats, the meaning of the term "infested" is set forth in Chapter VI of the Grain Inspection Manual

From *Federal Register*, 39 F.R. 32124, Sept. 5, 1974.

Table 234

SOME PHYSICAL CHARACTERISTICS OF THE OAT KERNEL AND RELATIVE WEIGHT DISTRIBUTION OF THE KERNEL FRACTIONS

Variety	Hulls (% of kernel)	1000 kwt. (g)	Embyronic axis (% of groat)	Scutellum (% of groat)	Starchy endosperm (% of groat)	Bran (% of groat)	Bran thickness (mm)
					Groat		
Orbit	28.1	21.5	1.2	1.8	68.3	28.7	0.063
Lodi	31.8	20.7	1.1	2.0	63.3	33.6	0.058
Garland	28.0	14.4	1.4	2.1	62.7	33.9	0.065
Froker	25.1	22.7	1.0	1.8	67.0	30.2	0.079
Portal	26.6	16.7	1.2	1.6	57.4	39.8	0.075
Dal	23.6	19.5	1.1	2.6	61.8	34.4	0.087
Goodland	26.6	20.3	1.1	1.7	55.8	41.4	0.101

[a] Values for groat fractions obtained by hand-dissection.

From Youngs, V. L., Protein distribution in the oat kernel, *Cereal Chem.*, 49, 407, 1972. With permission.

Table 235
PERCENT PROTEIN AND AMINO ACID IN OATS AND OAT FRACTIONS[a]

	Commercial samples[b]				Groats av, 286 varieties[c]	One cultivar (Orbit)[b]			
	Heavy oats	Flakes	Light oats	Hulls		Embryonic axis	Scutellum	Bran	Endosperm
Protein	13.4	17.6	9.6	5.7[c]	17.1	44.3	32.4	18.8	9.6
Lysine	4.2	4.1	5.2	4.9	4.2	8.2	6.9	4.1	3.7
Histidine	2.4	2.4	2.7	2.4	2.2	3.9	3.6	2.2	2.2
Ammonia	3.3	3.2	3.8	3.6	2.7	1.9	1.8	2.5	2.9
Arginine	6.4	6.0	6.3	6.8	6.9	8.3	9.0	6.8	6.6
Aspartic acid	9.2	9.0	11.1	10.5	8.9	10.2	9.7	8.6	8.5
Threonine	3.3	3.1	4.1	4.1	3.3	5.0	4.7	3.4	3.3
Serine	4.0	4.0	4.5	4.6	4.2	4.8	5.0	4.8	4.6
Glutamic acid	21.6	22.7	20.0	20.3	23.9	14.2	14.9	21.1	23.6
Cystine	1.7	1.7	0.4	0.5	1.6	0.5	1.0	2.4	2.2
Methionine	2.3	2.4	1.5	1.5	2.5	2.2	2.1	2.1	2.4
Glycine	5.1	5.0	6.0	6.1	4.9	6.3	6.2	5.4	4.7
Alanine	5.1	5.0	5.5	5.4	5.0	7.2	6.9	5.1	4.5
Valine	5.8	5.8	6.2	6.4	5.3	6.0	6.2	5.5	5.5
Proline	5.7	6.1	3.1	2.4	4.7	3.3	3.6	6.2	4.6
Isoleucine	4.2	4.3	4.5	4.5	3.9	3.9	3.8	3.8	4.2
Leucine	7.5	7.5	7.6	7.8	7.4	7.1	7.1	7.4	7.8
Tyrosine	2.6	2.1	2.4	2.9	3.1	2.9	3.0	3.5	3.3
Phenylalanine	5.4	5.6	5.3	5.3	5.3	4.2	4.4	5.1	5.6
Tryptophan		1.7							

[a] Protein = nitrogen × 6.25, dry basis. Amino acid values are percentages of amino acids recovered by weight.

[b] Data from Pomeranz, Y., Youngs, V. L., and Robbins, G. S., Protein content and amino acid composition of oat species and tissues, *Cereal Chem.*, 50, 702, 1973; except for tryptophan, from Tkachuk, R. and Irving, G. N., Amino acid compositions of cereals and oilseed meals, *Cereal Chem.*, 46, 206, 1969. Thus, total is over 100%.

[c] In another study hull protein from seven hand-dissected varieties was 1.6%.

Table 236
COMPOSITION OF WILD-RICE AND OTHER SELECTED CEREAL PRODUCTS

Component	Wild-rice (%)	Brown rice (%)	Polished white rice (%)	Oats (%)	Hard red winter wheat (%)	Corn (%)
Moisture	7.9—11.2	12.0	12.0	8.3	12.5	13.8
Protein	12.4—15.0	7.5	6.7	14.2	12.3	8.9
Fat	0.5—0.8	1.9	0.4	7.4	1.8	3.9
Ash	1.2—1.4	1.2	0.5	1.9	1.7	1.2
Crude fiber	0.6—1.1	0.9	0.3	1.2	2.3	2.0
Total carbohydrate	72.3—75.3	77.4	80.4	68.2	71.7	72.2

Table 237
MINERALS IN WILD-RICE AND OTHER CEREALS

Mineral	Wild-rice	Brown rice	Polished rice	Oats	Hard red winter wheat	Corn
Calcium	17—22	32	24	53	46	22
Iron	4.2	1.6	0.8	4.5	3.4	2.1
Magnesium	80—161	—	28	144	160	147
Potassium	55—344	214	92	352	370	284
Phosphorus	298—400	221	94	405	354	268
Zinc	3.3—6.5	—	1.3	3.4	3.4	2.1

ᵃ   Mg/100 g.

Table 238
VITAMIN CONTENT OF WILD-RICE AND OTHER CEREALS

Vitamin	Wild-rice	Brown rice	Polished rice	Oats	Hard red winter wheat	Corn
			mg/100 g			
Thiamine	0.45	0.34	0.07	0.60	0.52	0.37
Riboflavin	0.63	0.05	0.03	0.14	0.12	0.12
Niacin	6.2	4.7	1.6	1.0	4.3	2.2
Vitamin C	0	0	0	0	0	0
			I.U.			
Vitamin A	0	0	0		0	490

SORGHUM

Table 239
MARKET CLASSES FOR SORGHUM IN U.S. GRAIN STANDARDS

Brown sorghum	Sorghum with brown pericarps or brown subcoats with less than 10% sorghums of other colors
White sorghum	Sorghum with white pericarps with less than 2% of sorghums with pericarps or subcoats of other colors
Yellow sorghum	Sorghum with yellow, salmon-pink, red, or white pericarps, or white with spotted pericarp with less than 10% brown sorghums or those with subcoat, not meeting standard for white sorghums
Mixed sorghum	Sorghum which does not meet the requirement for any of the classes brown, yellow, or white sorghums

Table 240
GRADES, GRADE REQUIREMENTS, AND GRAIN DESIGNATIONS FOR U.S. SORGHUM

			Maximum limits of		
			Damaged kernels		Broken kernels, foreign material, and other grains (%)
Grade	Minimum test weight per bushel (lb)	Moisture (%)	Total (%)	Heat-damaged kernels (%)	
U.S. No. 1	57.0	13.0	2.0	0.2	4.0
U.S. No. 2	55.0	14.0	5.0	0.5	8.0
U.S. No. 3ᵃ	53.0	15.0	10.0	1.0	12.0
U.S. No. 4	51.0	18.0	15.0	3.0	15.0
U.S. Sample grade	U.S. Sample grade shall be sorghum which:				

(a) Does not meet the requirements for the grades U.S. Nos. 1, 2, 3, or 4.

(b) Contains more than 7 stones which have an aggregate weight in excess of 0.2 percent of the sample weight or more than 2 crotalaria seeds (*Crotalaria* spp.) per 1,000 grams of sorghum.

(c) Has a musty, sour, or commercially objectionable foreign odor (except smut odor) or

(d) Is badly weathered, heating, or distinctly low quality

SPECIAL GRADES

Smutty sorghum —	Sorghum which is covered with smut spores or which contains 20 or more smut masses in 100 grams of sorghum.
Weevily sorghum —	Sorghum which is infested with live weevils or other insects injurious to stored grain.

ᵃ   Sorghum which is distinctly discolored shall not be graded higher than U.S. No. 3.

Table 241
PROXIMATE ANALYSES OF SORGHUM AND MILLETS[a]

	Protein (N × 6.25)		Ether extract		Crude fiber		Ash		Nitrogen free extract	
	Mean	Range	Mean	Range	Mean	Range	Mean	Range	Mean	Range
Sorghum (*Sorghum bicolor*)	11.6	8.1—16.8 (1463)	3.4	1.4—6.2 (1462)	2.7	0.4—7.3 (1383)	2.2	1.2—7.1 (1436)	79.5	65.3—81.0 (1372)
Pearl millet (*Pennisetum americanum*)	14.5	6.9—20.9 (409)	5.3	0.3—6.9 (216)	2.2	0.9—3.6 (29)	2.1	0.3—5.1 (33)	72.5	59.8—80.6 (26)
Proso millet (*Panicum milacium*)	11.6	11.3—12.7 (20)	4.2	3.8—4.9 (20)	12.0	7.9—19.2 (17)	3.6	3.1—4.2 (20)	70.4	66.3—73.4 (20)
Japanese millet (*Echinochloa crusgalli*)	11.6	11.2—11.8 (6)	5.8	5.5—6.3 (6)	14.7	11.6—16.3 (6)	4.7	3.0—5.6 (6)	63.2	60.9—67.8 (6)
Foxtail millet (*setaria* spp.)	13.0	12.6—14.0 (6)	4.8	4.6—5.0 (6)	10.0	9.3—10.6 (6)	4.0	3.7—4.3 (6)	68.2	66.1—69.7 (6)
Finger millet or Ragi (*Eleusine coracana*)	8.0	4.9—11.3 (35)	2.0	0.9—7.7 (36)	3.6	0.7—8.0 (35)	3.0	2.0—5.0 (35)	72.4	69.3—73.4 (35)

Note: References included in the original.

[a] All values are expressed as a percentage based on oven dry weight of the sample. Numbers in parenthesis give the number of samples comprising the mean and range.

Table 242

AVERAGE ESSENTIAL AMINO ACID COMPOSITION OF SORGHUM AND MILLETS AS % PROTEIN

	Lys	Leu	Phe	Val	Trp	Met	Thr	His	Ileu
Sorghum	2.1	14.2	5.1	5.4	1.0	1.0	3.3	2.1	4.1
	(640)	(582)	(582)	(582)	(92)	(582)	(582)	(582)	(582)
Pearl millet	2.6	17.4	4.9	5.7	2.3	2.5	4.9	2.1	4.3
Proso	1.5	12.2	5.5	5.4	0.8	2.2	3.0	2.1	4.1
Ragi (finger millet)	2.0	6.4	3.2	5.8	1.5	2.3	2.7	0.3	3.7
Italian or foxtail	1.9	10.4	4.9	5.9	2.5	4.0	3.3	1.2	6.1
FAQ/ WHO Suggested Pattern	2.2	2.5	2.5	1.8	0.65	2.4	1.3	—	1.8

Note: Numbers in parenthesis give the number of samples used in determining means. References included in the original.

Table 243

NUTRITIVE VALUE OF SORGHUM COMPARED TO CORN

	Feed efficiency (% of corn)	
Animals	Yellow sorghum	Brown sorghum
Swine	95.8	—
	95.0	85.0
Poultry	97.4	91.4
	99.4	96.6
	98.5	94.0
	99.1	95.8
Beef cattle	—	80.4
	92.8	68.4

Note: References included in the original.

Table 244
A SUMMARY OF DRY MILLING PROCEDURES FOR SORGHUM

	Process description	Products	Composition
Pearling and degermination	The grain is tempered and pearled via an abrasive process. The pearled kernels are tempered and degerminated by impaction or pin milling. Fractions are separated by gravity separators and sieving	16, 7, 4, 4, 23, and 50% yields of feed, germ, flour, +20 grits, and +14 grits are obtained respectively, from yellow sorghum	Fat content was 9.0, 15.0, 3.0, 0.5, and 0.8% dry weight of the feed, germ, flour, +20 grits, and +14 grits, respectively
Pearling or decortication followed by grinding	The grain is tempered and subjected to milling by an abrasive milling procedure. Rice milling equipment works well for sorghum. Sometimes sorghum is pearled by hand using a mortar and pestle. Tempering permits greater retention of the germ tissue with the pearled kernel	Pearled whole kernels and broken kernels, flour and/or meal depending upon method of grinding the grits. Normally around 10 to 12% of the weight of the kernel is removed	Composition depends upon the amount of material removed by pearling. Usually a significant amount of germ remains with the endosperm so the products contain 2.0% fat or more. Degermination does not occur. Keeping properties of the flour are poor
Roller milling	The whole grain is tempered and subjected to roller milling similar to wheat milling although procedures vary	High extraction flours, 90% yield, and lower extraction flours (70%)	The 70 and 90% extraction flours have 2.0 and 2.8% fat, respectively. Ash content is 0.5 and 1.0% for the 70 and 90% extraction flour

Note: References included in the original.

Table 245
COMMON METHODS OF PROCESSING SORGHUM FOR USE IN LIVESTOCK FEEDS

Category	Method	Procedure	Characteristics
Mechanical action	Grinding	Particle size reduction by use of impaction, abrasion, attrition	Improves feed efficiency markedly. Least expensive processing method. Most commonly used
	Dry and steam rolling	Pass grain through roller mill with or without steam. Steam produces less fines and better flake. Grain is subjected to steam for only 2—3 min.	Increased digestibility, similar to dry grinding
Wet processing	Reconstitution	Enough water added to bring dry grain to 25—30% moisture level. Wet grain is anaerobically stored 15—21 days prior to grinding and feeding	Improves feed efficiency 15—20% over dry ground grain. Increases both protein and starch digestibility
	Early harvesting	Grain harvested at moisture level of 20—30%, stored anaerobically or with organic acids to prevent spoilage. Grinding is done prior to or after storage	Similar to reconstituted ground grain
	Soaking	Soak grain in water for 12 to 24 hr prior to rolling or grinding	Tendency for grain to ferment or sour. Only limited use
Heat and moisture	Steam rolling	Grain subjected to live steam 3—5 min at approx. 180°F then rolled	Little increase over dry rolling process
	Steam flaking	Grain exposed to high moisture steam. Moisture content of grain 18—20%. Then grain is rolled to desired flake thickness	Sorghum requires longer steaming than corn. Grain must be dried if to be stored. Thin flaking of grain increases utilization
	Pelleting	Ground grain is conditioned with steam, forced through a die and cooked	Reduces dust, improves palatability and improves handling of feeds
	Exploding	Grain exposed to high pressure steam, the starch is gelatinized, the pressure is decreased and rapid expansion of the kernel occurs	It is similar to puffing of cereals for breakfast foods. Feed efficiency is similar to steam flaked grain
Hot dry heat	Popping	Hot, dry air expansion of grain. Bulk density is low. Density is increased by spraying with water and rolling sometimes	Ruptures endosperm increasing starch availability. Feed efficiency is similar to steam flaking
	Micronizing	Heat grain with gas fired infrared burners to the point of eversion followed by rolling through a roller mill	Feed efficiency similar to steam flaking, dry product can be stored

Note: References included in the original.

CORN

Table 246
UNITED STATES GRADING STANDARDS FOR CORN[a]

Grade	Minimum test weight per bushel (lb)	Moisture (%)	Broken corn and foreign material (%)	Damaged kernels	
				Total (%)	Heat-damaged kernels (%)
1	56	14.0	2.0	3.0	0.1
2	54	15.5	3.0	5.0	0.2
3	52	17.5	4.0	7.0	0.5
4	49	20.0	5.0	10.0	1.0
5	46	23.0	7.0	15.0	3.0
Sample grade[b]	—	—	—	—	—

The "Maximum Limits" heading spans the Moisture, Broken corn and foreign material, and Damaged kernels columns.

[a] Grades and grade requirements for the classes Yellow Corn, White Corn, and Mixed Corn.

[b] Sample grade shall be corn which does not meet the requirements for any of the grades from No. 1 to No. 5, inclusive; or which contains stones; or which is musty, or sour: heating, or which has any commercially objectionable foreign odor; or which is otherwise of distinctly low quality.

Source: Official Grain Standards of the United States Federal Grain Inspection Service, U.S. Department of Agriculture, Washington, D.C., 1970.

Table 247
CLASSIFICATION OF KERNEL TYPES IN CORN

Type name	Genes involved	Characteristics
Polygenic differences		Extractable Oil, % d.b.
Normal oil		4.0—5.0
Low oil		4.0—0.75
High oil		5.0—20
		Crude Protein (N×6.25), % d.b.
Normal protein		8.0—11.0
Low protein		8.0—4.0
High protein		11.0—28
		Linoleic Acid, % of TFA
Normal oil composition		50—60
Low linoleic		19—50
High linoleic		60—71
		Total Carotenoids, mg/kg
Yellow endosperm	(3 gene pairs)	0.6; 8; 22; 50
Red, blue or purple pericarp or aleurone		
Monogenic differences	Recessive Mutant Gene	
Floury endosperm	fl_1	High lysine (low zein)
Floury endosperm	fl_2	Normal composition
Floury endosperm	h	Normal composition
Opaque endosperm	o_1	Normal composition
Opaque endosperm	o_2	High lysine (low zein)
Flinty endosperm	Dominant genes	Normal composition
Popcorn	Dominant genes	Normal composition
Dent endosperm	Dominant genes	Normal composition
Sugary endosperm (Sweet corn)	su_1	Low starch, med. sugar high phytoglycogen (PG)
Waxy endosperm	wx	0% amylose starch
High amylose endosperm	ae	60% amylose starch
Shrunken endosperm	sh_1, sh_2	Low starch, high sugar, low PG
Dull endosperm	du	Medium starch

Note: References included in the original.

Table 248
PHYSICAL PROPERTIES OF CORN, WHEAT, AND GRAIN SORGHUM

	Dent corn	Hard wheat	Grain sorghum
Kernels per ear	800	35	1,500
	(500—1,200)	(20—50)	(800—3,000)
Kernels per gram	3	35	50
Kernels per pound	1,200	16,000	20,000
	(900—1,500)	(8,000—24,000)	(13,000—24,000)
Kernel specific gravity	1.19	1.30	1.22
Bulk density, lbs/bu	56	60	57
Bulk density, lb/ft³	44.5	48	44.8
	(40—48)		
Air space in bulk, %	40	42.6	37
Specific heat, Btu/lb/°F	0.484	0.460	—
	(14.7% H_2O)	(13.6% H_2O)	
Equilibrium relative humidity, % (15% moisture, 70°F)	75	73	74
Angle of repose, degrees	27	27	—

Note: References included in the original.

Table 249

WEIGHT AND COMPOSITION OF COMPONENT PARTS OF DENT CORN KERNELS FROM 12 MIDWEST HYBRIDS

Part		Percent dry weight of whole kernel	Composition of kernel parts (% d.b.)				
			Starch	Fat	Protein	Ash	Sugar
Germ	Range	10.5—13.1	5.1—10.0	31.1—38.9	17.3—20.0	9.38—11.3	10.0—12.5
	Mean	11.5	8.3	34.4	18.5	10.3	11.0
Endosperm	Range	80.3—83.5	83.9—88.9	0.7—1.1	6.7—11.1	0.22—0.46	0.47—0.82
	Mean	82.3	86.6	0.86	8.6	0.31	0.61
Tip-cap	Range	0.8—1.1	—	3.7—3.9	9.1—10.7	1.4—2.0	—
	Mean	0.8	5.3[a]	3.8	9.7	1.7	1.5
Pericarp	Range	4.4—6.2	3.5—10.4	0.7—1.2	2.9—3.9	0.29—1.0	0.19—0.52
	Mean	5.3	7.3	0.98	3.5	0.67	0.34
Whole kernels	Range	—	67.8—74.0	3.9—5.8	8.1—11.5	1.27—1.52	1.61—2.22
	Mean	100	72.4	4.7	9.6	1.43	1.94

[a] Composite.

From Earle, F. R., Curtis, J. J., and Hubbard, J. E., Composition of the component parts of the corn kernel, *Cereal Chem.*, 44, 601, 1946. With permission.

Table 250
MOISTURE, CARBOHYDRATE, AND NITROGENOUS COMPONENTS OF DENT CORN GRAIN

	Range	Average
Moisture (% wet basis)	7—23	16.7
Carbohydrates		
Starch (% dry basis)	64—78	71.3
Total fiber (neutral detergent residue) (% d.b.)	8.3—11.9	9.5
Acid detergent fiber (cellulose) (% d.b.)	3.0—4.3	3.3
Crude fiber (% d.b.)	2.0—5.5	2.8
Pentosans (as xylose) (% d.b.)	5.9—6.6	6.2
Water soluble polysaccharides, (% d.b.)	0.3—0.5	0.4
Sugars, total (as glucose) (% d.b.)	1.0—3.0	2.58
Sucrose (% d.b.)	0.5—3.3	2.0
Raffinose (% d.b.)	0.08—0.30	0.19
Glucose (% d.b.)	0.07—0.17	0.10
Fructose (% d.b.)	0.05—0.13	0.07
Phytate, (% d.b.) (expressed as TriMg — hexa-phosphoinositol)	0.7—1.0	0.8
Nitrogenous		
Crude protein (Nx6.25), (% d.b.)	8—14	9.9
Albumins (% d.b.)	—	0.8
Globulins (% d.b.)	—	0.3
Prolamins (zein) (% d.b.)	1.1—7.3	3.8
Glutelins (% d.b.)[a]	3.9—7.3	4.1
Nonprotein Nitrogen	—	0.9
		Mg/kg (d.b.)
Nonprotein nitrogen (soluble)		277
21 Amino acids and ammonia		147
Peptides		23
Quarternary nitrogen compounds		
Choline		17
Trigonelline		2.4
Betaine		1.1
Heterocyclic compounds		
Adenine		2.8
Adenosine		2.6
Uridine		1.1
Uracil		0.7
Indoleacetic acid[37]		0.016
Not characterized		79

Note: References included in the original.

[a] Assumes nitrogen unextracted by alkali is glutelin.

Table 251
AMINO ACID ANALYSIS OF WHOLE DENT CORN GRAIN

Amino acid	% of dry matter	% of total protein[b]	
		Range	Average
Nutritionally essential[a]			
Methionine	0.19	1.0—2.1	1.9
Cystine	0.15	1.2—1.6	1.5
Lysine	0.25	2.0—3.8	2.5
Tryptophan	0.10	0.5—1.2	1.0
Threonine	0.39	2.9—3.9	3.8
Isoleucine	0.42	2.6—4.0	4.2
Histidine	0.22	2.0—2.8	2.1
Valine	0.48	2.1—5.2	4.7
Leucine	1.14	7.8—15.2	11.2
Arginine	0.59	2.9—5.9	5.8
Phenylalanine	0.50	2.9—5.7	4.9
Glycine	0.38	2.6—4.7	3.7
Nonessential			
Alanine	0.78	6.4—9.9	7.8[15]
Aspartic acid	0.68	5.8—7.2	6.8
Glutamic acid	1.77	12.4—19.6	17.7
Proline	.84	6.6—10.3	8.4
Serine	.46	4.2—5.5	4.6
Tyrosine	.47	2.9—4.7	4.7

Note: References included in the original.

[a] Listed in approximate order of decreasing importance in feed formulation.

[b] 10.1% total protein (N × 6.25), dry basis.

Table 252
LIPID COMPONENTS OF DENT CORN GRAIN

	Range	Dent corn average
Fats		
Free fat (hexane extractable) (% d.b.)	0.7—18.2	4.45
Total fat (free plus bound lipids) (% d.b.)	1.2—18.8	5.3
Triglycerides (% of TF)[a]	75—92	81.0
Palmitic acid, 16:0 (% of TGFA)[b]	7—19	11.5
Palmitoleic acid, 16:1 (% of TGFA)	1	1.0
Stearic acid, 18:0	1—3	2.0
Oleic acid, 18:1	20—46	24.1
Linoleic acid, 18:2	35—70	61.9
Linolenic acid, 18:3	0.8—2	0.7
Arachidic acid, 20:2	0.1—2	0.2
Diglycerides (% of TF)	1.0—2.9	2.9
Sterols, free (% of TF)	1.3—5.5	4.6
Sterylesters (% of TF)	1.1—2.9	2.8
Fatty acids, free (% of TF)	0.3—0.9	0.9
Polar lipids, (% of TF)	3.9—12.0	8.1
Tocopherols, mg/kg of corn, d.b.	42—87	
Carotenoid Pigments		
Xanthophyll pigments (mg/kg) (d.b.)	0.2—50	24
Zeinoxanthin	0.1—7.8	1.7
Lutein	0.1—33	15.5
Zeaxanthin	0.1—33	6.7
Carotene pigments (mg/kg) (d.b.)	0.2—7	6
α-carotene	0.1—0.6	0.2
β carotene	0.3—5.4	2.3
Zeta carotene	0.2—4.2	1.3
Cryptoxanthin	1.0—5.2	2.2
Polyoxy compounds (mg/kg) (d.b.)	0.1—7.5	1.2

Note: References included in the original.

[a] TF = Total Fat

[b] TGFA = Triglyceride Fatty Acids

Table 253
INORGANIC COMPONENTS OF DENT CORN GRAIN

	Range	Average
Total ash, (oxide), % d.b.	1.1—3.9	1.42
Phosphorus,[a] total, % d.b.	0.26—0.75	0.27
Inorganic P, % d.b.	—	0.08
Sulfur, % d.b.	0.01—0.22	0.14
Selenium, mg/kg	—	0.045
Iodine, mg/kg	73—810	385
Sodium, % d.b.	0.0—0.15	0.01
Potassium, % d.b.	0.32—0.72	0.38
Magnesium, % d.b.	0.09—1.0	0.17
Calcium, % d.b.	0.01—0.1	0.03
Iron, mg/kg	1—100	40
Zinc, mg/kg	12—30	22.4
Copper, mg/kg	0.9—10	3.8
Manganese, mg/kg	0.7—54	4.7
Cobalt, mg/kg	0.003—0.34	0.025
Chromium, mg/kg	0.06—0.16	0.07
Lead, mg/kg	0.2—0.3	0.27
Cadmium, mg/kg	0.04—0.15	0.07
Mercury, mg/kg	0.002—0.006	0.003

Note: References included in the original.

[a]   Largely phytin phosphorus.

Table 254
VITAMIN CONTENT OF DENT CORN (ON DRY BASIS)

	Range	Average
Vitamin A, IU/gm[a]	—	2.5
Vitamin E, mg/kg[b]	3.0—12.1	25
Thiamin, mg/kg	3.0—8.6	4.2
Riboflavin, mg/kg	0.25—5.6	1.25
Pantothenic acid, mg/kg	3.5—14	6.5
Biotin, mg/kg	50—600	90
Folic acid, mg/kg	100—683	426
Choline, mg/kg	—	500
Niacin, mg/kg	9.3—70	24.4
Pyridoxine, mg/kg	—	9.6

Note: References included in the original.

[a]   Assumes complete conversion of β-carotene to vitamin A. Most animals convert only 20—30%; poultry may convert 30—70%.

[b]   Most vitamin E activity resides in α-tocopherol, but γ-tocopherol is one-tenth as active.

Table 255

TYPICAL NUTRIENT ANALYSIS OF FEED PRODUCTS DERIVED FROM WET MILLING OF CORN

	Corn gluten feed	Corn gluten meal (41% protein)	Corn gluten meal (60% protein)	Corn germ meal	Condensed fermented corn extractives
Proximate analysis (%)					
Protein (N × 6.25)	22.3	42.2	62.0	22.6	25.0
Fat	2.5	2.5	2.5	1.9	0
Starch	14.0	NA	14.0	24.0	NA
Crude fiber	8.0	4.0	1.2	9.5	0
Ash	7.3	4.0	1.8	3.8	7.8
Moisture	10.0	10.0	10.0	10.0	47.0
NFE	50.0	35.0	20.0	48.0	18
Amino acids (% of sample)					
Alanine	1.5	3.3	5.2	1.4	1.8
Arginine	1.0	1.5	1.9	1.3	1.1
Aspartic acid	1.2	2.4	3.6	1.4	1.4
Cystine	0.5	0.8	1.1	0.4	0.8
Glutamic acid	3.4	8.6	13.8	3.2	3.5
Glycine	1.0	1.3	1.6	1.1	1.1
Histidine	0.7	0.9	1.2	0.7	0.7
Isoleucine	0.6	1.5	2.3	0.7	0.7
Leucine	1.9	6.0	10.1	1.8	2.0
Lysine	0.6	0.8	1.0	0.9	0.8
Methionine	0.5	1.2	1.9	0.6	0.5
Phenylalanine	0.8	2.3	3.8	0.9	0.8
Proline	1.7	3.6	5.5	1.3	2.0
Serine	1.0	2.0	3.1	1.0	1.0
Threonine	0.9	1.4	2.0	1.1	0.9
Tryptophan	0.1	0.2	0.3	0.2	0.1
Tyrosine	0.6	1.8	2.9	0.7	0.5
Valine	1.0	1.9	2.7	1.2	1.2
Vitamins (mg/lb)					
Choline	690	545	400	800	320
Niacin	33	29	25	19	40
Pantothenic acid	7.8	4.5	1.3	2.0	6.8
Pyridoxine	6.8	4.8		2.7	4.0
Riboflavin	1.1	1.0	1.0	1.7	2.7
Thiamine	0.9	0.5	0.1	2.8	1.3
Minerals					
Copper (mg/lb)	21.7	16.7	12.1	8.5	517
Manganese (mg/lb)	10.8	6.4	2.0	7.0	15.3
Iron (%)	0.03	0.01	0.002	0.3	0.01
Magnesium (%)	0.5	0.25	0.001	0.08	1.0
Phosphorus (%)	0.80	0.65	0.5	0.5	1.8
Calcium (%)	0.46	0.23	0.002	0.3	0.06
Potassium (%)	1.1	0.6	0.001	0.17	2.4
Energy					
Metabolizable energy (ME) (Cal/lb)	—	1260	1690	770	707
Total digestible nutrients (TDN) (%)	75	80	86	69	40
Estimated net energy (ENE) (therms/cwt)	66-70	—	78-80	—	—

From Shroder, J. D. and Heiman, V., Feed products from corn processing, in *Corn: Culture, Processing, Products,* Inglett, G. E., Ed., AVI, Westport, Conn., 1970, chap. 12. With permission.

330 *Practical Handbook of Agricultural Science*

Table 256
CAROTENOID COMPONENTS OF CORN GLUTENMEAL

	Concentration (mg/kg)			
Pigment	Gluten 1	Gluten 2	Gluten 3	Typical range
Total carotenes	88	49.3	37.8	20—88
Monohydroxy xanthophylls	—	39.4	48.6	—
Dihydroxy xanthophylls	—	274.0	151.8	—
Total xanthophylls	352	313.0	200.4	198—55

Note: References included in the original.

Table 257
NUTRIENT ANALYSIS—HOMINY FEED

Proximate analysis (%)
- Protein (N × 6.25) 10.7
- Fat 6.5
- Crude fiber 5.0
- Moisture 9.4
- Ash 2.5
- NFE 65.9

Amino acids (% of sample)
- Alanine 0.83
- Arginine 0.62
- Cystine 0.09
- Glutamic acid 2.17
- Glycine 0.61
- Histidine 0.37
- Isoleucine 0.39
- Leucine 1.23
- Lysine 0.47
- Methionine 0.22
- Phenylalanine 0.41
- Proline 1.35
- Serine 0.57
- Threonine 0.50
- Tryptophan 0.17
- Tryosine 0.34
- Valine 0.60

Vitamins
- Vitamin A (equiv. IU/lb) 6946
- Niacin (mg/lb) 23.2
- Pantothenic acid (mg/lb) 3.4
- Riboflavin (mg/lb) 0.91
- Thiamine (mg/lb) 3.6

Minerals
- Calcium (%) 0.05
- Iron (%) 0.006
- Magnesium (%) 0.24
- Phosphorus (%) 0.53
- Potassium (%) 0.67
- Sulfur (%) 0.03
- Cobalt (mg/lb) 0.027
- Copper (mg/lb) 6.6
- Manganese (mg/lb) 6.6

Energy
- ME (Cal/lb)
  - Cattle 1415
  - Swine 1530
  - Chickens 1303
- TDN (%)
  - Cattle 86
  - Sheep 81
  - Swine 82

Note: References included in the original.

PHYSICAL PROPERTIES AND DRYING

Table 258
GRAIN EQUILIBRIUM MOISTURE CONTENT (% WET BASIS)

Material	Temp °F	Relative humidity (%)									
		10	20	30	40	50	60	70	80	90	100
Barley	77	4.4	7.0	8.5	9.7	10.8	12.1	13.5	15.8	19.5	26.8
Buckwheat	77	5.0	7.6	9.1	10.2	11.4	12.7	14.2	16.1	19.1	24.5
Flaxseed	77	3.3	4.9	5.6	6.1	6.8	7.9	9.3	11.4	15.2	21.4
Oats	77	4.1	6.6	8.1	9.1	10.3	11.8	13.0	14.9	18.5	24.1
Rye	77	5.2	7.6	8.7	9.9	10.9	12.2	13.5	15.7	20.6	
Shelled corn YD	20				10.4	11.8	13.3	15.0	16.6		
	32				11.0	12.5	14.0	15.8	18.0	21.8	
	40	6.3	8.6	9.8	11.0	12.4	13.8	15.7	17.6	21.5	
	50	6.6	8.0	9.3	10.8	12.2	13.8	15.2	17.5	21.8	
	60	7.5	7.8	9.0	10.3	11.3	12.4	13.9	16.3	19.8	
	77	5.1	7.0	8.3	9.8	11.2	12.9	14.0	15.6	19.6	23.8
	86	4.4	7.4	8.2	9.0	10.2	11.4	12.9	14.8	17.4	
	90	4.9	6.6	7.7	9.3	10.8	12.4	14.0	16.2	19.3	
	100	4.0	6.0	7.3	8.7	9.0	11.0	12.5	14.2	16.7	
	120				8.6	10.0	11.2	13.1	14.9		
	155				7.4	8.4	10.0	11.5	12.2		
	160	3.9	6.2	7.6	9.1	10.4	11.9	13.9	15.2	17.9	
Sorghum	30		8.2	10.1	11.2	12.3	13.5	14.5	15.8		
	60		7.5	9.5	10.7	11.8	12.9	14.0	15.5		
	77	4.4	7.3	8.6	9.8	11.0	12.0	13.8	15.8	18.8	21.9
	90		7.0	8.7	10.2	11.8	12.2	13.1	14.8		
	120		6.6	8.0	9.4	10.7	11.6	12.7	14.3		
Soybeans	77		5.5	6.5	7.1	8.0	9.3	11.5	14.8	18.8	
Wheat											
Soft red winter	20				11.3	12.8	14.1	15.3	17.0		
	32				11.0	12.2	13.5	14.7	16.2		
	50				10.2	11.7	13.1	14.4	16.0		
	70				9.7	11.0	12.4	14.0			
	77	4.3	7.2	8.6	9.7	10.9	11.9	13.6	15.7	19.7	25.6
Hard red winter	77	4.4	7.2	8.5	9.7	10.9	12.5	13.9	15.8	19.7	25.0
Hard red spring	77	4.4	7.2	8.5	9.8	11.1	12.5	13.9	15.9	19.7	25.0
White	77	5.2	7.5	8.6	9.4	10.5	11.8	13.7	16.0	19.7	26.3
Durum	77	5.1	7.4	8.5	9.4	10.5	11.5	13.1	15.4	19.3	26.7
Wheat	30		6.9	9.2	10.4	11.8	13.2	14.5	16.3		
	50		7.9	9.2	10.6	12.7	14.2	15.0	17.3		
	60		6.1	7.8	9.6	10.7	12.7	13.8	15.3		
	68	5.6	7.1	8.3	9.6	10.9	12.2	13.5	15.0		
	77	5.8	7.6	9.1	10.7	11.6	13.0	14.5	16.8	20.6	
	90		5.3	7.0	8.6	10.3	11.5	12.9	14.3		
	104	4.6	6.2	7.4	8.6	10.0	11.3	12.3	14.2		
	120			6.2	7.4	9.6	10.4	11.9	13.6		
	122	4.0	5.8	6.7	8.1	10.0	10.8	12.6	15.1	19.4	
	176	2.5	3.7	4.8	5.7	6.7	8.0	9.8	11.5		

From Brooker, D. B., Bakker-Arkema, F. W., and Hall, C. W., *Drying Cereal Grains,* AVI Publishing, Westport, Conn., 1974. With permission.

Table 259
OVEN TEMPERATURE AND HEATING PERIOD FOR MOISTURE CONTENT DETERMINATIONS

Seed	Oven temp ± 1°C	Heating time hr	Heating time min	Seed	Oven temp ± 1°C	Heating time hr	Heating time min
Barley	130	20	0				
Corn	103	72	0	Rye	130	16	0
Fescue	130	3	0	Ryegrass	130	3	0
Flax	103	4	0	Safflower	130	3	0
Oats	130	22	0	Sorghum	130	18	0
Rape	130	4	0	Wheat	130	19	0

Note: References included in the original.

Table 260
THE PHYSICAL DIMENSIONS OF GRAIN KERNELS

Product	Equiv sphere diameter* (in.)	% Sphericity	Major (in.)	Interm. (in.)	Minor (in.)	Moisture % w.b.
			Ellipsoid dimensions*			
Barley	0.155 (0.008)	44.5	0.411 (0.035)	0.137 (0.010)	0.108 (0.058)	7.8
	0.141 (0.005)					8.4
	0.180 (0.011)					12.0
Flax	0.088 (0.002)	46.4	0.194 (0.008)	0.092 (0.005)		6.4
Milo	0.137	84.7	0.170 (0.010)	0.160 (0.007)	0.110 (0.012)	9.2
Navy bean	0.266 (0.016)					12.0
Oats	0.142 (0.004)	34.2	0.508 (0.058)	0.114 (0.012)	0.090 (0.011)	8.7
	0.136 (0.004)					9.5
	0.232 (0.022)					12.0
Rice	0.132 (0.004)	46.7	0.333 (0.017)	0.123 (0.004)	0.089 (0.004)	8.9
Rice						
Medium	0.179	46.2				15.0
Long	0.180	37.5				15.0
Rye	0.122					7.8
Shelled corn	0.315	99.4	0.640 (0.036)	0.798 (0.042)	0.504 (0.028)	6.7
	0.305 (0.006)					10.0
	0.386 (0.053)					12.0
Sorghum	0.138 (0.003)					10.2
Soybeans	0.249 (0.005)					<7.0
	0.253 (0.14)					14.6

Table 260 (continued)
THE PHYSICAL DIMENSIONS OF GRAIN KERNELS

Product	Equiv sphere diameter[a] (in.)	% Sphericity	Ellipsoid dimensions[a]			Moisture % w.b.
			Major (in.)	Interm. (in.)	Minor (in.)	
Wheat						
Hard	0.152	61.5	0.258	0.126	0.122	7.7
	(0.008)		(0.017)	(0.010)	(0.010)	
Soft	0.141					10.6
	(0.004)					
	0.158					12.0
	(0.010)					
	0.175					12.0
	(0.013)					

Note: References included in the original.

[a] Values in parentheses are standard deviations.

Table 261
WEIGHT, VOLUME, AND DENSITY OF GRAIN KERNELS

Product	Weight[a] ($\times 10^5$ lb)	Volume[a] ($\times 10^4$ in.³)	Unit density (lb/ft³)	Bulk density[a] (lb/ft³)	Moisture (% wb)
Barley	9.93	20.32	88.44	37.8	7.8
	(1.47)	(2.10)	—	(2.0)	
	7.56	—	—	38.6	12.0
Flax	1.46	3.47	72.71	43.6	6.4
	(0.16)	(0.45)	—	(0.5)	
Milo	6.34	13.43	81.57	48.1	9.2
	(0.21)	(0.44)	—	(0.2)	
Navy bean	46.60	—	—	38.6	12.0
	—	—	—	—	
Oats	7.50	15.18	85.38	29.0	8.7
	(0.73)	(1.22)	—	(3.8)	
	9.05	—	—	25.6	12.0
	—	—	—	—	
Rice	5.94	11.99	85.61	36.1	8.9
	(0.45)	(1.07)	—	(0.7)	
Rice, medium	9.80	20.48	82.67	37.4	12.0
	10.20	21.12	83.47	38.6	14.0
	10.70	21.87	84.53	39.6	16.0
	11.70	23.61	85.64	40.5	18.0
Rice, long	11.20	22.76	85.05	36.6	12.0
	11.30	22.81	85.59	36.7	14.0
	11.70	23.51	86.00	37.8	16.0
	12.00	24.02	86.34	38.4	18.0
Shelled corn	74.20	—	—	41.2	12.0
Wheat					
hard	9.48	18.47	88.69	50.0	7.7
	(1.44)	(2.78)	—	(0.7)	
soft	8.17	—	—	48.0	12.0
	(0.78)	—	—	—	

Note: References included in the original.

Table 262
RELATIONSHIP BETWEEN THREE AXIAL DIMENSIONS AND VOLUME, IN 50-KERNEL SAMPLES OF SHELLED CORN

Sample and statistics	Major diameter (mm)	Intermediate diameter (mm)	Minor diameter (mm)	Weight (g)
Single variety middle of ear				
Mean	10.78	7.58	4.57	0.27
Variance	0.475	0.214	0.151	0.001
Correlation with volume-r	0.66	0.59	0.27	—
Regression equation $\ln V = -7.14 + 1.18 \ln a + 1.04 \ln b + 0.48 \ln c$				
Single variety whole ear				
Mean	11.15	7.30	4.86	0.27
Variance	1.071	0.670	0.586	0.003
Correlation with volume-r	0.75	0.78	0.30	—
Regression equation $\ln V = -6.19 + 0.62 \ln a + 1.13 \ln b + 0.6 \ln c$				
Mixture of varieties				
Mean	12.66	8.5	5.24	0.35
Variance	1.72	0.77	0.5	0.004
Correlations with volume-r	0.35	0.41	0.32	—
Regression equation $\ln V = -6.18 + 0.9 \ln a + 0.71 \ln b + 0.68 \ln c$				

From Mohsenin, N. N., *Physical Properties of Plant and Animal Materials,* Gordon and Breach, New York, 1970.

Table 263
RELATIVE DRYING RATES OF DIFFERENT PRODUCTS AS COMPARED TO WHEAT AT 18% W.B. and 35°C (95°F)

	% Moisture content w.b.				
	22	20	18	16	14
Peas	95	55	35	23	15
Corn	135	90	50	30	15
Wheat	210	150	100	65	35
Rye	225	175	115	70	35
Oats	450	250	150	100	60
Sugar beet seed	1650	1200	800	500	300
Rapeseed	—	1500	1150	800	500

Source: Kreyger, J., *Drying and Storing Grains, Seeds and Pulses in Temperate Climates,* Inst. Stor. Proc. Agric. Prod., Wageningen, The Netherlands, 1972.

Table 264
RESULTS OF FOUR QUALITY FACTORS OF SOFT WHEAT DRIED AT DIFFERENT INLET AIR TEMPERATURES IN A ONE-STAGE CONCURRENT-FLOW DRYER

Drying temp °C	Germination %	Ash %	Protein %	AWR[a] %
121	71	0.37	8.2	51.5
149	71	0.38	8.6	52.9
204	51	0.39	9.0	59.0
232	28	0.40	8.9	66.3
Control	93	0.37	8.2	51.6

[a] AWR = Alkaline water retention.

Source: Ahmadnia, A., Quality of Soft Wheat Dried in a Concurrent-Counter Flow Dryer, M. S. thesis, Michigan State University, East Lansing, Mich., 1977.

Table 265
CRITICAL KERNEL TEMPERATURE (°C) OF SOME SMALL GRAINS AS A FUNCTION OF EQUILIBRIUM RELATIVE HUMIDITY (CRITERION: LESS THAN 5% VIABILITY DECREASE)

	% Relative Humidity			
	60	70	80	90
Oats	59	55	50	—
Wheat	63	62	58	52
Corn	52	51	48	46
Rye	53	50	45	41

Source: Ahmadnia, A., Quality of Soft Wheat Dried in a Concurrent-Counter Flow Dryer, M. S. thesis, Michigan State University, East Lansing, Mich., 1977.

Table 266
RECOMMENDED AERATION AIRFLOW RATES

Purpose of aeration	Airflow rate (cfm/bu)
Maintaining dry grain	1/20 to 1/10
Holding moist grain	1/4 to 1/2
Cooling hot grain (from dryer)	1/2 to 1

Table 267
FAN HORSEPOWER FOR AERATING SHELLED CORN[a] — REQUIREMENTS PER 100 ft² OF BIN CROSS-SECTION OR FLOOR AREA

Corn depth (ft)	Aeration air flow rate (cfm/bu)					
	1/20	1/10	1/4	1/2	3/4	1
	(hp/100 ft²)					
10	0.007	0.015	0.04	0.10	0.19	0.31
15	0.011	0.025	0.06	0.24	0.54	1.0
20	0.016	0.037	0.10	0.48	1.2	2.3
25	0.021	0.05	0.21	1.0	2.3	4.3
30	0.029	0.07	0.36	1.7	4.3	8.0
40	0.048	0.14	0.83	4.3	—[b]	—[b]
50	0.076	0.26	1.6	1.00	—[b]	—[b]
60	0.12	0.40	3.0	—[b]	—[b]	—[b]
80	0.24	0.99	7.8	—[b]	—[b]	—[b]
100	0.45	1.88	—[b]	—[b]	—[b]	—[b]

Note: Fan horsepower values shown for shelled corn may be used for soybeans.

[a] Horsepower based on fan static efficiency of 47%.
[b] Above recommended maximum of 10 hp/100 ft².

Table 268
FAN HORSEPOWER FOR AERATING WHEAT[a]
— REQUIREMENTS PER 100 ft² OF BIN CROSS-
SECTION OR FLOOR AREA

Wheat depth (ft)	Aeration air flow rate (cfm/bu)					
	1/20	1/10	1/4	1/2	3/4	1
	(hp/100 ft²)					
10	0.009	0.018	0.068	0.21	0.42	0.72
15	0.015	0.040	0.12	0.61	1.32	2.31
20	0.024	0.071	0.34	1.31	3.01	5.88
25	0.039	0.12	0.65	2.71	6.42	—[b]
30	0.059	0.19	1.11	4.41	—[b]	—[b]
40	0.12	0.42	2.49	—[b]	—[b]	—[b]
50	0.21	0.76	5.15	—[b]	—[b]	—[b]
60	0.34	1.33	8.62	—[b]	—[b]	—[b]
80	0.79	3.21	—[b]	—[b]	—[b]	—[b]
100	1.45	6.03	—[b]	—[b]	—[b]	—[b]

Note: Fan horsepower values shown for wheat may be used for sorghum, rough rice, barley and oats.

[a] Horsepower based on a fan static efficiency of 47%.
[b] Above recommended maximum of 10 hp/100 ft.

Table 269
CROSS-SECTIONAL OR FLOOR AREAS OF ROUND BINS

Bin diameter (ft)	Floor area (ft²)	Multiplier[a]
18	250	2.5
21	346	3.5
24	452	4.5
27	572	5.7
30	706	7.1
33	855	8.6
36	1017	10.2
40	1256	12.5
42	1385	13.9
48	1808	18.1

[a] Multiplier times horsepower values from Tables 267 and 268 is the fan horsepower required for the bin size selected.

Table 270
ENERGY REQUIRED TO DRY CORN TO 14% WET BASIS
(kWh/t)[a]

Drying from (% wet basis)	High speed to 14%, 90°C		High speed to 16%, 90°C, dryeration		High speed to 20%, 90°C, bin finish		Batch-in-bin, 60°C
	Fan	Heat	Fan	Heat	Fan	Heat	Fan and heat
30	4.6	365	3.8	310	16.0	240	370
28	4.1	315	3.3	260	15.5	190	315
26	3.6	270	2.8	210	15.0	145	260
24	3.2	225	2.2	170	14.6	100	210
22	2.7	180	1.8	130	14.1	60	165
20	2.2	135	1.4	90	—	—	120

[a]   Combination high temperature/low temperature. Values assume natural air only in the final stage for 30 days.

Table 271
PERFORMANCE OF HIGH-TEMPERATURE DRYERS WITH DESIGNS FOR ENERGY SAVING: ENERGY REQUIRED TO DRY CORN TO 14% WET BASIS (kWh/t)

Drying from (% wet basis)	Cross flow[a]				Concurrent flow[c]	
	Reverse cooling		Recycling[b]			
	Fan	Heat	Fan	Heat	Fan	Heat
30	4.8	353.0	4.6	298.2	7.7	287.7
28	4.3	303.1	4.1	255.8	6.5	244.7
26	3.7	256.1	3.5	215.8	5.4	204.1
24	3.2	211.2	3.0	177.6	4.4	165.6
22	2.6	166.9	2.5	140.0	3.5	129.1
20	2.1	123.6	2.0	103.2	2.5	94.3

[a]   Cross-flow dryer includes cooling section.
[b]   Recycling 50% of the exhausted drying air.
[c]   Drying bed 2 m deep, with airflow of 32 m³/min/t.

Table 272
ENERGY REQUIRED TO DRY CORN TO 14% WET
BASIS (kWh/t): LAYER METHOD

Drying from (% wet basis)	In-bin, 20°C Fan and heat	In-bin low-temperature ΔT = 4°C Fan and heat	Low-temperature solar, maximum ΔT = 10°C Fan and heat	Low-temperature heat pump, ΔT = 5.6°C Heat	Fan
30	—[a]	—[a]	—[a]	—[a]	—[a]
28	305	275	240[b]	190	35
26	230	190	120[b]	75	30
24	185	120	80[c]	35	25
22	145	80	50[c]	15	20
20	105	60	35[c]	10	15

Note: In-bin or layer-drying method — bin partially filled, dried, then second and third layer added and dried. An airflow of 1 m³/m/t is typical.

[a] Not recommended.
[b] Includes supplemental heat — solar heat not included.
[c] Fan only — solar heat not included.

SEED CONDITIONING

Table 273
STANDARD
SPECIFICATIONS FOR
WOVEN-WIRE CLOTH
SIEVES

U.S. standard sieve no.	Nominal sieve opening	
	mm	in.
4	4.76	0.187
6	3.36	0.132
8	2.38	0.0937
12	1.68	0.0661
16	1.19	0.0469
20	0.841	0.0331
30	0.595	0.0234
40	0.420	0.0165
50	0.297	0.0117
70	0.210	0.0083
100	0.149	0.0059
140	0.105	0.0041
200	0.074	0.0029
270	0.053	0.0021

Table 274
RECOMMENDED SIEVES FOR DETERMINING DOCKAGE AND FOREIGN MATERIAL IN GRAINS

Common name of sieve	Shape of perforation[a]	Description of sieve		Approximate number of perforations/ft²
		Size of perforation		
		Common designation (in.)	Decimal equivalent of diameter or width (in.)	
Grain sorghums dockage	Round hole	2 ½/64 diam.	0.0391	27,970
Corn sieve	Round hole	12/64 diam.	0.1875	2,640
Fine seed	Round hole	1/12 diam.	0.0833	9,820
Flaxseed dockage	Round hole	4 ½/64 diam.	0.0703	13,795
Soybean				
Foreign material	Round hole	8/64 diam.	0.1250	4,736
Splits[b]	Slotted	8/64 diam. × 3/4	0.1250	638
		9/64 diam. × 3/4	0.1406	590
		10/64 diam. × 3/4	0.15625	557
Small chess	Slotted	0.064 × 3/8	0.0640	2,633
Large chess	Slotted	0.070 × 1/2	0.0700	1,655
Barley sizing				
Class I, barley	Rectangular	5/64 × 3/4	0.0781	865
Class III, western	Rectangular	5 ½/64 × 3/4	0.0858	856
Small buckwheat	Equilateral triangle	Inscribed circle 5/64 diameter	0.0781	2,845
Flaxseed sieve	Slotted	3/64 × 3/8	0.0469	2,690
Flaxseed dockage	Wire mesh	4 × 16 meshes to in.²		

[a] Perforations in each row are staggered with respect to perforations in adjacent rows.

[b] The soybean splits sieves may be used in making a partial separation of the "splits" in soybeans.

Source: Grain Inspection Manual, Instruction 918(GR)-6 Agricultural Marketing Service, U.S. Department of Agriculture, Washington, D.C. 1974.

Table 275
REPRESENTATIVE VALUES OF SPECIFIC GRAVITY FOR SELECTED GRAINS

Product	Specific gravity
Barley	1.39
White hulless	1.33
Coast	1.13
Buckwheat	1.10
Corn	1.29
	1.19
Flax	1.14
Millet	1.11
Milo	1.32
Oats	1.37
	0.99
Rice	1.37
	1.12
Rye	1.23
Sorghum, grain.	1.24
Soybeans	1.18
Wheat	1.42
	1.30

Note: References included in the original.

Table 276

SCREEN PERFORATION AND MESH SIZES AVAILABLE FOR AIR-SCREEN CLEANERS

Finished screens made only in 9 in. and 8 in. model widths; sheet sizes 26 in. × 61½ in. and 16 in. × 53¾ in.

Perforated metal sheet								Wire cloth			
Round holes		Oblong holes									
Fractions	64ths	Fractions	64ths	Triangles 64ths	Oblong cross slot	Round hole half sizes	Oblong half sizes	Square openings	Oblong openings		
1/25	5½	1/24×½	5×¾	5	6×¾	6½	8½×¾	3×3	2×8	4×8½	6×14
1/24	6	1/22×½	5½×¾	8	7×¾	7½	9½×¾	4×4	2×9	4×15	6×15
1/23	7	1/22×½ Diag.	6×¾	9	8×¾	8½	10½×¾	5×5	2×10	4×16	6×16
1/22	8	3/64×5/16	6½×¾	10	9×¾	9½	11½×¾	7×7	2×11	4×18	6×18
1/21	9	1/20×½	7×¾	11	10×¾	10½	12½×¾	8×8	2×12	4×19	6×19
1/20	10	1/18×¼	8×¾-D		11×¾	11½	13½×¾	9×9	3×14	4×20	6×20
1/19	11	1/18×¾	9×¾		12×¾	12½	14½×¾	10×10	3×16	4×22	6×21
1/18	12	1/16×¼-A	10×¾-E		13×¾	13½		12×12	3×16 SP.	4×24	6×22
1/17	13	1/16×½	11×¾-F		14×¾	14½		14×14	3×18	4×24 SP.	6×23
1/16	14	1/15×½	12×¾-G		15×¾	15½		15×15	3×20	4×26	6×24
1/15	15	1/14×¼-B	13×¾-H		16×¾	16½		16×16	3×21	4×28	6×25
1/14	16	1/14×½	14×¾-I		18×¾	17½		17×17		4×30	6×26
1/13	17	1/13×½	15×¾-J		10½×¾	18½		18×18		4×32	6×28
1/12	18	1/12×½-C	16×¾-K		11½×¾	19½		20×20		4×34	6×30
	19		17×¾		12½×¾	20½		22×22		4×36	6×32
	20		18×¾			21½		24×24			6×34
	21		19×¾			22½		26×26			6×36
	22		20×¾					28×28			6×38
	23		21×¾					30×30			6×40
	24		22×¾					32×32			6×42
	25		24×¾-L					34×34			6×50
	26		32×1					36×36			6×60
	27							38×38			18×20
	28							40×40			20×22
	29							45×45			
	30							50×50			
	31							60×60			
	32										
	34										
	36										
	38										
	40										
	42										
	44										
	48										
	56										
	64										
	72										
	80										

From Ferrell-Ross Company, Saginaw, Mich. With permission.

Table 277

SCREEN OPENING SIZES COMMONLY USED FOR CONDITIONING VARIOUS KIND OF SEEDS

Seed	Col. 1	Col. 2	Col. 3	Col. 4
Alfalfa	1/14 or 1/13	6×26	1/16 or 1/14	6×24
Sorrel from	1/18	6×34	6×20 6×21, 6×22	6×32
Mustard from	1/18	6×34	3/64×6/16, 6/20, 6/21	6/32
Yellow trefoil from	1/18	6×34	6×22, 1/21, 1/22	6×32
Night flowering catchfly from	1/18	6×34	6×20 or 6×21	6×32
Allspice, whole	21	6	20	8/64×¾
Anise	14	1/20	12	1/18
Barley	11/64×¾ or 24	1/14×½	22, 21 or 20	1/13×½
Beans				
Cranberry	32	14/64×¾	30	16/64×¾
Yellow eye	24	11/64×¾	22	12/64×¾
Lima	56	16	48	20
Baby Lima	32	17	30	19
Red kidney	80	13/64×¾	28	14/64×¾
Navy or pea	22	10/64×¾	20	11/64×¾
Pinto	26	9/64×¾	24	10/64×¾
Great Northern	26	10/64×¾	24	11/64×¾
Mung	14	7	13	8/64×¾
Beet	24	7	22 or 20	8
Bent grass	26×26	60×60	28×28	50×50
Bermudagrass	6×28	6×42	6×30	6×40
Bromegrass	10	9 tri or 6×26	1/13×½	6×24
Blue grass				
Canada	1/18	6×42	1/20	6×40
Kentucky	6×28	6×42	26×26	6×40
Bulbous	10	4×24	7	4×22
Buckwheat	16	7	14	6/64×¾
Cabbage	10	1/18	8	8/64×5/16
Canary grass	9	6×26	3×16	6×24
Cantaloupe	20	7	16	9
Caraway	1/14×¼	6×24	1/16×½	6×22
Carrot	7	6×28	1/12	6×26
Carpet grass	12	1/14×½	10	1/13×½
			6×28	6×38
Celery	1/14	30×30	1/16	28×28
Clover				
Hop	1/20	6×36	1/25	6×34
Ladine	1/18	6×34	1/20 or 1/21	6×32
Alsike	1/18	6×34	1/19	6×32
Crimson	6	6×24	1/13	6×22
Bur hulled	1/12	6×24	1/13	6×22
Dalea or woods	6	6×23	1/13	6×22
Red	1/15	6×24 or 6×26	8/64×5/16	6×22 or 6×24
Red, hare's ear mustard from	3/64×5/16	6×24	3/64×5/16	6×24
White sweet	1/12	6×26	1¼	6×24
White sweet un-hulled	10	6×26	8	6×24
White	1/16−1/18	6×34	1/18, 1/19, 1/20	6×32
White, dock from	1/16	6×34	6×22 or 6×23	6×32
Cumin	14	1/25	12	1/22
Coriander	15	1/13×1/2	13	1/12×1/2
Corn, cleaning only	32	14	28, 30	16

Table 277 (continued)
SCREEN OPENING SIZES COMMONLY USED FOR CONDITIONING VARIOUS KIND OF SEEDS

Seed	Col. 1	Col. 2	Col. 3	Col. 4
Cotton seed				
Delinted	20	9/64×¾	18	10/64×¾
Not delinted	40	12/64×¾	36	13/64×¾
Cowpea	22	12 or 10/64×¾	21	11/64×¾
Crested wheat grass	1/16×½	6×30	1/18×¼	5 tri.
Crotalaria	12	6	1/12×½	7
Cucumber	18	8	17 or 16	9
Dallas grass	8	1/15	3×14	1/14
Fescue				
Meadow	1/13×1½	6×21	1/14×¼	4×22 or 5 tri.
Chewings	3/64×5/16	6×32	1/22×½	6×30 or 5 tri.
Ky. 31	3/64×5/16	6×32	1/22×½	5 tri. or 6×30
Fennel	14	1/16	12	1/14
Flax				
Bison	1/16×½	1/14	1/18×¾	1/13
Golden	1/13×½	1/12	1/14×¼	6
Small	3×16	1/14	3/64×5/16	1/12
Walsh & viking same as bison				
Garbanzo (chick pea)	80	11/64×¾	26	12/64×¾
Hemp	14	1/13×½	12	1/12×½
Johnson grass	10	1/18	8	1/16
Lentil	18	12	7/64×¾	12 or 13
Lespedeza				
Korean	6	1/17, 1/18	6×15 or 1/18×¼	1/16
Kobe	8	1/16	6×14 or 1/18×¾	1/14
Sericea, un-hulled	7	1/15	1/14×¼	6×26
Sericea, hulled	1/15	6×28	1/16 or 3/64×5/16	6×26
Lettuce	4×18	24×24	6×20	20×20
Milo, maize	14	1/13×½	13	1/12×½
Millet				
Siberian	7	1/20	1/12	1/15
Proso	9	3/64×5/16	7	1/18×¼
Mustard				
White	9	1/22×½	7	3/64×5/16
Brown	7	1/22	1/12	1/20
Oats	24	1/16×½ or 11 tri.	11/64×¾	1/14×½
Oats, corn from	10/64×¾ or 9/64×¾	1/16×½	9/64×¾ or 8/64×¾	1/14×½
Oat grass, tall	1/12×½	6×30	1/14×½	6×28
Okra	16	3/64×5/16	14	1/18×¾
Onion	11	1/15	9.8	1/14
Onion sets	1/¼	3×3	1 in.	8×3
Orchard grass	3/64×5/16	6×32	1/22×½ or 1/24×½	6x32 or 5 tri.
Peas				
Austrian winter	20	12 or 9/64×¾	17	10/64×¾
Field	22	10 or 9/64×¾	20	10/64×¾
Pepper, red	14	1/12	12	7
Pepper, black,	18	6/64×¾	16	7/64×¾
Pinion nuts	32	12/64×¾	30	14/64×¾
	10	4×24	7	4×22
Popcorn, pearl	21	10	20	11
Psyllium	4×18	6×38	6×20	26×26
Pumpkin	36	18	82	20
Radish	10	1/14	9 or 8	1/12

Table 277 (continued)
SCREEN OPENING SIZES COMMONLY USED FOR CONDITIONING VARIOUS KIND OF SEEDS

Seed	Col. 1	Col. 2	Col. 3	Col. 4
Rape, dwarf essex	7	1/23	1/12	1/22
Rape, German	9	1/18	7	1/16
Redtop	26×26	60×60	28×28	50×50
Redtop, timothy from	26×26	60×60	6×34	50×50
Rice, paddy	22	6 or 7	20 or 21	1/14×½, 1/18×½
Rice, hulled	14	1/12	12	14×14
Ryegrass	3/64×5/16	6×32	1/22×½	6×32 or 5 tri.
Rye	14	1/16×½	12	1/14×½
Rye, vetch from (100% Impossible)	7/64×¾	8	7/64×¾	9
Sesbania	10	1/18×¼	8 or 9	1/18×¾
Sesame	6	4×26	1/18×¼	1/16
Sorghum				
Grain	12	1/14×½	10	1/13×½
Sweet	8	1/22×½	6	3/64×5/16
Soybeans				
Large	22	10/64×¾	20	11/64×¾
Small	20	6/64×¼	18	7/64×¾
Black flat	18	7/64×¾	16	8/64×¾
Morninglory from	18	7/64×¾	16	11 or 12
Spinach	14	1/14	11 or 12	1/12
Squash	34	20	32	23
Sudangrass	12	1/22×½	10	3/64×5/16
Sunflower	24 to 32	10	24 to 32	11
Timothy	1/18	6×36	1/19 or 1/20	6×34 or 6×32
Timothy				
Sorrel from	6×20 or 6×21	6×36	6×20 or 6×21	6×34 or 6×32
Red clover from	6×24	6×36	6×26	6×34
Peppergrass from	1/19	6×34	1/20	6×32
Black plantain from	1″19	6×32	1/20	6×30
Redtop from	1/19	6×36	1/20	6/34
Buckhorn plantain from	1/20	6×30	1/22	6×30 or 6×28
Alsike clover from	1/20	6×34	1/25 or 6×26	6×32
Tobacco	30×30	50×50	32×32	40×40, 50×50
Tomato	12	1/14	10	1/12
Trefoil, birdsfoot	1/14	6×26	1/16	6×24
Turnip	1/12	1/22	1/14	1/20
Vetch	16	6/64×¾	14	6/64×¾ or 7/64×¾
Watermelon	28	14	24	16
Wheat	16	1/18×½ or 8 or 9	14	1/18×½ or 10 tri.
Wheat oats from	12	1/18×½	11	1/18×½
Wheatgrass, western	1/18×½	4×24	1/14×½	4×22

Note: Four-screen cleaners — Use Column 1 for top screen in top shoe, Column 2 for bottom screen in top shoe, Column 3 for top screen in bottom shoe, and Column 4 for bottom screen in bottom shoe. Three-screen cleaners — Use Column 1 for top screen, Column 3 for middle screen, and Column 4 for bottom screen.

From Ferrell-Ross Co., Catalogue of Seed Cleaning Equipment, Saginaw, Michigan, 1978. With permission.

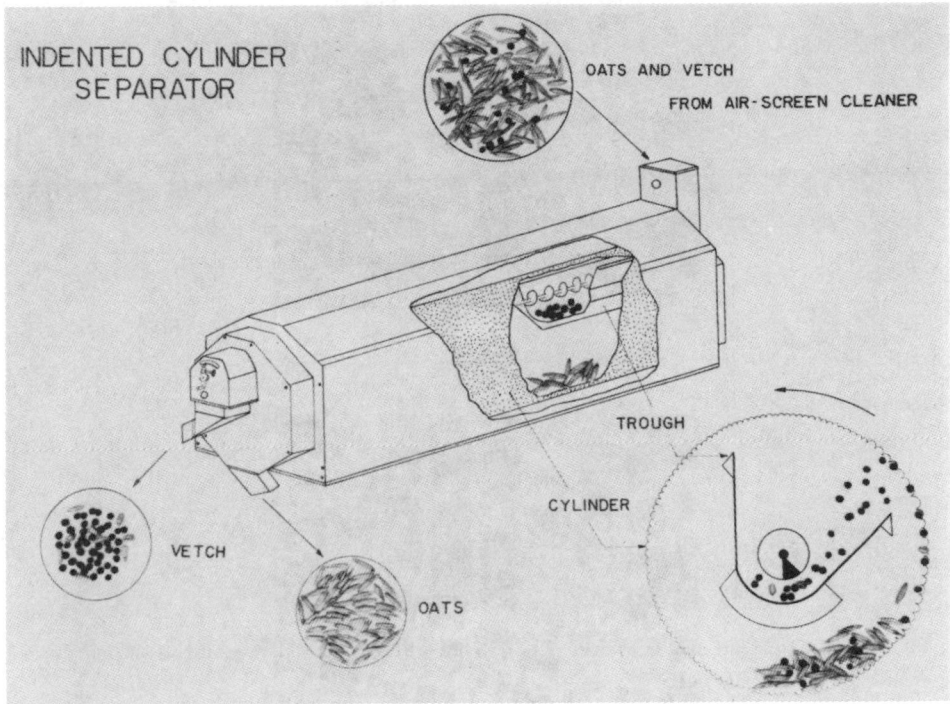

FIGURE 16. Indented cylinder separator. (Courtesy of the Seed Technology Laboratory, Mississippi State University, Mississipi State, 1979.)

A

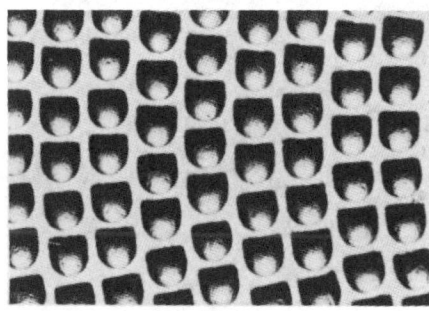

B

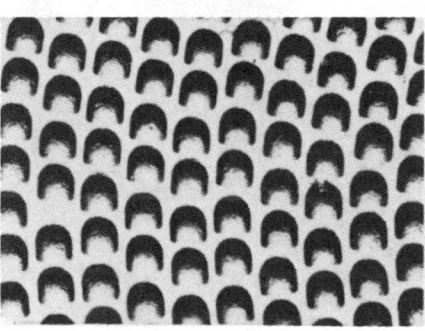

C

FIGURE 17.   Types of disc indents. (A) Square
pocket; (B) V pocket; (C) R pocket. (Courtesy of
the Carter-Day Company, Minneapolis, 1979.)

Table 278
INDENT CYLINDER SIZES USED FOR SEPARATING GRASSES, GRAINS, AND SEEDS

Indent number	Will lift	Will reject
2	Small pigweed, Alsike clover, small dodder, mullenweed, sand, etc.	Buckhorn, timothy, black medic clover, bluegrass (all varieties), alfalfa, crimson clover, lespedeza, etc.
3	Small sweet clover, pigweed, dodder, white clover, alsike, etc.	Thistles, buckhorn, sticks, alfalfa, red clover, etc.
4	Timothy, small clover (red and white), dodder, hulled water grass, mustard, sheep sorrel	Canada thistle, quackgrass, sticks, alfalfa, bluegrass, etc.
5	Red clover, alfalfa, small flax, water grass, mustard, bluegrass, etc.	Meadow fescue, wild brome, large buckhorn, quackgrass, cheat, chess, sticks, etc.
6 1/2	Small broken grain, small wild buckwheat, small vetch and cockle, wild mustard	Fescue, wheat, ryegrass, wheat grass, hulled orchard grass, flax, etc.
8 1/2	Buckwheat, cockle, vetch, sudangrass, small sugar beet seed, etc.	Wheat, rye, fescue, ryegrass, orchard grass, etc.
11	Broken grain, vetch, small onion and garlic, wild peas, coffee weed, etc.	Spring wheat, rye, rice, alta fescue
13	Spring wheat, small or broken durum, pearled and broken barley, flax	Durum, large spring wheat, barley, pin oats
19	Spring wheat, small durum	Oats, wild oats, barley, etc.
22	Wheat, winter wheat, hulled oats, rye, etc.	Oats, wild oats
24	Barley	Oats, wild oats, barley with tails, etc.
26		
28	Used primarily in length grading of seed corn and similar sized material	
32		
S-3	Equal to #22 Indent but has flat bottom; used primarily on corn	

From Carter-Day Company, Minneapolis, Minn. With permission.

Table 279
DISC POCKET SIZES USED FOR SEPARATING GRASSES, GRAINS, AND SEEDS

Type of pocket	Will lift	Will reject
V2 1/2	Alsike, timothy, ladino, white dutch, etc.	Buckhorn, Canada thistle
V3	Red clover, small sweet clover, etc.	Thistle, sticks, etc.
V3 1/2	Alfalfa, sweet clover, water grass, etc.	Hulled quack, large thistle, flax, etc.
V3 3/4		
V4	Water grass, mustard, small cockle, small cracked grain, etc.	Wheat, barley, pin oats, large flax, etc.
V4 1/2	Wild buckwheat, large cockle, vetch, cracked grain, onion bulblets	Wheat, oats, barley, rye, etc.
V5		
V5 1/2	Flax, extra large cockle, vetch, small or broken wheat, large wild buckwheat, etc.	Unhulled quack, pin oats, wheat, barley, oats, rye, etc.
V5 3/4		
V6		
V6 1/2		
R3 1/2	Water grass, mustard, smartweed, bentgrass, buckhorn	Flax, fescue, ryegrass, orchard grass, etc.
R3 3/4		
R4		
R4 1/2	Small seeds, cracked grain, wild buckwheat, bluegrass, buckhorn	Wheat, barley, oats, pin oats, rye, alta fescue, meadow brome, etc.
R5		
R5 1/2		
R6		
K	Large wild buckwheat, broken wheat, broken barley, etc.	Wheat, barley, oats, pin oats, rye, etc.
L		
M	Spring wheat, small or broken durum, pearled or broken barley, fescue, orchardgrass (Kentucky 31, Alta fescue — may use J& A)	Durum, large spring wheat, barley, oats, pin oats, etc.
AC		
EE		
J	Spring wheat, small durum, Kentucky 31 fescue, alta fescue, etc.	Oats, wild oats, barley, etc.
A	Wheat, winter wheat, durum, small barley, hulled oats, rye, safflower	Oats, wild oats, ragged barley, etc.
MM		
B	Barley	Oats, wild oats, sticks, etc.
RR-SS	Tailless barley	Sticks, stems, etc.
DD	Oats and all shorter grains	
AE-AD	Peanuts	Sticks, stems, etc.
SS-DD		

From Carter-Day Company, Minneapolis, Minn. With permission.

Table 280

COMMON AND SCIENTIFIC NAMES OF INSECTS AND
MITES COMMONLY FOUND ATTACKING CEREALS,
FEED GRAINS, AND SOYBEANS

Common name	Scientific name
Weevils	
Granary weevil[a]	*Sitophilus granarius*(L.)
Rice weevil[a]	*Sitophilus oryzae*(L.)
Maize weevil[a]	*Sitophilus zeamais*(Motschulsky
Broadnosed grain weevil	*Caulophilus oryzae*(Gyllenhal)
Coffee bean weevil	*Araecerus fasiculatus*(De Geer)
Bean weevil[b]	*Acanthoscelides obtectus*(Say)
Cowpea weevil[b]	*Callosobruchus maculatus*(F.)
Borers	
Lesser grain borer[a]	*Rhyzopertha dominica*(F.)
Larger grain borer[a]	*Prostephanus truncatus*(Horn)
Moths	
Angoumois grain moth[a]	*Sitotroga cerealella*(Olivier)
Pink scavenger caterpillar[a]	*Pyroderces rileyi*(Walsingham)
European grain moth[a]	*Nemapogen granella*(L.)
Rice moth[a]	*Corcyra cephalonica*(Stainton)
Indian meal moth[c]	*Plodia interpunctella*(Hubner)
Mediterranean flour moth[c]	*Anagasta kuehniella*(Zeller)
Meal moth[c]	*Pyralis farinalis* L.
Almond moth[c]	*Cadra cautella*(Walker)
Tobacco moth[c]	*Ephestia elutella*(Hubner)
Raisin moth[c]	*Cadra figuliella*(Gregson)
Beetles	
Cadelle	*Tenebroides mauritanicus*(L.)
Sawtoothed grain beetle[c]	*Oryzaephilus surinamensis*(L.)
Merchant grain beetle[c]	*Oryzaephilus mercator*(Fauvel)
Squarenecked grain beetle[c]	*Cathartus quadricollis*(Guerin-Meneville)
Foreign grain beetle[c]	*Ahasverus advena*(Waltl)
Mexican grain beetle[c]	*Pharoxonotha kirschi*(Reitter)
Siamese grain beetle[c]	*Lophocateres pusillus*(Klug)
Flat grain beetle[c]	*Cryptolestes pusillus*(Schonherr)
Rusty grain beetle[c]	*Cryptolestes ferrugineus*(Stephens)
Confused flour beetle[c]	*Tribolium confusum* Jacquelin duVal
Red flour beetle[c]	*Tribolium castaneum*(Herbst)
American black flour beetle[c]	*Tribolium audax* Halstead
Longheaded flour beetle[c]	*Latheticus oryzae* Waterhouse
Slenderhorned flour beetle[c]	*Gnatocerus maxillosus*(F.)
Smalleyed flour beetle[c]	*Palorus ratzeburgi*(Wissmann)
Depressed flour beetle[c]	*Palorus subdepressus*(Wollaston)
Broadhorned flour beetle[c]	*Gnatocerus cornutus*(F.)
Larger black flour beetle	*Cynaeus angustus*(LeConte)
Yellow mealworm[a]	*Tenebrio molitor* L.
Dark mealworm[a]	*Tenebrio obscurus* F.
Lesser mealworm[c]	*Alphitobius diaperinus*(Panzer)
Redhorned grain beetle[c]	*Platydema ruficorne*(Strum)
Black carpet beetle[c]	*Attagenus megatoma*(F.)
Warehouse beetles[a]	*Trogoderma* spp.
Hairy spider beetle[a]	*Ptinus villiger*(Reitter)
Whitemarked spider beetle[a]	*Ptinus fur*(L.)
Brown spider beetle[a]	*Ptinus clavipes* Panzer
Twobanded fungus beetle[c]	*Alphitophagus bifasciatus*(Say)
Hairy fungus beetle[c]	*Typhaea stercorea*(L.)
Corn sap beetle[a]	*Carpophilus dimidiatus*(F.)
Cigarette beetle[a]	*Lasioderma serricorne*(F.)
Drugstore beetle[a]	*Stegobium paniceus*(L.)

Table 280 (continued)
COMMON AND SCIENTIFIC NAMES OF INSECTS AND
MITES COMMONLY FOUND ATTACKING CEREALS,
FEED GRAINS, AND SOYBEANS

Common name	Scientific name
Mites	
Flour mite	*Acarus siro* L.
	Tyrophagus spp.
	Glycyphagus spp.
Insect parasites	*Anisopteromalus calandrae* (Howard)
	Choetospila elegans Westwood
	Bracon hebetor Say
	Venturia canescens (Gravenhost)
	Scenopinus fenistralis (L.)
	Scenopinus glabrifions Meigen
Insect predators	*Xylocoris flavipes* (Reuter)

[a]   Primary feeder.
[b]   Pest of stored soybeans primarily.
[c]   Secondary feeder.

FORAGE

HAY GRADES

Table 281
TERMS FOR DESCRIBING STAGES OF MATURITY

Preferred term	Definition	Comparable terms
	For Plants That Bloom	
Germinated	Stage in which the embryo in a seed resumes growth after a dormant period	Sprouted
Early vegetative	Stage at which the plant is vegetative and before the stems elongate	Fresh new growth, before heading out, before inflorescence emergence, immature, prebud stage, very immature, young
Pre-bloom	Stage at which stems are beginning to elongate to just before blooming; first bud to first flowers	Before bloom, bud stage, budding plants, heading to in bloom, heads just showing, jointing and boot (grasses), prebloom, preflowering, stems elongated
Early bloom	Stage between initiation of bloom and stage in which one tenth of the plants are in bloom; some grass heads are in anthesis	Early anthesis, first flower, headed out, in head, up to one tenth bloom
Midbloom	Stage in which one tenth to two thirds of the plants are in bloom; most grass heads are in anthesis	Bloom, flowering, flowering plants, half bloom in bloom, mid anthesis
Full bloom	Stage in which two thirds or more of the plants are in bloom	Three fourths to full bloom, late anthesis
Late bloom	Stage in which blossoms begin to dry and fall and seeds begin to form	15 days after silking, before milk, in bloom to early pod, late to past anthesis
Milk stage	Stage in which seeds are well formed, but soft and immature	After anthesis, early seed, fruiting, in tassel, late bloom to early seed, past bloom, pod stage post anthesis, post bloom, seed developing, see forming, soft, soft immature
Dough stage	Stage in which the seeds are of dough-like consistency	Dough stage, nearly mature seeds, dough, seeds well developed, soft dent
Mature	Stage in which plants are normally harvested to seed	Dent, dough to glazing, fruiting, fruiting plants, in seed, kernels ripe, ripe seed
Post ripe	Stage that follows maturity; seeds are ripe, and plants have been cast, and weathering has taken place (applies mostly to range plants)	Late see, over ripe, very mature
Stem cured	Stage in which plants are cured on the stem; seed have been cast, and weathering has taken place (applies mostly to range plants)	Dormant, mature and weathered, seeds cast
Regrowth early vegetative	Stage in which regrowth occurs without flowering activity, vegetative crop aftermath; regrowth in stubble (applied primarily to fall regrowth in temperate climates); early dry season regrowth	Vegetative recovery growth

Table 281 (continued)

TERMS FOR DESCRIBING STAGES OF MATURITY

Preferred term	Definition	Comparable terms
	For Plants That Bloom Intermittently[a]	
1 to 14 days' growth	A specified length of time after plants have started to grow	2 weeks' growth
15 to 28 days' growth	A specified length of time afer plants have started to grow	4 weeks' growth
29 to 42 days' growth	A specified length of time after plants have started to grow	6 weeks' growth
43 to 56 days' growth	A specified length of time after plants have started to grow	8 weeks' growth
57 to 70 days' growth	A specified length of time after plants have started to grow	10 weeks' growth

[a] These classes are for species that remain vegetative for long periods and apply primarily to the tropics. When the name of a feed is developed, the age classes form part of the name (e.g., alfalfa, aerial part, 15 to 28 days' growth). For plants growing longer than 70 days, the interval is increased by increments of 14 days.

From Fonnesbeck, P. V., Harris, L. E., and Kearl, L. C., 1st Int. Symp. Feed Composition, Animal Nutrient Requirements and Computerization of Diets, International Feedstuffs Institute, Logan State University, Logan, Utah, 1977. With permission.

Table 282
CRUDE PROTEIN CONCENTRATION IN MARKET HAY GRADES FOR LEGUMES, LEGUME GRASS MIXTURES, AND GRASSES

Grade	Description[a] Vegetative[b]	CP%
Prime	Leg. Pre-Bl.	>19
1	Leg. E.Bl., 20% grass-Veg.	17—19
2	Leg. M.Bl., 30% grass-EH	14—16
3	Leg. F.Bl., 40% grass-head	11—13[c]
4	Leg. F.Bl., 50% grass-head	8—10
Fair	Grass-head, and/or rain damaged	<8
Sample		—

Source: AFGC Hay Marketing Task Force, A Continuum from Legume Pre-bloom to Grass Headed and/or Heavily Weathered Forage, Descriptions adopted by the National Hay Quality Committee.

[a] Description adopted by National Alfalfa Hay Quality Committee.
[b] Pre.Bl., Pre bloom; E.Bl., early bloom; M.Bl., mid-bloom; F.Bl., mid- to full-bloom; veg, vegetative; EH, early head.
[c] Reference hay is mid- to full-bloom alfalfa; DDM = 54.2; DMI = 120.2 g/week;[0.75] DDMI = 65.2 g/week;[0.75] and RFV = 100%. (Abbreviations defined in Table 283.)

From Rohweder, D. A., Relative Feed Value (RFV) — What It Means to Producers, W-L Research, Inc., Franchisers Meeting, Carmel, Calif., July 1988.

Table 283
ADF AND NDF CONCENTRATION (%) IN MARKET HAY GRADES FOR LEGUMES, LEGUME-GRASS MIXTURES, AND GRASSES[a]

Grade[b]	Description ADF	NDF	DDM	DMI[c]	DDMI	RFV[d]
Prime	<30	<39	>65.5	>3.0	>1.97	>152.4
1	31—35	40—46	62—65	2.93—2.53	1.90—1.56	147.4—121.1
2	36—40	47—53	58—61	2.45—2.29	1.50—1.35	116.3—104.4
3	40—42	53—60	56—57	2.26—2.05	1.29—1.16	100—90[e]
4	43—45	61—65	53—55	2.04—1.85	1.15—0.99	89—77.1
Fair	>46	>65	<53	<1.84	<.98	<76

Source: AFGC Hay Marketing Task Force, A Continuum from Legume Pre-bloom to Grass Headed and/or Heavily Weathered Forage.

[a] Description and DDM adopted by National Alfalfa Hay Quality Committee.
[b] Prime = legume prebloom; 1 = legume early bloom, 20% grass, vegetative; 2 = legume mid bloom, 30% grass, early head; 3 = legume full bloom, 40% grass-head; 4 = legume full bloom, 50% grass-head; Fair = grass-head, and/or rain damaged.
[c] Percent body weight
[d] RFV (relative feed value) currently involves the following: (1) DDM (dry matter digestibility) 88.9 − 0.779 ADF% (ADF = acid detergent fiber); (2) DMI (dry matter intake) = 1.2/(NDF%/100) as a percent of body weight [NDF = neutral detergent fiber]; and (3) $RFV = \dfrac{DDM \times DMI}{1.29}$
[e] Reference hay is CP = 14.5, ADF = 41, NDF = 53, DDM = 54.2, DMI = 2.26, and RFV = 100%.

From Rohweder, D. A., Relative feed value (RFV) — What It Means to Producers, W-L Research, Inc., Franchisers' Meeting, Carmel, Calif., July 1988.

HARVESTING, COMPOSITION, AND FEEDING

Table 284
CUTTING SCHEDULE INFLUENCE OF FORAGE AND ANIMAL PERFORMANCE, ARLINGTON, WISCONSIN

Cutting schedule	Yield T/A	Forage quality CP	ADF	NDF	Milk (lb T[d]) 83—84	84—86	Milk (lb/A) 83—84	84—86
4x + A[a,c]	4.2	27.3	25.9	35.1	2152	—	9038	—
4x[a]	4.2	26.4	28.0	36.6	2091	—	8782	—
Early[a]	4.0	23.9	30.2	36.8	—	2364	—	9457
Late[a]	4.4	21.1	31.9	38.4	—	2212	—	9732
3x + A[b,c]	5.1	24.0	28.5	38.2	1851	—	9440	—
3x[b]	4.8	23.0	31.1	40.6	1718	—	8246	—
	4.7	16.6	38.5	44.5	—	1615	—	7590

[a] Cutting started near May 23rd; A = autumn cut.
[b] Cutting started near June 1.
[c] Fall cutting after a 45-day rest period.
[d] Determined from TMR ration for a 1,350-lb cow producing 80 lb 4.0% fat-corrected milk.

From Rohweder, D. R., Relative Feed Value (RFV) — What It Means to Producers, W-L Research, Inc., Franchisers' Meeting, Carmel, Calif., July, 1988.

Table 285
BASIC DIFFERENCES IN LEAF AND STEM
COMPOSITION IN ALFALFA HAY

Portion of plant	Crude protein (%)	Carotene (mg/lb)	Crude fiber (%)
Leaves	25.3	35.5	17.1
Stems	11.3	5.5	44.6

Source: High Quality Hay, Special Rep. ARS-22-52, Agricultural Research Services, U.S. Department of Agriculture, Washington, D.C., 1960.

Table 286
MILK PRODUCTION OF DAIRY COWS FED ALFALFA (AS SOLE RATION) HARVESTED AT THREE STAGES OF GROWTH

Growth stage at harvest	DM[a] intake/ day/cow (lb)	Average weight gain/ day/cow (lb)	Average milk production		
			FCM[b]/ day/cow (lb)	FCM/ 100 lb DM (lb)	FCM/ harvested acre (lb)
Initial bloom	34.9	0.08	27.9	79.9	6194
Half bloom	34.8	0.17	23.6	67.9	5145
Full bloom	33.6	0.0	20.8	62.0	3814

[a]   DM = dry matter.
[b]   FCM = fat corrected milk (quantity of milk produced converted to a 4% fat basis.

Source: High Quality Hay, Special Rep. ARS-22-52, Agricultural Research Service, U.S. Department of Agriculture, Washington, D.C., 1960.

Table 287
DIGESTION COEFFICIENTS (SHEEP) AND TOTAL DIGESTIBLE NUTRIENTS OF ALFALFA CUT AT VARIOUS INTERVALS

Cutting interval (weeks)	Organic matter (%)	Crude protein		Crude fiber (%)	Nitrogen- free extract (%)	Total digestible nutrients (dry basis) (%)
		Apparent digestibility (%)	True digestibility[a] (%)			
3	71.1	84.3	94.4	51.4	76.1	63.4
4	65.6	81.3	94.0	45.5	73.4	58.9
5	64.0	78.3	93.9	48.0	72.0	58.1
6	61.0	76.4	92.3	40.7	71.8	55.6

[a]   Corrected for metabolic protein in feces.

From Weir, W. C., Jones, L. G., and Meyer, J. H., Effect of cutting interval and stage of maturity on the digestibility and yield of alfalfa, *J. Anim. Sci.*, 19, 5, 1960. With permission.

Table 288
COMPOSITION OF ALFALFA CUT
ACCORDING TO STAGE OF MATURITY
AT SECOND CUTTING

	Composition (dry basis)		
Vegetative stage	Protein (%)	Crude fiber (%)	Lignin (%)
Prebud	30.9	15.2	5.3
1% bud	26.9	20.9	6.4
62% bud	25.2	21.7	6.6
11% bloom	21.3	27.3	8.2
46% bloom	19.1	28.5	8.2
96% bloom	16.9	31.4	8.3

From Weir, W. C., Jones, L. G., and Meyer, J. H., Effect of cutting interval and stage of maturity on the digestibility and yield of alfalfa, *J. Anim. Sci.,* 19, 5, 1960. With permission.

Table 289
PROTEIN DIGESTIBILITY (SHEEP) AND TOTAL
DIGESTIBLE NUTRIENTS OF SECOND CUTTING
ALFALFA HARVESTED AT SIX STAGES OF
MATURITY

	Organic matter digestion coefficient (%)	Crude protein		Total digestible nutrients (dry basis) (%)
Vegetative stage		Apparent digestion coefficient (%)	True digestion[a] coefficient (%)	
Prebud	73.1	78.8	87.9	66.1
1% bud	68.7	75.8	85.9	61.9
62% bud	64.2	73.4	84.4	58.9
11% bloom	61.9	72.8	85.4	57.2
46% bloom	60.1	70.3	84.8	54.7
96% bloom	60.2	69.2	84.4	55.7

[a] Corrected for metabolic protein in the feces.

From Weir, W. C., Jones, L. G., and Meyer, J. H., Effect of cutting interval and stage of maturity on the digestibility and yield of alfalfa, *J. Anim. Sci.,* 19, 5, 1960. With permission.

Table 290

YIELD OF ALFALFA CUT AT FOUR STAGES OF MATURITY FOR THREE GROWING SEASONS

Stage of maturity and year	Number of cuttings	Seasonal yield (lb/acre)		Protein yield (lb/acre)	
		Including first cutting	Without first cuttings[a]	Including first cutting	Without first cuttings[a]
Year 1					
Prebud	6	11,289	11,289	2,918	2,918
Bud	5	12,988	12,988	3,086	3,086
One tenth bloom	4	14,716	14,716	2,948	2,948
One half bloom	4	14,796	14,796	2,795	2,795
Year 2					
Prebud	8	17,373	12,277	4,291	3,347
Bud	7	19,892	15,130	4,436	3,543
One tenth bloom	6	22,731	17,881	4,755	3,645
One half bloom	5	21,867	16,785	4,240	3,107
Year 3					
Prebud	9	14,796	11,921	3,964	3,216
Bud	8	18,920	15,863	4,654	3,833
One tenth bloom	7	21,686	18,593	4,806	3,971
One half bloom	7	23,551	19,950	4,795	3,873
Total for 3 years					
Prebud	23	43,458	35,487	11,173	9,841
Bud	20	51,800	43,981	12,176	10,462
One tenth bloom	17	59,133	51,190	12,509	10,564
One half bloom	16	60,214	51,531	11,830	9,775

[a]   Includes second and third years only.

From Weir, W. C., Jones, L. G., and Meyer, J. H., Effect of cutting interval and stage of maturity on the digestibility and yield of alfalfa, *J. Anim. Sci.*, 19, 5, 1960. With permission.

Table 291
PERCENT DIGESTIBILITY AND COMPOSITION OF TEMPERATE GRASSES[a]

Grass	IRN	IVDMD	Protein	Ether extract	Ash	NDF	ADF	Hemicellulose	Holocellulose	PML
Kenhy—4 week		65.6	13.2	5.6	8.3	58.2	33.6	24.7	41.6	3.2
Ken-Blue—4 week	2-00-778	58.1	15.5	4.0	7.1	54.0	30.6	23.5	43.5	4.3
Brome — 4 week	2-00-956	64.2	14.3	4.5	8.8	56.2	34.3	21.9	51.1	4.8
Orchard—4 week	2-03-440	62.8	14.8	4.4	8.2	57.9	33.3	24.6	43.9	4.1
Kentucky-31—4 week	2-01-900	62.7	14.2	3.9	8.8	58.4	31.4	25.5	45.2	3.4
Timothy—4 week	2-04-902	66.8	13.4	4.3	8.8	55.6	34.7	20.9	42.0	4.1
Kentucky-31—4 week (Fall)	2-01-900	55.0	12.6	3.5	8.4	59.8	31.6	28.2	46.0	5.6
Kenhy—4 week (Fall)	—	60.1	12.6	3.4	8.0	57.3	30.7	26.6	43.8	4.2
Orchard—4 week (Fall)	2-03-440	58.8	17.8	4.4	9.7	54.0	29.4	24.6	45.5	4.8
Ken-Blue—4 week (Fall)	2-00-778	61.0	17.3	6.6	6.3	57.6	27.5	30.1	40.3	4.1
Average temperate		61.5	14.6	4.5	8.2	56.9	31.7	25.1	44.3	4.3

* IRN = international reference number; IVDMD = in vitro dry matter digestibility; NDF = neutral detergent fiber; ADF = acid detergent fiber; PML = permanganate lignin.

From Barton, F. E., II, Amos, H. E., Burdick, D., and Wilson, R. L., Relationship of chemical analysis to *in vitro* digestibility of selected tropical and temperate grasses, *J. Anim. Sci.*, 43, 504, 1976. With permission.

Table 292
PERCENT OF DRY MATTER
LOSSES IN FIELD-CURED HAY

Cause	Loss of dry matter
Plant respiration	4—15
Leaf shattering	
Grasses	2—5
Legumes	3—35
Rain (leaching[a])	5—14

[a]   Laboratory experiments indicate that leaching by water can remove 20 to 40% of the dry matter, 20% of the crude protein, 35% of the nitrogen-free extract, 30% of the phosphorus, and 65% of the potash.

Source: Camburn, O. M., Ellenberger, M. B., Jones, C. H., and Crooks, G. C., The Conservation of Alfalfa, Red Clover, and Timothy Nutrients as Silages and Hays (Part 2), Vermont Agricultural Experiment Station Bull. 494, University of Vermont, Burlington, 1942.

Table 293
TOTAL DRY MATTER AND NUTRIENT LOSSES FOR HAY THAT WAS FIELD-CURED UNDER VARIOUS WEATHER CONDITIONS

		Losses (%)		
Type hay	Location/weather	DM	CP	SE
Alfalfa	Colorado/favorable	5—26	—	—
Alfalfa	Vermont/good	13—24	21—29	—
Alfalfa	Germany/good	17	30	26
Alfalfa	Kansas/unfavorable	50	—	—
Alfalfa, 70%; grass, 30%	Switzerland/?	14	—	41
Clover	Vermont/unfavorable	36	40	—
Clover	Germany/unfavorable	41	—	64
Clover, 70%; grass, 30%	Denmark/very favorable	11—12	—	26—30
Clover, 70%; grass, 30%	Denmark/average	14—16	—	32—37
Clover, 70%; grass, 30%	Denmark/unfavorable	20—30	—	40—50
Clover,70—29%; grass, 30—71%	Switzerland/no rain	9	16	23
Clover, 70—29%; grass, 30—71%	Switzerland/5—6 showers	27	38	54
Grass hay	England/excellent	10—25	2—24	1—45
Grass hay	England/good	17	14	27
Grass hay	England/unfavorable	37	35	54
Coastal bermudagrass	Georgia/no rain	5	—	—
Coastal bermudagrass	Georgia/rain	11	1—2	—
Coastal bermudagrass	South Carolina/rain	—	12	—
Timothy, 45—46%; grass, 42—49%	New York/favorable	13	—	—
Timothy, 45—46%; grass, 42—49%	New York/poor	21	—	—
Timothy (early cut)	Vermont/excellent	3	19	—
Timothy (late cut)	Vermont/excellent	3	12	—
Alfalfa (first cutting)	Vermont/excellent	18	10	—
Alfalfa (second cutting)	Vermont/excellent	27	18	—
Red clover	Vermont/excellent	33	44[a]	—

Note: DM = dry matter; CP = crude protein; SE = starch equivalent (productive net energy in terms of digestible starch; 1 lb of starch equivalent is equal to 1.071 lb of net energy). References provided in original.

[a] High leaf shatter.

Table 294
YIELD AND DIGESTIBILITY OF ALFALFA HARVESTED WITH DIFFERENT MOWER-CONDITIONERS SYSTEMS

Machine	Yield of dry matter (%)		Digestibility (TDN)[a] (%)		Yield of digestible organic matter (%)	
	I[b]	II[b]	I	II	I	II
Sicklebar (control)	100	100	34.6	42.6	100	100
Flail mower	114	94	36.9	44.2	121	98
Sicklebar + steel roll conditioner	137	102	40.8	45.6	163	109
Sickle bar with experimental conditioning rotor	153	112	39.1	45.4	172	119

[a] TDN = total digestible nutrients.
[b] Drying time Experiment I, 8 to 12 days; Experiment II, 6 days, more favorable weather than I.

From Kliner, W. E., *Trans. ASAE*, 19, 237, 1976. With permission.

Table 295

EFFECT OF ACID TREATMENT AND PACKAGE TYPE ON DRY MATTER LOSSES OF ALFALFA HAY

Stage of losses	Rectangular bale		Round bale		Stacks	
	Acid treated (%)	Field cured (%)	Acid treated (%)	Field cured (%)	Acid treated (%)	Field cured (%)
Harvest moisture	27.2	17.0	24.5	16.0	27.4	15.0
Mowing and curing losses	7.1b	13.2a	7.1b	12.3a	7.1b	13.2a
Packaging loss	2.6d	3.5d	9.1b	10.2a,b	6.8c	13.4a
Storage loss	4.8d	7.8c,d	10.1b,c	11.7b	11.3b	16.6a
Feeding loss[a]	0.4c	5.2b	5.5b	14.3a	5.5b	16.3a
Total losses[b]	14.9d	29.7c	31.8c	49.4b	30.7c	59.5a

Note: Hay sprayed with 1% acetic to propionic acid (65:35). Round bales and stacks were stored outdoors for 6 months. Superscripts within rows show statistical grouping at 5% level.

[a] Acceptability trial with wintering brood cows.
[b] Based on average dry matter yield for five experiments.

From Nehrir, H., Kjelgaard, W. L., Anderson, P. M., Long, T. A., Hoffman, L. D., Washko, B., Wilson, L. L., and Mueller, J. P., *Trans. ASAE,* 21, 217, 1978. With permission.

Table 296

MOISTURE CONTENT OF COASTAL BERMUDAGRASS CONDITIONED WITH DIFFERENT EQUIPMENT[a]

Treatment	Moisture[b] content (%)
Sickle bar mower	20.3a[c]
Sickle bar mower and tedder at 1:00 p.m.	17.3b
Sickle bar mower and semicrusher	16.4b
Sickle bar mower and crimper	16.1b
Sickle bar mower and crusher	16.0b
Rotary mower with side up 90°	13.8c
Rotary mower with side down	12.3c

[a] Moisture content at 1:00 p.m., July 31, 1963, Tifton, Ga.
[b] Wet basis.
[c] Values followed by the same letter are not significantly different (p <0.05).

From Hellwig, R. E., Butler, J. L., Monson, W. G., and Utley, P. R., *Trans. ASAE,* 20, 1029, 1977. With permission.

Table 297

FIELD DRYING TIMES OF CRUSHED AND UNCRUSHED HAY

Time	Moisture (%)		Number of trials
	Uncrushed	Crushed	
First cutting			
First day			
At time of cutting	68	68	8
4 hr after cutting	52	41	8
6 hr after cutting	45	31	8
8 hr after cutting	35	24	4
Second day			
24 hr after cutting	35	26	3
26 hr after cutting	30	22	3
28 hr after cutting	26	18	3
Second day (rain overnight)			
24 hr after cutting	68	62	3
30 hr after cutting	41	36	2
Second cutting			
First day			
At time of cutting	76.0	76.0	4
At 2 p.m.	49.5	43.0	4
At 5 p.m.	39.0	29.5	4
Second day			
At 9 a.m.	37.5	29.7	4
At 2 p.m.	24.0	13.2	4

Time	Moisture (%)		Number of trials
	Uncrushed	Crushed	
Morning cuttings			
Baled first day			
As cut	61	61	2
As raked	33	28	2
As baled	24	22	2
Baled second day			
As cut	64	65	3
As raked	35	30	3
As baled	27	17	2
Baled third day			
As cut	69	70	2
As raked	35	26	2
As baled	24	15	2
Afternoon cuttings			
Baled second day			
As cut	61	62	4
As raked	34	29	3
As baled	23	17	3
Baled third day			
As cut	67	66	2
As raked	32	26	2
As baled	21	14	2

a   Timothy with 15 to 30% legumes (alfalfa, alsike, and red clover).

Source: Turk, K. L., Morrison, S. H., Norton, C. L., and Blaser, R. E., Effect of Curing Methods upon the Feeding Value of Hay, Cornell University Agricultural Experiment Station Bulletin 874, Cornell University, Ithaca, N.Y., 1950.

Table 298
STORAGE SPACE REQUIRED FOR HAY
AND SELECTED BEDDING MATERIALS

Material	wt/ft³ range (lb)	Space/ton range (ft³)
Hay — alfalfa		
Baled	6.0—10.0	200—330
Chopped	5.5—7.0	285—360
Loose	4.0—4.4	450—500
Hay — nonlegume		
Baled	6.0—8.0	250—330
Chopped	5.0—6.7	300—400
Loose	3.3—4.4	450—600
Hay — poor quality		
Baled	5.0—10.0	200—400
Chopped	6.0—9.0	225—325
Loose	3.0—5.0	400—600
Straw		
Baled	4.0—5.0	400—500
Chopped	5.7—8.0	250—350
Loose	2.0—3.0	670—1000
Sawdust		
Dry	8.0—12.0	175—250
Green	18.0—22.0	90—110
Shavings		
Baled	16.0—24.0	85—125
Loose	8.0—12.0	175—250

Adapted from various sources, including *Beef Housing and Equipment Handbook,* Midwest Plan Service, Iowa State University, Ames, 1968.

DEHYDRATED FORAGE

Table 299

PROXIMATE CONSTITUENTS OF ALFALFA MEALS, REGROUND PELLETED ALFALFA, AND GRASS AND GRASS-LEGUME MIXTURE MEALS

Sample	Moisture[a] (%)	Grit[a] (%)	Protein[b] (%)	Crude fat[b] (%)	Crude fiber[b] (%)	Ash[b] (%)	Sugar (as glucose)	
							Reducing[b] (%)	Total[b] (%)
Alfalfa meal								
N-1[c]	5.32	0.87	16.1	2.23	36.6	10.2	0.72	1.96
N-2	4.98	0.82	15.9	2.62	31.3	11.2	1.20	3.31
N-3	6.28	1.50	21.7	4.08	24.6	12.7	0.87	3.33
N-4	5.20	1.77	25.0	4.23	17.2	14.2	0.91	3.12
C-1	9.19	0.45	14.7	3.07	35.5	7.7	1.44	4.54
C-2	8.22	0.20	18.7	2.02	29.1	8.0	1.40	3.45
C-3	7.16	0.20	23.6	4.13	22.1	9.7	0.78	1.65
C-4	9.43	0.12	21.4	3.76	26.7	8.9	1.59	3.32
O-1	6.70	0.40	15.1	2.46	35.6	7.3	1.36	2.60
O-2	8.28	0.15	20.8	3.61	24.7	8.4	1.31	2.87
O-3	8.23	0.15	20.3	3.44	27.8	8.3	1.42	3.16
O-4	7.15	0.27	25.9	3.37	17.2	9.5	2.76	5.54
Reground pelleted alfalfa								
N-1-P	8.27	0.20	16.1	2.83	36.7	10.0	0.71	1.82
N-2-P	9.02	0.30	16.2	3.21	30.4	9.1	1.03	3.01
N-3-P	9.46	0.40	22.1	4.63	22.2	12.1	0.77	3.00
N-4-P	8.15	0.80	24.3	5.43	17.1	13.5	0.79	3.73
N-3-P-E-1	8.75	0.60	22.2	5.97	22.8	13.5	0.74	2.94
N-3-P-E-5	9.15	0.62	21.3	9.74	22.2	12.8	0.70	2.80
Grass and grass-legume mixture meals								
Reed canarygrass	7.69	0.20	24.8	5.11	20.7	8.2	2.39	5.76
Common ryegrass	8.75	0.83	27.4	5.93	18.4	11.5	3.55	7.08
Rye	7.20	2.35	31.3	7.24	16.0	12.6	2.24	5.34
Orchardgrass	9.92	0.13	25.0	5.76	22.6	10.3	1.86	4.30
Orchardgrass-ladino whiteclover	7.39	1.20	27.4	5.99	19.0	10.5	2.73	7.95

Table 299 (continued)
PROXIMATE CONSTITUENTS OF ALFALFA MEALS, REGROUND PELLETED ALFALFA, AND GRASS AND GRASS-LEGUME MIXTURE MEALS

Sample	Moisture[a] (%)	Grit[a] (%)	Protein[b] (%)	Crude fat[b] (%)	Crude fiber[b] (%)	Ash[b] (%)	Sugar (as glucose)	
							Reducing[b] (%)	Total[b] (%)
Orchardgrass-ladino whiteclover-fescue grass	8.23	0.31	24.1	6.67	21.4	8.9	3.64	8.17

[a] Percentage of air-dried material.

[b] Percentage of moisture-free and grit-free original material.

[c] Sample numbers.

Source: Binger, H. P., Thompson, C. R., and Kohler, G. O., Composition of Dehydrated Forages, *U.S. Dep. Agric. Tech. Bull.*, No. 1235, 1961.

Table 300
PROXIMATE CONSTITUENTS OF DEHYDRATED ALFALFA[a]

(%)	Crude protein grade (%)			
	15	17	20	22
Moisture	6.90	6.98	6.88	7.14
Crude protein	15.22	18.04	20.64	22.47
Crude fat	2.32	3.00	3.58	3.69
Crude fiber	26.41	24.27	20.15	18.54
Ash	8.40	8.96	10.34	10.32
Nitrogen free extract	40.77	38.81	38.42	37.85

[a] As fed basis.

From American Dehydrators Association, Dehydrated Alfalfa Assay Report, 3rd ed., Kansas City, Mo. With permission.

Table 301
VITAMIN CONTENT OF DEHYDRATED ALFALFA

	Crude protein grade (%)							
	15		17		20		22	
	ppm	mg/lb	ppm	mg/lb	ppm	mg/lb	ppm	mg/lb
β-Carotene	102.1	46.4	161.2	72.95	216.4	98.2	252.5	114.59
Xanthophylls[a]	175.0	79.7	257.0	116.9	328.0	149.0	401.0	182.0
Thiamine	3.0	1.37	3.5	1.5	3.9	1.76	4.2	1.91
Riboflavin	10.61	4.81	12.31	5.59	15.50	7.04	17.4	7.9
Pantothenic acid	20.91	9.49	30.0	13.61	32.8	14.9	33.0	14.99
Niacin	41.90	19.02	45.82	20.8	54.7	24.83	58.8	26.68
Pyridoxine	6.50	2.93	6.3	2.86	7.9	3.6	7.8	3.52
Choline	1550.	703.7	1518.	689.2	1618.	734.6	1853.	841.3
Proline betaine	6060.	2751.	6380.	2896.	7560.	3432.	8550.	3869.
Folic acid	1.54	0.70	2.1	0.95	2.67	1.21	3.0	1.36
Vitamin K	9.9	4.48	8.7	3.95	14.7	6.66	8.5	3.86
α-Tocopherol	98.0	44.5	128.0	58.1	147.0	66.73	151.0	68.55

[a] No vitamin activity.

From American Dehydrators Association, Dehydrated Alfalfa Assay Report, 3rd ed., Kansas City, Mo. With permission.

Table 302

VITAMIN A CONTENT (IU/LB) OF DEHYDRATED ALFALFA FOR SEVERAL SPECIES OF ANIMALS

Species	Conversion factor[a]	Crude protein grade (%)			
		15	17	20	22
Standard (poultry and laboratory rats)	1,667	77,349	121,608	163,699	191,022
Beef and dairy cattle	400	18,560	29,180	39,280	45,836
Sheep	400—600	18,560—27,840	29,180—43,770	39,280—58,920	45,836—68,754
Swine	500	23,200	36,475	49,100	57,295
Horses					
Growth	555	25,752	40,487	54,501	63,597
Pregnancy	333	15,451	24,292	32,701	38,158
Dogs	833	38,651	60,767	81,801	95,453

[a] To obtain (IU/lb.) of Vitamin A, multiply the amount of β-carotene (mg) for each level crude protein (see Table 121) by conversion factor.

From American Dehydrators Association, Dehydrated Alfalfa Assay Report, 3rd ed., Kansas City, Mo. With permission.

Table 303

DIGESTION COEFFICIENTS OF DEHYDRATED ALFALFA[a]

	Crude protein grade (%)			
	15	17	20	22
Crude protein	61.1	67.4	71.3	73.5
Crude fiber	33.5	38.0	37.9	42.4
Gross energy	53.0	56.6	60.0	65.0

[a] These data were determined with sheep, on an as-fed basis.

From American Dehydrators Association, Dehydrated Alfalfa Assay Report, 3rd ed., Kansas City, Mo. With permission.

LEAF PROTEIN

Table 304
EXTRACTION OF PROTEIN FROM VARIOUS PLANTS[a]

		Age and state of leaves	Dry matter (%)	N (%)	N[b] extracted (%)	Protein N extracted (%)
Egyptian or hyacinth bean	*Lablab atropurpureus*	85 days (dry season)	18.2	2.68	84.6	63.3
Egyptian or hyacinth bean	*L. atropurpureus*	105 days (dry season)	20.5	2.66	77.4	70.5
Egyptian or hyacinth bean	*L. atropurpureus*	136 days (before flowering)	19.7	2.60	74.2	69.6
Egyptian or hyacinth bean	*L. atropurpureus*	147 days flowering	18.3	2.68	57.8	61.8
Cowpeas	*Vigna sinensis*	45 days (early stage wet season)	18.3	2.16	84.3	61.3
Cowpeas	*V. sinensis*	60 days (wet season before flowering)	18.2	2.04	79.5	64.8
Cowpeas	*V. Sinensis*	75 days (wet season flowering)	20.2	2.05	65.5	59.1
Lucerne	*Medicago sativa*	45 days (new growth on old plants, dry season)	16.2	2.56	86.3	66.5
Oats	*Avena sativa*	74 days (wet season, before flowering)	21.1	2.4	83.1	63.8
Cauliflower	*Brassica alevacia*	After harvesting the flower head	12.0	3.44	82.1	62.3

[a] Plants were macerated with water (45 g/kg) and protein extracted at pH 11.
[b] Dry basis, except for dry matter.

From Nanda, C. L., Kondos, A. C., and Ternouth, J. H., An improved technique for plant protein extraction, *J. Sci. Food Agric.,* 26, 1917, 1975. With permission.

Table 305
EFFECT OF GRINDING ON THE PRESSING OPERATION AND YIELD OF LEAF PROTEIN CONCENTRATE FROM ALFALFA[a]

| Process parameter | Alfalfa[b] | | Net change |
	Chopped (%)	Chopped and ground (%)	(%)
Leaf protein concentrate	9.46	15.3	+ 61.7
Juice yield	52.5	63.8	+ 21.5
Juice dry matter	9.32	10.8	+ 15.9
Press cake dry matter	33.1	37.5	+ 13.3
Press cake protein	16.8	13.3	− 20.8
Press cake fiber	33.7	38.1	+ 13.1
Solids extracted	23.8	33.5	+ 40.8
Crude protein extracted	37.3	58.0	+ 55.5
Crude protein recovered	26.4	42.3	+ 60.2

[a] Means of nine runs; analyses on a dry basis except for dry matter content and juice yield which are on an as is basis.

[b] Composition before processing: 20.5% dry matter, 20.9% protein, 26.4% fiber.

Source: Kohler, G. O. et al., Means of increasing yields of leaf protein concentrate, Annual Meeting, American Society of Agricultural Engineering, Raleigh, N.C., June, 1977.

Table 306
AMINO ACID COMPOSITION (g/16 g N) OF ALFALFA LEAF PROTEIN CONCENTRATE (LPC)

	Whole LPC (Pro-Xan)	Green fraction LPC (Pro-Xan II)	White fraction LPC	Brown juice
Alanine	5.55	6.28	6.35	4.00
Ammonia	1.30	1.39	1.07	7.13
Arginine	5.99	6.14	7.98	1.56
Aspartic acid	9.43	10.27	10.04	9.69
Cystine	1.16	1.00	1.68	0.85
Glutonic acid	10.25	11.03	12.02	6.84
Glycine	4.88	5.71	5.52	2.61
Histidine	2.33	2.42	3.19	0.92
Isoleucine	4.94	6.19	5.28	1.97
Leucine	8.19	10.26	9.42	2.76
Lysine	6.18	5.88	6.43	2.81
Methionine	2.09	2.16	2.65	0.64
Phenylalanine	5.25	6.62	6.36	1.87
Proline	4.41	4.46	4.44	2.74
Serine	4.00	5.09	3.65	2.26
Threonine	4.81	4.71	5.70	2.38
Tryptophan	1.38	0.96	2.45	—
Tyrosine	4.09	4.54	5.63	1.36
Valine	6.25	6.99	6.91	2.66
N recovered (%)	85.2	92.6	97.7	75.0

Source: Kohler, G. O., de Fremery, D., and Edwards, R. H., Leaf Protein Research at Western Regional Center, paper presented at the 1975 Annual Meeting, American Society of Agricultural Engineering, Davis, Calif., June, 1975.

ROUGHAGES

Table 307
PROXIMATE COMPOSITION—DRY ROUGHAGES

Commodity	Dry matter (%)	Crude protein (%)	Ether extract (%)	Crude fiber (%)	Ash (%)	NFE[a] (%)
Barley hulls (range)	93.2	4.9—10.8	2.0—2.3	15.6—31.7	6.7—6.8	54.8—64.5
Mean value	93.2	7.9	2.2	23.7	6.8	59.4
Barley straw (range)	82.3—93.6	3.2—4.9	1.5—2.5	34.7—46.5	5.8—8.9	41.6—51.1
Mean value	88.2	4.1	1.8	34.7	6.6	45.1
Corncobs	68.2—97.0	1.8—6.5	0.3—0.9	27.9—42.4	1.0—2.6	50.4—64.7
Mean value	87.7	2.9	0.5	36.8	1.6	58.2
Corn stover (mature)	77.5—97.2	3.3—8.3	0.4—1.8	30.9—52.2	2.6—10.0	32.6—56.2
Mean value	87.2	5.9	1.2	37.1	7.1	48.7
Cotton stalks	90.0—92.6	4.3—7.2	0.1—1.2	44.5—48.6	3.8—22.9	34.6—43.3
Mean value	92.0	6.2	0.9	47.2	6.8	38.9
Cotton leaves	90.0—94.4	15.1—19.9	4.8—13.4	10.6—14.0	17.2—20.6	34.3—47.4
Mean value	92.0	16.8	7.8	11.2	17.6	46.6
Cottonseed hulls	88.8—96.0	2.4—6.4	0.3—2.9	39.0—53.5	2.2—4.0	38.1—53.5
Mean value	91.7	4.3	0.9	46.6	2.9	45.3
Flax hulls	90.0—91.1	4.6—13.2	1.0—2.0	29.6—33.0	8.7—11.4	46.1—50.5
Mean value	90.8	7.7	1.5	31.8	10.1	48.9
Flax straw	90.8—95.8	6.8—9.3	2.2—3.4	44.8—50.0	4.1—7.5	33.3—37.7
Mean value	92.9	7.8	3.3	46.2	7.2	35.5
Oat hulls	87.5—95.0	1.8—9.9	0.1—3.6	23.6—37.9	5.0—8.3	52.1—62.6
Mean value	92.6	4.1	1.4	32.4	6.6	55.5
Oat straw	83.9—95.2	2.0—8.6	0.8—3.2	33.3—54.0	4.9—12.9	38.2—57.2
Mean value	90.1	4.4	2.1	41.0	8.2	44.3
Peanut hulls	88.2—95.3	4.8—10.2	0.2—3.2	45.9—74.2	1.9—13.3	10.7—38.1
Mean value	92.3	7.4	1.2	65.4	4.7	21.3
Rye straw	85.7—92.9	2.0—8.2	1.0—2.0	39.7—59.5	3.4—8.0	32.7—50.2
Mean value	88.9	3.0	1.5	47.6	4.8	43.1
Pecan hulls	91.9	2.1	0.7	59.4	2.4	35.4
Rice hulls	90.7—93.9	1.7—3.8	0.2—1.5	38.4—47.9	14.6—24.6	26.0—37.6
Mean value	92.2	3.0	0.9	43.3	20.8	32.0
Rice straws	88.0—93.4	3.3—4.5	0.7—2.3	31.5—37.8	15.4—18.7	39.4—46.4
Mean value	90.3	3.8	1.3	35.6	16.7	42.6
Sorghum hulls	91.7—93.6	3.6—4.1	0.3—5.0	—	6.7—10.1	—
Mean value	92.7	3.9	0.4	—	8.2	—
Sorghum stover	69.0—94.6	2.8—7.1	1.0—3.2	23.8—41.6	6.0—12.4	44.9—60.8
Mean value	85.1	5.3	2.1	32.6	9.6	50.4
Soybean hulls	87.9—92.7	13.1—14.7	1.6—3.1	37.5—40.4	4.4—6.1	38.9—40.9
Mean value	91.3	13.7	2.3	38.9	5.0	40.1
Soybean straw	85.7—92.7	4.5—9.8	0.6—2.7	37.5—46.3	5.7—8.0	41.0—44.5
Mean value	87.6	5.5	1.4	44.1	6.4	42.6
Wheat straw	82.2—99.1	1.5—6.9	1.0—3.7	36.4—51.5	3.5—11.1	38.5—52.4
Mean value	90.1	3.6	1.7	41.5	8.1	45.1

[a] Nitrogen-free extract.

SILAGE CROPS — COMPOSITION AND FEEDING

Table 308
CHARACTERISTICS OF QUALITY SILAGE

Organoleptic
  Moisture and texture   Uniform moisture; tissues are firm, not slimy
  Odor   "Clean", pleasant vinegary odor
  Color   Light green to yellow (Dark brown, carmelized, or charred silage indicates excessive heat; black silage is rotten and should not be fed.)
  Taste   Pleasant; sharp taste indicative of acid pH
  Mold   No visible mold; not musty or slimy
  Animal acceptability   Palatable; animals consume it readily

Chemical
  pH <4.5
  Lactic acid (%)   1.5—2.5
  Acetic acid (%)   0.5—0.8
  Butyric acid (%)   Preferably <0.1
  Ammonia-N (% of total N)   Preferably <5—8

Note: Percentages are measured on a wet basis. References included in the original.

Table 309
COMPOSITION AND NUTRITIONAL VALUE OF CORN SILAGE

	Dry basis[a]		As fed
	Range	Average[b]	Average[b]
Proximate			
Dry Matter (% w.b.)	12.5—46.7	25.6	25.6
Protein, crude (N×6.25) (%)	3.5—15.9	8.4	2.2
Fat, crude, (ether extr.) (%)	0.7—6.7	3.3	0.8
Fiber, crude, %	19—42	26.0	6.6
Fiber, neutral detergent (%)	—	51	13
Fiber, acid detergent (%)	—	30	7.7
Nitrogen, free extract (%)	46.0—81.9	56.3	14.4
Lignin (%)	—	5	1.3
Mineral			
Ash (%)	1.3—10.5	6.0	1.5
Calcium (%)	0.2—0.6	0.33	.09
Phosphorus (%)	0.15—0.55	0.23	.06
Copper (mg/kg)	3.3—9.0	5.1	1.3
Potassium (%)	0.72—2.35	1.17	0.3
Magnesium (%)	0.10—0.40	0.21	.05
Iron (mg/kg)	40—550	180	47
Manganese (mg/kg)	5.1—81.6	67	17
Vitamins			
Niacin (mg/kg)	15.0—60.7	43	11
Carotene (mg/kg)	1.1—155	45.5	12
Vitamin D, IU/kg	363—495	438	114
Energy and Carbohydrate			
Starch (% d.b.)	55.9—59.2	57.6	15
Sugars (total)	0.0—11.5	1.5	0.4
Energy (calories/kg)	4334—4545	4382	1139
Energy, digestible, (cattle) (%)		2910	749
Protein, digestible, (cattle) (%)		3.9	1.0
Total digestible nutrients, (cattle) (%)		66	17

Note: References included in the original.

[a]   Except for dry matter.
[b]   Dough stage.

Table 310

MINERAL CONTENT OF CORN SILAGE AND
NATIONAL RESEARCH COUNCIL
RECOMMENDATIONS FOR MILK COWS
PRODUCING MORE THAN 30 kg OF MILK
PER DAY

| | Mineral content[a] | | NRC |
	Mean	Range	recommendations
Ca (%)	0.29	0.10— 1.51	0.53
P (%)	0.24	0.13— 0.66	0.39
K (%)	1.05	0.43— 2.77	0.70
Mg (%)	0.20	0.08— 0.74	0.10
S (%)	0.16	0.11— 0.21	0.20
Mn (ppm)	42	2 — 2.40	20
Fe (ppm)	235	31 —1800	100
Cu (ppm)	12	4 — 110	10
Zn (ppm)	34	13 — 213	40

[a]   Results of analysis of 396 samples; however, sulfur analysis
represents 29 samples.

From Hemken, R. W., High Silage Rations for Dairy Cows, papers
presented at Int. Silage and Research Conf., Washington, D.C., De-
cember 1971 (sponsored by National Silo Association, Inc., Waterloo,
Iowa). With permission.

Table 311

STAGE OF MATURITY FOR HARVESTING VARIOUS
SILAGE CROPS

Crop	Stage of maturity
Corn	Medium to hard dent
Sorghum	Dough or harvest before leaves and stalks deteriorate from disease
Alfalfa	Full bud
Red and alsike clover	One fourth to one half bloom
Birdsfoot trefoil	One fourth bloom
Smooth bromegrass	Medium head
Timothy and orchardgrass	Early head
Bermudagrass	15 to 18 in. high
Wheat, oats, millet, sudan	Late boot (heads begin to emerge)

Note: References included in the original.

Table 312
EFFECT OF STAGE OF MATURITY AT HARVEST ON POUNDS OF TDN IN CORN SILAGE

Maturity stage	Green weight (tons)		
	16	20	26
Milk	4,800	6,000	7,800
Dough	6,080	7,600	9,800
Full dent	7,040	8,800	11,440

Note: Milk = endosperm high in sugars and of milk-like consistency; dough = endosperm becoming firm, starch forming, dough-like consistency; full dent = endosperm consisting of dried starch which because of moisture loss has shrunk causing dent to appear in top of kernel.

Source: Cornell University, Cornell Recommends for Field Crops, New York State College of Agriculture, Ithaca, N.Y., 1963.

Table 313
CHANGES IN COMPOSITION AND QUALITY OF CORN HARVESTED FOR SILAGE AT DIFFERENT STAGES OF MATURITY

Harvest date and maturity stage	Dry matter content (%)	Forage yields		DM intake (% of dent stage)[a]
		Green (tons/A)	Dry (tons/A)	
August 18 (milk)	21	17.5	3.7	74
August 28	25	15.0	3.7	89
September 5	26	14.8	3.9	90
September 12 (dent)	35	12.3	4.3	100
September 20	38	10.0	3.8	95
September 26	41	9.2	3.7	89
October 2 (mature)	46	7.7	3.5	74

[a] Silage intake data were corrected for different body weights of the animal and expressed as dry matter intake on a relative basis; the highest value was 100.

Source: McCullough, M. E., Silage Research at the Georgia Station, University of Georgia College of Agricultural Experiment Station Research Report 75, 1970.

Table 314

EFFECT OF STAGE OF MATURITY OF WHEAT-RYEGRASS SILAGE ON INTAKE AND MILK PRODUCTION

Maturity stage	Dry matter intake (lb)	Milk production (lb/day)
Heading	33	37
Milk	28	22
Dough	27	20

Note: Heading = spikes or panicles emerging; milk = endosperm high in sugar and milk-like consistency; dough = endosperm becoming firm, starch forming, dough-like consistency.

From McCullough, M. E. and Sisk, R. S., Influence of stage of maturity at harvest and level of grain feeding on intake of wheat silage, *J. Dairy Sci.,* 50, 705, 1967. With permission.

Table 315

EFFECTS OF FILLING RATE ON CORN SILAGE QUALITY AND PRESERVATION

	Filled in 1 day	Filled in 5 days
Crude protein (%)	16.9	19.3[a]
pH	5.0	5.4
Butyric acid (%)	0.8	2.9
Lactic acid (%)	10.3	5.5
Fermentation losses	5.7	10.4
Silage dry matter consumption (percent of body weight)	1.21	1.06
Maximum temperature (°F)	94.0	99.0

[a] Increased protein is usually because of increased nitrogen because of higher temperature and does not indicate higher digestible protein.

From Miller, W. J., Clifton, C. M., and Cameron, N. W., Nutrient losses in silage quality as affected by rate of filling and soybean flakes, *J. Dairy Sci.,* 45, 403, 1962. With permission.

Table 316

EFFECT OF FINENESS OF CHOP OF CORN SILAGE ON AMOUNT DRY MATTER STORED PER CUBIC FOOT OF SILO CAPACITY

Dry matter of silage (%)	Dry matter (lb)/ft³ of silo		Increase (%)
	1/4—3/8 in. chop	5/8—3/4 in. chop	
28	13.40	11.14	20.3
31	12.32	11.55	6.7
43	13.15	12.13	8.4
48	11.99	9.96	20.4

Source: Sell, W. H., Silage Production, Harvesting, and Storing, Bulletin 716, Cooperative Extension Service, University of Georgia, Athens, 1975.

Table 317
PROXIMATE ANALYSIS OF CORN AND SORGHUM SILAGES AS ENSILED AND AS FED

| | Dry matter | Composition of dry matter (%) | | | | |
		Crude protein	Ether extract	N-free extract	Crude fiber	Ash
Experiment I						
Corn						
Ensiled	—	9.1	2.2	55.1	29.3	4.2
Fed	30.5	9.4	5.9	56.4	23.1	5.3
Sorghum						
Ensiled	—	6.9	1.6	65.7	20.6	5.3
Fed	25.4	6.8	6.3	56.7	23.9	6.2
Alfalfa-bromegrass hay	—	16.6	2.4	45.2	28.9	6.9
Experiment II						
Corn						
Ensiled	24.4	9.8	3.2	63.7	17.8	5.5
Fed	25.2	9.7	3.0	59.5	21.6	6.1
Sorghum						
Ensiled	23.0	8.6	1.9	61.5	22.3	5.7
Fed	22.1	9.7	2.6	55.0	26.0	6.7
Alfalfa-bromegrass hay	—	16.5	1.6	44.3	29.7	7.7

From Lance, R. D., Foss, D. C., Krueger, C. R., Baumgardt, B. R., and Neidermeier, R. P., Evaluation of corn and sorghum silages on the basis of milk production and digestibility, *J. Dairy Sci.*, 47, 254, 1964. With permission.

Table 318
RELATIVE ENERGY VALUES AND PERCENTAGE TDN FOR VARIOUS SILAGES

Silage	Relative energy value	TDN (%)
Corn	100	65—70
Ear corn	110—120	72—75
Sorghum head	110—120	72—75
Barley	95—100	62—65
Wheat	80—85	58—60
Oat	70—80	55—58
Alfalfa	65—80	53—58
Alfalfa-grass	60—70	50—57
Sudan grass	60—70	50—57
Sorghum stover	60—65	50—55
Cornstalk	50—60	45—50

From National Silo Association, Inc., Silage and Best-Cost Livestock Feeding Programs, Cedar Falls, Iowa, November 1976. With permission.

Table 319
EQUIVALENT ENERGY VALUES FOR
FEEDSTUFFS AS RELATED TO
PERCENTAGE TDN

TDN percent of DM	Approximate energy value/lb DM[a] (Mcal)			
	ME^a	$NE^a{}_{maint.}$	NE_{gain}	NE_{milk}
40	0.73	0.29	−0.03	0.28
42	0.76	0.33	0.01	0.32
44	0.80	0.37	0.04	0.36
46	0.82	0.40	0.08	0.40
48	0.84	0.44	0.12	0.44
50	0.87	0.47	0.16	0.48
52	0.90	0.50	0.19	0.51
54	0.93	0.53	0.22	0.54
56	0.96	0.56	0.25	0.58
58	0.98	0.59	0.28	0.61
60	1.01	0.62	0.31	0.64
62	1.04	0.64	0.34	0.67
64	1.06	0.67	0.36	0.70
66	1.09	0.69	0.38	0.72
68	1.12	0.72	0.41	0.75
70	1.15	0.74	0.44	0.78
72	1.17	0.76	0.46	0.80
74	1.20	0.78	0.48	0.82
76	1.23	0.80	0.51	0.85
78	1.26	0.82	0.53	0.87
80	1.28	0.84	0.55	0.90

[a] ME = metabolizable energy; NE = net energy; DM = dry matter.

Source: McCullough, M. E., *Fermentation of Silage — A Review,* McCullough, M. E., Ed., National Feed Ingredients Association, West Des Moines, Iowa, 1978, 1.

Table 320
SILAGE ADDITIVES AND PRESERVATIVES

Type	Purpose — desired results
Molasses	High in sugar; may increase acetic and lactic acid production, improve silage preservation, palatability and nutritive value; reduces dry matter losses and improves animal performance
Whey	Improves feeding value; may help to reduce nitrogen losses
Corn, barley, oats, milo, citrus pulp, beet pulp, other feed ingredients	Decreases moisture content and seepage; provides additional carbohydrates from which acids may be produced; improves palatability and feeding value; disadvantage is the quantity needed and labor required to handle and mix; citrus and beet pulp may not be available locally
Limestone	Neutralizes initial acids produced and thereby enhances production of lactic acid; provides additional calcium to low-calcium corn silage
Bacterial cultures (*Lactobacillus*)	Increases numbers of lactic acid producing bacteria to insure rapid fermentation; research results are variable
Antibiotics	Zinc bacitracin and other antibiotics have been added to silage, but their value as preservatives are of questionable value
Salt (NaCl)	At certain levels, inhibits certain microorganisms without inhibiting desirable-acid producing bacteria, but its value is questionable
Sodium metabisulfite, sulfur dioxide	Bacteriostatic agents; carotene content conserved; sulfur dioxide can be iritating to workers if care is not taken; proper handling and distribution of these compounds difficult and value is questionable
Urea, ammonia	Added to improve "protein" content of the low protein crops such as corn silage; with urea, fermentation of corn silage extended and feed efficiency reputed to be improved in some cases
Enzymes	Improve fermentation or digestibility; however, of questionable value
Mineral acids	Hydrochloric acid, sulfuric acid, and phosphoric acids have been used as preservatives primarily in Europe; substitute for acids produced by bacterial action; because of corrosive nature, use of mineral acids generally has more disadvantages than advantages
Organic acids, aldehydes	Bacteriostatic, fungicidal agents; formic acid, acetic acid, propionic acid, and salts of these acids and certain aldehydes (i.e., formaldehyde) are added to inhibit fermentation and proteolysis; formic acid and formaldehyde (paraformaldehyde) have not been approved as silage additives by the U.S. Food and Drug Administration; care must be exercised to protect workers handling certain of these chemicals, particularly formic acid and formaldehyde

Note: References included in the original.

Table 321
OPTIMUM MOISTURE LEVELS FOR ENSILING CROPS

Crop	Moisture levels
Grasses	Cut at stage that will make best hay; field wilt to 60—65% moisture (for low-moisture silage in air-tight structures, wilt to 45—60%)
Legumes	Harvest at bud to late bud (full bud) stage; maybe wilted to 45—60% low moisture air-tight silo storage
Corn	Harvest when kernels are in the late dent stage (generally when half leaves and all husks are brown), at a moisture level of about 60%; for conventional tower sile storage, 60—65% is ideal
Ear corn	For ground ear corn, 25—35% moisture; for whole grain, 25—32%; 25% moisture is ideal
Sorghum	Harvest when seeds are in dough stage, about 60—65% moisture; for grain sorghum, combine at about 25—30% moisture and ensile at 30% moisture
Small grain	Harvest at beading to milk stage of maturity and wilt to about 60—65% moisture

From the National Silo Association, *Operator's Manual,* Waterloo, Iowa, 1977. With permission.

Table 322
THE GRAB TEST FOR ESTIMATING MOISTURE CONDITION OF CROPS FOR SILAGE

Condition	Grab test
#1	Juice runs freely or shows between the fingers. The crop contains 75—84% moisture and is too wet to make high-quality silage without treatment. Silages made from crops in this condition will lose large quantities of juice. When possible, wilt these crops; if they must be ensiled without wilting, use an effective chemical preservative or 200 lb of ground grain per ton of crop.
#2	The ball holds its shape, and the hand is moist. The crop contains 70—75% moisture. Some juice will escape from tower silos. Additional wilting in the field is desirable. Where this is not done, use an effective chemical preservative or 125 lb of ground grain per ton of crop or layer with wilted crops. Odors will be strong without some treatment.
#3	The ball expands slowly; no dampness appears on the hand. The crop contains 60—70% moisture. This is the best condition for ensiling legumes without treatment. Treat grasses with a chemical perservative.
#4	The ball springs out in the opening hand. The crop contains less than 60% moisture. Only very young crops wilted to this condition can be safely ensiled. Others are likely to mold in the silo unless layered with wet crops.

Note: The grab test may be used to determine the moisture condition of crops standing in the field, lying in the swath or windrow, or chopped in the wagon. Squeeze tightly (with all your strength) for 90 sec a handful of fine-cut chop. Release the grip and note the condition of the ball of crop in the hand. The condition of this ball indicates the kind of treatment the crop must receive to make good silage.

From National Silo Association, Operator's Manual, Waterloo, Iowa, 1977. With permission.

Table 323
EXPECTED DRY MATTER LOSSES IN
HARVESTING AND STORING CORN AND HAY
CROP SILAGE

	Dry matter losses		
	Harvest (%)	Storage (%)	Total (%)
Moisture in corn silage			
70% or more	4.0	13.4	17.4
60—69%	5.0	6.3	11.3
Under 60%	16.2	6.3[a]	22.5[a]
Moisture in haycrop silage			
70% or more	2.0	21.2	23.2
60—69%	5.0	10.1	15.1
Under 60%	11.5	8.2	19.7

[a]   Estimated values

Source: Dum, S. A., Adams, R. S., Baylor, J. E., and Grant, A. R., Silage and Silos, Special Circular 223, The Pennsylvania State University, College of Agriculture, University Park, Pa.

Table 324
AMOUNT OF SILAGE IN SLICES FROM
TOWER SILOS OF DIFFERENT SIZES

Silo diameter (ft)	Volume/ft of depth (ft³)	Pounds of silage in slice[a]		
		1 in.	2 in.	3 in.
12	113.1	470	940	1,410
14	153.9	640	1,280	1,920
16	201.1	840	1,675	2,510
18	254.5	1,060	2,120	3,180
20	314.2	1,308	2,615	3,924
22	380.1	1,583	3,165	4,749
24	452.4	1,885	3,770	5,655
26	530.9	2,215	4,430	6,645
28	615.8	2,565	5,130	7,695
30	706.9	2,945	5,890	8,835
36	1,017.4	4,240	8,480	12,720

[a]   Based on 50 lb/ft³.

Source: Sell, W. H., Silage Production, Harvesting and Storing Bulletin 716, Cooperative Extension Service, University of Georgia, Athens, 1975.

Table 325
CAPACITY OF 12-FT-DEEP HORIZONTAL SILOS OF VARIOUS
DIMENSIONS AND THE AMOUNT OF SILAGE PER 4- AND 12-IN.-
THICK SLICE

		Capacity (tons)					Width (ft)	Amount of silage/slice (tons)	
60 ft[a]	80 ft	100 ft	120 ft	140 ft	160 ft	200 ft		4-in. slice	12-in. slice
288	384	480	576	672	768	960	20	1.6	4.8
432	576	720	864	1008	1152	1440	30	2.4	7.2
576	768	960	1152	1344	1536	1920	40	3.2	9.6
720	960	1200	1440	1680	1920	2400	50	4.0	12.0
864	1152	1440	1728	2016	2304	2880	60	4.8	14.4
1152	1536	1944	2292	2688	3072	3840	80	6.4	19.2

[a]   Length.

Source: Sell, W. H., Silage Production, Harvesting and Storing Bulletin 716, Cooperative Extension Service, University of Georgia, Athens, 1975.

SUGAR CROPS AND TOBACCO — COMPOSITION

Table 326
COMPOSITION OF CANE AND BEET MOLASSES

Constituents	Normal beet molasses (%)	Final cane molasses (%)	Cane refinery molasses (%)
Dry Substances	76—85	77—84	78—85
Total sugar as invert sugar	48—58	52—65	50—58
C	28—34		28—33
N	0.2—2.8	0.4—1.5	0.08—0.5
P_2O_5	0.02—0.07	0.6—2.0	0.009—0.07
CaO	0.15—0.7	0.1—1.1	0.15—0.8
MgO	0.01—0.1	0.03—0.1	0.25—0.8
K_2O	2.2—4.5	2.6—5.0	0.8—2.2
SiO_2	0.1—0.5		0.05—0.3
Al_2O_3	0.005—0.06		0.01—0.04
Fe_2O_3	9.001—0.02		0.001—0.01
Total ash	4—8	7—11	3.5—7.5

Table 327
COMPOSITION OF SWEET SORGHUM STALKS, SOUTH TEXAS (1977)

Variety	Rio (%)	MN1500 (%)
Water	64.8—66.0	63.6—67.5
Sugars		
Sucrose	12.78—13.77	6.89—7.88
Dextrose and levulose	0.54—0.72	1.68—1.94
Ash	1.3—1.4	1.3—1.4
Starch	1.0—2.0	0.6—1.2
Protein (Kjeldahl N × 6.25)	0.9—1.0	1.0—1.3
Fiber (ash-free)	12.5—14.8	16.0—17.2

Note: Range of values cited for Rio represent the high-and low-average data from 4 biweekly harvests; MN1500 values are average data similarly taken from 3 biweekly harvests.

From Smith, B. A. and Reeves, S. A., Jr., in *Sugar y Azucar,* 74, 25, 1979. With permission.

Table 328
SOLUBILITY OF SUCROSE IN WATER

Temperature, °C	Sucrose, wt %	Sucrose, g/100 g water
0	64.40	180.9
10	65.32	188.4
20	66.60	199.4
30	69.18	214.3
40	70.01	233.4
50	72.04	257.6
60	74.20	287.6
70	76.45	324.7
80	78.74	370.3
90	81.00	426.2

Table 329
RELATIVE SWEETNESS OF SUGARS IN 10% SOLUTIONS AT 20°C

Sugar	Rel. sweetness
Tructose	120
Sucrose	100
Glycerol	77
Glucose	69
Galactose	67
Inannitol	67
Lactose	39

Table 330
CHEMICAL ANALYSES OF PORTIONS OF FLUE-CURED AND AIR-CURED TOBACCOS

Constituents	Flue-cured (262 samples)		Maryland (32 samples)		Burley (72 samples)	
	Laminar	Stem	Laminar	Stem	Laminar	Stem
Total nitrogen (%)	1.99	1.46	2.26	1.08	3.26	1.97
Protein nitrogen (%)	0.77	0.42	1.24	0.62	1.49	0.65
α-Amino nitrogen (%)	0.179	0.163	0.193	0.103	0.254	0.119
Nicotine (%)	2.30	0.55	1.23	0.26	2.64	0.46
Petroleum ether extractive (%)	6.96	1.11	5.64	0.98	6.38	0.84
Soluble sugars (%)	17.98	17.21	—	—	—	—
Starch (%)	4.35	2.05	—	—	—	—
Nonvolatile acids[a]	14.00	16.45	16.79	18.97	26.65	23.84
Water-soluble acids[a]	4.15	3.88	1.80	0.81	1.82	0.29
pH	5.03	5.12	6.02	6.58	6.35	7.60

[a]   Milliliters of 0.1N alkali per gram tobacco.

Printed with permission from Darkis, F. R. and Hackney, E. J., Cigarette tobaccos: chemical changes that occur during processing, *Ind. Eng. Chem.*, 44, 284, 1952. Copyright 1952 American Chemical Society.

Table 331
CHEMICAL ANALYSES OF FLUE-CURED TOBACCO PARTS

Components (%)	Laminar	Total vein tissue	Midribs	Lateral veins	Veinules	Entire leaf	Stalks
Total nitrogen	2.72	1.76	1.57	2.03	2.60	2.40	1.03
Protein nitrogen	1.07	0.57	0.48	0.67	1.02	0.91	0.59
α-Amino nitrogen	0.277	0.210	0.196	0.284	0.315	0.254	0.060
Nicotine	2.80	0.87	0.61	1.15	0.221	2.15	0.37
Reducing sugars	11.33	11.26	11.62	8.79	11.53	11.31	4.36
Total sugars	12.37	12.16	12.46	9.90	12.60	12.25	6.13
Petroleum ether extractive	7.73	1.50	1.00	1.81	4.24	5.70	0.91
Soluble ash	10.32	1.414	1.519	13.31	9.75	11.59	4.49
CaO	3.05	3.05	3.01	2.74	2.33	3.07	0.94
K_2O	2.33	4.66	4.98	3.99	5.12	3.27	1.96
MgO	0.31	0.53	0.56	0.53	0.43	0.38	0.17
Cl	0.54	2.17	2.08	1.41	0.80	0.97	0.61
P_2O_5	0.66	1.05	1.05	1.14	1.01	0.78	0.45
S	0.64	0.53	0.54	0.54	0.63	0.61	0.39
Silica	1.05	0.87	0.61	0.96	1.97	0.99	0.55
Fe_2O_3 and Al_2O_3	0.18	0.12	0.10	0.12	0.21	0.16	0.12
pH	5.31	4.88	4.85	4.97	4.75	4.95	4.98
Total acidity[a]	15.83	17.90	18.25	17.05	15.29	16.51	10.32

[a] Milliliters of 0.1N acid/ per gram.

ANIMALS

GENERAL

REPRODUCTION

Table 332
DURATION OF ESTRUS, ESTROUS CYCLE, AND GESTATION PERIOD IN MAJOR FARM ANIMALS

	Cow	Ewe	Mare	Sow
Age at puberty (months)	4—14	7—10	15—24	5—8
Length of estrous cycle (d)				
Range	14—24	14—19	10—35	16—25
Mean	21	17	22	20
Duration of estrus				
Range	4—30 h	1—3 d	2—11 d	1—5 d
Mean	17 h	36 h	6 d	2 d
Gestation period (d)				
Range	262—359	140—159	316—363	102—128
Mean	279[a]	148	344[b]	114

[a] Holstein.
[b] Morgan.

From various sources.

Table 333

PLASMA TESTOSTERONE CONCENTRATION OF HEREFORD BULLS DURING CONFINEMENT IN ENVIRONMENTAL ROOM

Temperature (°C)	RH (%)	Week of exposure							Av.
		1	2	3	4	5	6	7	
21.0	50	3.10[a]±0.60	3.52±0.53	2.57±0.42	3.18±0.41	3.99±0.45	3.48±0.59	3.39±0.56	3.32±0.23
35.5	50	1.62±0.28	1.40±0.21	2.22±0.34	2.76±0.47	1.98±0.45	2.27±0.45	2.74±0.62	2.20±0.21

Note: Plasma testosterone concentration is in ng/m*l*.

[a] Each value represents the mean ± SE of eight bulls.

Source: Rhynes, W. E. and Ewing, L. L., *Endocrinology*, 92(2), 509, 1973.

Table 334
DEGREE OF THERMAL STRESS AND REPRODUCTIVE RESPONSES IN EWES

	Control[a]	Hot-room ewes (32.2°C, 60% RH)	
		Sheared	Unsheared
Rectal temperature (°C)	39.1±0.04	40.1±0.12	40.6±0.13
Ova fertilized (%)	92.6	64.0	40.7
Abnormal ova (%)	3.7	32.0	55.6

[a] Control = natural environment in October in Kentucky.

Source: Dutt, R. H., Ellington, E. F., and Carlton, W. W., *J. Anim. Sci.*, 18, 1308, 1959.

Table 335
EFFECT OF ENVIRONMENTAL TEMPERATURE ON SEMEN QUALITY OF RAMS

Semen quality (av)	Treatment			
	Cooled (20.6°C)	Number of observations	Heated (26.7°C)	Number of observations
Semen volume (mℓ)	0.92	11	1.20	8[a]
Motility[b]	3.5	14[c]	1.8	12
Live sperm (%)	67	14[a]	35	12
Normal sperm (%)	73	14[a]	43	12
Sperm concentration	3.48	14[c]	1.43	12

Note: Semen data of the last two collection periods.

[a] $p = 0.05$.
[b] Concentration expressed in millions per mm³.
[c] $p = 0.01$.

Source: Brooks, J. R. and Ross, C. V., *Mo. Agric. Exp. Stn. Res. Bull.*, 801, 1, 1962.

Table 336
EFFECTS OF ENVIRONMENTAL TEMPERATURE UPON SPERMATOGENESIS OF MATURING FOWL

Age in weeks		Number of birds per temperature		Spermatogenesis per age group (%)				
Trial 1	Trial 2	Trial 1	Trial 2	Trial 1		Trial 2		
				8°	30°	8°	19°	30°
—	10	—	10	—	—	0.0	0.0	0.0
11	11	8	10	0.0	0.0	0.0	20.0	0.0
13	12	8	10	0.0	37.5	0.0	20.0	10.0
15	13	8	10	25.0	100.0	40.0	60.0	50.0
17	14	8	10	50.0	100.0	40.0	70.0	90.0
19	15	8	10	87.5	100.0	50.0	80.0	80.0
Average				32.5	67.5[a]	23.0	42.0[a]	35.0[a]

[a]   Means significantly higher ($p < 0.5$) than that of the groups from 8°C.

From Houston, T. M., *Poult. Sci.,* 54(4), 1180, 1975. With permission.

Table 337
CAUSES OF LOSS OF POTENTIAL CALVES FOR BEEF AND DAIRY CATTLE

Beef cattle — bred naturally[a]		Dairy cattle — inseminated artificially[b]	
Status after breeding season	**Percentage of females**	**Outcome of first service**	**Percentage of females**
Cow weaned calf	71	Cow calved	50
Potential calves lost	29	Reproductive failure	50
Females not pregnant	17	Nonestrous cow inseminated	10
Females lost pregnancy	2	Anatomic abnormalities	2
Perinatal calf death	7	Ovulation failure	2
Calf deaths, birth to weaning	3	Lost or ruptured ova	5
Total	100	Fertilization failure	13
		Embryonic death	15
		Fetal death	3
		Total	100

[a]   A 14-year summary of data from the Livestock and Range Research Station, Miles City, Montana, including females 14 months to 10 years of age, with breeding seasons of 45 or 60 d.

[b]   Adapted from Hawk, H. W., in *Animal Reproduction,* Beltsville Symposia in Agriculture, No. 3, Hawk, H. W., Ed., Allan, Osmun, Montclair, NJ, 1979.

From Hawk, H. W. and Bellows, R. A., in *Reproduction in Farm Animals,* 4th ed., Hafez, E. S. E., Ed., Lea & Febiger, Philadelphia, 1980. With permission.

FEEDING

Table 338

NUTRIENT CONTENT OF REPRESENTATIVE ANIMAL FEEDS

Feed	Dry matter (%)	Net energy (Mcal/lb)			TDN (%)	Crude protein (%)	Crude fiber (%)	ADF (%)	Calcium (%)	Phosphorus (%)	Vitamin A equivalent (1,000 IU/lb)
		NE_l	NE_m	NE_g							
Alfalfa, fresh	27	0.621	0.594	0.268	61	19.0	28	35	1.72	0.31	36.1
Alfalfa hay, bud stage	89	0.669	0.640	0.372	65	21.7	24	31	2.12	0.30	90.9
Alfalfa hay, early bloom	90	0.590	0.562	0.268	58	17.2	31	38	1.25	0.23	15.4
Alfalfa hay, full bloom	88	0.544	0.522	0.200	54	15.0	35	42	1.28	0.20	2.2
Alfalfa dehydrated, 15% protein	93	0.621	0.594	0.313	61	16.3	33	41	1.32	0.24	19.9
Alfalfa silage, 25—40% DM (same values as for alfalfa hay of same maturity)											
Almond hulls, 13% crude fiber	91	0.579	0.561	0.358	57	4.4	14	27	0.23	0.11	—
Barley grain	89	0.866	0.889	0.594	83	13.9	6	7	0.05	0.37	—
Barley hay	87	0.576	0.558	0.249	57	8.9	26	—	0.21	0.30	—
Beet pulp, dried	91	0.812	0.812	0.540	78	8.0	22	34	0.75	0.11	—
Bermudagrass hay, "Coastal"	91	0.476	0.457	0.086	48	6.0	34	35	0.46	0.18	81.1
Bonemeal, steamed	95	0.122	0.259	0	16	12.7	2	—	30.51	14.31	—
Brewers grains, dried	92	0.680	0.653	0.390	66	26.0	16	23	0.29	0.54	—
Bromegrass hay, late bloom	90	0.544	0.522	0.200	54	7.4	40	44	0.30	0.35	—
Citrus pulp, dried	90	0.798	0.798	0.526	77	6.9	14	23	2.07	0.13	—
Clover hay, ladino	91	0.621	0.594	0.313	61	23.0	19	32	1.38	0.24	29.3
Coconut meal, mech-extd	93	0.844	0.857	0.572	81	21.9	13	—	0.23	0.66	—
Coconut meal, solv-extd	92	0.767	0.753	0.490	74	23.1	16	—	0.18	0.66	—
Corn fodder	82	0.667	0.640	0.372	65	8.9	26	33	0.43	0.23	0.9
Corn stover	87	0.599	0.571	0.281	59	5.9	34	39	0.60	0.09	—
Corn cobs, ground	90	0.467	0.458	0.068	47	2.8	35	35	0.12	0.04	0.1
Corn ears, ground	87	0.835	0.843	0.563	80	9.3	9	—	0.05	0.26	—
Corn grain, ground	89	0.921	0.975	0.644	88	10.0	2	3	0.03	0.31	0.4
Corn silage, well eared	25—40	0.721	0.699	0.440	70	8.0	24	31	0.27	0.20	8.2
Corn silage, average	25—40	0.694	0.670	0.406	68	8.0	24	31	0.27	0.20	8.2

Table 338 (continued)
NUTRIENT CONTENT OF REPRESENTATIVE ANIMAL FEEDS

Feed	Dry matter (%)	Net energy (Mcal/lb)			TDN (%)	Crude protein (%)	Crude fiber (%)	ADF (%)	Calcium (%)	Phosphorus (%)	Vitamin A equivalent (1,000 IU/lb)
		NE$_l$	NE$_m$	NE$_g$							
Corn silage, not well eared	25—40	0.667	0.640	0.372	65	8.4	32	41	0.34	0.20	2.3
Cottonseed hulls	90	0.367	0.390	0	38	4.3	50	71	0.16	0.73	—
Cottonseed, whole	93	1.034	1.175	0.730	98	24.9	18	29	0.15	0.73	—
Cottonseed meal, mech-extd, 41% protein	94	0.798	0.798	0.526	77	43.6	13	20	0.17	1.28	—
Cottonseed meal, solv-extd, 41% protein	92	0.780	0.766	0.503	75	44.8	13	—	0.17	1.31	0
Dicalcium phosphate	96	0	0	0	0	0	0	0	23.70	18.84	—
Fat, animal (not over 3% of ration)	99	2.381	2.381	1.188	182	0	0	0	—	—	—
Hominy feed	91	0.966	1.052	0.680	92	11.8	6	12	0.06	0.58	1.8
Limestone, ground	100	0	0	0	0	0	0	0	36.07	0.02	0
Milo grain	88	0.835	0.844	0.563	80	11.7	2	9	0.03	0.33	0.2
Molasses, beet, 79 deg. brix	77	0.780	0.767	0.503	75	8.7	0	0	0.21	0.04	—
Molasses, citrus	65	0.798	0.798	0.599	77	10.9	0	0	2.01	0.14	—
Molasses, sugarcane, dehydrated	96	0.698	0.676	0.417	68	10.7	5	—	0.87	0.29	—
Molasses, sugarcane, 79.5 deg. brix	75	0.744	0.726	0.467	72	4.3	0	0	1.19	0.11	—
Monosodium phosphate	87	0	0	0	0	0	0	0	0	25.80	0
Oats, grain	89	0.789	0.785	0.517	76	13.6	12	17	0.07	0.39	0
Oat hay	88	0.621	0.594	0.317	61	9.2	31	36	0.26	0.24	12.0
Oat silage, boot stage	20—30	0.635	0.603	0.331	62	—	—	—	—	—	—
Oat silage, dough stage	25—35	0.599	0.576	0.281	59	9.7	34	—	0.47	0.33	—
Oyster shells, ground	100	0	0	0	1.0	1.0	0	0	38.22	0.07	0
Phosphate rock, defluorinated	100	0	0	0	0	0	0	0	31.65	13.70	0
Pineapple bran	87	0.758	0.739	0.481	73	4.6	18	—	0.24	0.12	—
Rice bran	91	0.679	0.653	0.390	66	14.0	12	16	0.07	1.62	—
Safflower meal, solv-extd	92	0.558	0.531	0.218	55	23.9	34	—	0.37	0.80	—
Sodium tripolyphosphate	96	0	0	0	0	0	0	0	0	25.98	0
Sorghum silage, grain variety	25—40	0.558	0.531	0.218	55	8.3	26	—	0.32	0.18	—
Soybean meal, solv-extd, 44% protein	89	0.844	0.857	0.572	81	49.6	7	10	0.36	0.75	—

Soybean meal, solv-extd, 46% protein	89	0.844	0.857	0.572	81	51.8	5	8	0.36	0.75	—
Soybean meal, solv-extd, 48% protein	89	0.844	0.857	0.572	81	54.0	3	4	0.36	0.75	—
Sudangrass hay	89	0.599	0.571	0.281	59	11.0	29	42	0.56	0.31	10.7
Sudangrass silage	20—30	0.599	0.571	0.290	59	11.1	34	—	0.48	0.19	19.0
Timothy hay, early bloom	88	0.635	0.603	0.331	62	10.0	32	37	0.53	0.26	9.7
Timothy hay, late bloom	88	0.558	0.531	0.218	55	7.7	33	43	0.38	0.18	1.8
Timothy silage, 25—40% DM (same values as for Timothy hay of same maturity)											
Urea, 46% N	90	0	0	0	0	287.5	0	0	0	0	0
Vetch hay	88	0.635	0.603	0.331	62	19.0	31	43	1.18	0.34	79.1
Wheat bran	89	0.721	0.694	0.435	70	18.0	11	12	0.12	1.32	0.5
Wheat grain, soft	86	0.921	0.975	0.644	88	12.0	3	4	0.06	0.41	—
Wheat millrun	90	0.767	0.726	0.494	74	17.0	9	—	0.10	1.13	—

Note: Data from *Nutrient Requirements of Dairy Cattle*, 5th rev. ed., National Academy of Sciences — National Research Council, Washington, D.C., 1978.

From Bath, D. L., Dickinson, F. N., Tucker, H. A., and Appleman, R. D., *Dairy Cattle: Principles, Practices, Problems, Profits*, 3rd ed., Lea & Febiger, Philadelphia, 1985. With permission.

Table 339
SOLUBLE AND DEGRADABLE PROTEIN IN VARIOUS FEEDSTUFFS

| | | % of crude protein | | |
Ingredient	Crude protein	Soluble	ADF bound	Insoluble available[a]
High solubility feeds				
Corn sol. + germ meal and bran	29.4	63	2.8	35 (4)
Corn gluten feed	22.2	55	2.6	43 (1)
Rye middlings	18.3	48	2.0	50 (4)
Wheat middlings	18.4	37	2.3	61 (4)
Wheat flour	15.2	40	0.2	64 (4)
Intermediate solubility feeds				
Oats	12.9	31	4.8	64 (4)
Corn gluten feed w/corn germ	23.8	32	3.3	65 (2)
Cotton waste product	22.3	24	1.6	74 (?)
Hominy	11.0	24	2.8	73 (1)
Soybean meal	52.3	24	1.8	75 (4)
Soybean mill feed	15.2	22	14—20	57—63 (3)
Distillers DG w/sol.	29.1	19	15.3	65 (1)
Low solubility feeds				
Cottonseed meal	44.3	12	3.1	85 (?)
Corn	9.6	15	5.0	80 (1)
Brewer's dried grains	28.9	6	13.2	81 (1)
Distillers DG	26.7	6	18.8	76 (1)
Corn gluten meal	66.2	4	10.6	85 (2)
Beet pulp	8.5	3	10.9	86 (?)
Forages and silages				
Alfalfa hay	15—25	30	10	60 (4—5)
Alfalfa, dehydrated	17—25	25—30	10—30	40—75 (3—5)
Alfalfa silage[b]	17—25	30—60	15—40	0—50 (3—5)
Corn silage[b]	9	30—40	10—30	30—60 (1)

[a] Degree of degradability of insoluble available fraction: 1 — very resistant, 5 — very degradable.
[b] Higher moisture silages tend to have greater soluble NPN, while low DM silages may have greater heat damage.

Reprinted from: Peter J. Van Soest, *Nutritional Ecology of the Ruminant*. Copyright© 1982 by P. J. Van Soest. Used by permission of the publisher, Cornell University Press.

Table 340
DIGESTIBILITY OF CELLULOSE IN VARIOUS MATERIALS

Material	Nonruminant digestibility (%)	Ruminant digestibility (%)	Lignin/cellulose ratio[a]
Alfalfa	20—30	40—60	0.18—0.30
Temperate grass	0—20	48—90	0.08—0.20
Tropical grass	0—20	30—60	0.11—0.24
Straws	Negligible	40—60	0.10—0.26
Soybean hulls	40	94	0—0.03
Cottonseed hulls	Very low	50	0.55
Rice hulls	0	0	0.45
Common newsprints	0	23—37	0.34—0.43
All papers	Low	20—99	0—0.50
Wood	0	0—40	0.30—0.60
Vegetables	40—80	90—100	0—0.05

[a] Acid-detergent sulfuric acid lignin, including cutin fractions, expressed as a ratio to cellulose. A high ratio is associated with low digestibility.

Reprinted from: Peter J. Van Soest, *Nutritional Ecology of the Ruminant.* Copyright© 1982 by P. J. Van Soest. Used by permission of the publisher, Cornell University Press.

Table 341
MICROELEMENT PROBLEMS CAUSED BY DEFICIENCIES OR EXCESSES IN HERBAGE, AND REQUIREMENT LEVELS OF GROWING CATTLE

Element	Disease	Countries	Dietary requirement (ppm)	Condition
Cobalt	Bush sickness, enzootic marasmus, pining, nakuruitis, Grand Traverse disease, vosk, likzucht, hinsch, sukhotka	U.S., New Zealand, Australia, W. Europe, U.S.S.R., E. Africa	0.1	Co deficiency, resulting in impaired vitamin B_{12} synthesis
Copper	Enzootic ataxia, swayback, falling disease, pasture diarrhea (scouring disease), peat scours	Australia, New Zealand, W. Europe	8—14	Simple or conditioned (Mo,S) deficiency of Cu
	Teart, molybdenosis	U.K., western U.S.		High levels Mo in herbage
	Salt sick, coast disease	U.S., Australia		Dual Cu, Co deficiency
	Cu poisoning (toxemic jaundice)	Australia		Excess Cu or Cu/Mo imbalance
Iodine	Goiter	Worldwide	0.5	Simple iodine deficiency or goitrogens in plants
Selenium	White muscle disease, stiff lambs, ill thrift, infertility, neonatal mortality	Widespread, especially in New Zealand and U.S.	0.03—0.05	Deficiency of Se and/or vitamin E
	Alkali disease, blind staggers	U.S., Canada, S. America, Ireland, Israel, U.S.S.R., Australia		Poisoning due to excess Se in pasture
Manganese		Netherlands, U.K.	10—20	Conditioned deficiency

From Heath, M. E., Barnes, R. F., and Metcalfe, D. S., *Forages, The Science of Grassland Agriculture,* 4th ed., Iowa State University Press, Ames, 1985. With permission.

Table 342
METABOLIZABLE ENERGY REQUIRED BY DAM AND PROGENY MEAT ANIMALS AND POULTRY

| | Dam | | | | | Progeny[a] | | | | | Total Mcal ME |
| | | | | ME — Mcal | | | Livewt. | | | | |
	Livewt. (kg)	Time (days)	Progeny (number)	total	per capita	Age/market (days)	initial[b]	market	ME	per capita	kg/live-weight
Beef cow[c]	500	365	1	5900	500	600	150	500	7020	12920	25.8
Ewe	60	365	1	1188	1188	180	25	50	274	1458	29.1
Ewe	60	365	2	1278	639	200	20	45	320	959	21.3
Sow	135	200	7[d]	1649	235	180	12	100	1141	1376	13.8
Sow	170	365	14[e]	2963	212	180	12	100	1141	1353	13.53
Hen[f]	4	300	100	102	1.02	50	0.04	1.75	11.7	12.79	73.2
Turkey	5	240	80	117	1.44	90	0.045	5	39.27	40.71	8.0
Turkey	8	240	60	157	2.62	150	0.05	10	92.85	95.47	9.5

[a]   We have assumed 55% ME available for deposit in gain, 60% above maintenance in milk and wool, and 70% for eggs. The equation for cattle is $MEI = T (0.133W^{.75})$. For other species, $MEI = T (0.120W^{.75}) = ME$ for maintenance. For gain, ME content for cattle and pigs is assumed to be 5 Mcal/kg; wool 5 Mcal/kg; lamb 4 Mcal/kg; broilers 2 Mcal/kg; turkeys 3 Mcal/kg. For cows' milk 0.65 Mcal/kg; ewe's milk 1.1 Mcal/kg; sow's milk 1.2 Mcal/kg. For eggs 1.5 Mcal/kg. [5,12,31,45,57,99]

[b]   "Initial" weight produced by milk or egg.

[c]   Beef example assumes steer calf 150 days suckling, 300 days grass and harvested roughage, 150 days in feedlot; finished at choice.

[d]   One litter.

[e]   Two litters.

[f]   Broiler breeder.

Table 343
DAILY METABOLIZABLE ENERGY REQUIRED FOR MAINTENANCE OF LIVESTOCK AND POULTRY

Animal	Liveweight (kg)	A W$^{3/4a}$ (kg)	Estimate published
Cow	350	10.76	10.76
Cow	450	12.99	13
Cow	600	16.09	16.12
Cow	700	18.09	17.10
Bull	600	16.09	18.29
Bull	800	20.00	21.00
Bull	1000	23.67	26.83
Calf	150	5.69	5.60
Calf	200	7.08	7.00
Calf	300	9.80	9.50
Ewe	60	2.59	2.60
Sow	200	6.38	6.34
Hen	1.75	0.182	0.18[b]
Hen	2.5	0.242	0.252[b]
Hen	4.0	0.34	0.307[b]
Turkey hen	5.0	0.401	
Turkey hen	8.0	0.571	

[a] "A" is 0.133 for cattle, 0.12 for sheep, pigs, chickens and turkeys. See text.

[b] Calculated for diet containing 2.9 Mcal/ME/kg.

Note: References included in the original.

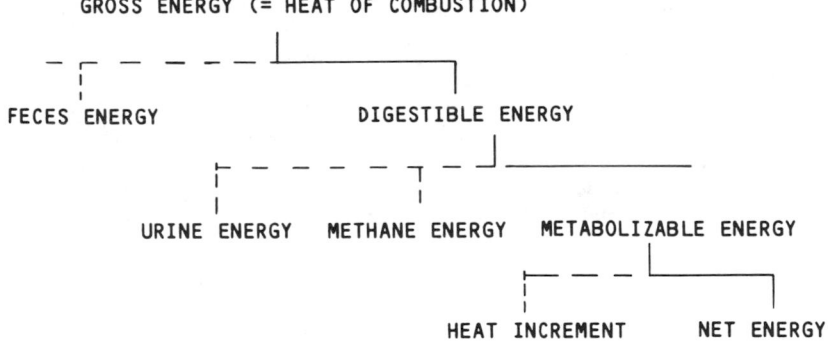

FIGURE 18.   Partition of food energy in the animal.

Table 344
APPROXIMATE PROPORTIONS OF THE TOTAL ENERGY REQUIREMENTS OF ANIMALS WHICH ARE CONTRIBUTED TO BY THEIR REQUIREMENTS FOR MAINTENANCE

	Daily values (kcal net energy) required for:		Maintenance as a percentage of total
	Maintenance	Production	
Dairy cow weighing 500 kg and producing 20 kg milk	8,000	18,400	30
Steer weighing 300 kg and gaining 1 kg	7,000	3,700	65
Pig weighing 50 kg and gaining 0.75 kg	1,730	2,320	43
Fowl weighing 1000 g and gaining 27 g	120	60	66

From McDonald, P., Edwards, R. A., and Greenhalgh, I.F.D., *Animal Nutrition,* 2nd ed., Oliver & Boyd, Edinburgh, 1973, 262. With permission.

Table 345
EFFICIENCY OF UTILIZATION OF METABOLIC ENERGY FOR MAINTENANCE AND PRODUCTION BY CATTLE, SWINE, AND POULTRY

Animal	$\dfrac{\text{Energy retention}}{\text{Metabolic energy intake}} \times 100$
Cattle	
Maintenance	0.60—0.75
Muscular work	0.20—0.30
Body-weight gain	
In veal calves	0.68
In growing animals	0.25—0.85
Fat storage in milk cows	0.75
Fetal tissues in pregnant cows	0.10—0.20
Milk production	
From food	0.60—0.75
From body fat	0.82
Sheep	
Maintenance	0.54
Body-weight gain	0.35—0.48
Swine	
Body-weight gain	0.59—0.78
Protein gain	0.36—0.77
Fat gain	0.70—0.75
Poultry	
Maintenance	0.71—0.92
Body-weight gain (growth)	0.60—0.80
Protein gain	0.51
Fat gain	0.70—0.84
Egg production	0.60—0.80
Egg protein	0.44
Egg fat	0.74

Note: References included in the original.

Table 346
PROTEIN REQUIREMENT — GRAMS PROTEIN (CP) PER MCAL ME[a]

Required for:	Beef cattle	Milk cows	Pigs	Sheep	Chickens	Turkeys
Growth	50	40	50	45	60	70
Finishing[b]	30	—	40	45	50	43
Late pregnancy	30	44	45	45	—	—
Lactation	50	50[c]	45	45	—	—
Egg production	—	—	—	—	46[d]	46
Adult males	40	—	45	40	42	42

[a] CP = crude protein.
[b] Castrates: steers, wethers, barrows. Young males: cocks and toms.
[c] 20 kg 3.5% fat milk from 600 kg cow. Calculated from data in *Nutrient Requirements of Domestic Animals*, No. 3, 5th rev. ed., National Academy of Sciences, Washington, D.C., 1978.
[d] 60% egg production.

Table 347
RELATIVE FOOD CAPACITY OF ANIMALS

		Maximum daily intake of food energy			
Animal	Body weight (kg)	kcal per animal	kcal per kg$^{3/4}$	Fasting energy loss (kcal/kg$^{3/4}$)	Relative food capacity
Chick	0.078	53.5	360	81	4.4
Rabbit	2.36	480	253	50	5.1
Sheep	50	5730	305	69	4.4
Swine	130	13980	363	64	5.7
Steer	427	42429	452	81	5.6
	444	36026	373	88	4.2

From Kleiber, M., *Fire of Life,* John Wiley & Sons, New York, N.Y., 1961. Copyright ©1961 by John Wiley & Sons. With permission.

Table 348
EXAMPLE OF THE INDEPENDENCE OF BODY SIZE AND FOOD UTILIZATION

	Animals tested	
	1 steer	300 rabbits
Total body weight	1,300 lb	1,300 lb
Food consumption per day	16⅔ lb	66⅔ lb
Duration of 1 ton of food	120 days	30 days
Heat loss per day	20,000 kcal	80,000 kcal
Gain in weight per day	2 lb	8 lb
Gain from 1 ton of food	240 lb	240 lb

From Kleiber, M., *Fire of Life,* John Wiley & Sons, New York, N.Y., 1961. Copyright ©1961 by John Wiley & Sons. With permission.

ENVIRONMENT

Table 349
TEMPERATURE, HUMIDITY, VENTILATION, AND SPACE REQUIREMENTS FOR CATTLE, SWINE, AND POULTRY

	Temperature range (°F)	Relative humidity (%)	Ventilation rate	Space
Dairy cattle				
Stall barns	35—75	40—80	W 36—47 cfm[a,b] S/F 142—190 cfm S 230—470 cfm	50—75 ft² per cow
Free-stall				75—100 ft² per cow
Swine barns				
Farrowing	50—68[c]	Up to 75 (maximum)	— [d]	34 ft² (stall)[e] 75 ft² (pens)[e]
Nursery	79—72[f]	—	3—38 cfm[g]	1.7—3.9 ft²[h]
Growing/gestation	55—72	75[i]	— [d]	5.8 ft² (75—150 lb)[j] 7.8 ft² (150—220 lb)[j] 14—23.7 ft² per sow 240 to 500 lbs
Poultry houses				
Broiler	60—80[k]	50—80	W 0.1 cfm/lb[l] S 1—2 cfm/lb[l]	0.6—1.0 ft² per bird[m]
Breeder (litter and slatted floors)	50—86	50—75	See broilers	2—3 ft² per bird
Laying (cages)	50—86	50—75	See broilers	50—75 in² per hen[n] (minimum)

[a]   W, winter; S/F, spring/fall; S, summer.
[b]   Beef cattle ventilation requirements are similar to those of dairy cattle on a unit weight basis.
[c]   Small areas warmed for pigs to 82—90°F.
[d]   See Table 350.
[e]   Per sow and litter.
[f]   79°F for first week after weaning. Lower 3°F per week to 72°F.
[g]   Per pig 13 to 79 lbs.
[h]   Per pig 13 to 31 lbs.
[i]   Maximum winter, no established limit in summer.
[j]   Per pig.
[k]   Temperature under brooder hover 86—91°F, reducing 5°F per week to reach room temperature.
[l]   Live weight.
[m]   Light minimum of 10 lux to 23 days of age, 1 to 20 lux for growout (in enclosed housing).
[n]   Controlled day length using light-controlled housing is generally practiced (January through June).

From Heating, Ventilating, and Air-Conditioning Systems and Applications, in *1987 ASHRAE Handbook*, American Society of Heating, Refrigerating and Air Conditioning Engineers, Inc., Atlanta, 1987. With permission.

Table 350
TOTAL PER HEAD VENTILATION RATES FOR CONFINEMENT SWINE BUILDING DURING VARIOUS TIMES OF THE YEAR (CFM)

Type of animal	Cold weather	Mild weather	Hot weather
Sow and litter	20	80	500
Prenursery pig (12—30 lb)	2	10	25
Nursery pig (30—75 lb)	3	15	35
Growing pig (75—150 lb)	7	24	75
Finishing hog (150—220 lb)	10	35	120
Gestation sow (325 lb)	12	40	150[a]
Boar (400 lb)	14	50	180[a]

[a] Use 300 per sow or boar in breeding facilities because of low animal density.

Source: Jones, D. D., Driggers, L. B., Fehr, R. L., and Stewart, W. R., Cooling swine, in *Pork Industry Handbook,* P1H-87, Cooperative Extension Service, Purdue University, West Lafayette, Ind., June 1983.

Table 351
AVERAGE DAILY TEMPERATURES FOR NOMINAL LOSSES IN PRODUCTION AND EFFICIENCY OF LIVESTOCK

Animal	Acceptable temperature range[a]
Dairy cattle —	
lactating or within 2 weeks of breeding	4—24°C
Calves	10—26°C
Beef cattle	4—26°C
Sheep	4—24°C
Hogs	
Weaning to market weight	10—24°C[b]
Farrowing sows	10—16°C
Poultry	
Growing (over 10 days old)	13—27°C[b]
Laying hens	7—21°C

[a] Acceptable average daily temperatures for long-term exposure (concurrent relative humidity less than 75%). The range should be shifted downward at least 3°C for high radiant heat loads (greater than 1 cal/cm²/min). Nutrition, management, housing, and other factors can also alter the acceptable range.

[b] Optimum temperature shifts downward within this range as weights increase.

From Hahn, G. L., in *Progress in Animal Biometeorology,* Johnson, H. D., Ed., Swets & Zeitlinger, Amsterdam, 1976, 496. With permission.

Table 352
OVERALL THERMAL CONDUCTANCE, COMPONENTS OF THERMAL INSULATION, MINIMUM EVAPORATIVE HEAT LOSS, AND THERMONEUTRAL HEAT PRODUCTION, AND THE CALCULATED CRITICAL TEMPERATURES OF VARIOUS DOMESTIC ANIMALS

| Species and breed | Weight (kg) | Surface area (m²) | Overall conductance (kcal/m² 24 hr) | Coat type | Insulation (°C × m² × 24 hr/Mcal) | | | Heat production (Mcal/m² × 24 hr) | | Minimum evaporative heat loss (Mcal/m² × 24 hr) | Calculated critical temperature (°C) | |
					Tissue	Coat	Air	At maintenance	Fully fed		At maintenance	Fully fed
Pig												
Large white	1.5	0.13	133	Normal	2.1	—	8.3ª	1.6	2.8	0.16	27	18
Large white	4	0.23	100	Normal	—	—	—	1.6	2.8	0.09	23	11
Large white	8	0.36	100	Normal	—	—	—	1.6	2.8	0.10	23	11
Large white	10	0.42	100	Normal	—	—	—	1.6	2.8	0.11	23	11
Landrace	23	0.71	98	Normal	4.6	—	7.2	1.6	2.8	0.11	23	11
Large white	150	2.31	36	Normal	—	—	—	1.4	2.5	—	2	−28
Sheep												
Merino	5	0.26	58	Fine	8.2	8.5		1.2	1.6	0.48	26	19
Merino	5	0.26	43	Coarse	7.5	14.7		1.2	1.4	0.48	21	17
Down	45	1.14	81	Clipped	3.5	nilᵇ		1.3	2.3	0.31	26	13
Down	45	1.14	31	5 cm	3.5	27.9	7.7	1.3	2.3	0.31	6	−26
Blackface	45	1.14	68	Clipped	5.7	nil		1.3	2.3	0.24	22	8
Blackface	45	1.14	29	5 cm	5.7	27.9	7.7	1.3	2.3	0.24	1	−33
Cattle												
Ayrshire	42	1.08	53	Normal	4.4	7.7	5.7	1.9	3.0	0.34	9	−12
Jersey	390	4.78	62	Normal	—	—	—	2.6	3.7	0.28	1	−17
Brahman	470	5.42	62	Normal	—	—	—	2.6	3.7	0.24	0	−18
Brown Swiss	550	6.02	52	Normal	—	—	—	2.6	3.7	0.21	−8	−29
Galloway	500	5.07	41	Normal	6.7	16.2		2.0	3.2	0.35	−2	−32
Hereford												
Steer	365	4.59	44	Severe winter	8.3	14.3		—	3.6	—	—	−48
Pregnant cow	500	5.78	36	Severe winter	12.7	15.3		2.2	—	—	−18	—
Calf	210	3.24	46	Severe winter	7.4	14.3			2.9ᶜ			−19ᶜ

Note: References included in the original.

ª   The data in this column apply for combined coat and air insulation.

ᵇ   nil = no coat.

ᶜ   The data in this column apply to moderately fed animals.

Table 353
APPARENT DIGESTIBILITY OF RATION DRY MATTER (DM)
IN WARM AND COLD ENVIRONMENTS

Animal	Ration	Exposure temp. (C)	Apparent DM dig. (C)	Change in % DM dig. per 1°C
Sheep	Alfalfa pellets	21	52.0	
		−6.5	44.5	−0.27
	Alfalfa-grass pellets	20	55.3	
		−8	48.4	−0.25
	Grain and alfalfa pellets	20	69.0	
		−8	57.9	−0.40
Calves	Grain and chopped alfalfa	18	70.4	
		−10[a]	65.1	−0.19
	Grain and chopped alfalfa	18	69.8	
		−9[a]	60.7	−0.34
Cows	Alfalfa-grass long hay	21	61.3	
		−11	61.6	−0.01
			mean ± SE	0.24 ± 0.06

[a] Average temperature of outdoor environment.

From Young, D. A. and Christopherson, R. J., in *Proc. Livestock Environment Symp. ASAE,* American Society of Agricultural Engineers, St. Joseph, Mich., 1974. With permission.

Table 354
HEAT-PRODUCTION VALUES FOR SELECTED
CATTLE UNDER NORMAL CONDITIONS

Species	Rectal temperature (°C)	Body weight (kg)	Heat production (kJ/hr)	Remarks
Brahman				
Lactating	38.1	331	2053	Resting
Dry	38.4	431	2095	Resting
Jersey				
Lactating	38.1	376	2597	Resting
Holstein				
Lactating	38.2	553	3561	Resting
Brown Swiss				
Lactating	38.0	608	3771	Resting
Dry	38.6	190	1466	Resting

Note: Normal conditions = average values for each breed between 4.4 and 15.6°C environmental temperature, the apparent comfort zone for these cows.

Source: Worstell, D. M. and Brody, S., *Mo. Agric. Exp. Stn. Res. Bull.,* No. 515, 1953, 1—42.

Table 355

SUMMARY OF STEER PERFORMANCE BY SEASONS AND MEAN SEASONAL CLIMATIC CONDITIONS

Month	Air temperature (°C)	Days per month below −23°C	Dew point (°C)	Wind chill (kcal/ m²/ hr)	Wind velocity (km/ hr)	Precip- itation (cm/ month)	Average weight (kg)	Average daily gain (kg)	Actual/ predicted gain[a]	Daily feed (kg)	Feed/ gain[b]	Daily kcal ME/ kg body weight
December to February	−17	11	−15	1302	15	2.2	419	1.03	0.99	8.95	9.8	56.7
March to May	2	1	−3	837	16	1.8	390	1.33	1.18	9.18	7.2	59.7
June to August	17	0	11	420	17	7.0	372	1.51	1.51	7.97	5.6	54.3
September to November	3	1	1	767	15	3.2	431	1.57	1.30	10.68	6.9	60.1

[a] Calculated on the basis of net energy intake (National Research Council, 1970).

[b] Feed per gain values are the average of results for individual pens and were not calculated from the overall means of feed and gain.

From Christison, G. I. and Milligan, J. D., in *Livestock Environment*, American Society of Agricultural Engineers, St. Joseph, Mich., 1974, 296. With permission.

Table 356

WEIGHT GAIN COMPARISONS OF CATTLE AT TWO ENVIRONMENTAL TEMPERATURES

Species	Air temperature (°C)	Weight at 4 months (kg)	Weight at 12 months (kg)	Total gain for 8-month period (kg)	Difference in gain from 10°C values (kg)
Santa Gertrudis	10	126.1	342.9	216.8	
	27	128.4	313.0	184.6	−32.2
Brahman	10	112.9	285.8	172.8	
	27	116.6	297.6	181.0	+ 8.2
Shorthorn	10	93.0	298.5	201.9	
	27	73.9	209.1	135.2	−66.7
Holstein	10	103.5	333.3	203.3	
	27	95.3	302.7	207.4	−22.9
Brown Swiss	10	74.2	303.3	229.1	
	27	89.0	310.2	221.2	− 7.9
Jersey	10	65.5	210.0	144.5	
	27	56.7	197.3	140.6	−3.9

Note: References included in the original.

Table 357

TEMPERATURE EFFECTS ON WEIGHT GAIN, FEED CONSUMPTION, AND FEED EFFICIENCY OF STEERS

Species	Air temperature (°C)	Rectal temperature (°C)	Weight gain (kg/day)	Feed consumption (kg/day)	Efficiency feed/gain (kg)	Remarks
Hereford × Angus steers[a]	32.6	39.4	1.16	7.56	6.54	Outside continuously, seasonal heat in Imperial Valley, Calif.
	23.9	39.1	1.29	8.46	6.77	Inside continuously, cooled 24 hr/day

[a] Average of four animals per treatment for 3 years.

Table 358
PERFORMANCE OF STEERS AND HEIFERS EXPOSED TO SEVERE COLD WITH VARYING DEGREES OF SHELTER

Animals	Month	Weight (kg)	Mean temperature (°C)	Average daily gain (kg)	Feed intake (kg)	Feed/ gain
Feedlot steers[a]	December to February	419	−17	1.03	8.95	9.8
	March to May	390	2	1.33	9.18	7.2
	June to August	372	17	1.51	7.97	5.6
	September to November	431	3	1.57	10.68	6.9
Heifers						
Control[b]	November to March	230	20	0.63	4.73	—
Shelter[c]	November to March	225	−16	0.59	5.25	—
Exposed[d]	November to March	218	−16	0.54	5.14	—

Note: References included in the original.

[a] Outdoors: porous fence windbreak, manure-pack bedding.
[b] Heated barn.
[c] Outdoors: roofed open-front shed and dry lot, straw bedding.
[d] Outdoors: dry lot, straw bedding.

PRODUCTION AND FEEDING

CATTLE

BEEF CATTLE

Table 359
DAILY TDN, PROTEIN, Ca, P, AND VITAMIN A REQUIREMENTS FOR GROWING-FINISHING STEER CALVES AND YEARLINGS (DRY MATTER BASIS)

Weight (lb)[a]	Daily gain (lb)	Total protein (lb)	Digestible protein (lb)	TDN (lb)	Ca (g)	P (g)	Vit. A (thousand IU)
220	0	0.40	0.22	2.6	4	4	5
	1.1	0.79	0.53	4.0	14	11	6
	1.5	0.88	1.62	4.4	19	13	6
	2.0	1.01	0.73	4.6	24	16	7
	2.4	1.08	0.79	5.1	28	19	7
331	0	0.51	0.29	3.5	5	5	6
	1.1	0.97	0.62	5.5	14	12	9
	1.5	1.08	0.73	6.0	18	14	9
	2.0	1.19	0.81	6.6	23	17	9
	2.4	1.28	0.90	6.8	28	20	9
441	0	0.66	0.37	4.2	6	6	8
	1.1	1.25	0.77	7.5	14	13	12
	1.5	1.34	0.86	7.9	18	16	13
	2.0	1.34	0.88	8.2	23	18	13
	2.4	1.39	0.95	8.6	27	20	13
551	0	0.77	0.44	5.1	8	8	9
	1.5	1.36	0.86	8.8	18	16	14
	2.0	1.52	0.97	9.9	22	19	14
	2.4	1.61	1.06	10.4	26	21	14
	2.9	1.67	1.12	11.5	30	23	14
661	0	0.88	0.51	5.7	9	9	10
	2.0	1.78	1.10	11.9	22	19	16
	2.4	1.80	1.14	12.3	15	22	16
	2.9	1.83	1.18	13.2	29	23	16
	3.1	1.91	1.25	13.7	31	25	16
772	0	1.01	0.57	6.4	10	10	12
	2.0	1.76	1.08	12.8	20	18	18
	2.4	1.83	1.14	13.7	23	20	18
	2.9	1.91	1.21	15.0	26	22	18
	3.1	1.98	1.25	15.4	28	24	18
882	0	1.12	0.64	7.3	11	11	13
	2.2	1.91	1.19	15.0	21	20	19
	2.6	1.91	1.19	15.4	23	21	19
	2.9	1.98	1.23	16.1	25	22	19
	3.1	2.07	1.30	17.0	26	23	19
992	0	1.19	0.68	—	12	12	14
	2.2	2.11	1.25	16.3	20	20	20
	2.6	2.13	1.28	17.4	23	22	20
	2.9	2.13	1.30	17.6	24	23	20
	3.1	2.16	1.32	18.5	25	23	20
1102	0	1.32	0.75	8.4	13	13	15
	2.0	2.09	1.23	16.5	19	19	23
	2.4	2.11	1.25	17.8	20	20	23
	2.6	2.11	1.28	18.1	21	21	23
	2.9	2.13	1.32	19.2	22	22	23

Note: Data from National Research Council (NRC), 1976.

[a] Weights taken from data in kilograms.

From Price, D. P., *Beef Production — Science and Economics. Application and Reality,* Southwest Scientific, Dalhart, Tex., 1981. With permission.

Table 360
DAILY TDN, PROTEIN, Ca, P, AND VITAMIN A REQUIREMENTS FOR GROWING-FINISHING HEIFER CALVES AND YEARLINGS (DRY MATTER BASIS)

Weight (lb)[a]	Daily gain (lb)	Total protein (lb)	Digestible protein (lb)	TDN (lb)	Ca (g)	P (g)	Vit. A (thousands IU)
220	0	0.40	0.22	2.6	4	4	5
	1.1	0.81	0.55	4.2	14	11	6
	1.5	0.92	0.64	4.6	19	14	6
	2.0	1.06	0.75	5.1	24	17	7
	2.4	1.17	0.86	5.5	29	19	7
331	0	0.53	0.31	3.5	5	5	6
	1.1	0.99	0.64	5.7	14	12	9
	1.5	1.10	0.73	6.2	18	14	9
	2.0	1.19	0.81	6.8	23	17	9
	2.4	1.32	0.92	7.5	28	20	9
441	0	0.66	0.37	4.2	6	6	8
	0.7	1.08	0.64	6.6	10	10	12
	1.1	1.28	0.77	7.7	14	13	13
	2.0	1.36	0.88	8.8	22	17	13
	2.4	1.41	0.97	9.5	25	19	13
551	0	0.77	0.44	5.1	7	7	9
	0.7	1.25	0.73	7.8	12	12	14
	1.1	1.36	0.81	8.6	13	13	14
	1.5	1.36	0.84	9.1	17	15	14
	2.0	1.43	0.92	10.1	21	17	14
	2.4	1.63	1.06	11.5	25	20	14
	2.6	1.65	1.08	11.9	27	21	14
661	0	0.88	0.51	5.7	9	9	10
	0.7	1.39	0.79	8.4	13	13	16
	1.1	1.47	0.88	9.9	14	14	16
	1.5	1.47	0.88	10.4	16	15	16
	2.0	1.54	0.97	11.5	19	17	16
	2.4	1.72	1.08	13.2	23	20	16
	2.6	1.74	1.10	13.7	24	20	16
772	0	1.01	0.57	6.4	10	10	12
	0.7	1.62	0.86	10.0	15	15	18
	1.1	1.61	0.92	11.2	15	15	18
	1.5	1.61	0.95	11.9	15	15	18
	2.0	1.69	1.01	13.2	17	17	18
	2.4	1.78	1.10	14.5	20	19	18
	2.6	1.78	1.10	15.2	21	20	18
882	0	1.12	0.64	7.3	11	11	13
	0.7	1.67	0.95	11.1	16	16	19
	1.1	1.72	0.95	11.9	15	15	19
	1.5	1.74	1.01	13.2	16	16	19
	2.0	1.74	1.03	14.3	17	17	19
	2.4	1.78	1.08	15.9	19	18	19
992	0	1.21	0.68	7.9	12	12	14
	0.4	1.63	0.90	10.6	16	16	19
	1.1	1.76	1.01	13.0	17	17	20
	1.8	1.80	1.06	15.0	16	16	10
	2.2	1.83	1.06	16.3	19	19	20

Note: Data from NRC, 1976.

[a] Weights taken from data in kilograms.

From Price, D. P., *Beef Production — Science and Economics. Application and Reality,* Southwest Scientific, Dalhart, Tex., 1981. With permission.

DAIRY CATTLE

Table 361
DAIRY BREEDS IN THE UNITED STATES

Breed	Estimated % of total U.S. population	Estimated no. of living registered cows in U.S.	Av no. animals registered annually	Av production, U.S. Official DHI herds (1981)						Mature cow body wt (lb)	Newborn calf wt (lb)
				No. herds	No. cows	Milk[a] (lb)	Fat[a] %	Fat[a] lb		
Ayrshire	1	85,000	10,500 females 475 males	427	18,456	12,226	3.9	481	1,200	75
Brown Swiss	1	100,000	12,500 females 1,100 males	561	26,427	12,665	4.1	516	1,500	95
Guernsey	2	164,000	20,500 females 750 males	985	54,954	11,222	4.7	523	1,100	75
Holstein	90	2,824,000	353,000 females 29,000 males	32,404	2,781,387	15,480	3.6	562	1,500	95
Jersey	5	370,000	46,200 females 2,000 males	1.486	105,193	10,608	4.9	516	1,000	60
Milking shorthorn	1	30,000	3,500 females 1,100 males	84	3,054	11,162	3.6	405	1,250	75

[a] Average per cow.

Source: Data compiled from various sources, including: 4-H Dairy: Calves & Heifers, University of Minnesota Agricultural Extension Service Publ. B-10, 1975; various breed association annual reports; DHI Letter, ARS-USDA, 59(1), 1983, and USDA-DHIA AI Sire Summary List.

Adapted from Bath, D. L., Dickinson, F. N., Tucker, H. A., Appleman, R. D., *Dairy Cattle: Principles, Practices, Problems, Profits*, 3rd ed., Lea & Febiger, Philadelphia, 1985. With permission.

Table 362
RECOMMENDED NUTRIENT CONTENT OF RATIONS FOR DAIRY CATTLE

Nutrients (conc in feed dry matter)	Lactating cow rations					Maximum conc (all classes)
	Cow wt (lb)	Daily milk yields (lb)				
	≤900 1,100 1,300 ≥1,550	<18 <24 <31 <40	18—29 24—37 31—46 40—57	29—40 37—51 46—64 57—78	>40 >51 >64 >78	
Crude protein (%)		13.0	14.0	15.0	16.0	—
Energy						
Ne₁ (Mcal/lb)		0.64	0.69	0.73	0.78	—
NEₘ (Mcal/lb)		—	—	—	—	—
NEₘ (Mcal/lb)		—	—	—	—	—
ME (Mcal/lb)		1.07	1.15	1.23	1.31	—
DE (Mcal/lb)		1.26	1.34	1.42	1.50	—
TDN (%)		63	67	71	75	—
Crude fiber (%)		17	17	17	17	—
Acid detergent fiber (%)		21	21	21	21	—
Ether extract (%)		2	2	2	2	—
Minerals						
Calcium (%)		0.43	0.48	0.54	0.60	—
Phosphorus (%)		0.31	0.34	0.38	0.40	—
Magnesium (%)		0.20	0.20	0.20	0.20	—
Potassium (%)		0.80	0.80	0.80	0.80	—
Sodium (%)		0.18	0.18	0.18	0.18	—
Sodium chloride (%)		0.46	0.46	0.46	0.46	5
Sulfur (%)		0.20	0.20	0.20	0.20	0.35
Iron (ppm)		50	50	50	50	1000
Cobalt (ppm)		0.10	0.10	0.10	0.10	10
Copper (ppm)		10	10	10	10	80
Manganese (ppm)		40	40	40	40	1000
Zinc (ppm)		40	40	40	40	500
Iodine (ppm)		0.50	0.50	0.50	0.50	50
Molybdenum (ppm)		—	—	—	—	6
Selenium (ppm)		0.10	0.10	0.10	0.10	5
Fluorine (ppm)		—	—	—	—	30
Vitamins						
Vitamin A (IU/lb)		1450	1450	1450	1450	—
Vitamin D (IU/lb)		140	140	140	140	—

	Nonlactating cattle rations					
	Dry pregnant cows	Mature bulls	Growing heifers and bulls	Calf starter concentrate mix	Calf milk replacer	Max.
	V	VI	VII	VIII	IX	
Crude protein (%)	11.0	8.5	12.0	16.0	22.0	—
Energy						
Ne₁ (Mcal/lb)	0.61	—	—	—	—	—
NEₘ (Mcal/lb)	—	0.54	0.57	0.86	1.09	—
NEₘ (Mcal/lb)	—	—	0.27	0.54	0.70	—
ME (Mcal/lb)	1.01	0.93	1.01	1.42	1.71	—
DE (Mcal/lb)	1.20	1.12	1.20	1.60	1.90	—
TDN (%)	60	56	60	80	95	—
Crude fiber (%)	17	15	15	—	—	—

Table 362 (continued)
RECOMMENDED NUTRIENT CONTENT OF RATIONS FOR DAIRY CATTLE

| | Nonlactating cattle rations | | | | | |
| | Dry pregnant cows | Mature bulls | Growing heifers and bulls | Calf starter concentrate mix | Calf milk replacer | |
	V	VI	VII	VIII	IX	Max.
Acid detergent fiber (%)	21	19	19	—	—	—
Ether extract (%)	2	2	2	2	10	—
Minerals						
Calcium (%)	0.37	0.24	0.40	0.60	0.70	—
Phosphorus (%)	0.26	0.18	0.26	0.42	0.50	—
Magnesium (%)	0.16	0.16	0.16	0.07	0.07	—
Potassium (%)	0.80	0.80	0.80	0.80	0.80	—
Sodium (%)	0.10	0.10	0.10	0.10	0.10	—
Sodium Chloride (%)	0.25	0.25	0.25	0.25	0.25	5
Sulfur (%)	0.17	0.11	0.16	0.21	0.29	0.35
Iron (ppm)	50	50	50	100	100	1000
Cobalt (ppm)	0.10	0.10	0.10	0.10	0.10	10
Copper (ppm)	10	10	10	10	10	80
Manganese (ppm)	40	40	40	40	40	1000
Zinc (ppm)	40	40	40	40	40	500
Iodine (ppm)	0.50	0.25	0.25	0.25	0.25	50
Molybdenum (ppm)	—	—	—	—	—	6
Selenium (ppm)	0.10	0.10	0.10	0.10	0.10	5
Fluorine (ppm)	—	—	—	—	—	30
Vitamins						
Vitamin A (IU/lb)	1450	1450	1000	1000	1720	—
Vitamin D (IU/lb)	140	140	140	140	270	—
Vitamin E (ppm)	—	—	—	—	300	—

Note: Data from *Nutrient Requirements of Dairy Cattle,* 5th rev. ed., National Academy of Sciences — National Research Council, Washington, D.C., 1978.

From Bath, D. L., Dickinson, F. N., Tucker, H. A., and Appleman, R. D., *Dairy Cattle: Principles, Practices, Problems, Profits,* 3rd ed., Lea & Febiger, Philadelphia, 1985. With permission.

Table 363
EFFECT OF ENVIRONMENTAL TEMPERATURE ON MILK PRODUCTION AND FEED INTAKE OF LACTATING DAIRY COWS

Breed	Mean ambient (°C)	Relative humidity (%)	Milk production (kg/day)	Digestible energy intake (Mcal/day)
Holstein[a]	−7	90—100	21.3	52.2
	−18	90—100	20.6	58.7
Holstein[b]	7	72—79	21.5	—
	−17	72—79	21.3	—
Holstein[c]	−10	64—67	15	49.3
	−13	60—68	14	45.8
Jersey[c]	10	64—67	7.6	30.0
	−13	60—68	3.5	32.6

Note: References included in the original.

[a] Unheated barn with manure pack. Concentrate fed for milk production, hay allowed *ad lib.*
[b] Same barn as [a], but with increased ventilation. Concentrate: fed for milk production, roughage fed for maintenance.
[c] Climatic laboratory: ambient temperature progressively decreased over a 2-month period. Concentrate fed for milk production, hay allowed *ad lib.*

Table 364
EFFECTS OF −13°C AS COMPARED TO 10°C ON MILK PRODUCTION, TDN INTAKE, AND THE RATIO OF CONVERSION[a]

Breed	Milk (kg/day)		TDN (kg/day)		Ration of conversion (feed/milk)	
	10°C	−13°C	10°C	−13°C	10°C	−13°C
Holstein	15	14	10.4	11.2	0.70	0.80
Brown Swiss						
Jersey	7.6	3.5	5.9	7.4	0.78	2.1
Brahman			4.5	6.2		

Adapted from Johnson, H. D., in *Progress in Biometeorology,* Tromp, S. W., Ed., Swets & Zeitlinger, Amsterdam, 1972, 43. With permission.

Table 365
EFFECTS OF 38°C AS COMPARED TO 10°C ON MILK PRODUCTION, TDN INTAKE, AND RATIO OF CONVERSION

	Milk (kg/day)		TDN (kg/day)		Ratio of conversion (feed/milk)	
	10°C	38°C	10°C	38°C	10°C	38°C
Holstein	18.6	4.9	12.1	2.6	0.65	0.53
Brown Swiss	20.4	9.5	12.2	2.5	0.60	0.26
Jersey	13.5	5.1	8.8	2.8	0.65	0.55
Brahman	3.6	3.2	5.2	3.3	1.44	1.03

Note: $\dfrac{\text{Feed kg/day}}{\text{Milk kg/day}}$ = kilograms of feed (TDN) required to produce 1 kg milk.

[a] Data from References 68 and 69.

Adapted from Johnson, H. D., in *Progress in Biometeorology,* Tromp, S. W., Ed., Swets & Zeitlinger, Amsterdam, 1972, 43. With permission.

Table 366
EFFECT OF WIND ON MILK PRODUCTION AT FOUR TEMPERATURES

Milk production (% normal)

Wind mi/hr (km/hr)	−8°C			10°C			27°C			35°C		
	Holstein	Jersey	Brown Swiss	Holstein	Jersey	Brown Swiss	Holstein	Jersey	Brown Swiss	Holstein	Jersey	Brown Swiss
0.4 (0.64)	76	36	72	100	100	100	85	100	100	63	74	83
5.0 (8.05)	85	39	74	100	100	100	95	100	100	79	94	90
9.0 (14.48)	84	35	75	100	100	100	95	100	100	79	94	90

Note: RH range is 60 to 70% at all four temperatures.

From Johnson, H. D., in *Progress in Biometeorology*, Tromp, S. W., Ed, Swets & Zeitlinger, Amsterdam, 1972, 43. With permission.

Table 367
EFFECT OF RADIATION ON MILK PRODUCTION AT THREE AIR TEMPERATURES

Milk production (% normal)

Level of radiation (cal/cm²-min)	8°C		21°C		27°C	
	Holstein	Jersey	Holstein	Jersey	Holstein	Jersey
Variable	100	100	99	99	84	94
5 (0.02)	100	100	100	100	93	99
40 (0.19)	100	100	100	99	92	94
90 (0.42)	100	100	93	95	77	93
130 (0.60)	100	100	90	97	69	87
180 (0.84)	100	100	88	95	57	79

From Johnson, H. D., in *Progress in Biometeorology*, Tromp, S. W., Ed., Swets & Zeitlinger, Amsterdam, 1972, 43. With permission.

Table 368
WATER ABSORPTION OF BEDDING

Material	Lb water absorbed per lb bedding
Wood	
Tanning bark	4.0
Dry fine bark	2.5
Pine chips	3.0
Sawdust	2.5
Shavings	2.0
Needles	1.0
Hardwood chips, shavings, or sawdust	1.5
Corn	
Shredded stover	2.5
Ground cobs	2.1
Straw, baled	
Flax	2.6
Oats	2.5
Wheat	2.2
Hay, chopped mature	3.0
Shells, Hulls	
Cocoa	2.7
Peanut	2.5
Cottonseed	2.5
Oats	2.0
Sand	0.25

Source: Dairy Housing and Equipment Handbook, Publ. No. 7, Midwest Plan Service, Iowa State University, Ames, 1976.

From Bath, D. L., Dickenson, F. N., Tucker, H. A., and Appleman, R. D., *Dairy Cattle: Principles, Practices, Problems, Profits,* 3rd ed., Lea & Febiger, Philadelphia, 1985. With permission.

Table 369
ESTIMATED DAILY BEDDING
REQUIREMENTS FOR DAIRY COWS

Type of housing	Type of bedding (lb per cow per d)		
	Straw	Shavings[a]	Sawdust[b]
Free stalls	3—5	3—6	8—10
Stanchion			
Stalls with rubber mats	2—5	6—8	8—12
Barn-concrete floor	5—7	8—15	20—42
Bedded pack			
Loose housing[c]	7—12	15—30	40—50

Note: Data from Ace, D. L., *Dairy Reference Manual,* College of Agriculture, Extension Service, Pennsylvania State University, University Park, 1970.

[a] Kiln dried 8% moisture, weight basis.
[b] Fifty percent moisture (green), weight basis.
[c] With separate lots for exercise.

From Bath, D. L., Dickinson, F. N., Tucker, H. A., and Appleman, R. D., *Dairy Cattle: Principles, Practices, Problems, Profits,* 3rd ed., Lea & Febiger, Philadelphia, 1985. With permission.

Table 370
VOLUME OF MILKHOUSE AND PARLOR WASTES

Weekly operation	Water volume
Bulk tank	
Automatic	50—60 gal per wash
Manual	30—40 gal per wash
Piperline — Volume increases with line diameter or length	75—125 gal per wash
Pail milkers	30—40 gal per wash
Miscellaneous equipment	30 gal/d
Cow prep	
Automatic	1—4$^{1}/_{2}$ gal per wash per cow (estimated average = 2 gal)
Manual	$^{1}/_{4}$—$^{1}/_{2}$ gal per wash per cow
Parlor floor	40—75 gal daily
Milkhouse floor	10—20 gal daily

Note: Data from *Livestock Waste Facilities Handbook,* MWPS-18, Midwest Plan Service, Iowa State University, Ames.

From Bath, D. L., Dickinson, F. N., Tucker, H. A., and Appleman, R. D., *Dairy Cattle: Principles, Practices, Problems, Profits,* 3rd ed., Lea & Febiger, Philadelphia, 1985. With permission.

SWINE

Table 371
A SUMMARY OF OBSERVATIONS ON SWINE OBTAINED IN CONTROLLED TEMPERATURE ROOMS AND OBSERVATIONS AT SLAUGHTER 25 DAYS AFTER BREEDING

	Dry-bulb temperature (°C)		
	26.7°	30.0°	33.3°
Average daily feed intake (kg)	2.11	2.03	1.86
Average daily gain (kg)	0.47	0.46	0.35
Average rectal temperature (°C)	38.4	39.3	39.9
Breeding and reproductive performance			
Number returning to estrus after breeding	2	8	8
Number failing to come into estrus in rooms	0	2	7
Number not pregnant at slaughter (did not return to heat after breeding)	5	2	3
Number pregnant (25 days)	67	67	62
Average number corpora lutea (25 days)	14.2	13.6	13.1
Average number live embyros (25 days)	10.3	9.7	9.8
Percent pregnant (at 25 days)	90.5	84.8	77.5

Note: Summary is adjusted for differences in age and initial weight.

From Teague, H. S., Roller, W. L., and Grifo, A. P., Jr., *J. Anim. Sci.,* 27(2), 408, 1968. With permission.

Table 372

RELATION BETWEEN METABOLIC RATE AND AMBIENT TEMPERATURE FOR NEWBORN AND 90-KG PIGS

Swine	Temperature (°C)	Metabolic rate	
		kcal/m², or hr	kJ/m², or hr
Newborn	5	—	—
90 kg	5	78	326
Newborn	10	139	582
90 kg	10	69	289
Newborn	15	134	561
90 kg	15	60	251
Newborn	20	120	502
90 kg	20	50	209
Newborn	25	104	435
90 kg	25	50	209
Newborn	30	79	331
90 kg	30	50	209

From Mount, L. E., in *Progress in Animal Biometeorology, The Effect of Weather and Climate on Animals,* Johnson, H. D., Ed., Swets & Zeitlinger, Amsterdam, 1976. With permission.

Table 373

HEAT PRODUCTION AND METABOLISM: SWINE

Age (months)	Body weight (kg)	Acclimation	Temperature (°C)	Duration	O₂ consumption		Remarks	Ref.
					l/kg/hr	kJ/kg/hr[a]		
43	98.0	Outdoors, seasonal	13.4	4—5 days	0.240	4.83	Fasting	113
46	139.0	Outdoors, seasonal	13.4	4—5 days	0.179	3.60	Fasting	113
48	169.0	Outdoors, seasonal	23.7	4—5 days	0.110	2.21	Fasting	113
	22.7	—	10.0	—	—	17.74	Resting	129
	90.7	—	10.0	—	—	8.30	Resting	129
	181.0	—	10.0	—	—	6.62	Resting	129
	22.7	—	22.0	—	—	15.41	Resting	129
	90.7	—	22.0	—	—	7.01	Resting	129
	181.0	—	22.0	—	—	5.15	Resting	129
	22.7	—	32.0	—	—	15.64	Resting	129
	90.7	—	32.0	—	—	6.43	Resting	129
	181.0	—	32.0	—	—	4.63	Resting	129

Note: References included in the original.

[a] Conversion: $l\,O_2$/kg/hr × 4.7 (TE) × 4.19 = kJ/kg/hr.

Table 374
FEED CONSUMPTION OF 8-WEEK-OLD PIGS

Air temperature (°C)	Body weight (kg)	Food consumption (kg/week)	Weight gain (kg/week)
10	29.9	11.15	4.03
15	33.4	12.84	5.39
20	37.5	12.17	5.10
25	38.4	12.77	4.72
30	29.2	9.11	4.56

Note: Data represent average values for feed consumption.

From Fuller, M. F., *Br. J. Nutr.,* 19, 531, 1965. With permission of Cambridge University Press.

Table 375
EFFECT OF AMBIENT AIR TEMPERATURE AND MEAN LIVEWEIGHT ON RATE OF GAIN IN SWINE

Average Daily Gain per Pig (kg)

Mean liveweight (kg)	Air temperature (°C)							
	4.4	10.0	15.6	21.1	26.7	32.2	37.8	43.3
45.36		0.62	0.72	0.91	0.89	0.64	0.18	−0.60
68.04	0.58	0.67	0.79	0.98	0.83	0.52	−0.09	−1.18
90.72	0.54	0.71	0.87	1.01	0.76	0.40	−0.35	
113.40	0.50	0.76	0.94	0.97	0.68	0.28	−0.62	
136.40	0.46	0.80	1.02	0.93	0.62	0.16	−0.88	
158.76	0.43	0.85	1.09	0.90	0.55	0.05	−1.15	

Adapted from Heitman, H., Jr., Kelly, C. F., and Bond, T. E., *J. Anim. Sci.,* 17, 62, 1958. With permission.

Table 376
SIMPLIFIED BALANCED RATIONS FOR SWINE[a]

Ingredient (lb)	Baby pigs 10—25 lb	Baby pigs 25—40 lb	Growing (40—125 lb)	Finishing (125 lb to market)	Sow[b]
Corn — yellow	975	1245	1558	1666	1635
Soybean meal (44%)	570	500	390	290	295
Dried whey	400	200	—	—	—
Calcium carbonate	13	15	15	17	20
Dicalcium phosphate	30	28	27	17	38
Salt	7	7	7	7	7
Vitamin-trace mineral mix[c]	5	5	3	3	5
Total (lb)	2000	2000	2000	2000	2000
Protein, %	19.20	17.70	15.40	13.70	13.70
Calcium, %	0.84	0.75	0.64	0.55	0.85
Phosphorus, %	0.71	0.64	0.56	0.45	0.65
Metab. energy, kcal/lb	1440	1447	1456	1464	1444

[a] Rations may be based on grain sorghum, barley, or wheat rather than yellow corn as the primary source of energy.
[b] This ration is designed for both bred and lactating sows.
[c] Vitamin and trace mineral mixes may be purchased separately. This is advisable if a premix is to be stored for more than 3 to 4 months.

Source: Luce, W. G., Hollis, G. R., Mahan, D. C., and Miller, E. R., Swine rations, in *Pork Industry Handbook,* PIH-23, Cooperative Extension Service, Purdue University, West Lafayette, Ind., 1982.

Table 377
TYPICAL SUMMERTIME WATER REQUIREMENTS FOR SWINE

Type of animal	Water per head per d (gal)[a]
Sow + litter	8
Nursery pig	1
Growing pig	3
Finishing hog	5
Gestation sow	6

[a] Includes water use for drinking and moderate water wastage. Water cooling systems may increase usage.

Source: Luce, W. G., Hollis, G. R., Mahan, D. C., and Miller, E. R., Swine rations, in *Pork Industry Handbook,* PIH-23, Cooperative Extension Service, Purdue University, West Lafayette, Ind., 1982.

SHEEP

Table 378
EXAMPLES OF LAMB FEED RATIONS

	Percentages	
Feed ingredients	Ration A	Ration B
Soybean meal	14	12
Corn meal	42	55
Stemmy alfalfa hay	44	—
Good quality alfalfa hay	—	33

	Approximate composition	
Nutrients	Ration A	Ration B
TDN (%)	66.0	70.0
Crude protein (%)	15.3	16.0
Calcium (%)	0.54	0.45
Phosphorus (%)	0.28	0.30
Vitamin A (IU/lb)	2600	5200

Source: Greene, D. L., *Sheep Management,* Cooperative Extension Service, Carroll County, Md., February, 1988.

Table 379
NUTRIENT REQUIREMENTS OF GROWING-FINISHING LAMBS

Lamb weight (lb)	Daily gain (lb)	Daily feed (lb)	TDN (%)	Crude protein (%)	Ca (%)	P (%)	Vitamin A (IU/lb)
30	0.55	1.8	66	15	0.44	0.26	600
45	0.60	2.4	66	15	0.44	0.26	700
60	0.60	3.4	66	13	0.44	0.26	800
75	0.55	3.7	65	11	0.31	0.20	300
90	0.55	4.1	63	10	0.27	0.18	300
105	0.50	4.4	63	10	0.27	0.18	300

From Greene, D. L., *Sheep Management,* Cooperative Extension Service, Carroll County, Md., February, 1988.

Table 380
NUTRIENT REQUIREMENTS OF BREEDING EWES

Ewe weight (lb) and stage of pregnancy	Daily feed (lb)	TDN (%)	Crude protein (%)	Ca (%)	P (%)	Vitamin A (IU/lb)
Dry, early pregnancy						
130	3.1	50	8	0.25	0.25	400
150	3.4	50	8	0.25	0.23	425
170	3.6	50	8	0.24	0.22	450
Last 6 weeks pregnancy						
150	5.0	52	8.2	0.20	0.18	1200
170	5.3	52	8.2	0.20	0.18	1275
190	5.6	52	8.2	0.20	0.18	1350
First 8 weeks lactation, twins						
130	6.3	60	10.4	0.40	0.30	800
150	6.8	60	10.4	0.39	0.28	850
170	7.3	60	10.4	0.38	0.27	900

From Greene, D. L., *Sheep Management,* Cooperative Extension Service, Carroll County, Md., February, 1988.

POULTRY

Table 381
HEAT PRODUCTION IN CHICKENS

Species	Temperature (°C)	Body weight (kg)	Heat production (kJ/hr)	Total vaporization (kJ/hr)	Remarks
Chickens	5	1.49	18.5	—	25 Day temperature exposure, 24-hr fast
	5—20 (cyclic)	1.50	19.1	—	
	20	1.44	16.3	—	
	20—34 (cyclic)	1.20	16.0	—	
	34	1.09	11.5	—	
Broilers	19.4	1.13	35.51	6.89	Resting
	19.4	2.04	41.50	7.16	Resting
	25.0	0.36	15.69	3.48	Resting
	25.0	0.72	24.25	5.43	Resting
	28.9	0.045	2.88	0.37	Resting
	28.9	0.13	6.52	1.65	Resting

Note: References included in the original.

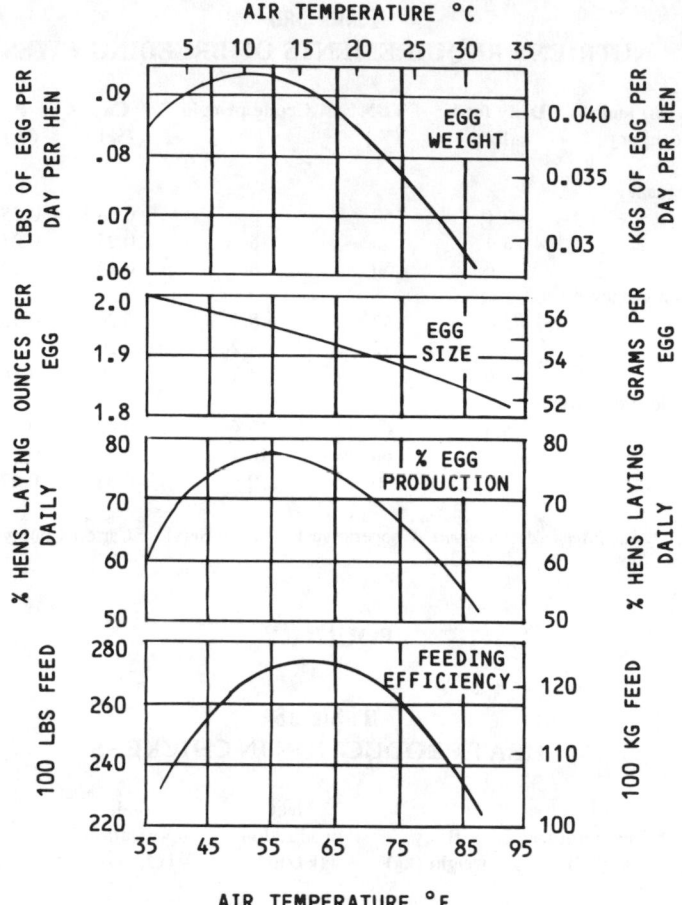

FIGURE 19.   The effect of temperature on egg weight, egg size, egg production, and feed efficiency for hens. (From Esmay, M. S., *Principles of Animal Environment,* AVI Publishing, Westport, Conn., 1969, 248. With permission.)

Table 382
EFFECT OF VARIATION OF TEMPERATURE ON GROWTH OF BROILER CHICKS FROM 2—10 WEEKS OF AGE

House temperature range (°F)	House temperature average (°F)	Average liveweight at 10 weeks (g)	Food conversion at 10 weeks	Liveweight increase 6—10 weeks (g)	Food conversion 6—10 weeks
69—75	71	1345[a]	2.94	586[a]	3.83
61—66	64	1638[b]	2.73	808[b]	3.41
54—66	58	1678[b]	2.79	866[b]	3.41
50—66	55	1665[b]	2.78	854[b]	3.38

[a,b]   Means in the same column with different superscripts differ ($p < .05$)

Source: Sorensen, P. H., in *Nutrition of Pigs and Poultry,* Morgan, J. T. and Lewis, D., Eds., Butterworths, London, 1962, 88.

Table 383
EGG PRODUCTION OF TURKEYS AND FERTILITY OF EGGS AT VARIOUS ENVIRONMENTAL TEMPERATURES

Temperature (°C)	Feed per bird (kg)		Egg production per bird		Settable egg production (%)		Egg weight (g)	Shell thickness (mm)	Fertile eggs (%)	Hatchable fertile eggs (%)
	0—12 weeks	0—24 weeks	12 weeks	24 weeks	12 weeks	24 weeks				
12.8	19.07	37.79	43.9	75.2	83.4	81.1	86.6	34.4	74.8	62.6
21.1	17.36	33.38	42.6	72.0	87.8	84.3	86.0	33.8	74.3	63.6
29.4	15.04	28.76	31.4	43.5	86.7	85.1	83.0	33.4	79.4	58.2

From Thomason, D. M., Leighton, A. T., Jr., and Mason, J. P., Jr., *Poult. Sci.*, 51, 1438—1449, 1972. With permission.

Table 384
EFFECT OF ENVIRONMENTAL TEMPERATURE ON MORTALITY, EGG PRODUCTION, FEED CONSUMPTION, AND BODY WEIGHT ON SINGLE-COMB WHITE LEGHORN PULLETS FROM 150 TO 435 DAYS OF AGE

	Temperature		
	Constant		Cycling
	32.2°C	12.8°C	12.8 to 32.2°C
Mortality (%)	17	3	2
Average egg production of survivors	140	179	190
Average egg production per pullet housed	129	177	188
Feed consumption per pullet per day (kg)	0.09	0.13	0.11
Feed per dozen eggs (kg)	2.36	2.45	2.00
Average body weight (kg)	1.77	2.00	2.00

From Mueller, W. J., *Poult. Sci.*, 40, 1562, 1961. With permission.

Table 385
CALCULATED COMPOSITION OF REPRESENTATIVE POULTRY DIETS[a]

	Broilers			
Nutrient	0—3 weeks	3—6 weeks	6 weeks to market	Layer-breeder
Protein (%)	24.00	21.00	19.00	17.00
Energy (M cal ME/kg)	2.90	2.97	3.11	2.62
Calcium (%)	1.00	1.01	1.00	3.05
Phosphorus (%) (total)	0.83	0.80	0.78	0.76
Vitamin A (IU/kg)	12,000	12,000	12,000	12,000
Vitamin D_3 (IU/kg)	3,000	3,000	3,000	3,000
Riboflavin (mg/kg)	10.00	10.00	10.00	10.00
Pantothenic acid (mg/kg)	20.00	20.00	20.00	20.00
Niacin (mg/kg)	100.00	100.00	100.00	100.00
Choline (g/kg)	2.00	2.00	2.00	2.00
Vitamin B_{12} (mcg)	100.00	100.00	100.00	100.00
Thiamine, (mg/kg)	10.00	10.00	10.00	10.00
Biotin (mg/kg)	2.00	2.00	2.00	2.00
Folic acid (mg/kg)	2.00	2.00	2.00	2.00
Vitamin E (IU/kg)	37.50	37.50	37.50	37.50
Vitamin K (mg/kg)	1.50	1.50	1.50	1.50

[a] Does not include vitamins supplied by dietary ingredients.

From *Diet Guidelines,* Agricultural Research Center, U.S. Department of Agriculture, Beltsville, Md., January, 1984.

HORSES

Table 386
ESTIMATING THE WEIGHT OF A HORSE FROM GIRTH LENGTH

Heart girth length (in.)	Weight (lb)	Heart girth length (in.)	Weight (lb)
30.0	100	64.5	800
40.0	200	67.5	900
45.5	300	70.5	1000
50.5	400	73.0	1100
55.0	500	75.5	1200
58.5	600	77.5	1300
61.5	700		

Source: Huff, A. N., Meacham, T. N., and Wahlberg, M. L., *Horse Nutrition: Feeds and Feeding,* Publ. 406-472, Virginia Cooperative Extension Service, reprinted February, 1987.

Table 387
NUTRIENT REQUIREMENTS OF HORSES (DAILY NUTRIENTS PER HORSE), 1100 lb MATURE WEIGHT

Type of animal	Weight (lb)	Daily gain (lb)	Digestible energy (M cal)	TDN (lb)[a]	Crude protein (lb)	Digestible protein (lb)	Ca (g)	P (g)	Vitamin A activity (1000 IU)	Daily feed (lb)[b]
Mature horses, maintenance	1100		16.39	8.20	1.39	0.64	23	14	12.5	16.4
Mares, last 90 d gestation		1.21	18.36	9.18	1.65	0.86	34	23	25.0	16.2
Lactating mare, first 3 months (15 kg milk per d)			28.27	14.14	2.99	1.85	50	34	27.5	22.2
Lactating mare, 3 months to weanling (10 kg milk per d)			24.31	12.16	2.42	1.36	41	27	22.5	20.6
Nursing foal (3 months of age)	341	2.64	13.66	6.83	1.65	1.19	33	20	6.2	9.2
Requirements above milk			6.89	3.45	0.90	0.68	18	13	0.0	4.9
Weanling (6 months of age)	506	1.76	15.60	7.80	1.74	1.14	34	25	9.2	11.0
Yearling (12 months of age)	715	1.21	16.81	8.41	1.67	0.99	31	22	12.0	13.2
Long yearling (18 months of age)	880	0.77	17.00	8.58	1.56	0.86	28	19	14.0	14.3
2 Years old (24 months of age)	990	0.33	16.45	8.23	1.39	0.72	25	17	13.0	14.5

[a] For horses heavier or lighter, add or subtract 8% of the TDN for each 100 lb difference.
[b] Dry matter basis.

Source: Huff, A. N., Meacham, T. N., and Wahlberg, M. L., *Horse Nutrition: Feeds and Feeding,* Publ. 406-472, Virginia Cooperative Extension Service, reprinted February, 1978; NRC, 1978.

Table 388
NUTRIENT CONCENTRATION IN DIETS FOR HORSES AND PONIES
EXPRESSED ON 90% DRY MATTER BASIS

Type of animal and/or activity	TDN (%)	Crude protein (%)	Ca (%)	P (%)	Vitamin A activity (IU/lb)
Mature horses and ponies at maintenance	45	7.7	0.27	0.18	650
Mares, last 90 d of gestation	50	10.0	0.45	0.30	1400
Lactating mare, first 3 months	55	11.0	0.40	0.25	1000
Creep feed	70	16.0	0.80	0.55	
Foal (3 months of age)	67	16.0	0.80	0.55	800
Weanling (6 months of age)	63	14.5	0.60	0.45	800
Yearling (12 months of age)	60	12.0	0.50	0.35	800
Long yearling (18 months of age)	55	10.0	0.40	0.30	800
2-Year old (light training)	60	9.0	0.40	0.30	800
Mature working horses					
Light work	50	7.7	0.27	0.18	650
Moderate work	60	7.7	0.27	0.18	650
Intense work	63	7.7	0.27	0.18	650

Source: Huff, A. N., Meacham, T. N., and Wahlberg, M. L., *Horse Nutrition: Feeds and Feeding,* Publ. 406-472, Virginia Cooperative Extension Service, reprinted February, 1987, NRC, 1978.

Table 389
RECOMMENDED CRUDE
PROTEIN PERCENTAGES IN
GRAIN RATIONS FOR HORSES

Type of animal	Percent[a]
Foal, creep feed	18—20
Weanlings	14—18
Yearlings	14—16
Mature horses	9—10
Pregnant mares	12—15
Lactating mares	12—16
Stallions	10—12
Hardworking horses	9—12

[a]   Horses fed legume hay will need less protein in the grain ration than those fed grass hay. Horses fed grass hay should receive upper protein percentage levels listed.

Source: Huff, A. N., Meacham, T. N., and Wahlberg, M. L., *Horse Nutrition: Feeds and Feeding,* Publ. 406-472, Virginia Cooperative Extension Service, reprinted February, 1987, NRC, 1978.

ENVIRONMENTAL CONTROL OF LIVESTOCK

Table 390
MAXIMUM AND MINIMUM VENTILATION RATES FOR VARIOUS CLASSES OF LIVESTOCK

Livestock	Weight (kg)	Ventilation rate Max (m³/sec kg × 10⁻⁴)	Min (m³/sec kg × 10⁻⁴)
Pullets and hens	1.8	16	1.3—2.0
	2.0	16	1.8
	2.5	15	1.7
	3.0	13	1.6
	3.6	12	1.6
Broilers	0.045	—	6.3
	0.36	—	3.9
	0.9	—	2.1
	1.4	—	1.7
	1.8	16	2.0
	2.3	16	2.1
Turkeys	0.45	37	4.2
	2.3	14	1.4
	4.5	9.4	0.94
	6.8	8.3	0.83
	11.3	6.7	0.67
Ducks	0.45	37	—
	1.4	20	—
	2.3	17	—
	4.5	14	—
Geese	0.45	42	—
	2.3	16	—
	4.5	10	—
	6.8	9.0	—
	9.1	8.8	—
Fattening pig	20—100	5.3	0.53
Sow and litter	150—200	5.3	0.53
Dry sow	125—200	5.3	0.53
Calf	45—200	5.3	1.1
Cattle	200 +	4.0	0.8
Sheep	20—100	11	2.2

Table 391
INSULATION "R" VALUES FOR SOME COMMON MATERIALS

Material	Insulation value[a]	
	Per inch thickness	For thickness listed
Batt or blanket insulation		
Glass wool, mineral wool, or fiberglass	3.50 approx, read label	
Fill-type insulation		
Glass or mineral wool	3.00	
Vermiculite (expanded)	2.13—2.27	
Shavings or sawdust	2.22	
Paper or wood pulp	3.70	
Rigid insulation		
Wood fiber sheathing	2.27—2.63	
Expanded polystyrene		
Extruded	4.00—5.26	
Molded	3.57	
Expanded polyurethane (aged)	6.25	
Glass fiber	4.00	
Ordinary building materials		
Concrete, poured	0.08	
Plywood, $3/_8''$	1.25	0.47
Plywood, $1/_2''$	1.25	0.62
Hardboard, $1/_4''$	1.00—1.37	
Cement asbestos board, $1/_8''$		0.03
Lumber (fir, pine), $3/_4''$	1.25	0.94
Wood beveled siding $1/_2'' \times 8''$		0.81
Asphalt shingles		0.44
Wood shingles		0.94
Window glass, includes surface condition		
Single-glazed		0.88
Single-glazed + storm windows		1.79
Double-pane insulating glass		1.45—1.73
Air space ($3/_4''$ or larger)		0.90
Surface conditions		
Inside surface		0.68
Outside surface (15 mph wind)		0.17

Note: Data from *ASHRAE Handbook of Fundamentals,* 1972.

[a] Mean temperature of 75°F.

From Bath, D. L., Dickinson, F. N., Tucker, H. A., and Appleman, R. D., *Dairy Cattle: Principles, Practices, Problems, Profits,* 3rd ed., Lea & Febiger, Philadelphia, 1985. With permission.

MEAT, MILK, AND EGGS — GRADES AND COMPOSITION

Table 392
BEEF QUALITY AND YIELD GRADES[a]

USDA Beef Quality Grades

Group	Approximate chronological age	Quality grades
Veal	9 weeks to 3 months	Prime, Choice, Good, Standard, Utility
Calf	3 months to 9 months	Prime, Choice, Good, Standard, Utility
Beef	9 months to 24 months	Prime, Choice, Good, Standard, Utility
Beef	42 months to over 96 months	Commercial, Utility, Cutter, Canner
Bullock	9 months to 42 months	Prime, Choice, Good, Standard, Utility

USDA Beef Yield Grades

Yield Grade	Percent of carcass weight (boneless four major cuts)
1	Over 52.3
2	52.3 to 50.0
3	50.0 to 47.7
4	47.7 to 45.4
5	Less than 45.4

Factors Used for Determination of USDA Beef Grades

	Quality Grades			Yield Grades	
Factor	Range	Average (ballpark) value	Factor	Range	Average (ballpark) value[b]
Marbling degree	"Devoid" to "Abundant"	"Small"	Fat thickness (12th rib)	0.00 to 1.6 in.	0.5 in.
Maturity score	"A" to "E"	"A"	Ribeye area (12th rib)	6 to 20 sq. in.	12 in.[a]
			Internal fat (% kidney, pelvic and heart)	0.5 to 6.5%	3%
			Carcass weight	350 to 1100 lb	675

[a] Official United States Standards for Carcass Beef (implemented in 1976) and veal and calf carcasses (1972).

[b] Average values represent a USDA Yield Grade 3.1 carcass which would yield 47.7 to 50.0% boneless, closely trimmed retail cuts from the round, rump, loin, rib and chuck.

Table 393
USDA GRADES FOR PORK CARCASSES

USDA Quality Grades

Group	Approximate live weight range	Quality grades
Slaughter pigs	Less than 120 lb	Choice, Good, Medium, Cull
Butcher hogs	120 to 270 lb	U.S., U.S. Utility
Packer sows	Over 270 lb	U.S., Medium, Cull

USDA Pork Yield Grades

Yield grade	Percent of carcass weight (bone-in four lean cuts)
NO. 1	Over 60[a]
NO. 2	55 to 60[a]
NO. 3	50 to 55[a]
NO. 4	Less than 50[a]

Factors Used for Determination of USDA Pork Carcass Grades

Quality grades

Factor	Range	Average (ballpark) value
Lean color	"Pale" to "dark red"	"Grayish pink"
Lean firmness	"Soft" to "very firm"	"Slightly firm"
Feathering	"None" to "abundant"	"Slight"
Fat condition	"Soft and oily" to "firm"	"Slightly firm"
Belly thickness	"Too thin for bacon production" to "thick"	"Slightly thick"

Yield grades

Factor	Range	Average (ballpark) value[b]
Average backfat thickness	0.8 to 2.5 in.	1.5 in.
USDA muscling score	"Very thin minus" to "Very thick plus"	"Moderately thick"
Carcass length or weight	27 to 34 in.	30.5 in.

[a] Factors affecting the desirability of barrow and gilt carcasses. A final report to USDA prepared by G. C. Smith and Z. L. Carpenter, page 251.
[b] Average values represent a U.S. NO. 2 pork carcass which would yield 55.0 to 60.0% bone-in trimmed ham, loin, and shoulder.

From USDA marketing bulletin NO. 49.

Table 394
LAMB AND MUTTON QUALITY AND YIELD GRADES[a]

USDA Quality Grades

Group	Approximate chronological age	Quality grades
Hothouse lambs	Less than 3 months	Extra Fancy, Fancy, Good, Fair, Plain
Lambs	3 months to 14 months	Prime, Choice, Good, Utility, Cull
Yearling mutton	14 months to 24 months	Prime, Choice, Good, Utility, Cull
Mutton	Over 24 months	Choice, Good, Utility, Cull

USDA Lamb Yield Grades

USDA yield grade	Percent of carcass weight (boneless four major cuts)
1	Over 47.3
2	47.3 to 45.5
3	45.5 to 43.7
4	43.7 to 41.9
5	Less than 41.9

Factors Used for Determination of USDA Lamb Grades

Quality grades

Factor	Range	Average (ballpark) value
Conformation	"Utility" to "prime"	"Choice"
Maturity score	"A" to "B"	"A"
Firmness of flank	"Slightly soft and watery" to "Extremely full and firm"	"Tends to be slightly full and firm"
Feathering in the ribcage	"Practically none" to "abundant"	"Slight"
Fat streaking in the flank	"None" to "abundant"	"Traces"

Yield grades

Factor	Range	Average (ballpark) value
Fat thickness (12th rib)	0.05 to 0.60 in.	0.20 in.
Leg conformation	"Cull minus" to "Prime plus"	"Choice"
Internal fat (% kidney and pelvic)	1 to 9%	4%

[a] Official United States Standards for grades of lamb, yearling mutton and mutton carcasses, 1960.
[b] Average values represent a USDA Yield Grade 3.4 carcass which would yield 43.7 to 45.5% boneless retail cuts from the leg, loin, rack, and shoulder.

Table 395

DRESSING PERCENTAGES OF VARIOUS KINDS OF LIVESTOCK, BY GRADES[a]

Livestock species/grade	Range	Average
Cattle		
Prime	62—67	64
Choice	59—65	62
Good	58—62	60
Standard	55—60	57
Commercial	54—62	57
Utility	49—57	53
Cutter	45—54	49
Canner	40—48	45
Calves/vealers[b]		
Prime	62—67	64
Choice	58—64	60
Good	56—60	58
Standard	52—57	55
Utility	47—54	51
Lambs/sheep[b]		
Prime	49—55	52
Choice	47—52	50
Good	45—49	47
Utility	43—47	45
Cull	40—45	42
Barrows/gilts[c]		
U.S. No. 1	68—72	70
U.S. No. 2	69—73	71
U.S. No. 3	70—74	72
U.S. No. 4	71—75	73
Utility	67—71	69

[a] All percentages are based on hot weights.
[b] Based on hide-off carcass weights.
[c] Based on packer-style dressing (ham facings, leaf fat, kidneys, and head removed).

From AMS-Livestock Division, U.S. Department of Agriculture, Washington, D.C.

Table 396
DISASSEMBLY PROCESS FOR MEAT ANIMALS[a]

	Weight (pounds) and Yield					
Species	LW[b]	DP (%)	CW[b]	RY[c] (%)	RCW	TY (%)
Choice steer	1146	61	699	74	517	45.1
Choice veal	202	61	123	83	102	50.5
Choice lamb	114	50	57	89	51	44.7
US No. 1 barrow	231	71	164	92	151	65.4

[a] Codes: LW = live weight; DP = dressing percent; CW = carcass weight; RY = retail yield; RCW = retail carcass weight; TY = total yield; DP = CW/LW × 100; RY = RCW/CW × 100; and, TY = RCW/LW × 100.
[b] Values (steer, lamb, hog) based on average weights for USDA inspected livestock slaughtered in 1979 (Livestock Slaughter, January, 1980). The average weights for Choice veal based on Livestock, Meat and Wool Market News (February, 1980), USDA.
[c] Agricultural Economic Report No. 138, 1977, USDA.

Table 397
COMPARISON OF RETAIL YIELD FOR BEEF SIDES REPRESENTING EACH US YIELD GRADE

USDA yield grade (YG) values for 300 lb sides

Retail cut class	Y.G. 1 side %	lb	Y.G. 2 side %	lb	Y.G. 3 side %	lb	Y.G. 4 side %	lb	Y.G. 5 side %	lb
Bone in										
Sirloin	9.1	27.3	8.7	26.1	8.3	24.9	7.9	23.7	7.5	22.5
Short loin	5.3	15.9	5.2	15.6	5.1	15.3	5.0	15.	4.9	14.7
Blade chuck	9.9	29.7	9.4	28.2	8.9	26.7	8.4	25.2	7.9	23.7
Rib (short cut)	6.3	18.9	6.2	18.6	6.1	18.3	6.0	18.0	5.9	17.7
Boneless										
Arm chuck	6.4	19.2	6.1	18.3	5.8	17.4	5.5	16.5	5.2	15.6
Rump	3.7	11.1	3.5	10.5	3.3	9.9	3.1	9.3	2.9	8.7
Inside round	4.9	14.7	4.5	13.5	4.1	12.3	3.7	11.1	3.3	9.9
Outside round	4.8	14.4	4.6	13.8	4.4	13.2	4.2	12.6	4.0	12.
Round tip	2.7	8.1	2.6	7.8	2.5	7.5	2.4	7.2	2.3	6.9
Brisket	2.5	7.5	2.3	6.9	2.1	6.3	1.9	5.7	1.7	5.1
Flank	0.5	1.5	0.5	1.5	0.5	1.5	0.5	1.5	0.5	1.5
Ground beef and lean trim	25.6	76.8	23.5	70.5	21.4	64.2	19.3	57.9	17.2	51.6
Bone, fat, kidney, other	18.3	54.9	22.9	68.7	27.5	82.5	32.1	96.3	36.7	110.1
Total	100.0	300.0	100.0	300.0	100.0	300.0	100.0	300.0	100.0	300.0

Table 398

COMPARISON OF YIELDS OF WHOLESALE CUTS FOR 68 KG PORK CARCASSES, BY GRADE[a] (% of carcass)

Cut	Wt. (lb)	U.S. No. 1	U.S. No. 2	U.S. No. 3	U.S. No. 4
Hams	14/DN		2.20	9.54	15.71
	14/17	20.40	17.56	9.10	1.82
	17/20	0.47			
	20/26				
Loins	8/14	14.75	16.41	15.60	14.52
	14/17	3.04	0.28		
	17/20				
Bellies	8/10	3.26	1.93	1.05	0.51
	10/12	10.06	10.91	10.83	9.82
	12/14	0.88	1.80	3.19	5.16
	14/16				
	16/18				
Spareribs	3/DN	3.30	3.37	3.16	2.92
	3/5	0.30			
Picnics	4/6	0.75	2.20	4.29	6.09
	6/8	7.94	6.12	3.66	1.47
	8/12				
Butts	4/8	7.15	6.72	6.31	5.88
Jowls		2.65	2.85	3.05	3.25
Neck bones		1.56	1.43	1.29	1.14
Fore feet		1.61	1.53	1.46	1.36
Tails		0.52	0.52	0.53	0.53
Lean trim (50% fat)		5.84	5.65	5.46	5.25
Lard		10.60	13.09	15.81	18.11
Total		95.08	94.57	94.33	93.54

[a]   The percentages under each grade represent the expected yields within each of the indicated weight categories. The product yield percentages do not total 100% because hind feet are excluded as being virtually valueless and because there is a loss in weight during the lard rendering process.

Source: Standards Branch, Meat Quality Division, Food Safety and Quality Service, U.S. Department of Agriculture, personal communication, 1979.

Table 399

COMPARISON OF YIELDS OF RETAIL CUTS FOR CHOICE LAMB CARCASSES, BY YIELD GRADE[a]

Retail cut	Yield grade (% of carcass)				
	1	2	3	4	5
Leg, short cut	23.6	22.2	20.8	19.4	18.0
Sirloin	6.7	6.4	6.1	5.8	5.5
Short loin	1.04	10.1	9.8	9.5	9.2
Rack	8.1	7.9	7.7	7.5	7.3
Shoulder	24.9	23.8	22.7	21.6	20.5
Neck	2.2	2.1	2.0	1.9	1.8
Breast	9.8	9.8	9.8	9.8	9.8
Foreshank	3.5	3.4	3.3	3.2	3.1
Flank	2.3	2.3	2.3	2.3	2.3
Kidney	0.5	0.5	0.5	0.5	0.5
Fat	4.6	8.2	11.8	15.4	19.0
Bone	3.4	3.3	3.2	3.1	3.0
Total	100.0	100.0	100.0	100.0	100.0

[a] The comparisons reflect average yields of retail cuts from lamb carcasses typical of the midpoint of each of the U.S. Department of Agriculture yield grades during April 1979.

Source: Report of cutting test. Standards Branch, Meat Quality Division, Food Safety and Quality Service, U.S. Department of Agriculture, Los Angeles, May 1965.

Table 400
COMPOSITION OF 100 g EDIBLE BEEF

Composite of Trimmed Retail Cuts, Separable Lean Only, All Grades

Nutrient (unit)	Raw, edible portion (mean)	Retention of nutrients on cooking (mean %)		
		Braised	Broiled	Roasted
Water (g)	71.60	50	63	65
Protein (g)	20.94	98	99	99
Total lipid (fat) (g)	6.33	114	120	129
Ash (g)	1.03	65	87	84
Calcium (mg)	6	84	103	91
Iron (mg)	2.27	102	94	99
Magnesium (mg)	23	66	86	83
Potassium (mg)	358	57	83	80
Sodium (mg)	63	59	85	83
Thiamin (mg)	0.111	45	70	58
Riboflavin (mg)	0.189	86	92	98
Niacin (mg)	3.605	61	78	75
Pantothenic acid (mg)	0.364	69	79	97
Vitamin B_6 (mg)	0.43	46	74	66
Folacin (mcg)	8	72	87	88
Vitamin B_{12} (mcg)	3.25	67	75	72
Cholesterol (mg)	60	103	106	103

All Grades, Cooked (g)

Fatty acids, total				
Saturated (g)	2.45		3.98	
Monounsaturated (g)	2.76		4.45	
Polyunsaturated (g)	0.26		0.38	

From Composition of Foods: Beef Products, *U.S. Dep. Agric. Agric. Handb.*, No. 8—13, rev. August 1986.

Table 401
COMPOSITION OF 100 g OF FRESH PORK

	Composite of Trimmed Leg, Loin, and Shoulder			
	Separable lean and fat		Without spareribs (separable lean only)	
Nutrient (unit)	Carcass Raw (mean)	With spareribs Raw (mean)	Raw (mean)	Cooked (mean)[a]
Water (g)	47.86	59.68	71.95	58.97
Protein (g)	13.35	16.74	20.22	27.04
Total lipid (fat) (g)	37.83	22.55	6.75	13.04
Ash (g)	0.72	0.86	1.04	1.10
Calcium (mg)	5	6	7	8
Iron (mg)	0.69	0.86	1.02	1.25
Magnesium (mg)	13	19	23	22
Phosphorus (mg)	152	188	224	258
Potassium (mg)	244	297	358	365
Sodium (mg)	44	55	64	69
Zinc (mg)	1.62	2.05	2.45	3.47
Copper (mg)	0.059	0.070	0.082	0.112
Thiamin (mg)	0.573	0.726	0.904	0.693
Riboflavin (mg)	0.212	0.233	0.275	0.356
Niacin (mg)	3.940	4.304	5.061	5.077
Pantothenic acid (mg)	0.877	0.663	0.795	0.783
Vitamin B_6 (mg)	0.28	0.38	0.47	0.42
Folacin (mcg)	4	5	6	8
Vitamin B_{12} (mcg)	0.66	0.69	0.79	0.83
Cholesterol (mg)	74	72	65	93
Fatty acids, total				
Saturated (g)	13.75	8.17	2.34	4.49
Monounsaturated (g)	17.63	10.43	3.06	5.86
Polyunsaturated (g)	4.16	2.38	0.71	1.58

[a] Roasted.

From Composition of Foods: Pork Products, Anderson, B. A., Principal Investigator, *U.S. Dep. Agric. Agric. Handb.*, No. 8—10, rev. August 1983.

Table 402
AVERAGE AMOUNT OF AMINO ACIDS IN 100 g OF SELECTED MEATS (EDIBLE PORTION)

Amino acids (g)	Beef[a,d]	Beef liver[a,e]	Pork, fresh[b,f]	Chicken[c,g]	Turkey[c,h]
Tryptophan	0.341	0.385	0.363	0.257	0.311
Threonine	1.329	1.222	1.271	0.922	1.227
Isoleucine	1.368	1.222	1.306	1.125	1.409
Leucine	2.404	2.513	2.197	1.653	2.184
Lysine	2.531	1.855	2.664	1.765	2.557
Methionine	0.779	0.675	0.666	0.591	0.790
Cystine	0.341	0.410	0.350	0.311	0.308
Phenylalanine	1.188	1.423	1.081	0.899	1.100
Tyrosine	1.022	1.060	0.964	0.732	1.066
Valine	1.480	1.650	1.449	1.100	1.464
Arginine	1.922	1.680	1.877	1.378	1.979
Histidine	1.042	0.731	1.371	0.655	0.845
Alanine	1.835	1.594	1.600	1.270	1.790
Aspartic acid	2.779	2.569	2.508	1.950	2.709
Glutamic acid	4.570	3.620	4.200	3.605	4.484
Glycine	1.660	1.530	1.232	1.378	1.682
Proline	1.343	1.410	1.025	1.178	1.304
Serine	1.163	1.282	1.111	0.832	1.240

[a] From Composition of Foods: Beef Products, *U.S. Dep. Agric. Agric. Handb.*, No. 8—13, rev. August 1986.
[b] From Composition of Foods: Pork Products, *U.S. Dep. Agric. Agric. Handb.*, No. 8—10, rev. August 1983.
[c] From Composition of Foods: Poultry Products, *U.S. Dep. Agric. Agric. Handb.*, No. 8—5, rev. August 1979.
[d] Composite of retail cuts, separable lean only, all grades, cooked.
[e] Pan fried.
[f] Composite of trimmed leg, loin, and shoulder, separable lean only, roasted.
[g] Broilers or fryers, flesh and skin, without neck, fried, batter dipped.
[h] All classes, flesh and skin, without neck, roasted.

Table 403

NUTRIENT COMPOSITION OF MILK, SKIM MILK, CREAM, AND
CERTAIN DAIRY PRODUCTS MADE THEREFROM (in 100 g portions)

Nutrient	Whole milk	Plain yogurt	Skim milk	Buttermilk	Cream	Sour cream
Calories	65.0	63.0	36.0	40.0	211.0	211.0
Protein	3.5	3.0	3.6	3.31	3.0	3.0
Fat (g)	3.5	1.6	0.1	0.9	20.6	20.4
Carbohydrate (g)	4.9	7.0	5.1	4.8	4.3	4.3
Ca (mg)	118.0	183.0	121.0	116.0	102.0	102.0
Fe (mg)	trace	0.08	trace	0.05	trace	0.04
Mg (mg)	13.0	17.0	14.0	11.0	9.0	11.0
P (mg)	93.0	144.0	95.0	89.0	80.0	77.0
K (mg)	144.0	234.0	145.0	151.0	122.0	56.0
Zn (mg)	0.38	0.89	0.40	0.42	0.27	0.27
Na (mg)	50.0	70.0	52.0	105.0	43.0	40.0
Ascorbic acid (mg)	1.0	0.80	1.0	0.98	1.0	1.0
Thiamin (mg)	0.03	0.04	0.04	0.03	0.03	30.0
Riboflavin (mg)	0.17	0.21	0.18	0.15	0.15	150.0
Niacin (mg)	0.1	0.11	0.1	0.06	0.1	0.07
Pantothenic acid (mg)	0.3	0.59	0.37	0.28	0.26—0.34	0.32—0.36
Vitamin B_6 (μg)	40.0	0.05	42.0	0.03	35.0	16.0
Folacin (μg)	0.6	3.9	1.0	—	2.0	11.0
Vitamin B_{12} (μg)	0.4	0.56	0.4	0.2	0.35	0.3
Vitamin A (IU)	140	66.0	tr.	33.0	840.0	839.0

Note: References included in the original.

Table 404

APPROXIMATE GROSS COMPOSITION
OF BUTTER

Compound	% (w/w)	Compound	mg/100 g
Fat	83	Ca	15
Water	15	P	20
Solids-not-fat	1[a]	Fe	0.1
Proteins	0.6[a]	Na	5[c]
Carbohydrates	0.5	K	15
Lactic acid	0.15[b]	Cu	0.002[d]
Citrates	0.02	Vitamin A	1
"Ash"	0.1[c]	Carotenoids	0.5
Linoleic acid	3	Vitamin D	0.001
Linolenic acid	0.6	Vitamin E	2
		Vitamin K	0.06
		Vitamin C	0.3
		Vitamin B_2	0.05

[a] If washing is deleted 1.5% solids-not-fat and 0.8% proteins.
[b] In the case of cultured cream butter (pH mostly ~4.7)
[c] Unsalted butter contains 0.01% NaCl, salted butter 0.2 to 2% NaCl, hence 0.3 to 2.1% "ash".
[d] If uncontaminated; in practice the content may be ten times higher.

Note: References included in the original.

Table 405

LACTOSE CONTENT OF MILK AND MILK PRODUCTS[a]

	Gram weight of one cup	Lactose Percentage[a]	Lactose Grams per cup
Whole milk	244	4.8	11.7
Lowfat milk (1% fat)	245	4.9	12.0
Skim or nonfat milk	245	5.1	12.5
Chocolate milk (carbohydrate other than lactose added)	250	11.0[b]	27.5[b]
Evaporated milk	252	9.7	24.4
Condensed milk	306	14.7	45.0
Sweetened condensed milk	306	11.4	34.9
Nonfat dry milk, reg. unreconstituted	120	52.3	35.6
Dry whole milk, unreconstituted	128	38.2	48.9
Unfermented acidophilus milk (lowfat)	245[c]	4.9	12.0
Buttermilk	244	4.3	10.5
Sour cream	230	3.4	7.8
Yogurt (plain)	227	4.1	9.3
Half-and-half	242	4.5	10.9
Light cream	240	4.2	10.1
Light whipping cream	239	3.6	8.6
Heavy whipping cream	238	3.2	7.6

[a]   Values from Webb, B. H., Johnson, A. H., and Alford, J. A., Eds., *Fundamentals of Dairy Chemistry,* AVI Publishing, Westport, Conn., 1974.
[b]   Carbohydrate other than lactose added.
[c]   Gram weight for unfermented acidophilus milk is the same as that for the fluid milk from which it is made.

From National Dairy Council, *Newer Knowledge of Milk,* Rosemont, Illinois, 1988, Table 9. Courtesy of the National Dairy Council.®

Table 406

MACROMINERALS IN MILK[a]

Macrominerals	Milk solids-not-fat (Mg/g)	Whole fluid milk (Mg/100 g)	Whole fluid milk (Mg/cup; 8 oz, 244g)
Calcium (mg)	13.8	119	291
Chlorine (mg)	11.5	103	244
Magnesium (mg)	1.6	13	33
Phosphorus (mg)	10.8	93	228
Potassium (mg)	17.5	152	370
Sodium (mg)	5.7	49	120
Sulfur (mg)	2.9	25	61

Note: The nutrient values are based on whole milk containing 8.67% solids-not-fat. The Handbook 8-1[15] includes standard error of the mean.

From National Dairy Council, *Newer Knowledge of Milk,* Rosemont, Illinois, 1988, Table 14. Courtesy of the National Dairy Council.®

Table 407
TRACE ELEMENTS IN MILK

Trace elements	Per gram milk solids-not-fat[a]	Per 100 g whole fluid milk	Per cup (8 oz 244g) whole fluid milk
Aluminum (mcg)	5.35	46.0	112.24
Arsenic[b] (mcg)	0.58	5.0	12.20
Barium (mcg)	—[c]	—[c]	—[c]
Boron[b] (mcg)	3.14	27.0	65.88
Bromine (mcg)	6.98	60.0	146.40
Bromine (coastal area) (mcg)	32.56	280.0	683.20
Cadmium (mcg)	0.30	2.6	6.34
Chromium[d] (mcg)	0.17	1.5	3.66
Cobalt[b] (mcg)	0.007	0.06	0.15
Copper[d] (mcg)	1.51	13.0	31.72
Fluorine[b,d] (mcg)	1.74	15.0	36.60
Iodine[b,d] (mcg)	0.50	4.3	10.49
Iron[b,d] (mg)	0.006	0.0492	0.12
Lead[b] (mcg)	0.47	4.0	9.76
Lithium (mcg)	—[c]	—[c]	—[c]
Manganese[b,d] (mcg)	0.26	2.2	5.37
Molybdenum[b,d] (mcg)	0.85	7.3	17.81
Nickel (mcg)	0.31	2.7	6.59
Rubidium[b] (mcg)	23.26	200.0	488.00
Selenium (nonseleniferous area)[b,d] (mcg)	0.47	4.0	9.76
Selenium (seleniferous area)[b,d] (mcg)	14.77	up to 127.0	up to 309.88
Silicon (mcg)	16.63	143.0	348.92
Silver (mcg)	0.55	4.7	11.47
Strontium (mcg)	1.99	17.1	41.72
Tin (mcg)	—[c]	—[c]	—[c]
Titanium (mcg)	—[c]	—[c]	—[c]
Vanadium (mcg)	0.0011	0.0092	0.02
Zinc[d] (mg)	0.044	0.3811	0.93

[a] Calculated for whole fluid milk containing 8.67% solids-not-fat.
[b] Effect of feed supplement.
[c] Dashes denote qualitative data, therefore difficult to assign a specific value.
[d] Trace elements for which the present state of knowledge allows an evaluation for human nutrition. An RDA has been established for iron, iodine, and zine.

From National Dairy Council, *Newer Knowledge of Milk,* Rosemont, Illinois, 1988, Table 15. Courtesy of the National Dairy Council.®

Table 408
CLASSES OF POULTRY

Chickens

Rock Cornish Game Hen or Cornish Game Hen
A Rock Cornish game hen or Cornish game hen is a
young immature chicken (usually 5-6 weeks of
age) weighing not more than 2 pounds ready-to-
cook weight, which was prepared from a Cornish
chicken or the progeny of a Cornish chicken
crossed with another breed of chicken.

Rock Cornish Fryer, Roaster, or Hen
A Rock Cornish fryer, roaster, or hen is the
progeny of a cross between a purebred Cornish
and a purebred Rock chicken, without regard to
the weight of the carcass involved. However, the
term "fryer", "roaster", or "hen" shall apply
only if the carcasses are from birds with ages and
characteristics that qualify them for such
designation under the next two paragraphs.

Broiler or Fryer
A broiler or fryer is a young chicken (usually under
13 weeks of age) of either sex that is tender-
meated with soft, pliable, smooth-textured skin
and flexible breastbone cartilage.

Roaster or Roasting Chicken
A bird of this class is a young chicken (usually 3 to
5 months of age), of either sex, that is tender-
meated with soft, pliable, smooth-textured skin
and breastbone cartilage that may be somewhat
less flexible than that of a broiler or fryer.

Capon
A capon is a surgically unsexed male chicken
(usually under 8 months of age) that is tender-
meated with soft, pliable, smooth-textured skin.

Hen, Fowl, or Baking or Stewing Chicken
A bird of this class is a mature female chicken
(usually more than 10 months of age) with meat
less tender than that of a roaster or roasting
chicken and nonflexible breastbone tip.

Cock or Rooster
A cock or rooster is a mature male chicken with
coarse skin, toughened and darkened meat, and
hardened breastbone tip.

Turkeys

Fryer-Roaster Turkey
A fryer-roaster turkey is a young immature turkey
(usually under 16 weeks of age), of either sex, that
is tender-meated with soft, pliable, smooth-
textured skin, and flexible breastbone cartilage.

Young Turkey*
A young turkey is a turkey (usually under 8 months
of age) that is tender-meated with soft, pliable,
smooth-textured skin and breastbone cartilage that
is somewhat less flexible than in a fryer-roaster
turkey. Sex designation is optional.

Yearling Turkey
A yearling turkey is a fully matured turkey (usually
under 15 months of age) that is reasonably tender-
meated and with reasonable smooth-textured skin.
Sex designation is optional.

Mature Turkey or Old Turkey (Hen or Tom)
A mature or old turkey is an old turkey of either
sex (usually in excess of 15 months of age) with
coarse skin and toughened flesh.

Ducks

Broiler Duckling or Fryer Duckling
A broiler duckling or fryer duckling is a young
duck (usually under 8 weeks of age), of either sex,
that is tender-meated and has a soft bill and soft
windpipe.

Mature Duck or Old Duck
A mature duck or an old duck is a duck (usually
over 6 months of age), of either sex, with
toughened flesh, hardened bill, and hardened
windpipe.

Roaster Duckling
A roaster duckling is a young duck (usually under
16 weeks of age), of either sex, that is tender-
meated and has a bill that is not completely
hardened and a windpipe that is easily dented.

Table 408 (continued)
CLASSES OF POULTRY

Geese

Young Goose
A young goose may be of either sex, is tender-meated, and has a windpipe that is easily dented.

Mature Goose or Old Goose
A mature goose or old goose may be of either sex and has toughened flesh and a hardened windpipe.

Guineas

Young Guinea
A young guinea may be of either sex, is tender-meated and has a flexible breastbone cartilage.

Mature Guinea or Old Guinea
A mature guinea or old guinea may be of either sex, has toughened flesh and a hardened breastbone.

Pigeons

Squab
A squab is a young immature pigeon of either sex, and is extra tender-meated.

Pigeon
A pigeon is a mature pigeon of either sex, with coarse skin and toughened flesh.

ᵃ For labeling purposes, the designation of sex within the class name is optional and the two classes of young turkeys may be grouped and designated as ''young turkeys''.

Table 409

SUMMARY OF SPECIFICATIONS OF QUALITY FOR INDIVIDUAL CARCASSES OF READY-TO-COOK POULTRY AND PARTS THEREFROM (MINIMUM REQUIREMENTS AND MAXIMUM DEFECTS PERMITTED)

Factor	A Quality	B Quality	C Quality
Conformation			
Breastbone	Normal; Slight curve or dent	Moderate deformities; Moderately dented, curved or crooked	Abnormal; Seriously curved or crooked
Back	Normal (except slight curve)	Moderately crooked	Seriously crooked
Legs and Wings	Normal	Moderately misshapen	Misshapen
Fleshing	Well fleshed, moderately long, deep and rounded breast	Moderately fleshed, considering kind, class and part	Poorly fleshed
Fat covering	Well covered — especially between heavy feather tracts on breast and considering kind, class and part	Sufficient fat on breast and legs to prevent distinct appearance of flesh through the skin	Lacking in fat covering over all parts of carcass
Pinfeathers			
Nonprotruding pins and hair	Free	Few scattered	Scattering
Protruding pins	Free	Free	Free

Exposed flesh[a]

Carcass Weight Minimum	Maximum	A Quality — Breast and Legs	Elsewhere	Part	B Quality — Breast and Legs[b]	Elsewhere	Part	C Quality
None	1½ lbs	None	¾ in.	Slight trim on edge	¾ in.	1½ in.	Moderate amount of the flesh normally covered	No limit
Over 1½ lbs	6 lbs	None	1½ in.		1½ in.	3 in.		
Over 6 lbs	16 lbs	None	2 in.		2 in.	4 in.		
Over 16 lbs	None	None	3 in.		3 in.	5 in.		

Discolorations[c]

Carcass Weight Minimum	Maximum	A Quality — Breast and Legs	Elsewhere	Part	B Quality — Breast and Legs	Elsewhere	Part	C Quality
None	1½ lbs	½ in.	1 in.	¼ in.	1 in.	2 in.	½ in.	No limit[d]

		A Quality			B Quality			C Quality
Over 1½ lbs	6 lbs	1 in.	2 in.	¼ in.	2 in.	3 in.	1 in.	No limit
Over 6 lbs	16 lbs	1½ in.	2½ in.	½ in.	2½ in.	4 in.	1½ in.	No limit
Over 16 lbs	None	2 in.	3 in.	½ in.	3 in.	5 in.	1½ in.	

	A Quality	B Quality	C Quality
Disjointed bones	1	2 disjointed and no broken or 1 disjointed and 1 nonprotruding broken	No limit
Broken bones	None	Wing tips, 2nd wing joint and tail	Wing tips, wings and tail
Missing parts	Wing tips and tail[e]	Back area not wider than base of tail and extending half way between base of tail and hip joints	Back area not wider than base of tail extending to area between hip joints
Freezing defects (when consumer packaged)	Slight darkening over the back and drumsticks. Few small 1/8 in. pockmarks for poultry weighing 6 lbs or less and ¼ in. pockmarks for poultry weighing more than 6 lbs. Occasional small areas showing layer of clear or pinkish ice.	Moderate dried areas not in excess of ½ in. in diameter. May lack brightness. Moderate areas showing layer of clear, pinkish or reddish colored ice.	Numerous pockmarks and large dried areas.

[a] Total aggregate area of flesh exposed by all cuts and tears and missing skin, not exceeding the area of a circle of the diameters shown.

[b] A carcass meeting the requirements of A quality for fleshing may be trimmed to remove skin and flesh defects, provided that no more than one third of the flesh is exposed on any part and the meat yield is not appreciably affected.

[c] Flesh bruises and discolorations such as blue back are not permitted on breast and legs of A quality birds. Not more than one half of total aggregate area of discolorations may be due to flesh bruises or blue back (when permitted), and skin bruises in any combination.

[d] No limit on size and number of areas of discoloration and flesh bruises if such areas do not render any part of the carcass unfit for food.

[e] In ducks and geese, the parts of the wing beyond the second joint may be removed, if removed at the joint and both wings are so treated.

Table 410
COMPOSITION OF 100 g EDIBLE TURKEY (ALL CLASSES)

Nutrient (unit)	Raw flesh and skin (mean)[a]	Raw flesh only (mean)[a]	Cooked flesh and skin (mean)[a,b]	Cooked flesh only (mean)[a,b]
Water (g)	70.40	74.16	61.70	74.88
Protein (g)	20.42	21.77	28.10	29.32
Total lipid (fat) (g)	8.02	2.86	9.73	4.97
Ash (g)	0.88	0.97	1.00	1.05
Calcium (mg)	15	14	26	25
Iron (mg)	1.43	1.45	1.79	1.78
Magnesium (mg)	22	25	25	26
Phosphorus (mg)	178	195	203	213
Potassium (mg)	266	296	280	298
Sodium (mg)	65	70	68	70
Zinc (mg)	2.20	2.37	2.96	3.10
Copper (mg)	0.103	0.109	0.93	0.094
Manganese (mg)	0.020	0.021	0.021	0.021
Thiamin (mg)	0.064	0.072	0.057	0.062
Riboflavin (mg)	0.155	0.168	0.177	0.182
Niacin (mg)	4.085	4.544	5.086	5.443
Pantothenic acid (mg)	0.807	0.907	0.858	0.943
Vitamin B_6 (mg)	0.41	0.47	0.41	0.46
Vitamin B_{12} (mcg)	0.40	0.43	0.35	0.37
Vitamin A (IU)	6	0	0	0
Cholesterol (mg)	68	65	82	76
Fatty acids, total				
Saturated (g)	2.26	0.95	2.84	1.64
Monounsaturated (g)	2.90	0.61	3.19	1.03
Polyunsaturated (g)	1.98	0.83	2.48	1.43

[a]   Without neck.
[b]   Roasted.

From Composition of Foods: Poultry Products, Posati, L. P., Principal Investigator, *U.S. Dep. Agric. Agric. Handb.*, No. 8—5, rev. August 1979.

Table 411
COMPOSITION OF 100 g EDIBLE CHICKEN (BROILERS OR FRYERS)

Nutrient (unit)	Raw flesh and skin (mean)[a]	Raw flesh only (mean)[a]	Fried, flesh and skin (mean)[a,b]	Fried, flesh only (mean)[a]
Water (g)	65.99	75.46	49.39	57.53
Protein (g)	18.60	21.39	22.54	30.57
Total lipid (fat)(g)	15.06	3.08	17.35	9.12
Ash (g)	0.79	0.96	1.30	1.09
Calcium (mg)	11	12	21	17
Iron (mg)	0.90	0.89	1.37	1.35
Magnesium (mg)	20	25	21	27
Phosphorus (mg)	147	173	155	205
Potassium (mg)	189	229	185	257
Sodium (mg)	70	77	292	91
Zinc (mg)	1.31	1.54	1.67	2.24
Copper (mg)	0.048	0.053	0.072	0.075
Manganese (mg)	0.019	0.019	0.057	0.028
Thiamin (mg)	0.060	0.073	0.115	0.085
Riboflavin (mg)	0.120	0.142	0.189	0.198
Niacin (mg)	6.801	8.239	7.043	9.663
Pantothenic acid (mg)	0.910	1.058	0.889	1.166
Vitamin B_6 (mg)	0.35	0.43	0.31	0.48
Vitamin B_{12} (mcg)	0.31	0.37	0.28	0.34
Vitamin A (IU)	140	52	93	59
Cholesterol (mg)	75	70	87	94
Fatty acids, total				
Saturated (g)	4.31	0.79	4.61	2.46
Monounsaturated (g)	6.24	0.90	7.09	3.35
Polyunsaturated (g)	3.23	0.75	4.10	2.15

[a] Without neck.
[b] Batter dipped.

From Composition of Foods: Poultry Products, Posati, L. P., Principal Investigator, *U.S. Dep. Agric. Agric. Handb.*, No. 8—5, rev. August 1979.

COMPARATIVE PRODUCTIVITY

Table 412
ANNUAL YIELDS FROM ANIMALS AND CROPS

	Energy (Mcal/ha)	Protein (kg/ha)
Dairy cows	2,500	115
Dairy and beef herds	2,400	102
Beef Cattle	750	27
Sheep	500	23
Pigs	1,900	50
Broilers	1,100	92
Eggs	1,150	88
Wheat	14,000	350
Peas	3,000	280
Cabbage	8,000	1,100
Potatoes	24,000	420

From Holmes, W., Animals for food, *Proc. Nutr. Soc.,* 29, 237—243, 1970. With permission of Cambridge University Press.

Table 413
EFFICIENCY WITH WHICH FARM ANIMALS PRODUCE FOOD PROTEIN

Food product	Level and/or rate of output	Protein production (g/Mcal of digestible energy)
Eggs	200 eggs/year	10.1
	250 eggs/year	13.7
Broiler	1.59 kg/12 wk; 3 kg feed/1 kg gain	11.9
	1.59 kg/10 wk; 2.5 kg feed/1 kg gain	13.7
	1.59 kg/8 wk; 2.1 kg feed/1 kg gain	15.9
Pork	91 kg/8.3 mo; 6 kg feed/1 kg gain	5.0
	91 kg/6.0 mo; 4 kg feed/1 kg gain	6.1
	91 kg/4.4 mo; 2.5 kg feed/1 kg gain	8.7
	91 kg/3.7 mo; 2.0 kg feed/1 kg gain	
Milk	3,600 kg/yr; No concentrates	10.5
	5,400 kg/yr; 25% of energy as concentrates	12.8
	9,072 kg/yr; 50% of energy as concentrates	16.3
	13,608 kg/yr; 65% of energy as concentrates	20.5
Beef	500 kg/15 mo; 8 kg feed/1 kg gain	2.3
	500 kg/12 mo; 5 kg feed/1 kg gain	3.2
	Highly intensive system; no losses	4.1

From Reid, J. T. and White, O. D., in *New Protein Foods,* Vol. 3, Altschul, A. M. and Wiecke, H. L., Eds., Academic Press, New York, 1978, 117.

Table 414
LAND REQUIRED TO PRODUCE 1 MILLION
CALORIES

Food source	Acres of land to produce 1 million calories
Sugar	0.15
Potatoes	0.44
Corn (as corn meal)	0.9
Wheat (as whole wheat flour)	0.9
Wheat (as refined wheat flour)	1.2
Hogs (pork and lard)	2.0
Whole milk	2.8
Eggs	7.8
Chickens	9.3
Steers	17.0

CONVERSION FACTORS

Table 415
VARIOUS CONVERSIONS AND CALCULATIONS

Unit	Equivalent

Area

Unit	Equivalent
1 in.^{2}	6.451626 cm^2
1 ft.^{2}	144 in.^{2} = 0.093 meter^2
1 yd^2	9 ft^2 = 0.836 meter^2
1 acre	43,560 ft^2 = 4,839 yd^2 = 0.404687 ha
1 mile^2	1 section = 640 acres = 258.998 ha
1 centimeter^2	100 mm^2 = 0.155 in.^{2}
1 meter^2	1 centare (ca.) = 1.196 yd^2
1 are	100 m^2
1 hectare (ha)	100 are = 10,000 m^2 = 2.47104 acres
1 kilometer^2	100 ha = 0.386 mile^2

Capacity Liquid (based on U.S. measures[a])

Unit	Equivalent
1 teaspoon	4.93 ml (ml = cm^3 [cc])
3 teaspoons	1 tablespoon = 14.79 ml
2 tablespoons (tb)	1 fluid oz = 29.578 ml
1 cup	8 fluid oz = 0.5 pt = 236.5835 ml
1 pint (pt)	473.167 ml = 0.473167 l
1 quart (qt)	2 pt = 0.946333 l
1 gallon (gal)	4 qt = 231 in.^{3} = 128 fluid oz = 3.785332 l
1 centiliter	10 ml = 0.021 pt
1 liter (l)	1000 ml or cm^3 = 2.11342 pt = 1.05671 qt

[a]   1 U.S. (liquid) pint, quart, or gallon = 0.833 British pint, quart, or gallon, respectively; and 1 British pint, quart, or gallon = 1.201 U.S. (liquid) pints, quarts, or gallons, respectively.

Capacity Dry (1 U.S. quart, dry measure = 1.1012 liters)

Unit	Equivalent
1 pint	33.6003125 in.^{3} = 0.550599 l
1 quart	2 pt = 1.101198 l
1 peck	8 qt = 8.80958 l
1 bushel	4 pk = 32 qt = 35.2383 l
1 liter	62.025 in.^{3} = 0.908102 qt
1 dekaliter	10 l = 0.28378 bu
1 hecktoliter	100 l = 2.8378 bu

Volume

Unit	Equivalent
1 in.^{3}	16.387162 cm^3
1 ft^3	1,728 in.^{3} = 28.317 decimeters^3
1 yd^3	27 ft^3 = 0.765 m^3
1 bushel (bu)	2,219.3 in.^{3} = 36.377 decimeters^3
1 cubic centimeter (cm^3 [cc] = ml)	0.061 in.^{3}
1 cubic decimeter (dm^3)	1000 cm^3 = 0.035 ft^3
1 cubic meter (m^3)	1000 dm^3 = 27.50 bu

Length

Unit	Equivalent
1 inch	25.4 mm = 2.540005 cm
1 link	7.92 in.

Table 415 (continued)
VARIOUS CONVERSIONS AND CALCULATIONS

Unit	Equivalent
1 foot	12 in. = 30.48006 cm
1 yard	3 ft = 36 in. = 0.9144 m
1 rod	16.5 ft = 5.5 yd = 25 links
1 chain	66 ft = 22 yd = 100 links
1 mile	5,280 ft = 1,760 yd = 1.609 km
1 millimeter (mm)	0.03937 in.
1 centimeter (cm)	10 mm = 0.3937 in.
1 meter (m)	39.37 in. = 3.28 ft
1 kilometer (km)	1000 m = 0.62137 mile

Weight

1 grain (gr)	64.798918 milligrams
1 ounce (oz)	437.5 gr = 28.349527 gm
1 pound (lb)	16 oz = 453.5924 gm = 0.454 kg
1 stone (st)	14 lb = 6.350 kg
1 (short) ton (U.S.)	2000 lb = 0.907 metric ton
1 (long) ton	2240 lb = 1.016 metric ton
1 milligram (mg)	0.015 grain = 0.001 gm
1 gram (gm)	15.432356 grain = 1000 mg
1 kilogram (kg)	1000 grams = 2.204622341 lb
1 metric quintal	100 kg = 220.46 lb
1 metric ton	1000 kg = 1.102 (short) ton, 0.984 (long) ton

Temperature

$1°C = 1.8°F$ $1°F = 5/9°C$

$°C = (°F - 32) \times 5/9$ $°F = 9/5°C + 32$

°C	°F	°C	°F	°C	°F
100	212	50	122	0	32
90	194	40	104	−10	14
80	176	30	86	−20	−4
70	158	20	68	−30	−22
60	140	10	50	−40	−40

Yield or Rate

Metric	English equivalent
ton/hectare	0.446 ton/acre
hectoliter/hectare	7.013 bushel/acre
kilogram/hectare	0.892 pound/acre
quintal/hectare	89.24 pound/acre

Light

lux (lx)	0.9029 foot candle (Fc)

Work and Energy

calorie (cal)	3.97×10^{-3} British thermal unit (Btu)

Table 415 (continued)
VARIOUS CONVERSIONS AND CALCULATIONS

Unit	Equivalant

Water

1 ft³	7.48 gal = 62.42 lb
1 acre-foot	43,560 ft³ = 323,136 gal
1 second foot	1 ft³ per second flow past a given point or 7.5 gal/sec (approx.)
1 second foot in 1 hour	1 acre-inch (approx.)

Dry Soil

1 ft³:muck	25—30 lb; clay and silt = 68—80 lb; loam = 80—95 lb: and sand = 100—110 lb.
1 acre foot (43,560 ft³)	2,000 tons approx.
Soil surface plow depth (6 2/3 in.)	1,000 tons/acre approx.

Amount of Formulated Chemical (Active Ingredient) Added to One Gallon to Apply One Pound Per Acre

Solid Formulations

Active ingredient content (%)	Grams in each gal for 1 lb/A at designated gallonage/A	
	10 gal/A	15 gal/A
10	453.39	302.39
20	226.80	151.20
25	181.44	120.96
50	90.72	60.48
75	60.48	40.32
100	45.36	30.24

Liquid Formulations

Pounds active ingredient/gal formulation	Milliliters needed for 1 gram	Milliliters/gal for 1 lb/A at designated gallonage/A	
		10 gal/A	15 gal/A
1	8.35	378.53	252.48
2	4.17	189.27	126.24
3	2.78	126.18	84.12
4	2.09	94.63	63.09
5	1.67	75.71	50.47
6	1.39	63.09	42.06

Source: Adapted from *Herbicide Handbook,* 5th Edition, Weed Society of America, Champaign, Ill., 1983.

Table 416
CONVERSION FACTORS FOR SI AND NON-SI UNITS

To convert Column 1 into Column 2, multiply by	Column 1 SI Unit	Column 2 non-SI Unit	To convert Column 2 into Column 1 multiply by

Length

0.621	kilometer, km (10^3 m)	mile, mi	1.609
1.094	meter, m	yard, yd	0.914
3.28	meter, m	foot, ft	0.304
1.0	micrometer, μm (10^{-6} m)	micron, μ	1.0
3.94×10^{-2}	millimeter mm (10^{-3} m)	inch, in	25.4
10	nanometer, nm (10^{-9} m)	Angstrom, Å	0.1

Area

2.47	hectare, ha	acre	0.405
247	square kilometer, km^2 (10^3 m)^2	acre	4.05×10^{-3}
0.386	square kilometer, km^2 (10^3 m)^2	square mile, mi^2	2.590
2.47×10^{-4}	square meter, m^2	acre	4.05×10^3
10.76	square meter, m^2	square foot, ft^2	9.29×10^{-2}
1.55×10^{-3}	square millimeter, mm^2 (10^{-6} m)^2	square inch, in^2	645

Volume

9.73×10^{-3}	cubic meter, m^3	acre-inch	102.8
35.3	cubic meter, m^3	cubic foot, ft^3	2.83×10^{-2}
6.10×10^4	cubic meter, m^3	cubic inch, in^3	1.64×10^{-5}
2.84×10^{-2}	liter, L (10^{-3} m^3)	bushel, bu	35.24
1.057	liter, L (10^{-3} m^3)	quart (liquid), qt	0.946
3.53×10^{-2}	liter, L (10^{-3} m^3)	cubic foot, ft^3	28.3
0.265	liter, L (10^{-3} m^3)	gallon	3.78
33.78	liter, L (10^{-3} m^3)	ounce (fluid), oz	2.96×10^{-2}
2.11	liter, L (10^{-3} m^3)	pint (fluid), pt	0.473

Mass

2.20×10^{-3}	gram, g (10^{-3} kg)	pound, lb	454
3.52×10^{-2}	gram, g (10^{-3} kg)	ounce (avdp), oz	28.4
2.205	kilogram, kg	pound, lb	0.454
0.01	kilogram, kg	quintal (metric), q	100
1.10×10^{-3}	kilogram, kg	ton (2000 lb), ton	907
1.102	megagram, Mg (tonne)	ton (U.S.), ton	0.907
1.102	tonne, t	ton (U.S.), ton	0.907

Yield and Rate

0.893	kilogram per hectare, kg ha^{-1}	pound per acre, lb acre^{-1}	1.12
7.77×10^{-2}	kilogram per cubic meter, kg m^{-3}	pound per bushel, lb bu^{-1}	12.87
1.49×10^{-2}	kilogram per hectare, kg ha^{-1}	bushel per acre, 60 lb	67.19
1.59×10^{-2}	kilogram per hectare, kg ha^{-1}	bushel per acre, 56 lb	62.71
1.86×10^{-2}	kilogram per hectare, kg ha^{-1}	bushel per acre, 48 lb	53.75
0.107	liter per hectare, L ha^{-1}	gallon per acre	9.35
893	tonnes per hectare, t ha^{-1}	pound per acre, lb acre^{-1}	1.12×10^{-3}
893	megagram per hectare, Mg ha^{-1}	pound per acre, lb acre^{-1}	1.12×10^{-3}

Table 416 (continued)
CONVERSION FACTORS FOR SI AND NON-SI UNITS

To convert Column 1 into Column 2, multiply by	Column 1 SI Unit	Column 2 non-SI Unit	To convert Column 2 into Column 1 multiply by
0.446	megagram per hectare, Mg ha^{-1}	ton (2000 lb) per acre, ton acre^{-1}	2.24
2.24	meter per second, m s^{-1}	mile per hour	0.447

Specific Surface

To convert Column 1 into Column 2, multiply by	Column 1 SI Unit	Column 2 non-SI Unit	To convert Column 2 into Column 1 multiply by
10	square meter per kilogram, m^2 kg^{-1}	square centimeter per gram, cm^2 g^{-1}	0.1
1000	square meter per kilogram, m^2 kg^{-1}	square millimeter per gram, mm^2 g^{-1}	0.001

Pressure

To convert Column 1 into Column 2, multiply by	Column 1 SI Unit	Column 2 non-SI Unit	To convert Column 2 into Column 1 multiply by
9.90	megapascal, MPa (10^6 Pa)	atmosphere	0.101
10	megapascal, MPa (10^6 Pa)	bar	0.1
1.00	megagram per cubic meter, Mg m^{-3}	gram per cubic centimeter, g cm^{-3}	1.00
2.09×10^{-2}	pascal, Pa	pound per square foot, lb ft^{-2}	47.9
1.45×10^{-4}	pascal, Pa	pound per square inch, lb in^{-2}	6.90×10^3

Temperature

To convert Column 1 into Column 2, multiply by	Column 1 SI Unit	Column 2 non-SI Unit	To convert Column 2 into Column 1 multiply by
1.00 (K -273)	Kelvin, K	Celsius, °C	1.00 (°C $+273$)
(9/5 °C) $+32$	Celsius, °C	Fahrenheit, °F	5/9 (°F -32)

Energy, Work, Quantity of Heat

To convert Column 1 into Column 2, multiply by	Column 1 SI Unit	Column 2 non-SI Unit	To convert Column 2 into Column 1 multiply by
9.52×10^{-4}	joule, J	British thermal unit, Btu	1.05×10^3
0.239	joule, J	calorie, cal	4.19
10^7	joule, J	erg	10^{-7}
0.735	joule, J	foot-pound	1.36
2.387×10^{-5}	joule per square meter, J m^{-2}	calorie per square centimeter (langley)	4.19×10^4
10^5	newton, N	dyne	10^{-5}
1.43×10^{-3}	watt per square meter, W m^{-2}	calorie per square centimeter minute (irradiance), cal cm^{-2} min^{-1}	698

Transpiration and Photosynthesis

To convert Column 1 into Column 2, multiply by	Column 1 SI Unit	Column 2 non-SI Unit	To convert Column 2 into Column 1 multiply by
3.60×10^{-2}	milligram per square meter second, mg m^{-2} s^{-1}	gram per square decimeter hour, g cm^{-2} h^{-1}	27.8
5.56×10^{-3}	milligram (H_2O) per square meter second, mg m^{-2} s^{-1}	micromole (H_2O) per square centimeter second, μmol cm^{-2} s^{-1}	180
10^{-4}	milligram per square meter second, mg m^{-2} s^{-1}	milligram per square centimeter second, mg cm^{-2} s^{-1}	10^4
35.97	milligram per square meter second, mg m^{-2} s^{-1}	milligram per square decimeter hour, mg dm^{-2} h^{-1}	2.78×10^{-2}

Table 416 (continued)
CONVERSION FACTORS FOR SI AND NON-SI UNITS

To convert Column 1 into Column 2, multiply by	Column 1 SI Unit	Column 2 non-SI Unit	To convert Column 2 into Column 1 multiply by
		Plane Angle	
57.3	radian, rad	degrees (angle), °	1.75×10^{-2}
		Electrical Conductivity, Electricity, and Magnetism	
10	siemen per meter, S m^{-1}	millimho per centimeter, mmho cm^{-1}	0.1
10^4	tesla, T	gauss, G	10^{-4}
		Water Measurement	
9.73×10^{-3}	cubic meter, m^3	acre-inches, acre-in	102.8
9.81×10^{-3}	cubic meter per hour, m^3 h^{-1}	cubic feet per second, ft^3 s^{-1}	101.9
4.40	cubic meter per hour, m^3 h^{-1}	U.S. gallons per minute, gal min^{-1}	0.227
8.11	hectare-meters, ha-m	acre-feet, acre-ft	0.123
97.28	hectare-meters, ha-m	acre-inches, acre-in	1.03×10^{-2}
8.1×10^{-2}	hectare-centimeters, ha-cm	acre-feet, acre-ft	12.33
		Concentrations	
1	centimole per kilogram, cmol kg^{-1} (ion exchange capacity)	milliequivalents per 100 grams, meq 100 g^{-1}	1
0.1	gram per kilogram, g kg^{-1}	percent, %	10
1	milligram per kilogram, mg kg^{-1}	parts per million, ppm	1
		Radioactivity	
2.7×10^{-11}	bequerel, Bq	curie, Ci	3.7×10^{10}
2.7×10^{-2}	bequerel per kilogram, Bq kg^{-1}	picocurie per gram, pCi g^{-1}	37
100	gray, Gy (absorbed dose)	rad, rd	0.01
100	sievert, Sv (equivalent dose)	rem (roentgen equivalent man)	0.01

Plant Nutrient Conversion

Elemental		Oxide	
2.29	P	P_2O_5	0.437
1.20	K	K_2O	0.830
1.39	Ca	CaO	0.715
1.66	Mg	MgO	0.602

Table 417
NATURE OF COMPUTATIONS INVOLVED IN SOME COMMON STATISTICAL TERMS

Chi (χ): the chi-square (χ^2) test is applied to determine the goodness of fit between the observed and expected values.

$$\chi^2 = S\left(\frac{x^2}{m}\right)$$

where x = difference between the observed and expected value in any one class; m = expected value in any one class; and S = summation over all classes.

Correlation Coefficient (r): mathematical measure of the degree of association between two variables, X and Y.

$$r = \frac{S(dx \cdot dy)}{\sqrt{Sdx^2 \cdot Sdy^2}} \quad \text{or} \quad \frac{S(XY) - (SX)\bar{y}}{\sqrt{[S(X^2) - (SX)\bar{x}] \cdot [S(Y^2) - (SY)\bar{y}]}}$$

where dx = deviation of the individual X variables from the mean ($\bar{x}$); dy = deviation of the individual Y variables from the mean ($\bar{y}$); S = summation over all classes; and X and Y = individual sample values (or variables).

Mean-square error:

$$s = \sqrt{\frac{S(X - \bar{x})^2}{n - 1}}$$

where X = single observation (sample value or variable); $\bar{x}$ = mean of all observations; S = summation of the series; and n = total number of observations.

Variance:

$$s^2 = \frac{S(X - \bar{x})^2}{n - 1}$$

Standard deviation ($\hat{\sigma}$ or s): the standard deviation is the square root of the variance.

$$s = \sqrt{\frac{S(d)^2}{n - 1}} \quad \text{or} \quad \sqrt{\frac{S(X^2) - (SX)\bar{x}}{n - 1}}$$

where S = summation; d = difference between an observation and the mean of all observations, i.e., $\bar{x}$; n = total number of observations; X = single observation.

Standard error of mean:

$$s_x = \frac{s}{\sqrt{n}}$$

Regression coefficient: the regression coefficient (b) is an expression of the change in Y per unit change in X.

$$b = \frac{S(X - \bar{x})(Y - \bar{y})}{S(X - \bar{x})^2}$$

Least significant difference (LSD): the least difference between the means of specific treatments and that of a predetermined control or check treatment that can be accepted as significant (normally at the 5% level, or at the 1% level for a highly significant difference).

$$LSD = t\sqrt{2s_{\bar{x}}^2}$$

Table 417 (continued)
NATURE OF COMPUTATIONS INVOLVED IN SOME COMMON
STATISTICAL TERMS

Duncan's multiple range test: standard error of the mean is derived from the error mean square in an analysis of variance:

$$s_{\bar{x}} = \sqrt{\frac{\hat{\sigma}^2}{r}}$$

where $\hat{\sigma}^2$ = mean square for error; r = number of replications.

Select appropriate size range (p); confidence level 5% or 1%; and enter table with appropriate degrees of freedom (n^2). Shortest significant range Rp = value from table $\times$ $s_{\bar{x}}$.

Each difference is significant when it exceeds the calculated shortest significant range; otherwise it is not significantly different. Note: no difference between two means can be judged significant when the two means in question are both contained in a subset of the means that has a non-significant range.

STATISTICAL TABLES

Table 418
TABLE OF χ^2

n.	P = .99	.98	.95	.90	.80	.70	.50	.30	.20	.10	.05	.02	.01
1	.000157	.000628	.00393	.0158	.0642	.148	.455	1.074	1.642	2.706	3.841	5.412	6.635
2	.0201	.0404	.103	.211	.446	.713	1.386	2.408	3.219	4.605	5.991	7.824	9.210
3	.115	.185	.352	.584	1.005	1.424	2.366	3.665	4.642	6.251	7.815	9.837	11.341
4	.297	.429	.711	1.064	1.649	2.195	3.357	4.878	5.989	7.779	9.488	11.668	13.277
5	.554	.752	1.145	1.610	2.343	3.000	4.351	6.064	7.289	9.236	11.070	13.388	15.086
6	.872	1.134	1.635	2.204	3.070	3.828	5.348	7.231	8.558	10.645	12.592	15.033	16.812
7	1.239	1.564	2.167	2.833	3.822	4.671	6.346	8.383	9.803	12.017	14.067	16.622	18.475
8	1.646	2.032	2.733	3.490	4.594	5.527	7.344	9.524	11.030	13.362	15.507	18.168	20.090
9	2.088	2.532	3.325	4.168	5.380	6.393	8.343	10.656	12.242	14.684	16.919	19.679	21.666
10	2.558	3.059	3.940	4.865	6.179	7.267	9.342	11.781	13.442	15.987	18.307	21.161	23.209
11	3.053	3.609	4.575	5.578	6.989	8.148	10.341	12.899	14.631	17.275	19.675	22.618	24.725
12	3.571	4.178	5.226	6.304	7.807	9.034	11.340	14.011	15.812	18.549	21.026	24.054	26.217
13	4.107	4.765	5.892	7.042	8.634	9.926	12.340	15.119	16.985	19.812	22.362	25.472	27.688
14	4.660	5.368	6.571	7.790	9.467	10.821	13.339	16.222	18.151	21.064	23.685	26.873	29.141
15	5.229	5.985	7.261	8.547	10.307	11.721	14.339	17.322	19.311	22.307	24.996	28.259	30.578
16	5.812	6.614	7.962	9.312	11.152	12.624	15.338	18.418	20.465	23.542	26.296	29.633	32.000
17	6.408	7.255	8.672	10.085	12.002	13.531	16.338	19.511	21.615	24.769	27.587	30.995	33.409
18	7.015	7.906	9.390	10.865	12.857	14.440	17.338	20.601	22.760	25.989	28.869	32.346	34.805
19	7.633	8.567	10.117	11.651	13.716	15.352	18.338	21.689	23.900	27.204	30.144	33.687	36.191
20	8.260	9.237	10.851	12.443	14.578	16.266	19.337	22.775	25.038	28.412	31.410	35.020	37.566
21	8.897	9.915	11.591	13.240	15.445	17.182	20.337	23.858	26.171	29.615	32.671	36.343	38.932
22	9.542	10.600	12.338	14.041	16.314	18.101	21.337	24.939	27.301	30.813	33.924	37.659	40.289
23	10.196	11.293	13.091	14.848	17.187	19.021	22.337	26.018	28.429	32.007	35.172	38.968	41.638
24	10.856	11.992	13.848	15.659	18.062	19.943	23.337	27.096	29.553	33.196	36.415	40.270	42.980
25	11.524	12.697	14.611	16.473	18.940	20.867	24.337	28.172	30.675	34.382	37.652	41.566	44.314
26	12.198	13.409	15.379	17.292	19.820	21.792	25.336	29.246	31.795	35.563	38.885	42.856	45.642
27	12.879	14.125	16.151	18.114	20.703	22.719	26.336	30.319	32.912	36.741	40.113	44.140	46.963
28	13.565	14.847	16.928	18.939	21.588	23.647	27.336	31.391	34.027	37.916	41.337	45.419	48.278
29	14.256	15.574	17.708	19.768	22.475	24.577	28.336	32.461	35.139	39.087	42.557	46.693	49.588
30	14.953	16.306	18.493	20.599	23.364	25.508	29.336	33.530	36.250	40.256	43.773	47.962	50.892

For larger values of n, the expression $\sqrt{2\chi^2} - \sqrt{2n-1}$ may be used as a normal deviate with unit variance.

Reprinted from Table III of Fisher: Statistical Methods for Research Workers, published by Oliver & Boyd, Ltd., Edinburgh, by permission of the publishers. From Fisher, R. A., and Yates, F., *Statistical Tables for Biological, Agricultural, and Medical Research*, 6th ed., Oliver & Boyd, Edinburgh, 1974.

Table 419
VALUES OF THE CORRELATION COEFFICIENT FOR DIFFERENT LEVELS OF SIGNIFICANCE

n	.1	.05	.02	.01	.001	n	.1	.05	.02	.01	.001
1	.98769	.99692	.999507	.999877	.9999988	16	.4000	.4683	.5425	.5897	.7084
2	.90000	.95000	.98000	.990000	.99900	17	.3887	.4555	.5285	.5751	.6932
3	.8054	.8783	.93433	.95873	.99116	18	.3783	.4438	.5155	.5614	.6787
4	.7293	.8114	.8822	.91720	.97406	19	.3687	.4329	.5034	.5487	.6652
5	.6694	.7545	.8329	.8745	.95074	20	.3598	.4227	.4921	.5368	.6524
6	.6215	.7067	.7887	.8343	.92493	25	.3233	.3809	.4451	.4869	.5974
7	.5822	.6664	.7498	.7977	.8982	30	.2960	.3494	.4093	.4487	.5541
8	.5494	.6319	.7155	.7646	.8721	35	.2746	.3246	.3810	.4182	.5189
9	.5214	.6021	.6851	.7348	.8471	40	.2573	.3044	.3578	.3932	.4896
10	.4973	.5760	.6581	.7079	.8233	45	.2428	.2875	.3384	.3721	.4648
11	.4762	.5529	.6339	.6835	.8010	50	.2306	.2732	.3218	.3541	.4433
12	.4575	.5324	.6120	.6614	.7800	60	.2108	.2500	.2948	.3248	.4078
13	.4409	.5139	.5923	.6411	.7603	70	.1954	.2319	.2737	.3017	.3799
14	.4259	.4973	.5742	.6226	.7420	80	.1829	.2172	.2565	.2830	.3568
15	.4124	.4821	.5577	.6055	.7246	90	.1726	.2050	.2422	.2673	.3375
						100	.1638	.1946	.2301	.2540	.3211

From Table VII of Fisher and Yates': *Statistical Tables for Biological, Agricultural, and Medical Research,* published by the Longman Group U.K. Ltd., London (previously published by Oliver & Boyd, Ltd., Edinburgh) and by permission of the publishers.

Table 420
VALUES OF F AND t

5% (Roman Type) and 1% (Bold Face Type) Points for the Distribution of F

Each cell: 5% (Roman) / 1% (Bold Face)

f_2	1	2	3	4	5	6	7	8	9	10	11	12	14	16	20	24	30	40	50	75	100	200	500	∞	Values of t
1	161 / **4,052**	200 / **4,999**	216 / **5,403**	225 / **5,625**	230 / **5,764**	234 / **5,859**	237 / **5,928**	239 / **5,981**	241 / **6,022**	242 / **6,056**	243 / **6,082**	244 / **6,106**	245 / **6,142**	246 / **6,169**	248 / **6,208**	249 / **6,234**	250 / **6,258**	251 / **6,286**	252 / **6,302**	253 / **6,323**	253 / **6,334**	254 / **6,352**	254 / **6,361**	254 / **6,366**	12.7 / **63.7**
2	18.51 / **98.49**	19.00 / **99.00**	19.16 / **99.17**	19.25 / **99.25**	19.30 / **99.30**	19.33 / **99.33**	19.36 / **99.34**	19.37 / **99.36**	19.38 / **99.38**	19.39 / **99.40**	19.40 / **99.41**	19.41 / **99.42**	19.42 / **99.43**	19.43 / **99.44**	19.44 / **99.45**	19.45 / **99.46**	19.46 / **99.47**	19.47 / **99.48**	19.47 / **99.48**	19.48 / **99.49**	19.49 / **99.49**	19.49 / **99.49**	19.50 / **99.50**	19.50 / **99.50**	4.30 / **9.92**
3	10.13 / **34.12**	9.55 / **30.82**	9.28 / **29.46**	9.12 / **28.71**	9.01 / **28.24**	8.94 / **27.91**	8.88 / **27.67**	8.84 / **27.49**	8.81 / **27.34**	8.78 / **27.23**	8.76 / **27.13**	8.74 / **27.05**	8.71 / **26.92**	8.69 / **26.83**	8.66 / **26.69**	8.64 / **26.60**	8.62 / **26.50**	8.60 / **26.41**	8.58 / **26.35**	8.57 / **26.27**	8.56 / **26.23**	8.54 / **26.18**	8.54 / **26.14**	8.53 / **26.12**	3.18 / **5.84**
4	7.71 / **21.20**	6.94 / **18.00**	6.59 / **16.69**	6.39 / **15.98**	6.26 / **15.52**	6.16 / **15.21**	6.09 / **14.98**	6.04 / **14.80**	6.00 / **14.66**	5.96 / **14.54**	5.93 / **14.45**	5.91 / **14.37**	5.87 / **14.24**	5.84 / **14.15**	5.80 / **14.02**	5.77 / **13.93**	5.74 / **13.83**	5.71 / **13.74**	5.70 / **13.69**	5.68 / **13.61**	5.66 / **13.57**	5.65 / **13.52**	5.64 / **13.48**	5.63 / **13.46**	2.78 / **4.60**
5	6.61 / **16.26**	5.79 / **13.27**	5.41 / **12.06**	5.19 / **11.39**	5.05 / **10.97**	4.95 / **10.67**	4.88 / **10.45**	4.82 / **10.27**	4.78 / **10.15**	4.74 / **10.05**	4.70 / **9.96**	4.68 / **9.89**	4.64 / **9.77**	4.60 / **9.68**	4.56 / **9.55**	4.53 / **9.47**	4.50 / **9.38**	4.46 / **9.29**	4.44 / **9.24**	4.42 / **9.17**	4.40 / **9.13**	4.38 / **9.07**	4.37 / **9.04**	4.36 / **9.02**	2.57 / **4.03**
6	5.99 / **13.74**	5.14 / **10.92**	4.76 / **9.78**	4.53 / **9.15**	4.39 / **8.75**	4.28 / **8.47**	4.21 / **8.26**	4.15 / **8.10**	4.10 / **7.98**	4.06 / **7.87**	4.03 / **7.79**	4.00 / **7.72**	3.96 / **7.60**	3.92 / **7.52**	3.87 / **7.39**	3.84 / **7.31**	3.81 / **7.23**	3.77 / **7.14**	3.75 / **7.09**	3.72 / **7.02**	3.71 / **6.99**	3.69 / **6.94**	3.68 / **6.90**	3.67 / **6.88**	2.45 / **3.71**
7	5.59 / **12.25**	4.74 / **9.55**	4.35 / **8.45**	4.12 / **7.85**	3.97 / **7.46**	3.87 / **7.19**	3.79 / **7.00**	3.73 / **6.84**	3.68 / **6.71**	3.63 / **6.62**	3.60 / **6.54**	3.57 / **6.47**	3.52 / **6.35**	3.49 / **6.27**	3.44 / **6.15**	3.41 / **6.07**	3.38 / **5.98**	3.34 / **5.90**	3.32 / **5.85**	3.29 / **5.78**	3.28 / **5.75**	3.25 / **5.70**	3.24 / **5.67**	3.23 / **5.65**	2.36 / **3.50**
8	5.32 / **11.26**	4.46 / **8.65**	4.07 / **7.59**	3.84 / **7.01**	3.69 / **6.63**	3.58 / **6.37**	3.50 / **6.19**	3.44 / **6.03**	3.39 / **5.91**	3.34 / **5.82**	3.31 / **5.74**	3.28 / **5.67**	3.23 / **5.56**	3.20 / **5.48**	3.15 / **5.36**	3.12 / **5.28**	3.08 / **5.20**	3.05 / **5.11**	3.03 / **5.06**	3.00 / **5.00**	2.98 / **4.96**	2.96 / **4.91**	2.94 / **4.88**	2.93 / **4.86**	2.31 / **3.36**
9	5.12 / **10.56**	4.26 / **8.02**	3.86 / **6.99**	3.63 / **6.42**	3.48 / **6.06**	3.37 / **5.80**	3.29 / **5.62**	3.23 / **5.47**	3.18 / **5.35**	3.13 / **5.26**	3.10 / **5.18**	3.07 / **5.11**	3.02 / **5.00**	2.98 / **4.92**	2.93 / **4.80**	2.90 / **4.73**	2.86 / **4.64**	2.82 / **4.56**	2.80 / **4.51**	2.77 / **4.45**	2.76 / **4.41**	2.73 / **4.36**	2.72 / **4.33**	2.71 / **4.31**	2.26 / **3.25**
10	4.96 / **10.04**	4.10 / **7.56**	3.71 / **6.55**	3.48 / **5.99**	3.33 / **5.64**	3.22 / **5.39**	3.14 / **5.21**	3.07 / **5.06**	3.02 / **4.95**	2.97 / **4.85**	2.94 / **4.78**	2.91 / **4.71**	2.86 / **4.60**	2.82 / **4.52**	2.77 / **4.41**	2.74 / **4.33**	2.70 / **4.25**	2.67 / **4.17**	2.64 / **4.12**	2.61 / **4.05**	2.59 / **4.01**	2.56 / **3.96**	2.55 / **3.93**	2.54 / **3.91**	2.23 / **3.17**
11	4.84 / **9.65**	3.98 / **7.20**	3.59 / **6.22**	3.36 / **5.67**	3.20 / **5.32**	3.09 / **5.07**	3.01 / **4.88**	2.95 / **4.74**	2.90 / **4.63**	2.86 / **4.54**	2.82 / **4.46**	2.79 / **4.40**	2.74 / **4.29**	2.70 / **4.21**	2.65 / **4.10**	2.61 / **4.02**	2.57 / **3.94**	2.53 / **3.86**	2.50 / **3.80**	2.47 / **3.74**	2.45 / **3.70**	2.42 / **3.66**	2.41 / **3.62**	2.40 / **3.60**	2.20 / **3.11**
12	4.75 / **9.33**	3.88 / **6.93**	3.49 / **5.95**	3.26 / **5.41**	3.11 / **5.06**	3.00 / **4.82**	2.92 / **4.65**	2.85 / **4.50**	2.80 / **4.39**	2.76 / **4.30**	2.72 / **4.22**	2.69 / **4.16**	2.64 / **4.05**	2.60 / **3.98**	2.54 / **3.86**	2.50 / **3.78**	2.46 / **3.70**	2.42 / **3.61**	2.40 / **3.56**	2.36 / **3.49**	2.35 / **3.46**	2.32 / **3.41**	2.31 / **3.38**	2.30 / **3.36**	2.18 / **3.05**
13	4.67 / **9.07**	3.80 / **6.70**	3.41 / **5.74**	3.18 / **5.20**	3.02 / **4.86**	2.92 / **4.62**	2.84 / **4.44**	2.77 / **4.30**	2.72 / **4.19**	2.67 / **4.10**	2.63 / **4.02**	2.60 / **3.96**	2.55 / **3.85**	2.51 / **3.78**	2.46 / **3.67**	2.42 / **3.59**	2.38 / **3.51**	2.34 / **3.42**	2.32 / **3.37**	2.28 / **3.30**	2.26 / **3.27**	2.24 / **3.21**	2.22 / **3.18**	2.21 / **3.16**	2.16 / **3.01**

f_1 degrees of freedom (for greater mean square)

Table 420 (continued)
VALUES OF F AND t

f_1 degrees of freedom (for greater mean square)

f_2	1	2	3	4	5	6	7	8	9	10	11	12	14	16	20	24	30	40	50	75	100	200	500	∞	Values of t
14	4.60 / 8.86	3.74 / 6.51	3.34 / 5.56	3.11 / 5.03	2.96 / 4.69	2.85 / 4.46	2.77 / 4.28	2.70 / 4.14	2.65 / 4.03	2.60 / 3.94	2.56 / 3.86	2.53 / 3.80	2.48 / 3.70	2.44 / 3.62	2.39 / 3.51	2.35 / 3.43	2.31 / 3.34	2.27 / 3.26	2.24 / 3.21	2.21 / 3.14	2.19 / 3.11	2.16 / 3.06	2.14 / 3.02	2.13 / 3.00	2.14 / 2.98
15	4.54 / 8.68	3.68 / 6.36	3.29 / 5.42	3.06 / 4.89	2.90 / 4.56	2.79 / 4.32	2.70 / 4.14	2.64 / 4.00	2.59 / 3.89	2.55 / 3.80	2.51 / 3.73	2.48 / 3.67	2.43 / 3.56	2.39 / 3.48	2.33 / 3.36	2.29 / 3.29	2.25 / 3.20	2.21 / 3.12	2.18 / 3.07	2.15 / 3.00	2.12 / 2.97	2.10 / 2.92	2.08 / 2.89	2.07 / 2.87	2.13 / 2.95
16	4.49 / 8.53	3.63 / 6.23	3.24 / 5.29	3.01 / 4.77	2.85 / 4.44	2.74 / 4.20	2.66 / 4.03	2.59 / 3.89	2.54 / 3.78	2.49 / 3.69	2.45 / 3.61	2.42 / 3.55	2.37 / 3.45	2.33 / 3.37	2.28 / 3.25	2.24 / 3.18	2.20 / 3.10	2.16 / 3.01	2.13 / 2.96	2.09 / 2.89	2.07 / 2.86	2.04 / 2.80	2.02 / 2.77	2.01 / 2.75	2.12 / 2.92
17	4.45 / 8.40	3.59 / 6.11	3.20 / 5.18	2.96 / 4.67	2.81 / 4.34	2.70 / 4.10	2.62 / 3.93	2.55 / 3.79	2.50 / 3.68	2.45 / 3.59	2.41 / 3.52	2.38 / 3.45	2.33 / 3.35	2.29 / 3.27	2.23 / 3.16	2.19 / 3.08	2.15 / 3.00	2.11 / 2.92	2.08 / 2.86	2.04 / 2.79	2.02 / 2.76	1.99 / 2.70	1.97 / 2.67	1.96 / 2.65	2.11 / 2.90
18	4.41 / 8.28	3.55 / 6.01	3.16 / 5.09	2.93 / 4.58	2.77 / 4.25	2.66 / 4.01	2.58 / 3.85	2.51 / 3.71	2.46 / 3.60	2.41 / 3.51	2.37 / 3.44	2.34 / 3.37	2.29 / 3.27	2.25 / 3.19	2.19 / 3.07	2.15 / 3.00	2.11 / 2.91	2.07 / 2.83	2.04 / 2.78	2.00 / 2.71	1.98 / 2.68	1.95 / 2.62	1.93 / 2.59	1.92 / 2.57	2.10 / 2.88
19	4.38 / 8.18	3.52 / 5.93	3.13 / 5.01	2.90 / 4.50	2.74 / 4.17	2.63 / 3.94	2.55 / 3.77	2.48 / 3.63	2.43 / 3.52	2.38 / 3.43	2.34 / 3.36	2.31 / 3.30	2.26 / 3.19	2.21 / 3.12	2.15 / 3.00	2.11 / 2.92	2.07 / 2.84	2.02 / 2.76	2.00 / 2.70	1.96 / 2.63	1.94 / 2.60	1.91 / 2.54	1.90 / 2.51	1.88 / 2.49	2.09 / 2.86
20	4.35 / 8.10	3.49 / 5.85	3.10 / 4.94	2.87 / 4.43	2.71 / 4.10	2.60 / 3.87	2.52 / 3.71	2.45 / 3.56	2.40 / 3.45	2.35 / 3.37	2.31 / 3.30	2.28 / 3.23	2.23 / 3.13	2.18 / 3.05	2.12 / 2.94	2.08 / 2.86	2.04 / 2.77	1.99 / 2.69	1.96 / 2.63	1.92 / 2.56	1.90 / 2.53	1.87 / 2.47	1.85 / 2.44	1.84 / 2.42	2.09 / 2.85
21	4.32 / 8.02	3.47 / 5.78	3.07 / 4.87	2.84 / 4.37	2.68 / 4.04	2.57 / 3.81	2.49 / 3.65	2.42 / 3.51	2.37 / 3.40	2.32 / 3.31	2.28 / 3.24	2.25 / 3.17	2.20 / 3.07	2.15 / 2.99	2.09 / 2.88	2.05 / 2.80	2.00 / 2.72	1.96 / 2.63	1.93 / 2.58	1.89 / 2.51	1.87 / 2.47	1.84 / 2.42	1.82 / 2.38	1.81 / 2.36	2.08 / 2.83
22	4.30 / 7.94	3.44 / 5.72	3.05 / 4.82	2.82 / 4.31	2.66 / 3.99	2.55 / 3.76	2.47 / 3.59	2.40 / 3.45	2.35 / 3.35	2.30 / 3.26	2.26 / 3.18	2.23 / 3.12	2.18 / 3.02	2.13 / 2.94	2.07 / 2.83	2.03 / 2.75	1.98 / 2.67	1.93 / 2.58	1.91 / 2.53	1.87 / 2.46	1.84 / 2.42	1.81 / 2.37	1.80 / 2.33	1.78 / 2.31	2.07 / 2.82
23	4.28 / 7.88	3.42 / 5.66	3.03 / 4.76	2.80 / 4.26	2.64 / 3.94	2.53 / 3.71	2.45 / 3.54	2.38 / 3.41	2.32 / 3.30	2.28 / 3.21	2.24 / 3.14	2.20 / 3.07	2.14 / 2.97	2.10 / 2.89	2.04 / 2.78	2.00 / 2.70	1.96 / 2.62	1.91 / 2.53	1.88 / 2.48	1.84 / 2.41	1.82 / 2.37	1.79 / 2.32	1.77 / 2.28	1.76 / 2.26	2.07 / 2.81
24	4.26 / 7.82	3.40 / 5.61	3.01 / 4.72	2.78 / 4.22	2.62 / 3.90	2.51 / 3.67	2.43 / 3.50	2.36 / 3.36	2.30 / 3.25	2.26 / 3.17	2.22 / 3.09	2.18 / 3.03	2.13 / 2.93	2.09 / 2.85	2.02 / 2.74	1.98 / 2.66	1.94 / 2.58	1.89 / 2.49	1.86 / 2.44	1.82 / 2.36	1.80 / 2.33	1.76 / 2.27	1.74 / 2.23	1.73 / 2.21	2.06 / 2.80
25	4.24 / 7.77	3.38 / 5.57	2.99 / 4.68	2.76 / 4.18	2.60 / 3.86	2.49 / 3.63	2.41 / 3.46	2.34 / 3.32	2.28 / 3.21	2.24 / 3.13	2.20 / 3.05	2.16 / 2.99	2.11 / 2.89	2.06 / 2.81	2.00 / 2.70	1.96 / 2.62	1.92 / 2.54	1.87 / 2.45	1.84 / 2.40	1.80 / 2.32	1.77 / 2.29	1.74 / 2.23	1.72 / 2.19	1.71 / 2.17	2.06 / 2.79
26	4.22 / 7.72	3.37 / 5.53	2.98 / 4.64	2.74 / 4.14	2.59 / 3.82	2.47 / 3.59	2.39 / 3.42	2.32 / 3.29	2.27 / 3.17	2.22 / 3.09	2.18 / 3.02	2.15 / 2.96	2.10 / 2.86	2.05 / 2.77	1.99 / 2.66	1.95 / 2.58	1.90 / 2.50	1.85 / 2.41	1.82 / 2.36	1.78 / 2.28	1.76 / 2.25	1.72 / 2.19	1.70 / 2.15	1.69 / 2.13	2.05 / 2.78

Table 420 (continued)
VALUES OF F AND t

f_1 degrees of freedom (for greater mean square)

f_2	1	2	3	4	5	6	7	8	9	10	11	12	14	16	20	24	30	40	50	75	100	200	500	∞	Values of t
27	4.21 / 7.68	3.35 / 5.49	2.96 / 4.60	2.73 / 4.11	2.57 / 3.79	2.46 / 3.56	2.37 / 3.39	2.30 / 3.26	2.25 / 3.14	2.20 / 3.06	2.16 / 2.98	2.13 / 2.93	2.08 / 2.83	2.03 / 2.74	1.97 / 2.63	1.93 / 2.55	1.88 / 2.47	1.84 / 2.38	1.80 / 2.33	1.76 / 2.25	1.74 / 2.21	1.71 / 2.16	1.68 / 2.12	1.67 / 2.10	2.05 / 2.77
28	4.20 / 7.64	3.34 / 5.45	2.95 / 4.57	2.71 / 4.07	2.56 / 3.76	2.44 / 3.53	2.36 / 3.36	2.29 / 3.23	2.24 / 3.11	2.19 / 3.03	2.15 / 2.95	2.12 / 2.90	2.06 / 2.80	2.02 / 2.71	1.96 / 2.60	1.91 / 2.52	1.87 / 2.44	1.81 / 2.35	1.78 / 2.30	1.75 / 2.22	1.72 / 2.18	1.69 / 2.13	1.67 / 2.09	1.65 / 2.06	2.05 / 2.76
29	4.18 / 7.60	3.33 / 5.42	2.93 / 4.54	2.70 / 4.04	2.54 / 3.73	2.43 / 3.50	2.35 / 3.33	2.28 / 3.20	2.22 / 3.08	2.18 / 3.00	2.14 / 2.92	2.10 / 2.87	2.05 / 2.77	2.00 / 2.68	1.94 / 2.57	1.90 / 2.49	1.85 / 2.41	1.80 / 2.32	1.77 / 2.27	1.73 / 2.19	1.71 / 2.15	1.68 / 2.10	1.65 / 2.06	1.64 / 2.03	2.04 / 2.76
30	4.17 / 7.56	3.32 / 5.39	2.92 / 4.51	2.69 / 4.02	2.53 / 3.70	2.42 / 3.47	2.34 / 3.30	2.27 / 3.17	2.21 / 3.06	2.16 / 2.98	2.12 / 2.90	2.09 / 2.84	2.04 / 2.74	1.99 / 2.66	1.93 / 2.55	1.89 / 2.47	1.84 / 2.38	1.79 / 2.29	1.76 / 2.24	1.72 / 2.16	1.69 / 2.13	1.66 / 2.07	1.64 / 2.03	1.62 / 2.01	2.04 / 2.75
32	4.15 / 7.50	3.30 / 5.34	2.90 / 4.46	2.67 / 3.97	2.51 / 3.66	2.40 / 3.42	2.32 / 3.25	2.25 / 3.12	2.19 / 3.01	2.14 / 2.94	2.10 / 2.86	2.07 / 2.80	2.02 / 2.70	1.97 / 2.62	1.91 / 2.51	1.86 / 2.42	1.82 / 2.34	1.76 / 2.25	1.74 / 2.20	1.69 / 2.12	1.67 / 2.08	1.64 / 2.02	1.61 / 1.98	1.59 / 1.96	2.04 / 2.74
34	4.13 / 7.44	3.28 / 5.29	2.88 / 4.42	2.65 / 3.93	2.49 / 3.61	2.38 / 3.38	2.30 / 3.21	2.23 / 3.08	2.17 / 2.97	2.12 / 2.89	2.08 / 2.82	2.05 / 2.76	2.00 / 2.66	1.95 / 2.58	1.89 / 2.47	1.84 / 2.38	1.80 / 2.30	1.74 / 2.21	1.71 / 2.15	1.67 / 2.08	1.64 / 2.04	1.61 / 1.94	1.59 / 1.94	1.57 / 1.91	2.03 / 2.73
36	4.11 / 7.39	3.26 / 5.25	2.86 / 4.38	2.63 / 3.89	2.48 / 3.58	2.36 / 3.35	2.28 / 3.18	2.21 / 3.04	2.15 / 2.94	2.10 / 2.86	2.06 / 2.78	2.03 / 2.72	1.98 / 2.62	1.93 / 2.54	1.87 / 2.43	1.82 / 2.35	1.78 / 2.26	1.72 / 2.17	1.69 / 2.12	1.65 / 2.04	1.62 / 2.00	1.59 / 1.94	1.56 / 1.90	1.55 / 1.87	2.03 / 2.72
38	4.10 / 7.35	3.25 / 5.21	2.85 / 4.34	2.62 / 3.86	2.46 / 3.54	2.35 / 3.32	2.26 / 3.15	2.19 / 3.02	2.14 / 2.91	2.09 / 2.82	2.05 / 2.75	2.02 / 2.69	1.96 / 2.59	1.92 / 2.51	1.85 / 2.40	1.80 / 2.32	1.76 / 2.22	1.71 / 2.14	1.67 / 2.08	1.63 / 2.00	1.60 / 1.97	1.57 / 1.90	1.54 / 1.86	1.53 / 1.84	2.02 / 2.71
40	4.08 / 7.31	3.23 / 5.18	2.84 / 4.31	2.61 / 3.83	2.45 / 3.51	2.34 / 3.29	2.25 / 3.12	2.18 / 2.99	2.12 / 2.88	2.07 / 2.80	2.04 / 2.73	2.00 / 2.66	1.95 / 2.56	1.90 / 2.49	1.84 / 2.37	1.79 / 2.29	1.74 / 2.20	1.69 / 2.11	1.66 / 2.05	1.61 / 1.97	1.59 / 1.94	1.55 / 1.88	1.53 / 1.84	1.51 / 1.81	2.02 / 2.70
42	4.07 / 7.27	3.22 / 5.15	2.83 / 4.29	2.59 / 3.80	2.44 / 3.49	2.32 / 3.26	2.24 / 3.10	2.17 / 2.96	2.11 / 2.86	2.06 / 2.77	2.02 / 2.70	1.99 / 2.64	1.94 / 2.54	1.89 / 2.46	1.82 / 2.35	1.78 / 2.26	1.73 / 2.17	1.68 / 2.08	1.64 / 2.02	1.60 / 1.94	1.57 / 1.91	1.54 / 1.85	1.51 / 1.80	1.49 / 1.78	2.02 / 2.70
44	4.06 / 7.24	3.21 / 5.12	2.82 / 4.26	2.58 / 3.78	2.43 / 3.46	2.31 / 3.24	2.23 / 3.07	2.16 / 2.94	2.10 / 2.84	2.05 / 2.75	2.01 / 2.68	1.98 / 2.62	1.92 / 2.52	1.88 / 2.44	1.81 / 2.32	1.76 / 2.24	1.72 / 2.15	1.66 / 2.06	1.63 / 2.00	1.58 / 1.92	1.56 / 1.88	1.52 / 1.82	1.50 / 1.78	1.48 / 1.75	2.02 / 2.69
46	4.05 / 7.21	3.20 / 5.10	2.81 / 4.24	2.57 / 3.76	2.42 / 3.44	2.30 / 3.22	2.22 / 3.05	2.14 / 2.92	2.09 / 2.82	2.04 / 2.73	2.00 / 2.66	1.97 / 2.60	1.91 / 2.50	1.87 / 2.42	1.80 / 2.30	1.75 / 2.22	1.71 / 2.13	1.65 / 2.04	1.62 / 1.98	1.57 / 1.90	1.54 / 1.86	1.51 / 1.80	1.48 / 1.76	1.46 / 1.72	2.01 / 2.68
48	4.04 / 7.19	3.19 / 5.08	2.80 / 4.22	2.56 / 3.74	2.41 / 3.42	2.30 / 3.20	2.21 / 3.04	2.14 / 2.90	2.08 / 2.80	2.03 / 2.71	1.99 / 2.64	1.96 / 2.58	1.90 / 2.48	1.86 / 2.40	1.79 / 2.28	1.74 / 2.20	1.70 / 2.11	1.64 / 2.02	1.61 / 1.96	1.56 / 1.88	1.53 / 1.84	1.50 / 1.78	1.47 / 1.73	1.45 / 1.70	2.01 / 2.68

Table 420 (continued)
VALUES OF F AND t

f_1 degrees of freedom (for greater mean square)

f_2	1	2	3	4	5	6	7	8	9	10	11	12	14	16	20	24	30	40	50	75	100	200	500	∞	Values of t
50	4.03 / 7.17	3.18 / 5.06	2.79 / 4.20	2.56 / 3.72	2.40 / 3.41	2.29 / 3.18	2.20 / 3.02	2.13 / 2.88	2.07 / 2.78	2.02 / 2.70	1.98 / 2.62	1.95 / 2.56	1.90 / 2.46	1.85 / 2.39	1.78 / 2.26	1.74 / 2.18	1.69 / 2.10	1.63 / 2.00	1.60 / 1.94	1.55 / 1.86	1.52 / 1.82	1.48 / 1.76	1.46 / 1.71	1.44 / 1.68	2.01 / 2.68
55	4.02 / 7.12	3.17 / 5.01	2.78 / 4.16	2.54 / 3.68	2.38 / 3.37	2.27 / 3.15	2.18 / 2.98	2.11 / 2.85	2.05 / 2.75	2.00 / 2.66	1.97 / 2.59	1.93 / 2.53	1.88 / 2.43	1.83 / 2.35	1.76 / 2.23	1.72 / 2.15	1.67 / 2.06	1.61 / 1.96	1.58 / 1.90	1.52 / 1.82	1.50 / 1.78	1.46 / 1.71	1.43 / 1.66	1.41 / 1.64	2.00 / 2.67
60	4.00 / 7.08	3.15 / 4.98	2.76 / 4.13	2.52 / 3.65	2.37 / 3.34	2.25 / 3.12	2.17 / 2.95	2.10 / 2.82	2.04 / 2.72	1.99 / 2.63	1.95 / 2.56	1.92 / 2.50	1.86 / 2.40	1.81 / 2.32	1.75 / 2.20	1.70 / 2.12	1.65 / 2.03	1.59 / 1.93	1.56 / 1.87	1.50 / 1.79	1.48 / 1.74	1.44 / 1.68	1.41 / 1.63	1.39 / 1.60	2.00 / 2.66
65	3.99 / 7.04	3.14 / 4.95	2.75 / 4.10	2.51 / 3.62	2.36 / 3.31	2.24 / 3.09	2.15 / 2.93	2.08 / 2.79	2.02 / 2.70	1.98 / 2.61	1.94 / 2.54	1.90 / 2.47	1.85 / 2.37	1.80 / 2.30	1.73 / 2.18	1.68 / 2.09	1.63 / 2.00	1.57 / 1.90	1.54 / 1.84	1.49 / 1.76	1.46 / 1.71	1.42 / 1.64	1.39 / 1.60	1.37 / 1.56	2.00 / 2.65
70	3.98 / 7.01	3.13 / 4.92	2.74 / 4.08	2.50 / 3.60	2.35 / 3.29	2.23 / 3.07	2.14 / 2.91	2.07 / 2.77	2.01 / 2.67	1.97 / 2.59	1.93 / 2.51	1.89 / 2.45	1.84 / 2.35	1.79 / 2.28	1.72 / 2.15	1.67 / 2.07	1.62 / 1.98	1.56 / 1.88	1.53 / 1.82	1.47 / 1.74	1.45 / 1.69	1.40 / 1.62	1.37 / 1.56	1.35 / 1.53	1.99 / 2.65
80	3.96 / 6.96	3.11 / 4.88	2.72 / 4.04	2.48 / 3.56	2.33 / 3.25	2.21 / 3.04	2.12 / 2.87	2.05 / 2.74	1.99 / 2.64	1.95 / 2.55	1.91 / 2.48	1.88 / 2.41	1.82 / 2.32	1.77 / 2.24	1.70 / 2.11	1.65 / 2.03	1.60 / 1.94	1.54 / 1.84	1.51 / 1.78	1.45 / 1.70	1.42 / 1.65	1.38 / 1.57	1.35 / 1.52	1.32 / 1.49	1.99 / 2.64
100	3.94 / 6.90	3.09 / 4.82	2.70 / 3.98	2.46 / 3.51	2.30 / 3.20	2.19 / 2.99	2.10 / 2.82	2.03 / 2.69	1.97 / 2.59	1.92 / 2.51	1.88 / 2.43	1.85 / 2.36	1.79 / 2.26	1.75 / 2.19	1.68 / 2.06	1.63 / 1.98	1.57 / 1.89	1.51 / 1.79	1.48 / 1.73	1.42 / 1.64	1.39 / 1.59	1.34 / 1.51	1.30 / 1.46	1.28 / 1.43	1.98 / 2.63
125	3.92 / 6.84	3.07 / 4.78	2.68 / 3.94	2.44 / 3.47	2.29 / 3.17	2.17 / 2.95	2.08 / 2.79	2.01 / 2.65	1.95 / 2.56	1.90 / 2.47	1.86 / 2.40	1.83 / 2.33	1.77 / 2.23	1.72 / 2.15	1.65 / 2.03	1.60 / 1.94	1.55 / 1.85	1.49 / 1.75	1.45 / 1.68	1.39 / 1.59	1.36 / 1.54	1.31 / 1.46	1.27 / 1.40	1.25 / 1.37	1.98 / 2.62
150	3.91 / 6.81	3.06 / 4.75	2.67 / 3.91	2.43 / 3.44	2.27 / 3.14	2.16 / 2.92	2.07 / 2.76	2.00 / 2.62	1.94 / 2.53	1.89 / 2.44	1.85 / 2.37	1.82 / 2.30	1.76 / 2.20	1.71 / 2.12	1.64 / 2.00	1.59 / 1.91	1.54 / 1.83	1.47 / 1.72	1.44 / 1.66	1.37 / 1.56	1.34 / 1.51	1.29 / 1.43	1.25 / 1.37	1.22 / 1.33	1.98 / 2.61
200	3.89 / 6.76	3.04 / 4.71	2.65 / 3.88	2.41 / 3.41	2.26 / 3.11	2.14 / 2.90	2.05 / 2.73	1.98 / 2.60	1.92 / 2.50	1.87 / 2.41	1.83 / 2.34	1.80 / 2.28	1.74 / 2.17	1.69 / 2.09	1.62 / 1.97	1.57 / 1.88	1.52 / 1.79	1.45 / 1.69	1.42 / 1.62	1.35 / 1.53	1.32 / 1.48	1.26 / 1.39	1.22 / 1.33	1.19 / 1.28	1.97 / 2.60
400	3.86 / 6.70	3.02 / 4.66	2.62 / 3.83	2.39 / 3.36	2.23 / 3.06	2.12 / 2.85	2.03 / 2.69	1.96 / 2.55	1.90 / 2.46	1.85 / 2.37	1.81 / 2.29	1.78 / 2.23	1.72 / 2.12	1.67 / 2.04	1.60 / 1.92	1.54 / 1.84	1.49 / 1.74	1.42 / 1.64	1.38 / 1.57	1.32 / 1.47	1.28 / 1.42	1.22 / 1.32	1.16 / 1.24	1.13 / 1.19	1.96 / 2.59
1000	3.85 / 6.66	3.00 / 4.62	2.61 / 3.80	2.38 / 3.34	2.22 / 3.04	2.10 / 2.82	2.02 / 2.66	1.95 / 2.53	1.89 / 2.43	1.84 / 2.34	1.80 / 2.26	1.76 / 2.20	1.70 / 2.09	1.65 / 2.01	1.58 / 1.89	1.53 / 1.81	1.47 / 1.71	1.41 / 1.61	1.36 / 1.54	1.30 / 1.44	1.26 / 1.38	1.19 / 1.28	1.13 / 1.19	1.08 / 1.11	1.96 / 2.58
∞	3.84 / 6.64	2.99 / 4.60	2.60 / 3.78	2.37 / 3.32	2.21 / 3.02	2.09 / 2.80	2.01 / 2.64	1.94 / 2.51	1.88 / 2.41	1.83 / 2.32	1.79 / 2.24	1.75 / 2.18	1.69 / 2.07	1.64 / 1.99	1.57 / 1.87	1.52 / 1.79	1.46 / 1.69	1.40 / 1.59	1.35 / 1.52	1.28 / 1.41	1.24 / 1.36	1.17 / 1.25	1.11 / 1.15	1.00 / 1.00	1.96 / 2.58

Based upon and reproduced by permission from George W. Snedecor, *Statistical Methods* (5th ed., 1956), copyright the Iowa State University Press, Ames, Iowa. Slightly modified to include values of t, obtained by extraction of the square root of values in Column 1. After LeClerg, E. L., Leonard, W. H., and Clark, A. G., *Field Plot Technique*, 2nd ed., Burgess Publishing Co., Minneapolis, Minnesota. With permission.

Table 421
SIGNIFICANT STUDENTIZED RANGES FOR A 5% LEVEL NEW[a] MULTIPLE RANGE TEST[b,c]

p \ n₂	2	3	4	5	6	7	8	9	10	12	14	16	18	20	50	100
1	17.97	17.97	17.97	17.97	17.97	17.97	17.97	17.97	17.97	17.97	17.97	17.97	17.97	17.97	17.97	17.97
2	6.085	6.085	6.085	6.085	6.085	6.085	6.085	6.085	6.085	6.085	6.085	6.085	6.085	6.085	6.085	6.085
3	4.501	4.516	4.516	4.516	4.516	4.516	4.516	4.516	4.516	4.516	4.516	4.516	4.516	4.516	4.516	4.516
4	3.927	4.013	4.033	4.033	4.033	4.033	4.033	4.033	4.033	4.033	4.033	4.033	4.033	4.033	4.033	4.033
5	3.635	3.749	3.797	3.814	3.814	3.814	3.814	3.814	3.814	3.814	3.814	3.814	3.814	3.814	3.814	3.814
6	3.461	3.587	3.649	3.680	3.694	3.697	3.697	3.697	3.697	3.697	3.697	3.697	3.697	3.697	3.697	3.697
7	3.344	3.477	3.548	3.588	3.611	3.622	3.626	3.626	3.626	3.626	3.626	3.626	3.626	3.626	3.626	3.626
8	3.261	3.399	3.475	3.521	3.549	3.566	3.575	3.579	3.579	3.579	3.579	3.579	3.579	3.579	3.579	3.579
9	3.199	3.339	3.420	3.470	3.502	3.523	3.536	3.544	3.547	3.547	3.547	3.547	3.547	3.547	3.547	3.547
10	3.151	3.293	3.376	3.430	3.465	3.489	3.505	3.516	3.522	3.526	3.526	3.526	3.526	3.526	3.526	3.526
11	3.113	3.256	3.342	3.397	3.435	3.462	3.480	3.493	3.501	3.509	3.510	3.510	3.510	3.510	3.510	3.510
12	3.082	3.225	3.313	3.370	3.410	3.439	3.459	3.474	3.484	3.496	3.499	3.499	3.499	3.499	3.499	3.499
13	3.055	3.200	3.289	3.348	3.389	3.419	3.422	3.458	3.470	3.484	3.490	3.490	3.490	3.490	3.490	3.490
14	3.033	3.178	3.268	3.329	3.372	3.403	3.426	3.444	3.457	3.474	3.482	3.484	3.485	3.485	3.485	3.485
15	3.014	3.160	3.250	3.312	3.356	3.389	3.413	3.432	3.446	3.465	3.476	3.480	3.481	3.481	3.481	3.481
16	2.998	3.144	3.235	3.298	3.343	3.376	3.402	3.422	3.437	3.458	3.470	3.477	3.478	3.478	3.478	3.478
17	2.984	3.130	3.222	3.285	3.331	3.366	3.392	3.412	3.429	3.451	3.465	3.473	3.476	3.476	3.476	3.476
18	2.971	3.118	3.210	3.274	3.321	3.356	3.383	3.405	3.421	3.445	3.460	3.470	3.474	3.474	3.474	3.474
19	2.960	3.107	3.199	3.264	3.311	3.347	3.375	3.397	3.415	3.440	3.456	3.467	3.472	3.474	3.474	3.474
20	2.950	3.097	3.190	3.255	3.303	3.339	3.368	3.391	3.409	3.436	3.453	3.464	3.470	3.473	3.474	3.474
24	2.919	3.066	3.160	3.226	3.276	3.315	3.345	3.370	3.390	3.420	3.441	3.456	3.465	3.471	3.477	3.477
30	2.888	3.035	3.131	3.199	3.250	3.290	3.322	3.349	3.371	3.405	3.430	3.447	3.460	3.470	3.486	3.486
40	2.858	3.006	3.102	3.171	3.224	3.266	3.300	3.328	3.352	3.390	3.418	3.439	3.456	3.469	3.504	3.504
60	2.829	2.976	3.073	3.143	3.198	3.241	3.277	3.307	3.333	3.374	3.406	3.431	3.451	3.467	3.537	3.537
120	2.800	2.947	3.045	3.116	3.172	3.217	3.254	3.287	3.314	3.359	3.394	3.423	3.446	3.446	3.585	3.601
∞	2.772	2.918	3.017	3.089	3.146	3.193	3.232	3.265	3.294	3.343	3.382	3.414	3.442	3.466	3.640	3.735

[a] Using protection levels based on degrees of freedom.

[b] Duncan, D. B., Multiple range and multiple F tests, *Biometrics*, 11, 1—42. 1955.

[c] Modified with corrections by Harter, H. L., Critical values for Duncan's New Multiple Range Test, *Biometrics*, 16, 671—685. 1960

From LeClerg, E. L., Leonard, W. H. and Clark, A. G., *Field Plot Technique*, 2nd ed., Burgess Publishing Co. Minneapolis, Minn. With permission of R. E. LeClerg.

Table 422
SIGNIFICANT STUDENTIZED RANGES FOR A 1% LEVEL NEW[a] MULTIPLE RANGE TEST[b,c]

n_2 \ p	2	3	4	5	6	7	8	9	10	12	14	16	18	20	50	100
1	90.03	90.03	90.03	90.03	90.03	90.03	90.03	90.03	90.03	90.03	90.03	90.03	90.03	90.03	90.03	90.03
2	14.04	14.04	14.04	14.04	14.04	14.04	14.04	14.04	14.04	14.04	14.04	14.04	14.04	14.04	14.04	14.04
3	8.261	8.321	8.321	8.321	8.321	8.321	8.321	8.321	8.321	8.321	8.321	8.321	8.321	8.321	8.321	8.321
4	6.512	6.677	6.740	6.756	6.756	6.756	6.756	6.756	6.756	6.756	6.756	6.756	6.756	6.756	6.756	6.756
5	5.702	5.893	5.989	6.040	6.065	6.074	6.074	6.074	6.074	6.074	6.074	6.074	6.074	6.074	6.074	6.074
6	5.243	5.439	5.549	5.614	5.655	5.680	5.694	5.701	5.703	5.703	5.703	5.703	5.703	5.703	5.703	5.703
7	4.949	5.145	5.260	5.334	5.383	5.416	5.439	5.454	5.464	5.472	5.472	5.472	5.472	5.472	5.472	5.472
8	4.746	4.939	5.057	5.135	5.189	5.227	5.256	5.276	5.291	5.309	5.316	5.317	5.317	5.317	5.317	5.317
9	4.596	4.787	4.906	4.986	5.043	5.086	5.118	5.142	5.160	5.185	5.199	5.205	5.206	5.206	5.206	5.206
10	4.482	4.671	4.790	4.871	4.931	4.975	5.010	5.037	5.058	5.088	5.106	5.117	5.122	5.124	5.124	5.124
11	4.392	4.579	4.697	4.780	4.841	4.887	4.924	4.952	4.975	5.009	5.031	5.045	5.054	5.059	5.061	5.061
12	4.320	4.504	4.622	4.706	4.767	4.815	4.852	4.883	4.907	4.944	4.969	4.986	4.998	5.006	5.011	5.011
13	4.260	4.442	4.560	4.644	4.706	4.755	4.793	4.824	4.850	4.889	4.917	4.937	4.950	4.960	4.972	4.972
14	4.210	4.391	4.508	4.591	4.654	4.704	4.743	4.775	4.802	4.843	4.872	4.894	4.910	4.921	4.940	4.940
15	4.168	4.347	4.463	4.547	4.610	4.660	4.700	4.733	4.760	4.803	4.834	4.857	4.874	4.887	4.914	4.914
16	4.131	4.309	4.425	4.509	4.572	4.622	4.663	4.696	4.724	4.768	4.800	4.825	4.844	4.858	4.892	4.892
17	4.099	4.275	4.391	4.475	4.539	4.589	4.630	4.664	4.693	4.738	4.771	4.797	4.816	4.832	4.874	4.874
18	4.071	4.246	4.362	4.445	4.509	4.560	4.601	4.635	4.664	4.711	4.745	4.772	4.792	4.808	4.858	4.858
19	4.046	4.220	4.335	4.419	4.483	4.534	4.575	4.610	4.639	4.686	4.722	4.749	4.771	4.788	4.845	4.845
20	4.024	4.197	4.312	4.395	4.459	4.510	4.552	4.587	4.617	4.664	4.701	4.729	4.751	4.769	4.833	4.833
24	3.956	4.126	4.239	4.322	4.386	4.437	4.480	4.516	4.546	4.596	4.634	4.665	4.690	4.710	4.802	4.802
30	3.889	4.056	4.168	4.250	4.314	4.366	4.409	4.445	4.477	4.528	4.569	4.601	4.628	4.650	4.772	4.777
40	3.825	3.988	4.098	4.180	4.244	4.296	4.339	4.376	4.408	4.461	4.503	4.537	4.566	4.591	4.740	4.764
60	3.762	3.922	4.031	4.111	4.174	4.226	4.270	4.307	4.340	4.394	4.438	4.474	4.504	4.530	4.707	4.765
120	3.702	3.858	3.965	4.044	4.107	4.158	4.202	4.239	4.272	4.327	4.372	4.410	4.442	4.469	4.673	4.770
∞	3.643	3.796	3.900	3.978	4.040	4.091	4.135	4.172	4.205	4.261	4.307	4.345	4.379	4.408	4.635	4.776

[a]   Using protection levels based on degrees of freedom.

[b]   Duncan, D. B., Multiple range and multiple F tests, *Biometrics*, 11, 1—42, 1955.

[c]   Modified with corrections by Harter, H. L., Critical values for Duncan's New Multiple Range Test, *Biometrics*, 16, 671—685, 1960.

Table 423

SETS OF RANDOM SAMPLE NUMBERS (NUMBERS 1 TO 60 INCLUSIVE)

I	II	III	IV	V	VI	VII	VIII	IX	X
39	21	17	01	60	02	43	25	29	39
15	02	40	57	06	57	42	46	48	38
34	42	15	30	37	22	50	08	59	15
11	47	45	33	22	27	22	53	28	10
49	33	26	52	26	05	23	30	15	27
33	08	31	56	01	38	29	50	25	51
16	07	36	03	25	59	56	54	19	34
19	53	01	39	02	54	35	23	04	14
12	44	42	24	09	21	19	40	09	05
56	20	05	37	53	29	36	02	30	56
48	39	38	50	56	11	55	60	22	02
26	41	13	32	08	20	38	44	01	20
32	15	25	58	03	32	39	59	41	49
41	03	10	43	50	58	15	13	11	52
59	31	14	27	27	18	09	42	42	59
60	57	11	29	20	45	02	14	02	26
50	59	06	10	42	31	28	35	23	48
27	55	56	11	31	39	46	17	26	04
28	01	52	41	44	30	03	49	47	37
40	29	34	22	21	23	05	22	37	12
37	35	09	13	29	01	53	06	34	55
36	36	49	34	41	14	04	16	44	33
08	48	60	04	18	48	21	55	33	16
57	12	07	51	10	15	59	07	27	08
22	56	30	08	57	06	12	47	13	53
24	05	59	20	54	52	07	45	39	35
55	10	04	48	43	40	32	31	03	09
31	06	19	44	51	55	08	21	55	23
14	14	53	25	05	25	51	52	·57	19
45	16	44	40	46	49	54	27	38	21
05	37	57	42	35	08	49	43	52	13
30	30	58	17	17	26	13	12	24	17
17	43	50	07	47	16	10	38	53	03
44	13	28	36	40	19	26	56	45	57
25	45	43	18	30	41	48	34	49	40
47	22	03	14	55	35	06	32	10	44
58	49	48	35	36	50	24	48	54	24
53	58	08	38	11	13	60	10	16	46
43	32	20	49	58	53	33	41	14	50
21	04	21	23	07	10	47	26	43	30
20	27	51	09	04	43	35	09	17	01
23	11	54	12	32	09	44	29	35	06
42	46	22	47	12	17	16	28	32	07
52	18	18	15	49	36	25	57	56	42
13	23	55	19	28	07	57	01	40	22
54	28	37	59	45	03	30	24	60	60
07	51	23	45	34	44	45	39	50	41
06	24	16	06	48	60	17	37	20	58
02	50	35	55	24	34	18	15	06	31
35	26	47	60	59	47	20	36	18	32
29	17	41	31	14	28	37	05	36	11
04	09	33	05	15	24	52	58	58	45
18	19	39	02	52	33	40	04	46	18
46	60	24	26	33	56	01	51	31	29
10	34	32	21	19	51	31	11	51	54
09	54	27	46	23	12	41	03	05	25
03	25	29	53	38	37	11	20	12	28
01	38	12	28	13	42	58	18	07	47
38	52	02	16	16	47.	27	33	21	36
51	40	46	54	39	04	14	19	08	43

From LeClerg, E. L., Leonard, W. H., and Clark, A. G., *Field Plot Technique*, 2nd ed., Burgess Publishing Co., Minneapolis, Minn. With permission.

GLOSSARY

Table 424
GLOSSARY OF SOME TERMS ENCOUNTERED IN THE IMPROVEMENT AND MANAGEMENT OF PLANTS, ANIMALS, AND SOILS

A horizon. The surface and subsurface soil that contains most of the organic matter.

Aberrant. Abnormal; differing from normal type but not readily assignable to another group.

Abiotic. Nonliving components of the environment.

Abortive. Imperfectly developed.

Abscission. Natural separation of leaves, flowers, and fruits from other plant parts by formation of thin-walled cells.

Absorption. Intake of water or other substance, by a cell, root, or other organ.

Acid equivalent. The theoretical yield of parent acid from the active ingredient content of a formulation.

Acidity, total. Total acidity in a soil or clay. Usually estimated by a buffered salt determination (cation exchange capacity − exchangeable bases = total acidity).

Active ingredient. The chemical in a formulation primarily responsible for its effectiveness when applied or used.

Additive factors. Non-allelic factors affecting the same character and enhancing the effects of each other.

Additive resistance. Resistance governed by more than one gene, each of which can be expressed independently.

Adenine. A purine base occurring in both DNA and RNA. Normally pairs with thymine in DNA.

Adjuvant. Any substance added to pesticides and herbicides to improve effectiveness or application characteristics.

Ad libitum. Free choice access to feed.

Adsorption. A surface phenomenon, in which water or other substances associate with a surface, e.g., a soil colloid.

Adult plant resistance. Horizontal resistance of adult plants, also known as mature plant resistance.

Adventitious. Arising from an unusual position on a stem (also at crown of grasses).

Aerobic. Pertaining to life in free oxygen or to conditions requiring the same.

Afterripening. A type of curing process required by seed, bulb, and related structures, before germination can occur.

Agglutination. Aggregation or clumping of cells in a fluid.

Agglutinin. An antibody that produces clumping of the antigenic structure.

Aggregate. A mass or cluster of soil particles or other small objects.

Air porosity. Proportion of bulk volume of soil that is filled with air at any given time or under a given condition.

Aleurone. Outer layer of the cells of the endosperm of the seed.

Alkali soil. Soil containing excessive amounts of alkaline salts, usually sodium carbonate (pH 8.5 or higher).

Alkaline soil. Any soil with a pH >7.

Alkaloids. A diverse group of secondary plant constituents containing nitrogen; many are toxic to animals or rumen bacteria.

Allele (also Allelomorph). One member of a pair or a series of genes that occur at a particular locus on homologous chromosomes.

Alleopathic substances. Chemical compounds produced by plants that affect interactions between different plants.

Alloploid or Allopolyploid. A polyploid with whole chromosome sets from different species.

Amendment, soil. Any substance added to soil that alters soil properties, e.g., gypsum, sawdust, fertilizers, etc.

Amino acid. The building block molecules of proteins. They contain both an amino (NH_2) group and a carboxyl (COOH) group.

Amphiploid or Amphidiploid. A tetraploid individual with two sets of chromosomes from each of two known ancestral species.

Anaerobic. Living in the absence of free oxygen.

Anemia. Deficiency in the quality or quantity of blood.

Aneuploidy. Variation in chromosome number by whole chromosomes, but not exact multiples of haploid complement, e.g., $2n+1$ (trisomic), $2n-1$ (monosomic).

Animal unit. One mature cow (454 kg) or the equivalent based upon average daily forage consumption of 12 kg dry matter per day.

Animal unit month (AUM). Amount of feed or forage required by an animal unit for 1 month.

Antagonism. An interaction of two or more chemicals that reduces the predicted effect of each separate chemical.

Anthocyanin. A water-soluble plant pigment that produces many of the red, blue, and purple colors in plants.

Antibiosis. A toxic or other direct detrimental effect of one organism upon another.

Antibody. A substance that acts to neutralize a specific antigen in a living organism.

Antifeedant. A substance that deters feeding by an insect but does not kill the insect directly.

Anthesis. The period of pollen distribution; the process of dehiscence of the anthers.

Antigen. Any substance, usually a protein, that stimulates the production of a specific antibody.

Antimetabolite. Prevents the formation or utilization of organic compounds in specific metabolic processes in living tissue.

Anoxia. Oxygen deficiency.

Apomixis. Reproduction from an unfertilized egg or from somatic cells associated with the egg.

Asexual reproduction. Any process of reproduction that does not involve the fusion of cells.

ATP. Adenosine triphosphate, an energy-rich compound participating in energy use and storage in the cell.

Auricle. Earlike lobe at the base of leaf blades of barley, wheat, and certain grasses grown for turf and forage.

Autogamy. The condition in which a flower is pollinated by its own pollen.

Autoploid or Autopolyploid. A polyploid with all sets of chromosomes from the same species.

Autosome. A chromosome not associated with the sex of the individual.

Auxins. Natural or synthetic substances that can regulate or stimulate growth and cell enlargement in plants.

Average daily gain (ADG). The average daily liveweight increase of an animal.

Awn. A ''beard'' or bristle extending from tip or back of the lemma of a grass flower.

Axil. The angle between leaf and stem.

Axillary tiller. A new shoot arising from junction of leaf and stem.

B horizon. Subsoil layer in which certain leached substances (e.g., iron) are deposited.

Backcross. Mating the result of a cross with its parent to reinforce or increase the gene frequency of a desired characteristic (BC_1, BC_2, etc., symbols for first and second backcross generations)

Backgrounding. Growing of replacement cattle, usually on high-forage systems. Can take place anytime during postweaning period, until animal goes to feedlot or to breeding herd.

Bacteriophage. See **Phage.**

Bacteriostatic. Preventing or hindering the growth and multiplication of bacteria.

Basal diet. A diet common to all groups of experimental animals to which the experimental substance(s) is added.

Basal metabolic rate (BMR). The heat produced by an animal during complete rest (but not sleeping) following fasting.

Base analog. A slightly modified purine or pyrmidine molecule that may substitute for the normal base in nucleic acid molecules.

Base-saturation percentage. The extent to which the adsorption complex of a soil is saturated with exchangeable cations other than hydrogen. Expressed as percentage of total cation-exchange capacity.

Basic number. The number of chromosomes in gametes of a diploid organism or a diploid ancestor of a polyploid plant.

BC soil. A soil profile with B and C horizons, and little or no A horizon.

Berry. A simple fruit derived from one flower, in which the parts remain succulent. The true berry is derived from an ovary, as in the grape, while the false berry is derived from an ovary plus receptacle tissue, as in the blueberry.

Bias. Average error in estimating the true value of a trait from a sample of data that results in an overestimate (positive bias) or underestimate (negative bias).

Bioassay. Measuring the effect on an organism of a given stress, e.g., toxic substance, extreme temperatures, and drought.

Biomass. The part of a given habitat consisting of living matter (weight/unit area, or volume/unit volume of habitat).

Biosynthesis. The production of a new material in living cells or tissues.

Biota. Animal and plant life of a particular region (considered as a total ecological entity).

Biotype. A population within a species that exhibits distinct genetic variation. Some biotypes are distinguished by criteria other than morphology.

Bivalent. A pair of synapsed homologous chromosomes.

BOD (biochemical oxygen demand). Quantity of oxygen used in the biochemical oxidation of organic matter in a specified time and specified temperature and conditions.

Bolus. Mass of food ready to be swallowed or regurgitated (also, a large pill for dosing animals).

Boot stage. Growth stage of grasses when the head is enclosed by the sheath of the uppermost leaf.

Bound water. Water held tenaciously as an extremely thin film surrounding very small solid particles.

Bract. A reduced or modified leaf that can be large and showy, or small and inconspicuous.

Breeder seed. Seed (or vegetative propagating material) increased by the originating or sponsoring plant breeder or institution (source for increase of foundation seed).

Breeding value. Genetic worth of an animal's genotype for a certain trait; twice the animal's genetic transmitting ability.

Browse. That part of leaf and twig growth of shrubs, woody vines, and trees available for animal consumption (to consume).

Buffer compounds, soil. The clay, organic matter, and compounds such as carbonates and phosphates that enable soil to resist appreciable change in pH.

Bulk density, soil. The mass of dry soil per unit bulk volume. Bulk volume determined before drying to constant weight at 105°C.

Caliche. A layer near the surface, more or less cemented by secondary carbonates of calcium or magnesium precipitated from the soil solution (may occur as soft thin soil horizon, as a hard thick bed, or as a surface layer exposed by erosion).

Calorie (gram calorie). Unit for measuring chemical energy (heat necessary to raise the temperature of 1 gram of water from 14.5 to 15.5°C at standard pressure).

Cannula. A tubular device (of metal, rubber, or glass) used to connect an internal structure with the outside of the animal.

Carbohydrate. A compound of carbon, hydrogen, and oxygen in the ratio of one atom each of carbon and oxygen to two of hydrogen (as in sugar, starch, and cellulose).

Carbohydrate, nonstructural. Photosynthetic products existing in plant tissue as a solute or as stored insoluble material and functioning as readily metabolizable compounds, e.g., fructose, glucose, sucrose, starch, fructosans, and hemicellulose.

Carbon cycle. The sequence of transformations whereby carbon dioxide is fixed in living organisms by photosynthesis or by chemosynthesis and then released by respiration and by the death and decomposition of the fixing organisms, and ultimately returned to its original state.

Carbon-organic nitrogen ratio. Ratio of the weight of organic carbon to the weight of organic nitrogen in soil, organic material, plants, or cells of microorganisms.

Carboxyl group. An acidic organic chemical group — COOH.

Carotene. A yellow pigment in green leaves and other plant parts (precursor of vitamin A).

Carrying capacity. The number of animal units that a property or area will carry on a year-round basis.

Caryopsis. The grain or fruit of grasses.

Catabolism. Metabolic breakdown of complex substances to simple substances.

Catheter. A hollow tube for insertion into various body cavities for withdrawal of fluids, or, with artificial insemination, for the deposition of semen.

Cation exchange. Interchange between a cation in solution and another cation on the surface of any surface-active material, e.g., clay colloid or organic colloid.

Cation-exchange capacity (CEC). The sum total of exchangeable cations that a soil can absorb (expressed in meg/100 g or per gram of soil, clay, or other exchangers).

Cell culture. A growth of cells *in vitro*.

Cell-free extract. A fluid extract of soluble materials of cells, obtained by rupturing cells and discarding particulate materials and intact cells.

Cellulose. Major skeletal material in the cell wall of plants (an anhydride of beta-D-glucose units).

Centromere. A specialized, complex region of the chromosome, consisting of one kinetochore for each sister chromatid.

Certified seed. Seed which is the progeny of breeder, select, foundation, or registered seed handled to maintain satisfactory genetic purity and identity, which is acceptable to the certifying agency.

Chasmogamy. Pollination that occurs only after the opening of the flower.

Chiasma. Visible connection or crossover between two chromatids seen at prophase 1 of meiosis.

Chimera. A mixture of tissues of different genetic constitution in the same organism.

Chi-square test. Statistical test for determining probability that a set of experimental values will be equalled or exceeded by chance alone for a given theoretical expectation.

Chloroplast. Cytoplasmic organelle containing several pigments, especially the light-absorbing chlorophylls and also DNA and polysomes.

Chlorosis. Yellowing or blanching of leaves and other parts of chlorophyll-bearing plants.

Chromatid. One of two identical longitudinal halves of a chromosome that shares a common centromere with a sister chromatid.

Chromatin. Nuclear material comprising the chromosomes.

Chromosome. The physical sites of nuclear genes arranged in linear order. They are nucleoprotein structures that appear more or less rodlike during nuclear division.

Classification score. A final point rating placed on animals in breed association conformation rating programs; pertains to ideal conformation of the breed.

Clay. A soil separate consisting of particles <0.002 mm in equivalent diameter.

Clay pan. Dense, compact layer in the subsoil having a much higher clay content than overlying material and separated by a well-defined boundary (usually hard when dry and sticky when wet).

Cleistogamy. Fertilization that occurs without opening of the flower.

Climax. Fully developed plant community with plant cover and environment at equilibrium.

Clone. A group of cells or organisms, derived from a single ancestral cell or individual and all genetically alike.

Close breeding. Intense inbreeding often applied to mating closely related animals, such as full sibs or parent-offspring.

Codon. Set of nucleotides specific for a particular amino acid in protein synthesis.

Coefficient of digestibility. The percentage value of a food nutrient that is absorbed.

Coleoptile. The sheath covering the first leaf of a grass seedling as it emerges from the soil.

Collagen. A white, papery transparent type of connective tissue that is protein in composition.

Combining ability, general. The average performance of a genetic strain in a series of crosses.

Combining ability, specific. The performance of specific combinations of genetic strains in crosses in comparison with the average performance of all combinations.

Complementary resistance. Resistance governed by two or more genes, where one alone is ineffective.

Complementary sire selection. Applying selection criteria to a group of bulls, rather than selecting bulls individually in order to achieve herd-breeding goals.

Complementation. The ability of linearly adjacent segments of DNA to supplement one another in phenotypic effect. Complementary genes interact to change the expression of a trait.

Compound fertilizer. A mixed fertilizer containing at least two of the primary plant nutrients (N, P_2O_5, and K_2O).

Conditioning. The mechanical handling of seed from harvest until marketing.

Confidence interval. The interval above and below an estimated value within which there is confidence that the true value lies with a specified level of probability.

Contact herbicide. One that causes localized injury to plant tissue where contact occurs.

Continuous distribution. A method of depicting the frequency of occurrence of different levels of traits that vary in increments too small to measure (often measured for convenience in arbitrary finite units, e.g., centimeters, kilograms, etc.).

Controlled-release fertilizers. Fertilizers in which one or more nutrients have limited solubility in soil, becoming available to growing plants over a period of time.

Convergent improvement. A system of crossing, backcrossing, and selfing with selection to improve inbred lines of cross-pollinated plants without changing their behavior in hybrid combinations.

Corm. A swollen underground storage stem that serves as resting phase in plants such as timothy.

Corpus luteum. A temporary structure formed on the ovary after ovulation. During the time it secretes progesterone, the animal does not undergo estrous cycles.

Correlation. Tendency of two or more traits to vary in the same direction (positive correlation) or in opposite directions (negative correlation) because of common forces. The degree of correlation (r) ranges from $+1$ to -1.

Cotyledons. The foliar portion or first leaves of the embryo as found in the seed.

Coumarin. A white, crystalline compound ($C_9H_6O_2$) with a vanilla-like odor.

Coumestrol. An estrogenic factor occurring naturally in forage species, especially in certain legumes.

Covariance. Variation that is common to two traits because of the same genetic and/or environmental effects.

Critical nutrient concentration. Nutrient concentration in the plant or its organs at the time when the nutrient becomes deficient for growth.

Crossbreeding. Mating of animals of different breeds.

Cross-fertilization. Fertilization secured by pollen from another plant.

Cross-inoculation. Inoculation of one legume species by the symbiotic bacteria from another.

Cross-pollination. The transfer of pollen from one plant genotype to another.

Crossing-over. A process in which genes are exchanged between nonsister chromatids of homologous chromosomes.

Crown. The base of the stem where roots arise.

Crude fiber. Coarse fibrous portions of plants, such as cellulose, partially digestible and relatively low in nutritional value.

Culling. Removal from a population of individuals of lower genetic or phenotypic merit.

Culm. The jointed stem of grasses.

Cultivar (derived from "cultivated variety"). Denotes an assemblage of cultivated individuals that are distinguished by any characters significant for the purposes of agriculture, forestry, or horticulture which when reproduced retain their distinguishing features.

Cuticle. The outer corky or waxy covering of the plant.

Cutin. Complex fatty or waxy substance in cell walls of plants, particularly in epidermal layers.

Cytogenetics. The study of cellular structures and mechanisms associated with genetics.

Cytokinins. A class of plant hormones that stimulate cell divisions and delay senescence.

Cytoplasm. Contents of a cell outside the nucleus.

Cytoplasmic inheritance. Inheritance dependent upon hereditary units in the cytoplasm.

Cytoplasmic male sterility. Male sterility caused by cytoplasm rather than nuclear genes: transmitted through female parent.

Day neutral plant. A plant capable of flowering under either long or short daylengths.

Deflocculate. To separate individual components of compound particles by chemical and/or physical means.

Dehiscence. Splitting open of a fruiting structure or anther.

Deletion (deficiency). Loss of a part of a chromosome, usually involves one or more genes.

Denitrification. Biochemical reduction of nitrate or nitrite to gaseous nitrogen either as molecular nitrogen or as an oxide of nitrogen.

Determinate. The flowering of plant species uniformly within certain time limits.

Diallel cross. Intercrossing of parents in all combinations of two.

Diapause. A period of dormancy in insects.

Dicoumarol. A chemical compound found in spoiled sweet clover hay or made synthetically (anticoagulant that can cause internal hemorrhages when eaten by animals).

Digestibility, apparent. Food consumed, less food in the feces.

Digestibility, true. Actual availability of a feed or nutrient as represented by the balance between intake and the feed residues in the feces exclusive of metabolic products.

Digestible dry matter (DDM). Feed-feces difference expressed in dry matter (DM).

Digestible energy (DE). Feed-feces difference expressed in calories.

Digestible nutrients. That portion of nutrients consumed that is digested and taken into the animal body (either apparent or true digestibility).

Dihybrid. The result of a cross between parents that differ by two specified genes.

Dimorphism. The presence of two forms of leaves, flowers, or other organs upon the same plant, or upon other plants of the same species.

Dioecious. Having male and female flowers borne on different plants.

Diploid. Having two sets of chromosomes. Each parent contributes one set of chromosomes.

Diuretic. Increasing urinary secretion.

DNA. Deoxyribonucleic acid; see **Nucleic acids.**

Dominance. Where one gene of an allelic series masks the phenotypic expression of another gene in the same allelic series (may be complete or incomplete).

Dormancy. Internal condition of a seed or bud that prevents its prompt germination or sprouting under normal conditions for growth.

Double cross. First generation hybrid between two foundation single crosses.

Dyspnea. Difficult or labored respiration.

Dystocia. Abnormal or difficult labor and/or birth.

Ecotype. A variety or strain adapted to a particular environment.

Edaphic. Pertaining to the influence of the soil upon plant growth.

Embryo. The period in the development of an individual between conception and the completion of organ formation. The rudimentary plant in a seed.

Embryo transplant. Removal of an embryo at early stage of development from one animal and insertion of that embryo into another animal for development and birth.

Endemic. A species native to a particular environment or locality.

Endocrine. Internal secretions elaborated directly into the blood or lymph that affect another organ or tissue in the body.

Endogenous. Developing or originating within the organism.

Endosperm. The starchy interior of a grain.

Endotrophic. Nourished or receiving nourishment from within (as fungi or their hyphae receiving nourishment from plant roots in a mycorrihizal association).

Environmental variance. That portion of the phenotypic variance that is caused by environmental factors.

Enzyme. Any substance, protein in whole or in part, that regulates the rate of a given biochemical reaction in living organisms.

Epicotyl. The stem of the embryo or young seedling above the cotyledons.

Epiphytotic. A sudden or abnormally destructive outbreak of a plant disease, usually over a wide geographic area.

Epistasis. The masking of the phenotypic effect of either or both members of one pair of alleles by a gene of a different pair.

Equilibrium frequency. That gene frequency which varies nondirectionally about the mean of a population, under conditions of unchanging selection pressure and mutation rate and no mixing with other populations.

Erosion. The wearing away of the land surface by running water, wind, ice, or other geological agents.

Erosion classes. A grouping of erosion conditions based on the degree of erosion or on characteristic patterns (applied to accelerated erosion, not to normal, natural, or geological erosion). Four classes recognized for water erosion and three for wind erosion.

Erosion index. A measure of the erosive potential of a specific rainfall event.

Estrogenic. Pertaining to any hormonal substance capable of producing estrus in a female animal.

Estrous. Referring to the entire cycle of reproductive changes in the nonpregnant female mammal.

Estrus. The period of sexual receptivity in female animals.

Ether extract. Fats, oils, waxes, and similar plant compounds extracted with dry ethyl ether in chemical analysis.

Etiolation. The condition of spindly white growth caused by reducing or excluding light.

Eukaryote. Any organism or cell with a structurally discrete nucleus.

Eutrophic. Having concentrations of nutrients optimal, or nearly so, for plant or animal growth (said of nutrient and soil solutions and bodies of water).

Evapotranspiration. The combined loss of water from a given area, and during a specific period of time, by evaporation from the soil surface and by transpiration from plants.

Exchange capacity. The total ionic charge of the adsorption complex active in the adsorption of ions.

Exogenous. Originating outside the organism.

F_1, F_2, etc. Symbols used to designate the first generation, the second generation, etc. after a cross.

Facultative anaerobes. Microorganisms capable of living in either the presence or absence of oxygen.

Fallow. Cropland left idle, usually for one growing season, while soil is cultivated to control weeds and conserve moisture.

Fasciation. Flattening of the stem due to multiple terminal buds growing in the same place.

Fat. Frequently used in a general sense to include both fats and oils, or a mixture of the two. Fats and oils have the same general structure and chemical properties but differ in physical characteristics.

Fat corrected milk (FCM). Adjustment of milk with different fat percentages to equivalent amounts on the basis of energy.

Fatty acid. One of a group of compounds that usually combine with glycerol to form glycerides or fats.

Feed efficiency. The ratio expressing the number of units of feed required for one unit of production (meat, milk, eggs) by an animal.

Feed unit. One pound (0.45 kg) of corn or its equivalent in other feed.

Fertilization. The union of an egg and a sperm (gametes) to form a zygote.

Fertilizer grade. Indicates the nutrient content expressed in percentages of N, P_2O_5, and K_2O, in that order.

Fertilizer salt index. The ratio of the increase in osmotic pressure by a fertilizer compound or mixture to that produced by the same weight of $NaNO_3$ ($NaNO_3$ = 100).

Field capacity (water). The amount of moisture remaining in soil after free water (gravitational) has drained away.

Field resistance. Resistance observed under field conditions as distinguished from resistance observed in the laboratory or greenhouse.

Fixation. The process of conversion of an element in the soil essential to plants from a readily available to a less available form.

Flocculate. To aggregate individual particles into small groups or granules.

Flow velocity (of water in soil).. The volume of water transferred per unit of time and per unit of area normal to the direction of the net flow.

Flushing. The practice of feeding females more generously 2 to 3 weeks before breeding in order to enhance conception.

Flux. The rate of transfer of a quantity (water, heat, etc.) across a surface.

Forage. Vegetable matter, fresh or preserved, gathered and fed to animals.

Formic acid. A colorless pungent liquid (HCOOH) that has been used to improve preservation of grass silage.

Foundation seed. Seed which is the progeny of breeder or foundation seed produced under the control of the breeder or sponsoring institution. It is a class of certified seed produced under procedures for the maintenance of genetic purity and identity.

Fragipan. A natural subsurface horizon with high bulk density — seemingly cemented when dry and moderately brittle when moist (low in organic matter and slowly permeable to water).

Frequency distribution. Method of depicting the number of variates having the same value or belonging to the same class at various points on a horizontal axis.

Fructosin. A polysaccharide yielding primarily fructose on hydrolysis.

Full feed. When animals are provided as much feed as they will consume safely without going off feed.

Full sibs. Individuals having the same maternal and paternal parents.

Fusiform. Feather shaped — enlarged in the middle and tapering at both ends.

Gametes. Cells formed in sexual organs or their equivalent and specialized for fertilization.

Gene. The blueprint of a hereditary trait, represented by a particular segment of a DNA molecule, generally located in a chromosome.

Gene expression. The process of protein synthesis which turns the genetic code into traits.

Gene interaction. Modification of gene action by a nonallelic gene.

Gene pool. The total of all genes in a population.

Genetic drift. A change in gene frequency in a population because of differences among genotypes in efficiency of reproduction (often found in small, isolated populations).

Genetic load. The average number of lethals per individual in a population (proportional reduction in fitness relative to the optimum).

Genome. A complete set of chromosomes inherited as a unit from one parent.

Genotype. The genetic make-up of an organism; the sum total of its genes; also, a group of organisms with the same genetic make-up.

Geotropic. Turning downward in response to a stimulus caused by the force of gravity.

Germplasm. The total genetic information in a population of a particular organism.

Gestation. The period of development of an individual between fertilization and birth.

Glabrous. Smooth (opposite of pubescent).

Glacial drift. Rock debris that has been transported by glaciers and deposited.

Glaucous. Covered with a whitish, waxy bloom.

Gluten. The protein in wheat flour that enables the dough to rise.

Glycoside. Any of a group of organic compounds that produce sugars and related substances on hydrolysis.

Grading up. Use of sires of superior genetic merit on cows of lesser genetic merit.

Granular pesticide. A dry formulation consisting of discrete particles, for application without a liquid carrier.

Granule. A natural soil aggregate or ped which is relatively nonporous.

Gras (generally recognized as safe). A designation of food additives that have been judged safe for human consumption.

Grass silage. Silage made from grasses, legumes, or mixtures.

Grass tetany (hypomagnesemia). Marked by staggers, convulsions and often death in cattle and sheep (animals have low level of blood magnesium).

Grazing pressure. Number of animals per unit area of available forage.

Guanine. A purine base occurring in DNA and RNA.

Guttation. Exudation of water from the ends of the vascular system at the margins of leaves that occurs under humid conditions.

Gynoecious. Producing only female flowers.

Half sib. Having one parent in common.

Haploid. Having a single set (genome) of chromosomes in a cell or an individual; the reduced number (n), as in a gamete.

Hardpan. Hardened soil layer, in lower A or in B horizon, caused by cementing of soil particles with organic matter, or with silica, sequioxides, or calcium carbonate.

Heavy metals. Metals that have densities >5.0.

Heliotropic. Turning towards the sun.

Hemicellulose. Heterogeneous polysaccharide fraction that exists largely in the secondary cell wall of the plant.

Herbicide resistance. The trait of a population of plants, or plant cells in tissue culture, of having tolerance for a particular herbicide greater than the average for the species.

Heritability. That portion of observed variability in a progeny that is due to heredity.

Hermaphrodite. Perfect, having both stamens and pistils in the same flower.

Heterochromatin. Condensed chromatin.

Heterosis. Increased vigor, growth, size, yield or function of a hybrid progeny over that of the parents.

Heterotrophic. Organisms capable of deriving energy from the oxidation of organic compounds.

Heterozygote. An individual whose chromosomes bear unlike genes of a given allelic pair or series.

Hexaploid. Having six sets (genomes) of chromosomes.

Homeostasis. Maintenance of constancy or a high degree of uniformity in functions of an organism.

Homologous chromosomes. Chromosomes occurring in pairs, one from each of two parents, normally morphologically alike and having the same gene loci.

Homozygous. Having like genes at corresponding loci on homologous chromosomes (organism may be homozygous for one, several, or all genes).

Horizontal resistance. A resistance trait not involving a gene-for-gene relationship.

Hormone. A chemical growth-regulating substance that can be or is produced by a living organism.

Humus. The more-or-less stable fraction of the soil organic matter remaining after major portion of added plant and animal residues have decomposed.

Hybrid. The first generation offspring of a cross between two individuals differing in one or more genes; also, progeny of a cross between different species.

Hybridize. To produce hybrids by crossing individuals that have different genotypes.

Hybrid vigor. (See Heterosis.)

Hydrocyanic acid (HCN). Poison produced as a glucoside by several plant species (also called prussic acid).

Hydrologic cycle. Fate of water from the time of precipitation until the water has been returned to the atmosphere by evaporation.

Hydrology. Science dealing with the distribution and movement of water.

Hydrolysis. Decomposition of a compound into other compounds by the chemical addition of the elements of water (H^+ and OH^-).

Hydromorphic soils. Soils formed under conditions of poor drainage in marshes, swamps or seepage areas.

Hydroponics. The growing of plants in aqueous chemical solutions.

Hygroscopic. The ability of a substance to attract and hold moisture from the air.

Hypocotyl. The stem below the cotyledons and above the root of a young seedling.

Imbibition. Process of absorption of liquids by the force exerted through surface tension.

Immune. Free from attack by a given organism.

Imperfect flower. One lacking either stamens or pistils.

Inbred. Resulting from successive self-fertilization.

Inbred line. A relatively true-breeding strain resulting from at least five successive generations of controlled self-fertilization or its equivalent.

Inbreeding coefficient. A percentage expressing the degree of added homozygosity in an inbred organism because of the relationship between its parents.

Inbreeding depression. Loss of vigor and often increased mortality in successive generations of inbreeding.

Incomplete dominance. Condition in heterozygotes where the phenotype is intermediate between the two homozygotes.

Increaser. A plant that spreads under existing management.

Indeterminate. Pertaining to the growth of plants, the flowers of which are borne on lateral branches with blooming continued over a relatively long period.

Indigenous. Produced or living naturally in a specific environment.

Infiltration rate. Rate at which water enters the soil.

Infiltrometer. Device for measuring rate of entry of fluid into a porous body.

Inflorescence. Flowering part of a plant.

Inoculate. To place an inoculum where it will produce a reaction in the host organism, or to introduce nitrogen-fixing bacteria into the soil.

Insolation. Solar radiation received by the earth.

Integuments. Coats or walls of an ovule that eventually form the seedcoat.

Interaction. Interacting of two sources of variation on a trait, so that the effect caused by either source is dependent in part on the influence of the other.

Interpolation. Estimation of points or values falling between known points, such as those in a table.

Interspecific hybrid. Plant or animal progeny that results from the successful cross between two species.

Introgression. The entry or introduction of a gene or genes from one gene complex into another.

Inversion. Reversal of the order of a block of genes in a given chromosome.

Inversion temperature. The condition in which cold air settles beneath warm air on a clear night.

In vitro. Occurring in an artificial environment, as in a test tube.

In vivo. Occurring in the living organism.

Involucre. A circle of bracts below a flower or a flower cluster.

Ion. An electrically charged element, group of elements, or particle.

Isotonic. Having the property of equal tension, especially osmotic pressure.

Kaolin. A group name for aluminum silicates. Kaolinite is a clay mineral that characterizes most kaolins.

Karyotype. The somatic chromosome complement of an individual, usually defined at mitotic metaphase by morphology and number (arranged in a sequence standard for the organism).

Ketone body. Any compound containing the carbonyl group, CO.

Ketosis. Metabolic disease characterized by excessive ketone body formation.

Key pests. Serious, perennially occurring persistent species that dominate control practices.

Kinetochore. The attachment region (within the centromere) of the microtubules of the spindle in cells undergoing either mitosis or meiosis.

Labile. A substance which readily undergoes transformation or is readily available to organisms.

Lamina. The blade of a leaf, particularly those portions between the veins.

Landrace. A strain or variety developed locally by indigenous people, presumably in absence of plant breeding.

Lateritic soils. Soils formed in warm, temperate, and tropical regions.

Latosol. Soils formed under forested, tropical, humid conditions — characterized by low base-exchange capacity, low content of most primary minerals, and usually red in color.

Leaching requirement. That fraction of applied water that should be passed through the root zone to control soil salinity at a specified level.

Lethal gene. A gene whose phenotypic effect is drastic enough to kill the bearer — death may occur at any stage in the life cycle.

Lignin. A complex noncarbohydrate strengthening material in the thickened cell walls of plants; practically indigestible.

Ligule. A membranous projection on inner side of leaf at top of sheath in many grasses.

Lime, agricultural. A soil amendment consisting of calcium carbonate but including magnesium carbonate and perhaps other materials used to neutralize soil acidity and furnish calcium and magnesium to plants.

Lime requirement. The mass of agricultural lime, or equivalent, required to raise the pH of the soil to a specified value under field conditions.

Line. A group of individuals from a common ancestry — more narrowly defined group than strain.

Line breeding. A form of inbreeding in which an attempt is made to concentrate genes of a superior ancestor in the progeny of later generations.

Linkage group. All genes located physically on a given chromosome.

Lithosols. Soils characterized by an incomplete solum or no clearly expressed soil morphology.

Locus (pl. loci). The position on a chromosome occupied by a particular gene or one of its alleles.

Loess. Material transported and deposited by wind and consisting of predominately silt-sized particles.

Luxury uptake. Absorption by plants of nutrients in excess of their need for growth.

Lysimeter. A device for measuring gains (irrigation, precipitation, and condensation) and losses (evapotranspiration) by a column of soil.

Male sterility. A condition in which pollen is absent or nonfunctional in flowering plants.

Mass selection. A system of breeding in which individuals are selected on the basis of phenotypes and used to produce the next generation (in plants, seed may be composited to grow the next generation).

Mean. Statistical term for the average value of a set of values.

Meiosis. The cell division that produces the gametes which contain half the number (haploid) of chromosomes found in the somatic cells of the species.

Meristem. An undifferentiated cellular region in plants characterized by repeated cell divisions.

Mesophyte. A plant that thrives under medium conditions of moisture and salt content of the soil.

Messenger RNA. Ribonucleic acid conferring amino acid specificity on ribosomes.

Metabolism. The sum of the physical and chemical processes whereby the living organism is produced and maintained.

Metabolite. Any substance produced in metabolism.

Metabolizable energy (ME). Food intake gross energy minus fecal energy, minus energy in gaseous products of digestion, minus urinary energy.

Metamorphic rock. Derived from preexisting rock but differing from them as a result of natural geological processes.

Metaxenia. The effect of pollen on tissue characteristics outside of the embryo.

Microclimate. Atmospheric environmental conditions in the immediate vicinity of an organism.

Microsporocyte. In plants, a cell destined to produce microspores by meiosis (also, pollen mother cell).

Mineralization. Conversion of an element from an organic form to an inorganic state as a result of microbial decomposition.

Mines. Deep holes or tunnels in plant parts caused by burrowing insects or their larvae.

Mitochondrion (pl. **mitochondria**). Small cytoplasmic organelle where cellular respiration occurs.

Mitosis. Cell division of somatic cells with each daughter cell receiving the same number (diploid) of chromosomes as the parent cell before division.

Modifying gene. One that causes minor changes in phenotype by interacting with the gene(s) that are the major determiners of the phenotype under consideration.

Moisture tension. The equivalent negative pressure in the soil water (equal to the equivalent pressure that must be applied to soil water to bring it to hydraulic equilibrium).

Monocotyledon. A plant whose embryo has one cotyledon, as in grasses.

Monoecious. Having male and female reproductive organs in separate flowers on the same plant.

Monophagous. Feeding on or using a single kind of food.

Monoploid. See **Haploid.**

Monosomic. An individual lacking one chromosome of a set $(2n - 1)$.

Multiline. A crop grown from a mixture of seed of several (usually related but with different genotypes) resistant lines.

Multiple alleles. A series of three or more alternative alleles, any one of which may occur at a particular locus on a chromosome.

Mutation. A spontaneous change in DNA, altering the genetic code (mutations may be gene mutations or chromosomal changes).

Mycorrhiza. Literally "fungus root". The association, usually symbiotic, of specific fungi with the roots of higher plants.

Mycotoxins. Toxic metabolites produced by molds during growth.

Net energy (NE). The difference between metabolizable energy and heat increment (includes energy for maintenance or maintenance plus production.)

Nitrate poisoning. A condition that may result when ruminants ingest nitrates (NO_3) that rumen bacteria convert to nitrite NO_2.

Nitrogen assimilation. Uptake of nitrate by living organisms and conversion to cell substances (e.g., protein).

Nitrogen cycle. Sequence of biochemical changes undergone by nitrogen wherein it is used by the living organism, liberated upon the death and decomposition of the organism, and converted to its original state of oxidation.

Nitrogen fixation. The conversion of atmospheric (free) nitrogen to nitrogen compounds.

Nitrogen-free extract (NFE). Obtained by subtraction of the crude fiber from the total carbohydrate analysis in the proximate system of feed analyses.

Nonpreference. Behavior of insects when they avoid or exhibit negative reactions to a plant or a host species (may describe trait inducing such behavior).

Nonprotein nitrogen (NPN). Nitrogen in feed from substances such as urea, ammonium salts, amines, amino acids, etc. but not from preformed protein.

Nonrecurrent parent. The parent that is not involved in a backcross.

Nontarget species. A species not intentionally affected by a pesticide.

Normal curve. A smooth, symmetrical, bell-shaped curve of distribution.

Normal distribution. Frequency distribution usually characteristic of biological traits (basis of many statistical procedures).

Nucleic acids. The two nucleic acids are DNA and RNA which are composed of long chains of molecules called nucleotides. DNA carries the genetic code, while there are three types of RNA, namely, ribosomal RNA (rRNA), messenger RNA (mRNA), and transfer RNA (tRNA).

Nucleotide. Portion of a DNA or RNA molecule composed of one deoxyribose phosphate unit (in DNA), or one ribose phosphate unit (in RNA), plus a purine or a pyrimidine.

Nucleus. A body of specialized protoplasm containing the chromosomes within a cell.

Nullisomic. A cell or organism lacking both members of a given chromosome pair $(2n - 2)$.

Nutrient balance. The approximate balance among concentrations of nutrients essential for plant growth that permits maximum growth rate and yield (imbalances may result from either a deficiency or an excess supply of one or more nutrients).

Nutrient requirement. Refers to meeting the animals minimum needs, without margins of safety, for maintenance, fitting, reproduction, lactation, and work.

Off-types (in crops). Any seed or plant not a part of the cultivar (variety) in that it deviates in one or more characteristics from the cultivar as described and may include: a seed or plant of another cultivar; a seed or plant not necessarily any cultivar; a seed or plant resulting from cross-pollination by another kind or cultivar; a seed or plant resulting from uncontrolled self-pollination during production of hybrid seed; or segregates from any of the above.

Open-pollination. Pollination whereby seed is produced as a result of natural pollination (as opposed to controlled pollination).

Operon. A system of cistrons, operators, and promotor sites, by which a given genetically controlled, metabolic activity is regulated.

Organelle. Subcellular structures that make up a cell.

Organic soil. One that contains a high percentage (>15 or 20%) of organic matter throughout the solum.

Organophosphate. Any of several derivatives of phosphoric acid (includes materials related to parathion, malathion, diazinon, etc.).

Osmosis. The passage of a solvent from one side of a membrane to another where the escaping tendency of the solvent on the two sides is unequal.

Outbreeding. A system of mating animals that are less closely related than the average relationship in the population.

Outcross. A cross to an individual that is not closely related.

Ovary. The female gonad in animals or the ovule-containing portion of the pistil of a flower.

Ovule. The structure that bears the female gamete and becomes the seed after fertilization.

Oxidation. Loss of electrons from an atom or its gain in positive charges (combination with oxygen, or the loss of a hydrogen, or the loss of an electron).

Pans. Horizons or layers in soils, that are strongly compacted, indurated, or very high in clay content.

Parent material. Unconsolidated mineral or organic matter from which soils develop.

Parthenocarpy. Production of fruits without fertilization and, normally, without seeds.

Parthenogenesis. The development of an individual from a gamete without fertilization.

Partial dominance. Lack of complete dominance.

Particle size. The effective diameter of a particle measured by sedimentation, sieving, or micrometric methods.

Pathogenicity. The ability of an organism to incite a disease.

Peat soil. An organic soil containing more than 50% organic matter.

Ped. A unit of soil structure such as an *aggregate, crumb, pism, block,* or *granule,* formed by natural processes (contrasts with *clod,* that is formed artificially).

Percolation, soil water. The downward movement of water through soil.

Perfect flower. A flower having both pistil and stamens.

Permanent wilting percentage. Water content of a soil when indicator plants growing in that soil wilt and fail to recover when placed in a humid chamber.

Phage. A virus that infects bacteria.

Phenol. Carbolic acid, C_6H_5OH.

Phenotype. Physical or external appearance of an organism as contrasted with its genetic constitution (genotype).

pH, soil. The negative logarithm of the hydrogen-ion activity of a soil.

Phosphoric acid. Usually a solution of P_2O_5 in water but frequently referred to as P_2O_5. Phosphoric anhydride, phosphorus pentoxide, or phosphoric oxide are more accurate terms for P_2O_5.

Photosensitization. Hypersensitivity to light.

Photosynthesis. The synthesis, in the presence of light, of carbohydrates and other organic compounds from simple molecules through the aid of chlorophyll.

Phytoalexin. A substance in plants that, in combination with antibodies, causes destruction of bacteria and other antigens.

Phytotoxic. Injurious to plant life or life processes.

Plasmagene. A self-replicating, cytoplasmically located gene.

Plasmid. A free-floating circular piece of DNA, not part of the main chromosome, in bacteria.

Plastic soil. A soil that may be molded or deformed continuously and permanently, by relatively modest pressure, into various shapes.

Pleiotrophy. The influencing of more than one trait by a single gene.

Ploidy. Referring to the number of sets of chromosomes in a cell.

Podzol. Soils formed in cool-temperate to temperate, humid climates, under coniferous or mixed coniferous-deciduous forest.

Pollination. Transfer of pollen from the anther to a stigma.

Polycross. An isolated group of plants or clones arranged to facilitate random interpollination.

Polygenes. Two or more different pairs of alleles, with a presumed cumulative effect governing quantitative traits.

Polymorphic. Having many forms.

Polyploid. An organism with more than two sets (genomes) of chromosomes in its body cells.

Polysome. A group of ribosomes joined by a molecule of messenger RNA.

Population. A group of organisms that are considered genetically as a unit for purposes of experimentation.

Porosity. The volume percentage of the total bulk not occupied by solid particles.

Potash (K_2O, potassium oxide). The trade term "potash" is used interchangeably with the word "potassium", and expresses the percentage of potassium oxide (K_2O) in potash salts and mixtures.

Probability. The likelihood that some event or outcome will occur.

Productivity, soil. The capacity of a soil, in its normal environment, for producing a specified plant or a sequence of plants under a specified management system.

Progeny test. A progeny, or groups of progenies, grown for the purpose of evaluating the genotype of the parent.

Prokaryote. A simple one-celled organism without a "true nucleus".

Protandrous. Having anthers that shed their pollen before the stigmas are receptive.

Protein. Complex combination of amino acids, always containing carbon, hydrogen, oxygen, and nitrogen, and sometimes phosphorus and sulfur.

Protein, crude. All nitrogenous substances contained in feedstuffs (% crude protein = % N × 6.25).

Protein digestible. That part of the diet that is made available for absorption from the gastrointestinal tract.

Protogynous. Having pistils ready for fertilization before anthers are matured.

Protoplasm. All the material in a cell, including the nucleus.

Protoplast. A cell without a cell wall (a "naked cell").

Proximate analysis. Analytical system that includes the determination of ash, crude fiber, crude protein, ether extract, moisture (dry matter), and nitrogen-free extract.

Pure line. A strain of organisms that is genetically pure (homozygous) because of continued inbreeding, self-fertilization, or other means.

Pure live seed (PLS). The percentage of the content of a seed lot that is pure and viable.

Race. A group of individuals having certain characteristics in common because of common ancestry (subdivision of species).

Raceme. A simple inflorescence of pediceled (short stems) flowers upon a more or less elongated axis, opening from the base.

Random drift. Changes in gene frequency in a population due to random causes rather than to identifiable causes.

Reaction, soil. Degree of acidity or alkalinity of a soil, usually expressed as a pH value.

Recessive. The member of a pair of genes that fails to express itself in the presence of its dominant allele (ordinarily expressed only in the homozygous state).

Reciprocal crosses. Two crosses between two plants in which the male parent of one cross is the female parent of the second cross.

Recombinant. An individual derived from a crossover gamete.

Recombinant DNA. A DNA molecule (often a bacterial plasmid) that has been cut enzymatically and DNA of another individual of the same or a different species inserted in the space so produced.

Recombination. The new association of genes in a recombinant individual, or formation of new gene combinations as a result of cross-fertilization between different genotypes.

Recurrent parent. The parent to which hybrid material is crossed in a backcross.

Recurrent selection. A breeding system designed to increase the frequency of favorable genes for specific characteristics by repeated cycles of selection and recombination.

Registered animal. An animal recorded in the herdbook of the breed.

Registered seed. The progeny of breeder or foundation seed handled under procedures acceptable to the certifying agency to maintain satisfactory genetic purity and identity. Used to produce certified seed.

Regression. Mathematical method for determining the amount of change in a trait that is due to change in another trait.

Regulator site. The specific segment of DNA responsible for the production of a regulatory protein (may serve as a repressor or an activator).

Residue. The quantity of a chemical remaining in or on the soil, plant parts, animal tissues, whole organisms and surfaces. Also, that which remains of any particular substance.

Rhizobia. Bacteria capable of living symbiotically in the roots of legumes, from which they receive energy and often utilize molecular nitrogen (collective common name, rhizobium).

Rhizome. An underground stem, usually horizontal and often elongated (has nodes, internodes, and sometimes scale-like leaves).

Rhizosphere. Soil zone where the microbial population is altered both quantitatively and qualitatively by the presence of plant roots.

Ribonucleic acid (RNA). A single-stranded nucleic acid molecule, synthesized principally in the nucleus from deoxyribonucleic acid. Functions to carry genetic information from nuclear DNA to the ribosomes.

Ribosome. A cytoplasmic structure which is the site of protein synthesis.

Rogue. To cull or weed out diseased or defective individuals or those considered inferior in one or more respects.

Rumination. The act of regurgitating previously eaten feed and chewing a coarse mass of coarse feed particles, called a bolus or cud.

Runoff. In soil science, "runoff" usually refers to the water lost from an area by surface flow, while in geology and hydraulics "runoff" usually includes both surface and subsurface flow.

Saline soil. A nonsodic soil containing sufficient soluble salt to impair its productivity.

Salinization. The process whereby soluble salts accumulate in soils.

Salt index. A measure of the relative tendency of a fertilizer to increase the osmotic pressure of the soil solution, as compared to the increase caused by an equal weight of sodium nitrate.

Saponin. Any of various plant glucosides that form soapy colloidal solutions when mixed and agitated in water.

Saturated fat. A completely hydrogenated fat — each carbon atom is associated with the maximum number of hydrogens; there are no double bonds.

Scabrous. Having a rough surface.

Secondary nutrient elements. Calcium, magnesium, and sulfur are secondary macronutrient elements, essential in lesser quantity for plant growth than primary macronutrients.

Seed. A seed consists of the seed coat, embryo, and in certain plants, an endosperm (a mature ovule plus coverings).

Seed, hard. With a seed coat impervious to the water or oxygen necessary for germination.

Segregation. The separation of homologous chromosomes (and genes) from different parents at meiosis.

Selection. Any process, natural or artificial, that permits an increase in the proportion of certain genotypes or groups of genotypes in succeeding generations (based on differential rate of reproduction).

Selection pressure. The intensity of selection for or against a trait in the selection process.

Self-fertilization. The functioning of a single individual as both male and female parent.

Self-incompatible. Inability of a plant to set seed as a result of fertilization with its own pollen.

Self-sterility. Failure to complete fertilization and obtain seed after self-pollination.

Siblings. Offspring of the same parental organisms.

Significance. Significant probability values indicate that the results, although possible, deviate too greatly from the expectancy to be acceptable when chance alone is operating.

Silica-aluminum ratio. The molecules of silicon dioxide (SiO_2) per molecule of aluminum oxide (Al_2O_3) in clay minerals or in soils.

Single cross. The first generation progeny of a cross between two different parents. Specifically, the first generation (F_1) of a cross between two inbred lines or their equivalent.

Sire summary. An estimate of the genetic transmitting ability of a bull.

Slow release. A term used interchangeably with delayed release, controlled release, controlled availability, slow acting, and metered release to designate a rate of dissolution (usually in water) much less than is obtained for completely water-soluble compounds.

Sodic soil. One containing sufficient exchangeable-sodium to interfere with the growth of most crop plants.

Soil amendment. See **Amendment, soil**.

Soil water. Water in soil is subject to several force fields originating from: the presence of the soil solid phase; the dissolved salts; the action of external gas pressure; and the gravitational field. These effects may be quantitatively expressed by assigning an individual component potential to each. The sum of these potentials = the total potential of soil water.

Solum. The upper and most weathered part of the soil profile; the A and B horizons.

Soma (adj. **somatic**). The body, cells of which in mammals and flowering plants normally have two sets of chromosomes, one derived from each parent.

Specific combining ability. The response in vigor, growth, or fruitfulness from crossing a given line with other particular lines (often inbred lines).

Standard deviation. A measure of the variation in a sample.

Stigma. The part of the pistil that receives the pollen.

Strain. A group of plants derived from a cultivar.

Suberize. To form a protective corky layer on a cut surface of plant tissue.

Surfactant. A material that improves the emulsifying, dispersing, spreading, wetting, or other surface properties of liquids.

Sward. The grassy surface of a pasture, lawn or playing field. "Turf" and "sod" are similar terms but refer to the stratum of earth filled with grass roots as well as to the surface.

Symbiotic nitrogen fixation. Fixation of atmospheric N by rhizobia growing in nodules on the roots of legumes.

Sympatric. Occurring in the same area without loss of identity through interbreeding.

Synapsis. The pairing of homologous chromosomes that occurs at meiosis.

Syndrome. A group of symptoms which occur together and characterize a disease.

Synergism. The combined action of two or more agents that produce a greater response than the sum of their individual effects.

Synthetic cultivar. Advanced generation (usually limited) progeny of a number of clones or lines obtained by open-pollination under isolation from possible sources of contamination.

Systemic chemical. A compound that is translocated throughout a plant and that may have an effect on pests or other agents associated with the plant system, may modify growth, or may kill the plant.

Tandem method of selection. Exertion of selection pressure first on one trait, then on a second, and so on, rather than selection for all traits simultaneously.

Tannin. A broad class of soluble polyphenols with common property of condensing with protein to form an insoluble substance of impaired digestibility.

Taxon. A taxonomic group of any rank.

Temperature inversion. See **Inversion temperature**.

Template. A model, mold, or pattern; DNA acts as a template for RNA synthesis.

Tensiometer. A device for measuring the soil-water matric potential (or tension) of water in soil *in situ*.

Teratogenic. Capable of causing birth defects.

Test cross. The cross of an individual, usually a dominant phenotype, with one having the recessive phenotype.

Tetany. A condition in an animal in which there are localized, spasmodic muscular contractions.

Tetraploid. A cell, tissue, or organism with four sets of chromosomes (4n).

Thermophilic organisms. Those that grow readily at temperatures above 45°C.

Three-way cross. In plants, the first generation of a cross of a foundation single cross and an inbred line or a foundation backcross.

Tissue culture. The growing of aggregates of cells in a solution of salts, sugars, vitamins, and hormones. (Regeneration of whole plants may be possible.)

Top cross. An outcross of selections, clones, lines, or inbreds, to a common pollen parent.

Total acidity. (See **Acidity, total**.)

Total digestible nutrients (TDN). Sum total of all digestible organic nutrients, i.e., proteins, nitrogen-free extract, fiber, and fat. An average of all feeds 1 g of TDN = 4.4 kcal.

Total milk solids. Primarily milk fat, proteins, lactose, and minerals.

Toxicity. The quality or potential of a substance to cause injury or illness.

Trace element. A chemical element used in minute amounts by organisms and held essential to their physiology, includes cobalt, copper, iodine, iron, manganese, selenium, and zinc.

Transcription. Synthesis of messenger RNA from a DNA template.

Transduction. Recombination in a bacterium whereby DNA is transferred by a phage from one cell to another.

Transformation. Genetic recombination, particularly in bacteria, whereby naked DNA from one individual becomes incorporated into that of another.

Transgressive segregation. The segregation of individuals, in the F_2 or a later generation of a cross, which show a more extreme development of a trait than either parent.

Translocation. Physical movement of water, nutrients, or chemicals, such as a herbicide or elaborated food, within a plant.

Transmitting ability. One-half of an organism's breeding value; the average genetic superiority or inferiority that is transmitted by an individual to its offspring.

Transpiration. Water loss by evaporation from the internal surface of the leaves and through the stomata.

Triploid. A polyploid cell, tissue, or organism having three sets of chromosomes (3n).

Tropism. An involuntary response in the form of movement of a plant, the direction of which is determined by the source of the stimulus.

True protein. A nitrogenous compound that will hydrolyze completely to amino acids.

Tuff. Volcanic ash usually more or less stratified and in various states of consolidation.

Ungulates. Mammals having hoofs.

Unit character. A hereditary trait that is transmitted by a single gene.

Unit, fertilizer. Represents percent or 20 pounds of nutrient (N, P_2O_5, or K_2O) in a 2000-lb ton of fertilizer. Thus a 5—20—10 fertilizer contains five units of N, 20 units of P_2O_5, and 10 units of K_2O per ton.

Unsaturated fat. A fat having one or more double bonds; not completely hydrogenated.

Urea. A white, crystalline substance found in urine, blood, and lymph, that is the final produce of protein metabolism. Synthetic urea is used as N source by plants, and as N source for protein synthesis by ruminant animals.

Vapor drift. The movement of chemical vapors from the area of application.

Variance. A statistic providing an unbiased estimate of population variability (the square of the standard deviation).

Variant. Any seed or plant which (a) is distinct within the cultivar (variety) but occurs naturally in the cultivar; (b) is stable and predictable with a degree of reliability comparable to other cultivars of the same kind, within recognized tolerances, when the cultivar is reproduced or reconstituted; and (c) was originally a part of the cultivar as released. A variant is not an off-type.

Variation. The occurrence of differences within a species, cultivar, or breed.

Variety. See **Cultivar.**

Vernalization. To treat seed, bulbs, or seedlings of a plant to shorten its vegetative period and induce earlier flowering and fruiting.

Virulence. The ability to produce disease.

Water-stable aggregate. A soil aggregate that is stable to the action of water.

Water table. The upper surface of ground water, or that level in the ground where the water is at atmospheric pressure.

Water use efficiency. Dry matter or harvested portion of crop produced per unit of water consumed.

Weed. Any plant that is objectionable or interferes with the activities or welfare of man.

Wetting agent. A substance that serves to reduce interfacial tensions and causes solutions and suspensions to make better contact with surfaces.

Wild type. Most frequently encountered phenotype in natural breeding populations (= ''normal'' phenotype).

Xenia. The physiological effect of foreign pollen on maternal fruit tissue.

Xerophytes. Plants that grow in or on extremely dry soils or soil materials.

Yield, sustained. A continual annual, or periodic, yield of plants or plant material from an area; implies management practices that will maintain productive capacity of the land.

Zygote. The protoplast resulting from the fusion of two gametes in sexual reproduction; a fertilized egg.

INDEX

A

Abelmoschus esculentus, see Okra
Abies
  balsamea, see Fir, balsam
  concolor, see Fir, white
  nobilis, see Fir, noble
  procera, see Fir, noble
Abiological nitrogen fixation, 78
Abutilon theophrasti, see Jute, China
Acacia senegal, see Gum arabic
Acari, 77
Acer saccharum, see Maple, sugar
Acetylene reduction, 80
Acid detergent fiber (ADF), 355, 359
Acid soils, 43, 46, 226
  beans and, 51
  corn and, 49—50
Acid treatment, 362
Actinidia chinensis, see Kiwi
Actinomycetes, 77
Additives, to silage, 380
ADF, see Acid detergent fiber
Aeration
  airflow rates for, 227
  of corn, 337
  of wheat, 338
Agave sisalana, see Sisal
Ageratum (*Ageratum* spp.), 124
Agropyron
  cristatum, see Wheatgrass, fairway
  desertorum, see Wheatgrass, crested
  elongatum, see Wheatgrass, tall
  intermedium, see Wheatgrass, intermediate
  repens, 219
  riparium, see Wheatgrass, streambank
  sibiricum, see Wheatgrass, Siberian
  smithi, see Wheatgrass, western
  spicatum, see Wheatgrass, bluebunch
  trachycaulum, see Wheatgrass, slender
  tricophorum, see Wheatgrass, pubescent
Agrostemma githago, see Corncockle
Agrostis
  alba, see Redtop
  canina, see Bentgrass, velvet
  palustris, see Bentgrass, creeping
  stolonifera, see Bentgrass, creeping
  tenuis, see Bentgrass, colonial
  vulgaris, 219
Air-screen cleaners, 343
Aleurites fordii, see Tung-oil tree
Alfalfa, 109, see also varieties by name
  annual productivity of, 208
  AOSCA certification of seeds of, 164, 169
  carbon-nitrogen ratio in, 75
  cellulose in, 397
  composition of, 357
  crop factors for, 190
  crop rotation and, 229, 251
  digestibility of, 361
  digestion coefficients of, 368
  diseases in, 250, 251, 257
  distribution factors for, 117
  dry matter losses in, 361, 362
  fungicides and, 246
  germination requirements for, 147
  harvesting of, 375
  insects in, 233
  irrigation of, 179, 182
  leaf composition in, 356
  lime rate for, 49
  lime response in, 45
  maximum yields of, 198
  meals, 365—366
  mean growth rate for, 208
  milk production in dairy cows and, 356
  moisture content of, 365, 367
  nematodes in, 259
  nitrogen fixation and, 79
  nutient amounts needed for production of, 88
  nutrient losses in, 361
  nutrients in, 356, 357, 393
  nutrients in above-ground portion of, 73
  nutrient uptake in, 89, 91, 97
  optimum pH for, 44
  photosynthesis in, 198
  protein in, 357, 365, 367, 368, 370
  proximate analyses of, 365—366, 367
  roots of, 220, 222
  salt and productivity of, 58
  salt tolerance of, 55
  seed conditioning in, 344
  seeding rates in, 173
  seed longevity in, 124
  silage, 378, 393
  soil pH for, 49
  stem composition in, 356
  transpiration ratio constants for, 186
  vitamins in, 367, 368
  water requirements of, 185
  water stress effects on, 179
  yields of, 49, 97, 358, 361
Alfalfa, Grimm, 186
Alfalfa, lucerne, 124
Alfalfa, Peruvian, 186
Alfalfa meals, 365—366
Alfalfa silage
  nutrients in, 393
  TDN for, 378
Alfilaria, 147
Alfisols, 31—33
  lime response in, 46
  phosphorus in, 95
  potassium in, 96
Algae
  biomass of, 77

maximum yields of, 198
photosynthesis in, 198
Alkali sacaton, 61
Allium
   ampeloprasum, see Leek
   cepa, see Onion
   fistulosum, see Onion, Welsh
   sativum, see Garlic
   schoenoprasum, see Chives
Allspice, 344
Almond, 109, 117
   amino acids in, 282
   conversion factors for shells of, 73
   nutrients in hulls of, 393
   protein in, 282
   salt and productivity of, 56
   toxic compounds from, 283
   weed losses in, 260
Aloe barbadensis (Aloe, Barbados), 109, 117
Alopecurus
   arundinaceus, see Foxtail, creeping
   myosuroides, see Foxtail, black grass
   pratensis, see Foxtail, meadow
Aluminum
   in beans, 51
   in corn leaves, 49—50
   in limestone, 47
   in milk, 445
   in sewage sludge, 85
   in slags, 48
Aluminum sulfate, 66
Alysicarpus vaginalis, see Clover, alyce
Amaranth, Chinese, 109, 117
Amaranth, feathered, 124
Amaranthus
   graicizans, see Tumbleweed
   retroflexus, see Pigweed
   tricolor, see Amaranth, Chinese; Tampala
Amino acids, 473
   in alfalfa leaf protein, 370
   in animal feeds, 307, 308, 329, 330
   in cereal grains, 297, 299
   in corn, 327
   in hominy feed, 330
   in meats, 442
   in millet, 320
   in nuts, 282
   in oats, 316
   in sorghum, 320
   in soybean meal, 290
   in soybean products, 287
   in sunflower meal, 290
   in triticales, 309
Ammonia, 106, 107
Ammonia-ammonium nitrate solution, 104
Ammonia-ammonium nitrate-urea solutions, 105
Ammonia-urea, 104
Ammonium nitrate, 106, 107
Ammonium sulfate, 106
Ammophila
   arenaria, see Beachgrass, European

breviligulata, see Beachgrass, American
Anacardium occidentale, see Cashew
Ananas comosus, see Pineapple
Andropogon
   furcatus, 219
   gerardii, see Bluestem, big
   hallii, see Bluestem, sand
   ischaemum, see Bluestem, yellow
   sorghum, 185, see also Sorghum
Anethum graveolens, see Dill
Animal feeds, see also specific types
   amino acids in, 307, 308, 329
   chicken consumption of, 427
   from corn, 329
   dry matter in, 393—395
   elements in, 397
   minerals in, 307, 308, 329, 393—395
   nutrients in, 307, 308, 329, 393—395
   protein in, 329, 393—395, 396
   proximate analysis of, 329
   sorghum in, 322
   soybean products in, 286, 287
   vitamins in, 307, 308, 329, 393—395
   wheat mill feed values in, 308
   whole wheat values in, 307
Animal reservoirs of infectious disease, 266
Animals, see also specific types
   farm, see Farm animals
   as vectors, 266
Anise, 344
Annual productivity, 207—211
Anthoxanthum odoratum, see Vernalgrass, sweet
Anthyllis vulneraria, see Vetch, kidney
Antirrhinum majus, see Snapdragon
AOSCA certification, 164—165, 169, 170
Aphids, 234, 235, 237, 239
Apium graveolens, see Celery
Apple, 109, 117
   chilling injury to, 278
   diseases in, 250
   freezing sensitivity of, 277
   fungicides and, 246
   heat killing temperature for, 177
   insects in, 233
   nutrients in, 74, 280
   nutrient uptake in, 92
   optimum pH for, 44
   salt and productivity of, 56
   seed longevity in, 124
   seed rates for, 218
   spacing of, 218
   storage requirements of, 273
   waterlogging tolerance in, 197
Apricot, 109, 117
   conversion factors for pits of, 73
   freezing sensitivity of, 277
   nutrients in, 280
   optimum pH for, 44
   salt and productivity of, 56
   seed longevity in, 124
   storage requirements of, 273

toxic compounds from, 283
Aqua ammonia, 104
Arable soils, 76
Arachis hypogaea, see Peanut
Arborvitae, 44, 58
Arctium lappa, see Burdock, great
Area, conversion factors, 455
Arenga pinnata, see Palm, sugar
Aridisols, 31—33
Armoracia rusticana, see Horseradish
Arrhenatherum, see Oatgrass, tall
Arsenic
    in milk, 445
    in sewage sludge, 85
    toxicity, 283
Arthropods, 77, 269, see also Insects
Artichoke
    germination requirements for, 159
    seed longevity in, 124
Artichoke, globe, 109
    distribution factors for, 117
    storage requirements of, 274
Artichoke, Jerusalem, 109
    distribution factors for, 117
    storage requirements of, 274
Artocarpus
    altilis, see Breadfruit
    heterophyllus, see Jackfruit
Ascorbic acid, see Vitamin C
Ash, European, 124
Asparagus, see also Asparagus, garden
    freezing sensitivity of, 277
    fungicides and, 246
    germination requirements for, 159
    nutrients in, 281
    optimum pH for, 44
    roots of, 175, 219
    salt tolerance of, 55
    seed longevity in, 124
    storage requirements of, 274
Asparagus, garden, 109, 117
Asparagus officinalis, see Asparagus; Asparagus,
    garden
Association of Official Seed Certifying Agencies, see
    AOSCA
Aster, 125
Astragalus spp., see Milkvetch
Avena
    byzantina, see Oats, red
    elatior, see Oatgrass, tall
    fatua, see Oats, wild
    sativa, see Oats
    strigosa, see Oats, bristle
Avocado, 109, 117
    chilling injury to, 278
    crop factors for, 190
    freezing sensitivity of, 277
    nutrients in, 280
    soil factors, 56, 191
    storage requirements of, 273
Axonopus affinis, see Carpetgrass

Azonal soils, 30

B

Bacteria, 77
Bagasse, 72
Bahiagrass, 109, 117
    germination requirements for, 147
    roots of, 220
    seeding rates in, 172
    seed longevity in, 125
Bahiagrass, Pensacola, 91
Balanites aegyptiaca, see Date, desert
Bamboo, 125
Bamboo, heavenly, 56
Bambusa arundinacea, see Bamboo
Banana, 109, 117
    chilling injury to, 278
    crop factors for, 190
    crop residue management and disease control in,
        256
    diseases in, 226
    freezing sensitivity of, 277
    lime response in, 45
    nutrients in, 280
    nutrient uptake in, 92
    seed rates for, 218
    storage requirements of, 273
    toxic compounds from, 283
    waterlogging tolerance in, 197
Barbarea verna, see Cress, upland
Barium
    in milk, 445
    in sewage sludge, 85
Barley, 109, 117, see also specific types
    annual productivity of, 211
    AOSCA certification of seeds, 164
    average composition of, 297
    carbohydrates in, 299
    carbon dioxide uptake in, 199
    composition of, 297, 311
    conversion factors for residues of, 72
    crop rotation and disease control in, 251
    digestion coefficients of, 305
    diseases in, 250
    fatty acids in, 303
    forage yields of, 311
    frost resistance in, 176
    germination requirements for, 147
    grain, size of, 295, 334
    growth rate for, 211, 213
    heat killing temperature for, 177
    humidity effects on, 177
    insects in, 233
    irrigation of, 182
    lime response in, 45
    minerals in, 301
    moisture content of, 331, 332
    nematodes in, 259
    nutrients in, 393
    nutrients in above-ground portion of, 73

nutrient uptake in, 90
optimum pH for, 44
physical properties of, 295
protein in, 300
proximate analysis of, 296, 371
radiation use efficiency in, 213
relative feeding values of, 305
residue production of, 70
roots of, 220
salt and productivity of, 61, 64
salt tolerance of, 55
seed conditioning in, 344
seeding rates in, 171
seed longevity in, 125
steer intake of, 312
TDN for, 378
waterlogging tolerance in, 197
water requirements of, 185
weed losses in, 260
yields of, 216
Barley, pearled
    carbohydrates in, 298
    vitamins in, 302
Barnyardgrass, 263
Basil, 109, 117
Bay rum tree, 109, 117
Beachgrass, American, 109, 117
Beachgrass, European, 109, 117
Bean, see also specific types
    carbon-nitrogen ratio in stems of, 75
    crop factors for, 190
    diseases in, 226
    frost resistance in, 176
    fungicides and, 246
    humidity effects on, 177
    insects in, 233
    nutrients in, 51, 74
    salt and productivity of, 56
    seed conditioning in, 344
    soil pH for, 51
    toxic compounds from, 283
    waterlogging tolerance in, 197
    yields of, 51
Bean, adzuki, 147
Bean, African locust, 109
Bean, asparagus, 159
Bean, baby lima, 344
Bean, broad, 109, 117
    germination requirements for, 159
    salt and productivity of, 58
    seeding rates in, 171
    seed longevity in, 126
    waterlogging tolerance in, 197
Bean, castor, 110, 118
    germination requirements for, 150
    salt tolerance of, 55
    seeding rates in, 171
    seed longevity in, 128
Bean, dry, see also specific variety
    crop rotation and disease control in, 251
    diseases in, 250

storage requirements of, 274
    weed losses in, 264, 265
Bean, dwarf, 175
Bean, Egyptian (hyacinth bean), 109, 117, 369
Bean, fava
    AOSCA certification of seeds of, 164
    seed longevity in, 126
Bean, field
    AOSCA certification of seeds of, 164
    carbon dioxide uptake in, 205
    germination requirements for, 147
    optimum pH for, 44
    salt tolerance of, 55
    seed rates for, 217
Bean, French, 214
Bean, garden
    AOSCA certification of seeds of, 164
    seed longevity in, 125
Bean, great northern, 344
Bean, green, 109, 117
    diseases in, 250
    salt tolerance of, 55
    seeding rates in, 171
    storage requirements of, 274
Bean, horse, 186
Bean, hyacinth (Egyptian bean), 109, 117, 369
Bean, kidney
    crop rotation and disease control in, 251
    seed rates for, 217
Bean, lablab, 109, 117, 369
Bean, lima, 109, 117
    crop rotation and disease control in, 251
    germination requirements for, 159
    nutrients in, 281
    seed conditioning in, 344
    seeding rates in, 171
    seed longevity in, 125
    storage requirements of, 274
Bean, Mexican, 186
Bean, mung, 109, 117
    AOSCA certification of seeds of, 165
    germination requirements for, 147
    seed conditioning in, 344
    seeding rates in, 171
Bean, navy
    physical dimensions of kernels of, 332, 333
    seed conditioning in, 344
    transpiration ratio constants for, 186
Bean, pinto, 344
Bean, pole, 175
Bean, runner, 159
Bean, scarlet runner, 109
    distribution factors for, 117
    seed longevity in, 125
Bean, snap
    chilling injury to, 278
    disease control in, 251, 256
    freezing sensitivity of, 277
    nutrients in, 281
    nutrient uptake in, 91
    weed losses in, 264

Bean, tepary, 109
  distribution factors for, 117
  seeding rates in, 171
Bean, velvet, 115, 123
  germination requirements for, 157
  seeding rates in, 174
Bean, winged, 109, 117
Bean, yard-long, 109, 117
Bean, yellow eye, 344
Bean sprouts, 274
Bedding, 417
Beef, see also Cattle, beef
  amino acids in, 442
  composition of, 440, 453
  grades of, 433, 437
  protein in, 453
  retail yields of, 437
  vitamins in, 440
Beef cows, see Cattle, beef
Beet, see also specific types
  fodder, 176
  freezing sensitivity of, 277
  germination requirements for, 159
  humidity effects on, 177
  nutrients in, 281
  nutrients in pulp of, 393
  seeding guide for, 175
  topped, 274
Beet, bunched, 274
Beet, field, 147
Beet, garden, 109
  distribution factors for, 117
  salt and productivity of, 61
  seed longevity in, 125
Beet, mangel
  seeding guide for, 175
  seeding rates in, 171
  seed longevity in, 125
Beet, sugar
  annual productivity of, 209
  carbon dioxide uptake in, 204
  crop factors for, 190
  crop rotation and disease control in, 251
  diseases in, 226, 250
  drying rate of seeds of, 335
  frost resistance in, 176
  fungicides and, 246
  germination requirements for, 147
  growth rate, 209, 213
  insects in, 239
  irrigation of, 182
  maximum yields of, 198
  nutrients in above-ground portion of, 74
  nutrient uptake in, 90
  optimum pH for, 44
  photosynthesis in, 198
  radiation use efficiency in, 213
  roots of, 219
  salt and productivity of, 64
  salt tolerance of, 55
  seeding guide for, 175

  seeding rates in, 171, 218
  seed longevity in, 125
  spacing of, 218
  transpiration ratio constants for, 185
  waterlogging tolerance in, 197
  water requirements of, 185
  weed losses in, 264
Beet, table
  crop residue management in, 256
  crop rotation in, 252
  nutrient uptake in, 92
  optimum pH for, 44
  roots of, 219
  salt tolerance of, 55
Beetles, 237, 239, 241
  in cereal grains, 351
  in grains, 351
  in soybean, 351
Beggarwood, Florida, 147
Bentgrass, see also specific types
  crop residue management in, 256
  crop rotation in, 252
  nutrient uptake in, 91
  salt and productivity of, 58
  seed conditioning in, 344
Bentgrass, colonial, 109, 117
  germination requirements for, 148
  seeding rates in, 172
  seed longevity in, 126
Bentgrass, creeping, 109, 117
  germination requirements for, 148
  seeding rates in, 172
Bentgrass, velvet, 109, 117
  germination requirements for, 148
  roots of, 219
  seeding rates in, 172
Bermudagrass, see also specific types, 109, 117
  annual productivity of, 209
  carbon dioxide uptake in, 203
  germination requirements for, 148
  growth rate, 209, 212
  harvesting of, 375
  lime response in, 45
  moisture content of, 362
  nutrients in above-ground portion of, 73
  nutrient uptake in, 89, 91
  radiation use efficiency in, 212
  roots of, 220
  salt and productivity of, 64
  salt tolerance of, 55
  seed conditioning in, 344
  seeding rates in, 172
Bermudagrass, African, 109, 117
Bermudagrass, coastal
  dry matter losses in, 361
  nutrients in, 361, 393
Bermudagrass, giant, 148
Berries, see also by names
  defined, 474
  freezing sensitivity of, 277
  storage requirements of, 273

toxic compounds from, 283
Berries, cane, 191
Bertholletia excelsa, see Brazil nut
Beta vulgaris, see Beet, field; Beet, garden; Beet,
  mangel; Beet, sugar; Chard, Swiss
Betula spp., see Birch
Beverages, 217
Bindweed, 126, 219
Bindweed, field, 263
Biochemical oxygen demand (BOD), 86
 defined, 474
Biogas, 84
Biomass values, 77
Biotin
 in animal feeds, 308
 in cereal grains, 302
 in corn, 328
 in poultry feeds, 426
 in triticales, 310
 in wheat, 310
Birch, 211
Blackberry, 109, 117
 nutrients in, 280
 salt and productivity of, 56
 storage requirements of, 273
 waterlogging tolerance in, 197
Blueberry, see also specific types
 nutrients in, 280
 storage requirements of, 273
Blueberry, highbush, 109
 distribution factors for, 117
 optimum pH for, 44
Blueberry, lowbush, 109, 117
Blueberry, rabbiteye, 109, 117
Bluegrass, see also specific types
 nutrients in above-ground portion of, 73
 nutrient uptake in, 91
Bluegrass, bulbous
 germination requirements for, 148
 seed conditioning in, 344
Bluegrass, Canada
 germination requirements for, 148
 seed conditioning in, 344
Bluegrass, glaucantha, 148
Bluegrass, Kentucky, 109, 117
 germination requirements for, 148
 optimum pH for, 44
 roots of, 221, 222
 seed conditioning in, 344
 seeding rates in, 172
 seed longevity in, 126
Bluegrass, Nevada, 149
Bluegrass, rough, 149
Bluegrass, roughstalk, 109, 117
 seeding rates in, 172
 seed longevity in, 126
Bluegrass, Texas, 149
Bluegrass, wood
 germination requirements for, 149
 seed longevity in, 126
Bluestem, Australian, 109, 117

Bluestem, big, 109, 117
 germination requirements for, 149
 nutrient uptake in, 91
 seeding rates in, 172
 seed longevity in, 126
Bluestem, Caucasian, 109, 117
Bluestem, diaz, 109, 117
Bluestem, little, 110, 117
 germination requirements for, 149
 nutrient uptake in, 91
 seeding rates in, 172
 seed longevity in, 126
Bluestem, sand, 110, 117
 germination requirements for, 149
 seeding rates in, 172
 seed longevity in, 126
Bluestem, yellow, 110
 distribution factors for, 117
 germination requirements for, 149
BOD, see Biochemical oxygen demand
Boehmeria nivea, see Ramie
Bonemeal, 393
Borax, 106
Borers, 351
Boric acid, 106
Boron, 87
 amounts needed for plant production, 88
 in corn, 93
 in fertilizers, 106
 in irrigation water, 190
 in limestone, 47
 in manure, 82
 in milk, 445
 in secondary effluents, 86
 in sewage sludge, 85
 in soils, 45
 uptake of, 97
Boswellia carteri, see Frankincense
Bothriochloa
 caucasica, see Bluestem, Caucasian
 intermedia, see Bluestem, Australian
 ischaemum, see Bluestem, yellow
Bottlebrush, 58
Bougainvillea, 64
Bougainvillea spectabilis, see Bougainvillea
Bouteloua
 curtipendula, see Grama, side-oats
 gracilis, see Grama, blue
Boxwood, 58
Boysenberry, 56
Brachiaria
 mutica, see Paragrass
 ramosa, see Millet, browntop
Brassica
 arvensis, 222
 campestris, see Kale; Mustard; Rape, bird; Rape,
  turnip
 caulorapa, 219
 chinensis, see Pak-choi
 juncea, see Mustard, India
 napobrassica, 219

napus, see Kale, Siberian; Rape; Rape, annual; Rape, winter; Rutabaga
nigra, see Mustard, black
oleracea, see Brussels sprouts; Cabbage, tronchuda; Cauliflower; Collard; Kale; Kohlrabi
oleracea var. *botrytis,* 219
oleracea var. *capitata,* 219, see Broccoli; Cabbage
pekinensis, see Cabbage, Chinese
perviridis, see Mustard, spinach
rapa, see Cabbage, Chinese; Mustard; Turnip
Brazil nut, 113
    amino acids in, 282
    distribution factors for, 121
    protein in, 282
Breadfruit, 110, 117
Brewers grains, nutrients in, 393
Broccoli, 110, 117
    crop rotation and disease control in, 252
    freezing sensitivity of, 277
    germination requirements for, 159
    nutrients in, 281
    optimum pH for, 44
    salt and productivity of, 61
    salt tolerance of, 55
    seeding guide for, 175
    storage requirements of, 274
Broilers, see Chickens
Brome, see Bromegrass
Brome, smooth, see Bromegrass, smooth
Bromegrass
    nutrients in, 393
    nutrient uptake in, 91
    salt and productivity of, 61
    seed conditioning in, 344
    water requirements of, 185
Bromegrass, field
    germination requirements for, 149
    seed longevity in, 126
Bromegrass, mountain
    germination requirements for, 149
    seed longevity in, 127
Bromegrass, smooth
    chromosome number for, 110
    distribution factors for, 117
    germination requirements for, 149
    harvesting, 375
    roots of, 219
    salt tolerance, 55
    seeding rates in, 172
    seed longevity in, 127
    target yield for, 110
Bromine, 445
Bromus
    arvensis, see Bromegrass, field
    catharticus, see Rescuegrass
    inermis, see Bromegrass, smooth; Bromegrass
    marginatus, see Bromegrass, mountain
    mollis, see Chess, soft
    polyanthus, see Bromegrass, mountain
Broomcorn, 149, 221
Browallia viscosa, 142

Brussels sprouts, 110, 117
    freezing sensitivity of, 277
    germination requirements for, 159
    nutrients in, 281
    seeding guide for, 175
    seed longevity in, 127
    storage requirements of, 274
Buchloe dactyloides, see Buffalograss
Buckwheat, 110, 117, 303
    carbohydrates in, 298
    flour, 298
    frost resistance in, 176
    germination requirements for, 149
    minerals in, 301
    moisture content of, 331
    nutrient uptake in, 90
    protein in, 300
    seed conditioning in, 344
    seeding rates in, 171
    seed longevity in, 127
    transpiration ratio constants for, 185
    vitamins in, 302
    water requirements of, 185
Buffalograss, 110, 117
    carbon dioxide uptake in, 203
    caryopses, 172
    germination requirements for, 149
    roots of, 219
    seeding rates in, 172
Buffelgrass, 110, 117
    germination requirements for, 149
    seed longevity in, 127
Bulls, see also Cattle; Steers
    metabolizable energy requirements of, 399
    residues of, 81
    testosterone in, 390
Burclover, California, 110, 117
Burdock, great, 159
Burnet, 110
    distribution factors for, 118
    salt and productivity of, 56
Burnet, little, 149
Burning, 226
Buttercup, creeping, 127
Buxus microphylla, see Boxwood

C

Cabbage, 110, 118, see also specific types
    crop rotation and disease control in, 252
    diseases in, 226
    early, 175
    frost resistance in, 176
    germination requirements for, 159
    insects in, 234
    late, 175, 274
    nutrients in, 281
    nutrients in above-ground portion of, 74
    nutrient uptake in, 91
    old, 277
    optimum pH for, 44

salt and productivity of, 58
salt tolerance of, 55
seed conditioning in, 344
seed longevity in, 127
seed rates for, 217
seeding guide, 175
sensitivity to freezing, 277
spacing of, 217
toxic compounds from, 283
transpiration ratio constants for, 186
yields of, 453
Cabbage, Chinese, 110, 118
  germination requirements for, 159
  roots of, 219
  seeding guide for, 175
Cabbage, tronchuda, 160
Cacao, 110
  diseases in, 226
  distribution factors for, 118
Cadmium
  in corn, 328
  in milk, 445
  in secondary effluents, 86
  in sewage sludge, 85
  in sludge, 87
  toxicity, 283
Cajanus cajan, see Pea, pigeon
Calcimorphic soils, 29, 30
Calcium, 87
  in above-ground portion of crops, 73—74
  amounts needed for plant production, 88
  in animal feeds, 329
  in beans, 51
  in beef, 440
  in cereal grains, 317
  in chickens, 451
  in corn, 93, 328
    leaves, 49—50
    silage, 374, 375
  in dairy cattle feed, 412, 413
  in dairy products, 443
  in fertilizers, 98—100, 106
  in fruits, 280
  in hominy feed, 330
  in horse feeds, 429
  in limestone, 47
  livestock requirements for, 409, 420, 428
  in manure, 82
  in milk, 444
  in pork, 441
  in poultry feeds, 426
  in root material, 75
  in secondary effluents, 86
  in sewage sludge, 85
  in slags, 48
  in soils, 41—42, 45
  in soybean products, 287, 290
  in sunflower meal, 290
  in turkey, 450
  uptake of, 97
  in vegetables, 281

Calcium carbonate, 106
Calcium hydroxide, 106
Calcium oxide, 106
Callistemon viminalis, see Bottlebrush
Callistephus chinensis, see Aster
Calopo *(Calopogonium muconoides),* 203
Calves, see Cattle
Camellia sinensis, see Tea
Camphor tree, 110, 118
Campion, white, 127
Canals, 192, 193
Canarygrass, 110, 118, see also specific types
  germination requirements for, 150
  salt and productivity of, 58
  seed conditioning in, 344
  seeding rates in, 172
  seed longevity in, 127
Canarygrass, reed, 110, 118
  germination requirements for, 150
  proximate analyses of, 365
  roots of, 220
  salt and productivity of, 59
  salt tolerance of, 55
  seed longevity in, 127
Cane trash, 75
Canna, edible, 118
Cannabis sativa, see Hemp; Marihuana
Canna edulis, see Canna, edible
Cantaloupe, 110, 118
  chilling injury to, 278
  crop rotation and disease control in, 252
  fungicides and, 246
  germination requirements for, 161
  irrigation of, 182
  nutrients in, 280
  nutrient uptake in, 92
  optimum pH for, 44
  salt and productivity of, 59
  seed conditioning in, 344
  seeding guide for, 175
  seed longevity in, 128
  storage requirements of, 274
  transpiration ratio constants for, 186
Capacity, liquid, conversion factors, 455
Capacity, dry, conversion factors, 455
Capsicum
  annuum, see Pepper
  frutescens, see Pepper, tabasco
Caraway, 110, 118
  seed conditioning in, 344
  seed longevity in, 128
Carbohydrates, 475, see also specific types
  in cereal grains, 298
  in corn, 326
  in corn silage, 374
  in dairy products, 443
Carbon, 87
  mineralization and release of, 76
  in sewage sludge, 85
Carbon dioxide
  effect on photosynthesis, 198

rate of uptake, 199—206
Carbon-nitrogen ratios, 75, 76
Cardoon, 160
Carica papaya, see Papaya
Carissa grandiflora, see Plum, natal
Carotene
  in alfalfa, 367
  in corn silage, 374
  in peanut, 289
Carotenoids, 330
Carpetgrass, 110, 118
  germination requirements for, 150
  seed conditioning in, 344
  seeding rates in, 172
Carrot, 110, 118
  crop rotation and disease control in, 252
  freezing sensitivity of, 277
  frost resistance in, 176
  fungicides and, 246
  germination requirements for, 160
  insects in, 234
  irrigation of, 182
  maximum growth rate for, 212
  nutrients in, 281
  optimum pH for, 44
  radiation use efficiency in, 212
  roots of, 220
  salt and productivity of, 56
  salt tolerance of, 55
  seed conditioning in, 344
  seeding guide for, 175, 217
  seed longevity in, 128
  spacing of, 217
  storage requirements of, 274
  weed losses in, 265
  yields of, 216
Carrot, bunched, 274
Carthamus tinctorius, see Safflower
Carum carvi, see Caraway
Carya illinoensis, see Pecan
Carya spp., see Hickory
Casaba melon, 278
Cashew, 110
  amino acids in, 282
  distribution factors for, 118
  protein in, 282
Cassava, 110, 118
  annual productivity of, 209
  growth rate for, 209, 214
  radiation use efficiency in, 214
  seed rates for, 217
  toxic compounds from, 283
Castanea
  dentata, see Chestnut, American
  mollissima, see Chestnut, Chinese
  sativa, see Chestnut, European
Castor plant, 176
Cations
  in acid soils, 43
  exchangeable, 41—42, 53
Cats, 292

Cattle, see also Bulls; Livestock; Steers; specific
    types
  arthropods in, 269
  beef, see Cattle, beef
  composition of manure from, 82
  dairy, see Cattle, dairy
  digestibility of dry matter in, 405
  digestion coefficients of cereal grains for, 305
  diseases in, 269
  dressing percentages of, 436
  estrus in, 389
  excreta production in, 81
  gestation period in, 389
  heat-production values for, 405
  infectious diseases in, 266
  metabolic energy use by, 400
  metabolizable energy requirements of, 399
  nutritional disorders in, 269
  parasites in, 269
  relative feeding values of cereal grains for, 305
  sorghum digestibility in, 313
  temperature requirements of, 403, 404, 407
  ticks on, 268
  triticale digestibility in, 313
  weight gain in, 407
Cattle, beef, see also Beef; Heifers; Steers
  corn nutritive value in, 320
  loss of calves in, 392
  metabolizable energy requirements of, 398
  nutrients in manure from, 82
  protein requirements of, 401
  sorghum nutritive value in, 320
  soybean products in feeds for, 286
  temperature requirements of, 403
  vitamin A in feeds for, 368
  yields of, 453
Cattle, dairy
  bedding requirements for, 417
  breeds of in U.S., 411
  diseases in, 269
  environmental conditions required by, 402
  food intake in, 414
  loss of calves in, 392
  milk production in, 356, 414, 415, 416
  mineral recommendations for, 375
  nutrients in feed for, 412—413
  nutrients in manure from, 82
  nutritional disorders in, 269
  parasites in, 269
  protein requirements of, 401
  soybean products in feeds for, 286
  space requirements for, 402
  TDN in, 414, 415
  temperature requirements of, 403
  total energy requirements of, 400
  vitamin A in feeds for, 368
  waste volume in, 418
  yields of, 453
Cauliflower, 110, 118
  crop rotation and disease control in, 252
  freezing sensitivity of, 277

germination requirements for, 160
nutrients in, 281
optimum pH for, 44
protein in, 369
salt and productivity of, 58
salt tolerance of, 55
seeding guide for, 175
storage requirements of, 274
Cedar, Port-Orford, 128
Cedar, western red, 128
Celeriac, see Celery
Celery, 110, 118
    crop rotation and disease control in, 252
    diseases in, 226, 250
    freezing sensitivity of, 277
    germination requirements for, 160
    insects in, 234
    nutrients in, 281
    nutrient uptake in, 91
    optimum pH for, 44
    salt and productivity of, 56
    salt tolerance of, 55
    seed conditioning in, 344
    seeding guide for, 175
    seed longevity in, 128
    storage requirements of, 274
Cellulose, 397
Celosia cristata, see Amaranth, feathered
Cenchrus ciliaris, see Buffelgrass
Centers of diversity, 117—123
Centipede grass, 112, 119
Cereal grains, see also specific types
    amino acids in, 297, 299
    average composition of, 297
    carbohydrates in, 298
    carbon dioxide uptake in, 199
    characteristics of, 311
    crop rotation and disease control in, 252
    digestion coefficients of, 305
    fatty acids in, 303—304
    forage yields of, 311
    grain size of, 295
    insects in, 351—352
    minerals in, 301
    mites in, 351—352
    nitrogen fixation and, 79
    physical properties of, 295
    protein in, 300
    proximate analysis of, 296
    relative feeding values of, 305
    seed rates for, 217
    vitamins in, 302, 317
Chaetochloa italica, see Millet
Chamaecyparis lawsoniana, see Cedar, Port-Orford
Chard, 274
Chard, Swiss, 110, 118
    germination requirements for, 160
    seeding guide for, 175
    seed longevity in, 125
Chemical elements, see Elements
Chemical oxygen demand (COD), 86

Chemicals, dilution tables for, 457
Chenopodium album, see Lambsquarters
Cherry, see also specific types
    conversion factors for pits of, 73
    diseases in, 250
    fungicides and, 246
    toxic compounds from, 283
    waterlogging tolerance in, 197
Cherry, sour, 110, 118
    optimum pH for, 44
    storage requirements of, 273
Cherry, sweet, 110, 118
    nutrients in, 280
    optimum pH for, 44
    storage requirements of, 273
Chess, soft
    germination requirements for, 150
    seed longevity in, 128
Chestnut, American, 110, 118
Chestnut, Chinese, 110, 118
Chestnut, European, 110, 118
Chi square, table of, 463
Chickens, see also Poultry
    amino acids in, 442
    composition of, 451
    egg production in, 426
    feed consumption in, 426
    food capacity of, 401
    heat production in, 423
    land requirements for, 454
    protein requirements of, 401
    relative feeding values of cereal grains for, 305
    residues of, 81
    temperature effects on, 424, 426
    ventilation rates for, 431
    yields of, 453
Chicory, 110, 171
    distribution factors for, 118
    germination requirements for, 160
Chilling injury, 278
Chives, 110, 118
    germination requirements for, 160
    seed longevity in, 129
    toxic compounds from, 283
Chlorella, maximum yields of, 198
Chloride
    in irrigation water, 191
    in secondary effluents, 86
    in soils, 191
Chlorine, 87, see also Chloride
    in corn, 93
    in milk, 444
    in soybean products, 287
Chloris gayana, see Rhodesgrass
Choline
    in alfalfa, 367
    in animal feeds, 308, 329
    in corn, 328
    in poultry feeds, 426
    in soybean meal, 290
    in sunflower meal, 290

Chromium
  in corn, 328
  in milk, 445
  in secondary effluents, 86
  in sewage sludge, 85
  toxicity, 283
Chromosome numbers, of selected economic plants,
    109—116
Chrysanthemum, 129
Chrysanthemum
  cinerariifolium, see Pyrethrum
  coronarium, see Chrysanthemum
Cicer arientinum, see Pea, chick
Cichorium
  endivia, see Endive
  intybus, see Chicory
Cinchona spp., see Quinine
Cinnamomum
  camphora, see Camphor tree
  verum, see Cinnamon
Cinnamon, 110, 118
Citron, 110
  distribution factors for, 118
  germination requirements for, 160
Citrullus
  lanatus, see Watermelon
  vulgaris, see Citron
Citrus, see also specific types
  carbon dioxide uptake in, 205
  chloride in soil of, 191
  diseases in, 226
  fungicides and, 246
  nutrients in pulp of, 393
  seed rates for, 218
  spacing of, 218
  storage requirements of, 273
  waterlogging tolerance in, 197
  weed losses in, 260
Citrus
  aurantifolia, see Lime; Lime, Rangepur
  aurantium, see Orange, sour
  limon, see Lemon
  medica, see Citron
  nobilis, see Mandarin, cleopatra
  paradisi, see Grapefruit
  reticulata, see Tangerine
  sinensis, see Orange
Clay, 37, 39—40, 193, 194
  beans and, 51
  corn and, 49, 50
  crop rotation tillage and, 224
  defined, 475
  design velocities in drain tile and, 197
  limestone and pH of, 47
  mean velocity values and, 192
  nitrogen in, 68
  nonerosive velocities and, 193
  phosphorus in, 68
  salt percentage in, 54
  soil-water characteristics in, 196
Climate groups, 1

Climatic zones, 1
Clove, 110, 118
Clover, see also specific types
  AOSCA certification of seeds of, 164
  carbon-nitrogen ratio in residues of, 75
  dry matter losses in, 361
  heat killing temperature for, 177
  lime response in, 45
  nitrogen fixation and, 79
  nutrient losses in, 361
Clover, alsike, 110, 118
  germination requirements for, 150
  harvesting of, 375
  salt and productivity of, 58
  salt tolerance of, 55
  seed conditioning in, 344
  seeding rates in, 173
Clover, alyce, 110, 118
  germination requirements for, 147
  seeding rates in, 173
Clover, arrowleaf, 110, 118
Clover, ball, 110, 118
Clover, barrel, 147
Clover, berseem, 110, 118
  germination requirements for, 150
  salt and productivity of, 61
Clover, bur
  germination requirements for, 149
  seed conditioning in, 344
  seeding rates in, 173
Clover, button, 110, 118
  germination requirements for, 150
  seeding rates in, 173
Clover, cluster, 150
Clover, crimson, 110, 118
  germination requirements for, 150
  seed conditioning in, 344
  seeding rates in, 173
  seed longevity in, 129
  transpiration ratio constants for, 186
Clover, dalea, 344
Clover, hop, 150, 344
Clover, Kenya, 150
Clover, kura, 111, 118
Clover, ladino, 344
  germination requirements for, 150
  nutrients in, 393
  salt and productivity of, 58
  salt tolerance of, 55
  seeding rates in, 173
  seed longevity in, 129
Clover, lappa, 150
Clover, large hop, 150
Clover, Persian, 150
Clover, red, 111, 118
  annual productivity of, 209
  carbon dioxide uptake in, 202
  dry matter losses in, 361
  germination requirements for, 150
  growth rate, 209, 214
  harvesting of, 375

nutrient losses in, 361
nutrients in above-ground portion of, 74
nutrients in roots of, 75
optimum pH for, 44
phosphorus in, 95
radiation use efficiency in, 214
roots of, 221
salt and productivity of, 58
salt tolerance of, 55
seed conditioning in, 344
seeding rates in, 173
seed longevity in, 129
transpiration ratio constants for, 186
water requirements of, 185
yields of, 95
Clover, rose, 150
Clover, small hop
    germination requirements for, 151
    seed longevity in, 129
Clover, sour, 156
Clover, strawberry
    germination requirements for, 151
    salt and productivity of, 58
    salt tolerance of, 55
Clover, subterranean, 111, 118
    carbon dioxide uptake in, 202
    germination requirements for, 151
    maximum growth rate for, 213
    radiation use efficiency in, 213
Clover, sweet, see also Clover, sweet, white; Clover,
        sweet, yellow
    carbon-nitrogen ratio in, 75
    crop rotation and, 229
    optimum pH for, 44
    salt and productivity of, 59
    salt tolerance of, 55
    transpiration ratio constants for, 186
    water requirements of, 185
Clover, sweet, white, 115, 122, see also Clover, white
    germination requirements for, 157
    roots of, 220
    seed conditioning in, 344
    seeding rates in, 174
    seed longevity in, 144
Clover, sweet, yellow, 115, 122
    germination requirements for, 157
    seeding rates in, 174
    seed longevity in, 144
Clover, white, 111, 118, see also Clover, sweet, white
    carbon dioxide uptake in, 202
    germination requirements for, 151
    maximum growth rate for, 214
    nutrients in roots of, 75
    optimum pH for, 44
    radiation use efficiency in, 214
    roots of, 221, 222
    salt tolerance of, 55
    seed conditioning in, 344
    seeding rates in, 173
    seed longevity in, 129
Clover, woods, 344

Clovergrass, 89, 91
Coastal grass, 73
Cobalt, 87
    in animal feeds, 397
    in corn, 328
    in dairy cattle feed, 412, 413
    in hominy feed, 330
    in limestone, 47
    in milk, 445
    in secondary effluents, 86
    in sewage sludge, 85
    toxicity, 283
Cocksfoot
    annual productivity of, 209
    carbon dioxide uptake in, 201
    maximum growth rate for, 213
    mean growth rate for, 209
    roots of, 220
Cocoa
    crop factors for, 190
    nutrient uptake in, 92
    seed longevity in, 129
    seed rates for, 217
Coconut, 111, 118
    nutrient uptake in, 92
    seed rates for, 218
    storage requirements of, 273
Coconut meal, 393
Cocos nucifera, see Coconut
COD, see Chemical oxygen demand
Coffea
    arabica, see Coffee
    canephora, see Coffee, robusta
Coffee, 111, 118, see also specific types
    crop factors for, 190
    lime response in, 45
    seed rates for, 217
Coffee, robusta, 111, 118
Cold temperatures, 278, 408, see also Freezing
Cole
    fungicides and, 246
    insects in, 234
Collard, 111, 118
    germination requirements for, 160
    insects in, 235
    seeding guide for, 175
    storage requirements of, 274
Collembola, 77
Colocasia esculenta, see Taro
Columbus grass, 203, see also Sorghum, almum
Combustion, 78
Comfrey, 111, 118
Concrete, 193
Conductivity, 54
Conringia orientalis, 222
Controlled burning, 226
Conversion factors, 455—457, 458—460, see also
        specific types
    for fertilizer materials, 106—107
    for fruit pits, 73
    for nut shells, 73

Convolvulus arvensis, see Bindweed
Cool soil planting, 226
Copper, 87
  in above-ground portion of crops, 73—74
  amounts needed for plant production, 88
  in animal feeds, 307, 308, 329, 397
  in chickens, 451
  in corn, 93, 328
  in corn silage, 374, 375
  in dairy cattle feed, 412, 413
  in hominy feed, 330
  in limestone, 47
  in manure, 82
  in milk, 445
  in pork, 441
  in secondary effluents, 86
  in sludge, 85, 87
  in soils, 45
  in soybean products, 287
  toxicity, 283
  in turkey, 450
  uptake of, 97
Corchorus olitorius, see Jute, tussa
Coriander
  frost resistance in, 176
  seed conditioning in, 344
Corn, see also specific types
  aeration of, 336
  amino acids in, 327
  animal feeds from, 329
  annual productivity of, 208, 210
  AOSCA certification of seeds of, 164
  average composition of, 297
  as barrier crop, 226
  bran, 301
  carbohydrates in, 326
  carbon dioxide uptake in, 200
  carbon-nitrogen ratio in stalks of, 75
  chromosome number for, 111
  composition of, 317, 376
  conversion factors for residues of, 72
  crop factors for, 190
  crop residue management and disease control in,
    256
  crop rotation and, 224, 228, 229, 252
  dent, 325—328
  digestion coefficients of, 305
  disease
    control, 226, 252, 257
    losses from, 248, 250
  distribution factors for, 118
  drying of, 338, 339
  drying rate of, 334
  earthworms and, 225
  fats in, 303, 327
  frost resistance in, 176
  fungicides and, 246
  grading of, 323
  growth rate
    maximum, 212
    mean, 208, 210

  harvesting of, 375
  heat killing temperature for, 177
  heterocyclic compounds in, 326
  humidity and, 177, 336
  hybrids, 164
  inorganic components of, 328
  irrigation of, 180, 181
  kernel dimensions, 295, 332—334
  kernel types classified, 324
  land requirements for, 454
  leaf composition of, 49
  lime response in, 45
  lipids in, 327
  maximum yields of, 198
  minerals in, 301, 317
  moisture content of, 326, 331, 332, 381
  nematodes in, 259
  nitrogen deficiency in, 94
  nitrogen fixation and, 79
  nitrogenous components of, 326
  nutient amounts needed for production of, 88
  nutrient content of, 49, 393
    above-ground portion of plant, 73, 93
    cobs, 393
    ears, 393
    roots, 75
  nutrient uptake in, 89, 90, 96
  nutritive value of, 320
  open-pollinated, 164
  optimum pH for, 44
  phosphorus in, 95
  photosynthesis in, 198
  physical properties of, 295, 324
  products, see products by names
  protein in, 300
  proximate analysis of, 296, 371
  quality of, 376
  radiation use, efficiency in, 212
  relative feeding values of, 305
  residue production of, 70
  roots of, 221, 222
  salt and productivity of, 58
  salt tolerance of, 55
  seed conditioning in, 344, 346
  seeding rates in, 171, 217
  soil acidity and, 49—50
  spacing of, 217
  target yield for, 111
  tocopherols in, 327
  toxic compounds from, 283
  transpiration ratio constants for, 185
  vitamins in, 302, 317, 328
  waterlogging tolerance in, 197
  water requirements of, 185
  water use of, 180
  weed losses in, 260, 264
  weight and composition of component parts
    of, 325
  wet milling of, 329
  yields of, 49, 95, 96, 180, 216
Corn dent, 325—328, see also Corn; Corn, field

Corn, field
  carbohydrates in, 298, 299
  germination requirements for, 151
  insects in, 235
  seed longevity in, 130
Corn, opaque—2, 300
Corn, pop
  carbohydrates in, 298
  germination requirements for, 151
  seed conditioning in, 345
  seeding rates in, 171
Corn, sweet
  AOSCA certification of seeds of, 164
  carbohydrates in, 298
  fungicides and, 246
  germination requirements for, 160
  insects in, 235
  nutrients in, 281
  nutrient uptake in, 92
  salt and productivity of, 58
  salt tolerance of, 55
  seeding guide for, 175
  seeding rates in, 171
  silage, see Corn silage
  storage requirements of, 274
  volunteer, as pest, 263
  weed losses in, 264
Corncockle, 212
Corn flour, 301
  carbohydrates in, 298
  fatty acids in, 303
Corn grits, 303
Cornmeal
  carotenoids in, 330
  fatty acids in, 303
Corn oil
  fatty acids in, 303
  tocopherols in, 293
Cornsalad (fetticus), 160
Corn silage, 311
  composition of, 374
  filling rate of, 377
  fineness of chop of, 377
  minerals in, 374, 375
  moisture content of, 382
  nutrients in, 393, 394
  nutrient uptake in, 91
  nutritional value of, 374
  protein in, 377
  proximate analysis of, 374, 378
  seeding rates in, 171
  storage of, 382
  TDN for, 376, 378
Corn stover
  nutrients in, 393
  proximate analysis of, 371
Coronilla varia, see Crownvetch
Correlation coefficient, 461, 464
Corylus
  avellana, see Filbert, European
  chinensis, see Hazel, Chinese

cornuta, see Hazel, beaked
Cotton, see also specific types
  AOSCA certification of seeds of, 164
  carbon dioxide uptake in, 205
  conversion factors for residues of, 72
  crop residue management in, 256
  crop rotation and, 229, 252
  disease control in, 252, 256
  diseases in, 226, 248, 250
  frost resistance in, 176
  fungicides and, 246
  germination requirements for, 151
  humidity effects on, 177, 178
  insects in, 235
  irrigation of, 182
  lime response in, 45
  maximum growth rate for, 213
  nematodes in, 259
  nutrients in above-ground portion of, 74
  nutrient uptake in, 89, 90
  photosynthesis in, 198
  proximate analyses, 371
  radiation use, efficiency in, 213
  residues, 70, 256
  salt and productivity of, 64
  salt tolerance of, 55
  seed, see Cottonseed
  seed rates for, 218
  spacing of, 218
  transpiration ratio constants for, 185
  waterlogging tolerance in, 197
  weed losses in, 260, 264
  yields of, 216
Cotton, Pima, 171
Cotton, Sea Island, 111, 118
  seeding rates in, 171
  seed longevity in, 130
Cotton, tree, 111, 118
Cotton, upland, 111, 118
  root length of, 222
  seeding rates in, 171
  seed longevity in, 130
Cotton, wild, 111, 118
Cottonseed
  cellulose in hulls of, 397
  composition of, 291
  conversion factors for residues of, 72
  nutrients in, 394
  protein in, 285
  proximate analysis of hulls of, 371
  seed conditioning in, 346
Cottonseed meal
  nutrients in, 394
  toxicity of, 292
Cottonseed oil, 293
Cover and management factors, 229—230
Cows, see Cattle
Crabapple, 44
Crambe, 111, 118
  AOSCA certification of seeds of, 164
  germination requirements for, 151

seeding rates in, 171
seed longevity in, 130
Crambe abyssinica, see Crambe
Cranberry, 111, 119
    chilling injury to, 278
    freezing sensitivity of, 277
    nutrients in, 280
    seed conditioning in, 344
    storage requirements of, 273
Cranberry, large, 44
Crenshaw melon
    chilling injury to, 278
    storage requirements of, 275
Cress, garden, 111, 119
    germination requirements for, 160
    seed longevity in, 130
Cress, upland, 160
Cress, water, 116, 123
    germination requirements for, 161
    storage requirements of, 275
Cropping treatments, 227, see also specific types
Crop residues, 70, see also specific types
    conversion factors for, 72
    decomposition of, 69
    management of in disease control, 256
    N-P-K in, 71
Crop rotation, 226, 228, 229—230
    disease control and, 251—255
    flooding and, 226
    tillage and, 224
Cross-pollinated grasses, 164
Crotalaria, see also specific types
    nematodes in, 259
    seed conditioning in, 345
    seed longevity in, 131
Crotalaria, lance, 151
Crotalaria, showy, 111, 119
    germination requirements for, 151
    seeding rates in, 173
Crotalaria, slenderleaf, 151
Crotalaria, striped, 151
Crotalaria, sunn, 151
Crotalaria
    intermedia, see Crotalaria, slenderleaf
    juncea, see Crotalaria; Crotalaria, sunn
    lanceolata, see Crotalaria, lance
    mucronata, see Crotalaria, striped
    spectabilis, see Crotalaria, showy
Crownvetch, 111, 119
    AOSCA certification of seeds of, 164
    germination requirements for, 151
    seeding rates in, 173
Crucifers, 252, see also specific crops
Cucumber, 111, 119
    chilling injury to, 278
    crop rotation and disease control in, 252
    frost resistance in, 176, 277
    fungicides and, 246
    germination requirements for, 161
    nutrients in, 281
    nutrient uptake in, 91

optimum pH for, 44
roots of, 219
salt and productivity of, 58
salt tolerance of, 55
seed conditioning in, 345
seeding guide for, 175
seed longevity in, 131
storage requirements of, 274
transpiration ratio constants for, 186
Cucumis
    anguria, see Gherkin, West Indian
    melo, see Cantaloupe
    sativus, see Cucumber
Cucurbita
    maxima, see Marrow, Boston; Squash
    mixta, see Squash, mixta
    moschata, see Pumpkin; Squash
    pepo, see Marrow, vegetable; Pumpkin
Cultural practices, see also specific types
    corn disease control and, 257
    disease control and, 226
Cumin, 344
Curcuma domestica, see Turmeric
Currant, 273
    black, 111, 119
    red, 111, 11
Cutting schedule, 355
Cyamopsis tetragonoloba, see Guar
Cydonia oblonga, see Quince
Cynara
    cardunculus, see Cardoon
    scolymus, see Artichoke; Artichoke, globe
Cynodon
    dactylon, see Bermudagrass; Bermudagrass, giant
    plectostachyus, see Stargrass
    transvallensis, see Bermudagrass, African
Cynosurus cristatus, see Dogtail, crested

D

Dactylis glomerata, see Cocksfoot; Orchardgrass
Dairy cows, see Cattle, dairy
Dairy products, 443, 444, see also specific types
Daisy, African, 124
Dallisgrass, 111, 119, 346
    germination requirements for, 151
    roots of, 220
    salt and productivity of, 61
    salt tolerance of, 55
    seeding rates in, 172
Dandelion, 161
Date, desert, 111, 119
    crop factors for, 190
    freezing sensitivity of, 277
    nutrients in, 280
    salt and productivity of, 64
    storage requirements of, 273
Datura (jimson weed), 131
Datura
    ferox, see Datura
    innoxia, see Datura

metal, see Datura
stramonium, see Datura
Daucus carota, see Carrot
Daylengths, 28
Daytime hours, 28
Degree days, 21—27
Delphinium (*Delphinium* spp.), 131
Design velocities in drain tile, 197
Desmodium tortuosum, see Beggarweed, Florida
Dicalcium phosphate, 394
Dichanthium annulatum, see Bluestem, diaz
Dichondra (*Dichondra repens),* 151
Digestibility
    of alfalfa, 361
    of cellulose, 397
    defined, 477
    of dry matter, 405
    of grasses, 359
Digestible nutrients, 356, 357
    total, see Total digestible nutrients (TDN)
Digestion coefficients
    of alfalfa, 368
    of cereal grains, 305
    for sheep, 356
Digitaria decumbens, see Pangolagrass
Dill, 111
    distribution factors for, 119
    seed longevity in, 131
Dimorphotheca spp., see Daisy, African
Dioscorea
    alata, see Yam, winged
    composita, see Yam, composite
    rotundata, see Yam, eboe
Diospyros
    kaki, see Persimmon, Japanese
    virginiana, see Persimmon, common
Diptera, 77, see also insects
Disease control in plants
    crop rotation, 251—256
    cultural practices, 257
    residue management, 256
Diseases
    in animals, 266—267, 269, 397
    in plants, see Insects; Plant pathogens; see also
        affected crop by name
Distichlis stricta, see Saltgrass
Distribution of economic plants, 117—123
Disturbed soils, 224
Dock, curly, 131
Dodonea viscosa (Dodonea), 58
Dogs
    toxicity of cottonseed meals in, 292
    vitamin A in feeds for, 368
Dogtail, crested
    germination requirements for, 151
    seed longevity in, 131
Dolichos lablab, 109, 117, 369
Donkeys, 81
Douglas fir, 132, 222
Dracaena (*Dracaena endivisa),* 61
Drain tile, design velocities, 197

Dropseed, sand, 151
Drying, 336
    of corn, 338, 339
    field, 363
    of hay, 363
    rates of, 334
Dry matter, 379, see also specific types
    in alfalfa, 362
    in animal feeds, 393—395
    digestibility of, 405
    in hay, 360, 361
    losses of, 360—362
    in roughages, 371
    in silages, 378
    storage of, 377
Dry milling process, 321
Ducks
    classes of, 446
    composition of manure from, 82
    ventilation rates for, 431
Dwarf milo, 185

E

Earthworms, 77, 225
Echinochloa crusgalli, see Millet, Japanese
Effluents, properties of, 86
Eggplant, 111, 119
    chilling injury to, 278
    freezing sensitivity of, 277
    germination requirements for, 161
    nutrients in, 281
    seeding guide for, 175
    seed longevity in, 131
    storage requirements of, 274
    toxic compounds from, 283
Eggs
    chicken, 426
    land requirements for, 454
    protein in, 453
    turkey, 425
    yields of, 453
Ehrharta calycina, see Veldtgrass
Einkorn wheat, see wheat, einkorn
Elaeagnus pungens, see Silverberry
Elaeis guineensis, see Palm, oil
Elder, American, 111, 119
Electrical resistance, 54
Elements, see also Metals; Minerals; Nutrients;
        specific types
    in animal feeds, 397
    deficiencies in, 397
    defined, 485
    essential, 87
    in milk, 445
    in secondary effluents, 86
    as toxic, 283
Elephant grass, 112, 119
Eleusine coracana, see Millet, finger; Millet, Ragi
Elm, 226
Elm, American, 131

Elymus
  angustus, see Wildrye, altai
  canadensis, see Wildrye, Canada
  condensatus, see Wildrye, giant
  glaucus, see Wildrye, blue
  psathyrostachys juncea, see Wildrye, Russian
  triticoides, see Wildrye, beardless
Emmer, see Wheat, emmer
Enchytraeidae, 77
Endive, 111, 119
  germination requirements for, 161
  seeding guide for, 175
  storage requirements of, 274
Entisols, 31, 32
  land surface percentage for, 33
  potassium in, 96
Eragrostis
  curvula, see Lovegrass, weeping
  lehmanniana, see Lovegrass, Lehmann
  spp., see Lovegrass
  tef, see Teff
  trichodes, see Lovegrass, sand
Eremochloa ophiuroides, see Centipede grass
Erodium cicutarium, see Alfilaria
Esparto, 111, 119
Estrus, 389, 477
Eucalyptus *(Eucalptus* spp.), 111, 119
  distribution factors, ll9
  maximum yields of, l98
Euchalena mexicana, see Teosinte
Euonymus japonica, (Euonymus), 61
Evaporation, 187—188, 189
Evaporative demand stress, 184
Evapotranspiration, 190
Evening primrose, 131
Ewes, see Sheep
Exchangeable cations, 41—42, 53
Excreta production, 81
Extraterrestrial radiation, 183

F

Fagopyrum esculentum, see Buckwheat
F and t, values of, 465—468
Farina, 303
Farm animals, see also specific types
  body size and food utilization in, 401
  diseases in, 397
  estrous cycle in, 389
  food capacity of, 401
  gestation period in, 389
  humidity requirements of, 402
  metabolizable energy requirements of, 398, 399
  protein production efficiency of, 453
  protein requirements of, 401
  space requirements of, 402
  total energy requirements of, 400
  yields of, 453
Feed, see Animal feeds
Feed efficiency, 407, see also Animal feeds

Feedlot performance, 311
Feijoa sellowiana, see Guava, pineapple
Fennel, 346
Fertilizers, see also specific types
  abbreviations for, 98—100
  acidifying effects of, 48
  acidity of, 102
  analyses of, 98—100
  basicity of, 102
  bulk densities of, 103
  chemical composition of, 104—105
  conversion factors for materials in, 106—107
  crop residue composition and, 71
  grades of, 98—100
  macronutrients in, 98—100
  manufacture of, 98—100
  manure as source of elements in, 84
  nitrogen fixation and, 78
  nitrogen sources for, 48
  phosphorous sources for, 48
  physical properties of, 104—105
  salt indices of, 101
  water solubility of, 98—100
Fescue, see also specific types
  lime response in, 45
  moisture content of, 332
  roots of, 220
  salt and productivity of, 61
Fescue, chewing
  germination requirements for, 151
  seed conditioning in, 345
Fescue, creeping red, 132
Fescue, hair, 152
Fescue, hard, see also Fescue, sheep
  germination requirements for, 151
  roots of, 220
  seed longevity in, 132
Fescue, Kentucky, 345
Fescue, meadow, 111, 119
  germination requirements for, 152
  roots of, 220
  salt tolerance of, 55
  seed conditioning in, 345
  seeding rates in, 173
  seed longevity in, 132
Fescue, red, 111, 119
  germination requirements for, 152
  seeding rates in, 173
Fescue, sheep, 111, 119, see also Fescue, hard
  germination requirements for, 152
  roots of, 220
  seeding rates in, 173
Fescue, tall, 111, 119
  annual productivity of, 209
  carbon dioxide uptake in, 201
  germination requirements for, 152
  growth rate for, 209, 213
  nutrient uptake in, 91
  radiation use efficiency in, 213
  salt tolerance of, 55
  seeding rates in, 173

seed longevity in, 132
Festuca
  arundinacea, see Fescue, tall
  capillata, see Fescue, hair
  elatior, see Fescue; Fescue, meadow
  ovina, see Fescue, hard; Fescue, sheep
  pratensis, see Fescue, meadow
  rubra, see Fescue, chewing; Fescue, creeping red;
    Fescue, red
Fetticus (cornsalad), 160
Fiber crops, see also specific crops
  carbon dioxide uptake in, 205
  nutrient uptake in, 90
Ficus carica, see Fig, common
Field crops, see also specific crops
  diseases in, 250
  optimum pH for, 44
  residues of, 72
  salt tolerance of, 55
Field drying, 363
Fig, common, 111, 119
  nutrients in, 280
  salt and productivity of, 61
  storage requirements of, 273
Filbert
  amino acids in, 282
  conversion factors for shells of, 73
  protein in, 282
Filbert, European, 111, 119
Filling rate, 377
Findi, 295
Fir, balsam, 132
Fir, Douglas, 132, 222
Fir, noble, 132
Fir, white, 133
Flax, 111, 119, see also Flaxseed
  AOSCA certification of seeds of, 164
  conversion factors for residues of, 72
  crop factors for, 190
  diseases in, 226, 250
  frost resistance in, 176
  germination requirements for, 152
  moisture content of, 332
  nutrient uptake in, 90
  optimum pH for, 44
  physical dimensions of kernels of, 332, 333
  proximate analyses of hulls of, 371
  salt and productivity of, 58
  salt tolerance of, 55
  seeding rates in, 171
  seed longevity in, 133
  straw, 371
  transpiration ratio constants for, 186
  varieties, 346
  water requirements of, 185
  weed losses in, 264
Flax, bison, 345
Flax, golden, 345
Flax, small, 345
Flaxseed, 331
Flooding, 226

Flowers, see also specific plants
  evaporative demand stress and, 184
  optimum pH for, 44
  storage requirements of, 276
Fluid conductivity, 224
Fluorine
  in dairy cattle feed, 412, 413
  in limestone, 47
  in milk, 445
Fodder, 217, see also specific types
Folacin
  in beef, 440
  in dairy products, 443
  in pork, 441
  in triticales, 310
  in wheat, 310
Folic acid, see Vitamin B
Food capacity of farm animals, 401, see also Animal
    feeds
Forages, see also specific types
  carbon dioxide uptake in, 201
  cutting schedule for, 355
  diseases in, 250
  nutrient uptake in, 91
  seed rates for, 217
  spacing of, 217
  weed losses in, 260
  yields of, 311
Forest plants, 68, see also specific plants
  diseases in, 250
  nitrogen fixation and, 78
  optimum pH for, 44
Fortunella spp., see Kumquat
Foxtail, 171, see also specific types
Foxtail, black grass, 133
Foxtail, creeping, 111
  distribution factors for, 119
  seed longevity in, 133
Foxtail, giant, 263
Foxtail, meadow, 111, 119
  germination requirements for, 153
  salt and productivity of, 59
  salt tolerance of, 55
  seed longevity in, 133
Fragaria
  ananassa, see Strawberry, garden
  chiloensis, see Strawberry, Chilean
  spp., see Strawberry
  virginiana, see Strawberry, Virginia
Frankincense, 111
Fraxinus excelsior, see Ash, European
Freeze-free periods, 16—19
Freezing
  of fruits, 273, 277
  of vegetables, 277
Frost resistance, 176
Fruit, see also specific types
  carbon dioxide uptake in, 205
  chilling injury to, 278
  chloride in soil of, 191
  citrus, see Citrus

conversion factors for pits of, 73
diseases in, 250
evaporative demand stress and, 184
freezing points of, 273
freezing sensitivity of, 277
fungicides and, 246
nutrients in, 74, 280
nutrient uptake in, 92
optimum pH for, 44
properties of, 273
seed rates for, 218
spacing of, 218
stone, 191
storage requirements of, 273
toxic compounds from, 283
water content of, 273
weed losses in, 260
Fungi, 77, see also Plant pathogens
Fungicides, 246—247
Furcraea foetida, see Hemp, Mauritius

G

Garbonzo, see Pea, chick
Garlic, 111, 119
  roots of, 219
  toxic compounds from, 283
Gaultheria procumbens, see Wintergreen
Geese
  classes of, 447
  composition of manure from, 82
  ventilation rates for, 431
Geranium, cutleaf *(Geranium dissectum),* 133
Geranium, dovefoot *(Geranium molle),* 133
Germination requirements, 147—163
  agricultural seeds, 147—159
  vegetable seeds, 159—163
Germination times, 171—174
Gestation period in farm animals, 389
Gherkin, West Indian, 133
Ginger, 111, 119
Gladiola *(Gladiolus* spp.) 133
Glossary, 473—486
Glycine
  javanica, see Siratro
  max, see Soybean
  wightii, see Soybean, perennial
Glycyrrhiza glabra, see Licorice, common
Gooseberry, 111, 119, 273
Gooseberry, European, 119
Gossypium
  anomalum, see Cotton, wild
  arboreum, see Cotton, tree
  barbadense, see Cotton, Sea Island
  hirsutum, see Cotton, upland
Grab test, 381
Graded yields, 216
Grains, 49, see also specific crops
  annual productivity of, 210
  cereal, see Cereal grains
  crop factors for, 190

density of kernels of, 333
foreign material in, 342
fungicides and, 246
humidity and, 335
insects in, 351—352
mean growth rate for, 210
mites in, 351—352
moisture content of, 331, 381
nematodes in, 259
nutrient uptake in, 90
photosynthesis in, 198
physical dimensions of kernels of, 332, 333
separation of grasses and seeds from, 349, 350
volume of kernels of, 333
weight of kernels of, 333
Grama, 186, see also specific types
Grama, blue, 111, 119
  germination requirements for, 152
  seeding rates in, 173
Grama, side-oats, 111, 119
  germination requirements for, 152
  seeding rates in, 173
Grape
  carbon dioxide uptake in, 206
  chloride in soil of, 191
  diseases in, 249, 250
  freezing sensitivity of, 277
  fungicides and, 246
  heat killing temperature for, 177
  insects in, 236
  nutrients in, 280
  nutrient uptake in, 92
  salt and productivity of, 58
  toxic compounds from, 283
Grape, American, 273
Grape, muscadine, 112, 119
Grape, vinifera, 273
Grape, wine, 112, 119
Grapefruit, 112, 119
  chilling injury to, 278
  crop factors for, 190
  freezing sensitivity of, 277
  fungicides and, 246
  nutrients in, 280
  salt and productivity of, 56
  seed longevity in, 133
  storage requirements of, 273
Grapefruit, white, 280
Grasses, see also by name
  age of, 224
  AOSCA certification of seeds of, 164—165
  composition of, 359
  crop rotation and, 229, 252
  detergent fiber in, 355
  digestibility of, 359
  earthworms and, 225
  moisture content of, 365, 381
  nitrogen fixation and, 79
  protein in, 354, 359, 365
  proximate analysis of, 365—366
  seeding rates in, 172—173

separation of grains and seeds from, 349, 350
waterlogging tolerance in, 197
weed losses in, 260
Grassland diseases, 250
Gravel
design velocities in drain tile and, 197
mean velocity values and, 192
Great soil groups, 29—30, 31, 32
Greens, see also specific types
nutrients in, 281
storage requirements of, 274, 275
Grinding, 370
Groundnut
nitrogen fixation and, 79
seed rates for, 217
Groundnut, Bambarra, 112, 119
Growing degree days, 21—27
Growth rates of crops
maximum, 212—215
mean, 207—211
Guar, 112, 119
germination requirements for, 152
seeding rates in, 171
seed longevity in, 133
transpiration ratio constants for, 186
Guava, pineapple, 56
Guayule, 112, 119
seeding rates in, 171
seed longevity in, 133
Guineagrass, 112, 119
annual productivity of, 208
germination requirements for, 152
mean growth rate for, 208
nutrient uptake in, 91
seeding rates in, 173
Guineas, 447
Gum arabic, 112, 119
Gutta percha, 112, 119
Gypsum, 66, 106

H

Halomorphic soils, 29, 30
Hardiness zones for plants, 20
Harding grass, 61, 112, 119
germination requirements for, 152
seeding rates in, 173
Hay, see also specific types
acid detergent fiber in, 355
carbon-nitrogen ratio in, 75
diseases in, 250
drying of, 363
dry matter losses in, 360, 361
fungicides and, 246
moisture content of, 363
neutral detergent fiber in, 355
nutrient losses in, 361
nutrients in above-ground portion of, 73
protein in, 354
silage from, 382
storage space required for, 364

Hazel, beaked, 112, 119
Hazel, Chinese, 112, 119
Heat killing temperatures, 177
Hedera canariensis, see Ivy, Algerian
Heifers, see also Cattle, beef
cold temperature effects on, 408
dietary requirements, 410
performance of, 408
residues of, 81
Helianthus
annuus, see Sunflower
tuberosus, see Artichoke, Jerusalem
Heliothis, 236
Hemarthria altissima, see Limpo grass
Hemlock, western, 134
Hemp
frost resistance in, 176
germination requirements for, 152
seed conditioning in, 345
seed longevity in, 134
Hemp, Mauritius, 112, 119
Hens, see also Chickens; Poultry
composition of manure from, 82
excreta production in, 81
metabolizable energy requirements of, 398, 399
ventilation rates for, 431
Heterocyclic compounds, see also specific types
in corn, 326
Hevea brasiliensis, see Rubber tree
Hibiscus, 56
Hibiscus
cannabinus, see Kenaf
rosa-sinensis, see Hibiscus
sabdariffa, see Roselle
Hickory, 112, 119
Hilaria belangeri, see Mesquite, curly
Histosols, 31— 33
Hogs, see Swine
Holcus lanatus, see Velvetgrass
Holly, burford, 56
Hominy feed, 330, 394
Honeydew melon
chilling injury to, 278
storage requirements of, 275
Hops, 112
crop rotation and disease control in, 252
distribution factors for, 120
Hordeum
distichum, 222
vulgare, see Barley
Horizontal silos, 383
Horseradish, 112
distribution factors for, 120
storage requirements of, 274
Horses
composition of manure from, 82
infectious diseases in, 266
nutrient requirements of, 428
nutrients in feed for, 429
residues of, 81
vitamin A in feeds for, 368

weight estimates of, 427
Host resistance, 257
Humidity, 177, 178, 188, 189, 335
    farm animals, requirements for, 402
    fruits and, 273
    transpiration ratio constants and, 185—186
    vegetables and, 274—275
Humulus lupulus, see Hops
Humus
    carbon-nitrogen ratio in, 75
    formation of, 67
    model of, 69
Hydrogen, 87
    in soils, 41—42
Hydromorphic soils, 29, 30
Hyparrhenia rufa, see Jaragua grass

I

Ilex
    cornuta, see Holly, burford
    paraguariensis, see Mate
Inceptisols, 31—33
Incipient scour, 192
Indented cylinder separators, 347, 349
Indiangrass, 112, 120
    nutrient uptake in, 91
    seeding rates in, 173
    seed longevity in, 134
Indiangrass, yellow, 152
Indigo, 112, 120
Indigo, hairy, 112, 120
    germination requirements for, 152
    seeding rates in, 174
Indigofera
    hirsuta, see Indigo, hairy
    tinctoria, see Indigo
Infectious diseases in livestock, 266, 267
Insects, see also Beetles; Mites
    in cereal grains, 351—352
    density of, 233—240
    Diptera, 77
    as disease vectors, 266, 267
    in grains, 351—352
    plant losses caused by, 233—245, see also specific
        crop
    planting time and, 241
    in soybean, 351—352
Insulation values, 432
Intrazonal soils, 29, 30
Iodine
    in animal feeds, 397
    in corn, 328
    in dairy cattle feed, 412, 413
    in milk, 445
Ipomoea batatas, see Sweet potato
Iris, 177
Iron, 48, 85—87, 97
    amounts needed for plant production, 88
    in animal feeds, 307, 308, 329, 330
    in beef, 440

    in cereal grains, 317
    in chickens, 451
    in corn, 93, 328
    in corn silage, 374, 375
    in dairy cattle feed, 412, 413
    in dairy products, 443
    in fruits, 280
    in limestone, 47
    in manure, 82
    in milk, 445
    in pork, 44
    in soils, 45
    in soybean products, 287
    in turkey, 450
    in vegetables, 281
Iron sulfate, 66
Irrigation, 178, 191—195
    boron and, 190
    chloride and, 191
    crop productivity and, 182
    delay of, 179
    energy requirements for pumping water in, 195
    lengths for borders in, 193
    maximum allowable sprinkling rates in, 194
    out of season, 226
    seepage in unlined canals, 192
    sprinkler system efficiencies in, 194
    velocity and, 192, 193
    water-use efficiency and, 181
Ivy, Algerian, 56

J

Jackfruit, 112, 120
Jaragua grass, 112, 119, 208
Jasmine, star, 57
Jimson weed, 131
Job's tears, 300
Johnsongrass, 112, 120
    germination requirements for, 152
    seed conditioning in, 345
    seeding rates in, 173
    seed weight relationships and, 263
Jojoba, 112, 120
Juglans nigra, see Walnut, black
Juniper, 58
Juniper, alligator, 134
Juniper, one-seed, 134
Juniperus
    chinensis, see Juniper
    deppeana, see Juniper, alligator
    monosperma, see Juniper, one-seed
    osteosperma, see Juniper, one-seed
Jute, China, 112, 120
Jute, tussa, 112, 120

K

Kale, 112, 120, 161
    annual productivity of, 209
    freezing sensitivity of, 277

germination requirements for, 161
growth rate for, 209, 214
radiation use efficiency in, 214
salt and productivity of, 62
salt tolerance of, 55
seeding guide for, 175
seed rates for, 217
storage requirements of, 274
yields of, 216
Kale, Chinese, 161
Kale, Siberian, 161
Kenaf, 112
distribution factors for, 120
seed longevity in, 134
Kikuyugrass, 112, 120
annual productivity of, 207
carbon dioxide uptake in, 203
Kiwi, 112
distribution factors for, 120
storage requirements of, 273
Kleingrass, 112, 120
Kochia, 112
distribution factors for, 120
roots of, 220
Kochia scoparia, see Kochia
Kohlrabi, 112, 120
freezing sensitivity of, 277
germination requirements for, 161
seeding guide for, 175
storage requirements of, 274
Kok-saghyz, 176
Kudzu, 112, 120
germination requirements for, 152
seeding rates in, 174
Kudzu, tropical, 112, 120
Kumquat, 112, 120

L

Lablab atropurpureus, see Bean, Egyptian
Lactuca sativa, see Lettuce
Ladysthumb, 134
Lambs, see also Sheep
disassembly process for, 437
dressing percentages of, 436
feed for, 422
grades of, 435
predator losses in, 272
residues of, 81
retail yields of, 439
Lambsquarters
seed longevity in, 134
seed production, 263
transpiration ratio constants for, 185
Land requirements, 454
Lantana camara, 58
Larch (*Larix* spp.), 134
Lateritic soils, 29, 30
Lathyrus
hirsutus, see Pea, rough
odoratus, see Pea, sweet

Latuca sativa, see Lettuce
Lawngrass, Japanese, 112
distribution factors for, 120
germination requirements for, 152
Lead
in corn, 328
in milk, 445
in secondary effluents, 86
in sludge, 85, 87
Leek, 112, 120
germination requirements for, 161
seeding guide for, 175
seed longevity in, 135
storage requirements of, 274
Legumes, see also specific types
acid detergent fiber in, 355
carbon dioxide uptake in, 204
crop rotation and, 229
earthworms and, 225
moisture content of, 365, 381
neutral detergent fiber in, 355
nitrogen fixation and, 78, 79
protein in, 354, 365
proximate analysis of, 365—366
seeding rates in, 173—174, 217
toxic compounds from, 283
transpiration ratio constants for, 186
weed losses in, 260
Lemon, 112, 120
chilling injury to, 278
diseases in, 250
freezing sensitivity of, 277
fungicides and, 246
nutrients in, 280
salt and productivity of, 56
seed longevity in, 134
storage requirements of, 273
Lemon, rough, 134
Length, conversion factors, 455
Lentil (*Lens culinaris*), 113, 120
frost resistance in, 176
germination requirements for, 153
nitrogen fixation and, 79
seed conditioning in, 345
seeding rates in, 171
seed longevity in, 135
Lepidium sativum, see Cress, garden
Lespedeza
cuneata, see Lespedeza, sericea
hedysaroides, see Lespedeza, Siberian
stipulacea, see Lespedeza, Korean
striata, see Lespedeza; Lespedeza, Kobe;
Lespedeza, 113, 120, see also specific types
AOSCA certification of seeds of, 165
germination requirements, 153
lime response in, 45
nematodes in, 259
nutrient uptake in, 91
roots of, 220
seeding rates, 174
Lespedeza, Chinese, 153

Lespedeza, Kobe
  germination requirements for, 153
  seed longevity in, 135
  soil conditioning in, 345
Lespedeza, Korean, 113, 120
  germination requirements for, 153
  seed conditioning in, 345
  seeding rates in, 174
  seed longevity in, 135
Lespedeza, sericea, 113, 120
  germination requirements for, 153
  seed conditioning in, 345
  seeding rates in, 174
  seed longevity in, 135
Lespedeza, Siberian, 153
Lespedeza, striate, 153
Lespedeza, Tennessee, 153
Lettuce, 120
  chromosome number, 113
  crop rotation and disease control in, 252
  diseases in, 250
  freezing sensitivity of, 277
  fungicides and, 246
  germination requirements for, 161
  insects in, 236
  irrigation of, 182
  nutrients in, 281
  nutrient uptake in, 91
  optimum pH for, 44
  roots of, 220, 222
  salt and productivity of, 58
  salt tolerance of, 55
  seed conditioning in, 345
  seeding guide for, 175
  seed longevity in, 135
  storage requirements of, 274
Licorice, common, 113, 120
Lightning, 78
Ligustrum lucidum, see Privet, Texas
Lilium regale (Lily, regal), 135
Lime (compound), see also Fertilizers
  rate for alfalfa, 49
  response to, 45, 46
  soil, 226
  -sulfur solution, 66
Lime (tree)
  chilling injury to, 278
  chromosome number for, 113
  defined, 480
  distribution factors for, 120
  freezing sensitivity of, 277
  nutrients in, 280
  salt and productivity of, 62
  storage requirements of, 273
Lime, Rangepur, 62
Limestone, 47, 394
Limpo grass, 112, 119
Linum usitatissimum, see Flax
Liquid nitrogen materials, 104—105
Litchi *(Litchi sinensis)*, 113, 120
Lithium, 445

Livestock, see also Cattle; specific animals
  arthropods in, 269
  diseases in, 269
  dressing percentages of, 436
  infectious diseases in, 266, 267
  losses in, 269—271
  metabolizable energy requirements of, 399
  nutritional disorders in, 269
  parasites in, 269
  poisonous plants and, 271
  temperature requirements of, 403, 404, 407
  toxic conditions in, 270
  ventilation rates for, 431
Loam, 194
  crop rotation tillage and, 224
  design velocities in drain tile and, 197
  limestone and pH of, 47
  mean velocity values and, 192
  nitrogen in, 68
  nonerosive velocities and, 193
  organic matter in, 68
  phosphorus in, 68
  salt percentage in, 54
  soil-water characteristics in, 196
Lobelia cardinalis (lobelia; cardinal flower), 135
Loganberry, 273
Lolium
  multiflorum, see Ryegrass, annual; Ryegrass, Italian
  perenne, see Ryegrass, perennial
  rigidum, see Ryegrass, wimmera
Lotus
  acutangula, see Luffa, angled
  aegyptiaca, see Luffa, vegetable sponge
  corniculatus, see Trefoil, birdsfoot
  pedunculatus, see Trefoil, big
  spp., see Trefoil
  tenuis, see Trefoil, narrowleaf
  uliginosus, see Trefoil, big
Lovegrass, see also specific types
  germination requirements for, 153
  salt and productivity of, 58
Lovegrass, Lehmann, 113, 120, 173
Lovegrass, sand, 153
Lovegrass, weeping, 113, 120
  germination requirements for, 153
  seeding rates in, 173
Low-pressure liquid nitrogen materials, 104—105
Lucerne
  carbon dioxide uptake in, 202
  maximum growth rate for, 213
  nitrogen fixation and, 79
  protein in, 369
  roots of, 220, 222
Lucerne, Townsville, growth rate, 215
Luffa, angled, 137
Luffa, vegetable sponge, 137
Lupine, 79, see also specific types
  frost resistance in, 176
  nematodes in, 259
  transpiration ratio constants for, 186
Lupine, blue, 153, 174

Lupine, European blue, 113, 120
Lupine, white, 113, 120
  germination requirements for, 153
  seeding rates in, 174
Lupine, yellow
  frost resistance in, 176
  germination requirements for, 153
  seeding rates in, 174
  seed longevity in, 137
Lupinus
  albus, see Lupine, white
  angustifolius, see Lupine, blue; Lupine, European
    blue
  luteus, see Lupine, yellow
Lychee, 273
Lychnis alba, see Campion, white
Lycopersicon spp., see Tomato

M

Macadamia (*Macadamia* spp.)113
  distribution factors for, 120, 121
  seed longevity in, 137
Macronutrients, see Major nutrients
Magnesia, 106
Magnesium, 87
  in above-ground portion of crops, 73—74
  amounts needed for plant production, 88
  in animal feeds, 307, 308, 329
  in beans, 51
  in beef, 440
  in cereal grains, 317
  in chickens, 451
  in corn, 93, 328
    leaves, 49—50
    silage, 374, 375
  in dairy cattle feed, 412, 413
  in dairy products, 443
  in fertilizers, 98—100, 106
  in hominy feed, 330
  in limestone, 47
  in manure, 82
  in milk, 445
  in pork, 441
  in root material, 75
  in secondary effluents, 86
  in sewage sludge, 85
  in slags, 48
  in soils, 41—42, 45
  in soybean products, 287, 290
  in sunflower meal, 290
  in turkey, 450
  uptake of, 89, 90—92, 97
Magnesium carbonate, 106
Magnesium sulfate, 106
Maize, see Corn
Major nutrients, 87, see also Nutrients; specific types
  in fertilizers, 98—100
Mallow, venice, 263
Malt, 298
Malus sylvestris, see Apple

Mandarin, 273
Mandarin, Cleopatra, 62
Manganese, 87
  in above-ground portion of crops, 73—74
  amounts needed for plant production, 88
  in animal feeds, 307, 308, 329, 397
  in beans, 51
  in chickens, 451
  in corn, 93, 328
    leaves, 49—50
    silage, 374, 375
  in dairy cattle feed, 412, 413
  in fertilizers, 106
  in hominy feed, 33
  in limestone, 47
  in manure, 82
  in milk, 445
  in secondary effluents, 86
  in sewage sludge, 85
  in slags, 48
  in soils, 45
  in soybean products, 287, 290
  in sunflower meal, 290
  in turkey, 450
  uptake of, 97
Manganese sulfate, 106
Mangifera indica, see Mango
Mango, 113, 120
  chilling injury to, 278
  storage requirements of, 273
Manihot
  esculenta, see Cassava
  utilissima, see Tapioca
Manilagrass, 153
Manure
  carbon-nitrogen ratio in, 75
  composition of, 82
  fertilizer elements from, 84
  methane from, 84
  nitrogen losses from, 83
  spreading method for, 84
  swine, 83, 84
Maple, sugar, 113, 120
Marbut classification of soils, 29
Mares, 389
Marigold, 137
Marihuana, 113, 120
Marl, 48
Marrow, Boston, 113, 120
Marrow, vegetable, 113, 120
Maté, 113, 120
Matricaria inodora, see Mayweed, scentless
Matthiola incana, see Stock
Maturity stages, of hay, 353—354
Maximum allowable sprinkling rates in irrigation, 194
Maximum growth rates, 212—215
Maximum photosynthetic productivity, 198
Maximum yields, 198
Mayweed, scentless, 137
Mean growth rates, 207—211
Mean velocity values at incipient scour, 192

Meat animals, see also specific types
  disassembly process for, 437
  metabolizable energy requirements of, 398
Meats, see also specific types
  amino acids in, 442
Medic, black, 113, 120
  germination requirements for, 153
  seed longevity in, 137
Medicago
  arabica, see Clover, bur, spotted
  hispida, see Clover, bur, California
  lupulina, see Medic, black
  orbicularis, see Clover, button
  polymorpha, see Burclover, California
  sativa, see Alfalfa; Alfalfa, Grimm; Alfalfa,
    Peruvian; Lucerne
  tribuloides, see Clover, barrel
Melilotus
  alba, see Clover, sweet, white
  indica, see Clover, sour
  officinalis, see Clover, sweet, yellow
  spp., see Clover, sweet
Melinis minutiflora, see Molassesgrass
Melon, see also specific types
  chilling injury to, 278
  frost resistance in, 176
  storage requirements of, 274
Melon, Persian, 278
Mentha
  arvensis, see Mint, field
  piperita, see Peppermint
  spicata, see Spearmint
Mercury
  in corn, 328
  in secondary effluents, 86
  in sewage sludge, 85
Mesquite, 113
Mesquite, curly, 120
Mesquitegrass, vine, 137
Metals, see also Elements; specific metals
  in secondary effluents, 86
  sludge, 87
Methane, 84
Metric units, conversion factors, 455—460
Microbial tissue, 75
Micronutrients, see Trace nutrients
Mignonette, common, 137
Milk, 311, 356, 377
  elements in, 445
  environmental temperature effect on, 414, 415
  lactose in, 444
  land requirements for, 454
  minerals in, 444
  nutrients in, 443
  protein in, 453
  radiation effects on, 416
  wind effects on, 416
Milkvetch
  AOSCA certification of seeds of, 165
  salt and productivity of, 62
Milkweed, common, 263

Millet, see also specific types
  amino acids in, 320
  AOSCA certification of seeds of, 165
  frost resistance in, 176
  germination requirements for, 153
  harvesting of, 375
  nematodes in, 259
  proximate analysis of, 296, 319
  rogi, see Millet, finger
  vitamins in, 302
  water requirements of, 185
Millet, browntop, 113, 120
  germination requirements for, 153
  seeding rates in, 171
Millet, bulrush
  annual productivity of, 207
  carbon dioxide uptake in, 203
  growth rate for, 207, 212
  physical properties of, 295
  radiation use efficiency in, 212
Millet, finger, 113, 120, 319
  amino acids in, 320
  physical properties of, 295
  protein in, 300
  proximate analysis of, 319
Millet, foxtail, 113, 120
  amino acids in, 320
  frost resistance in, 176
  germination requirements for, 153
  protein in, 300
  proximate analyses of, 319
  salt and productivity of, 59
Millet, German, 153
Millet, golden, 153
Millet, Hungarian, 153
Millet, Italian, see Millet, foxtail
Millet, Japanese
  germination requirements for, 153
  proximate analysis of, 319
Millet, kursk, 185
Millet, pearl, 113, 120
  amino acids in, 320
  carbohydrates in, 299
  fatty acids in, 303
  germination requirements for, 153
  protein in, 300
  proximate analysis of, 319
  seeding rates in, 171
Millet, proso, 113, 120
  amino acids in, 320
  carbohydrates in, 298
  germination requirements for, 154
  minerals in, 301
  protein in, 300
  proximate analysis of, 319
  seed conditioning in, 345
  seeding rates in, 171
  transpiration ratio constants for, 185
Millet, ragi, see Millet, finger
Millet, Siberian
  germination requirements for, 153

seed conditioning in, 345
Millet, Turkestan, 185
Millet, white wonder, 153
Milling fractions, 310
Milo
  digestion coefficients of, 305
  nutrients in, 394
  physical dimensions of kernels of, 332, 333
  relative feeding values of, 305
  seed conditioning in, 345
Mineralization, 76
Minerals, see also Elements; specific minerals
  in animal feeds, 307, 308, 329, 393—395
  in beef, 440
  in cereal grains, 301
  in chickens, 451
  in corn silage, 374, 375
  in dairy cattle feed, 412, 413
  in dairy products, 443
  in hominy feed, 330
  in milk, 444
  in pork, 441
  in secondary effluents, 86
  in soybean products, 287
  in turkey, 450
  in wild rice, 317
Mint, field, 113, 120
Mites, 234, 238, 241
  in cereal grains, 351—352
  in grains, 351—352
  in soybean, 351—352
Moisture availablity, 196
Moisture content, see also Water content
  of alfalfa, 365, 367
  of bermudagrass, 362
  of corn, 326
  of corn silage, 382
  ensiling crops and, 381
  of grains, 331
  of grasses, 365
  of hay, 363
  of hay crop silage, 382
  heating for determination of, 332
  of legumes, 365
Moisture treatments, see Irrigation
Molasses
  beet
    composition of, 385
    nutrients in, 394
  citrus, 394
  sugarcane, 385, 394
Molassesgrass, 112, 119
  germination requirements for, 154
  seeding rates in, 173
Mollisols, 31—33
  lime response in, 46
  phosphorus in, 95
  potassium in, 96
Molluscs, 77
Molybdenum, 87
  amounts needed for plant production, 88

in corn, 93
in dairy cattle feed, 412, 413
in limestone, 47
in milk, 445
in sewage sludge, 85
uptake of, 97
Monosodium phosphate, 394
Morningglory
  seed conditioning in, 345
  seed weight relationships and, 263
Morus rubra, see Mulberry, red
Mucuna deeringiana, see Velvetbean
Mulberry, red, 113, 120
Mules, 81
Multiple range test, signigicant ranges, 469, 470
Musa × paradisiaca, see Banana
Mushroom
  storage requirements of, 275
  toxic compounds from, 283
Muskmelon, see Cantaloupe
Mustard, see also specific types
  AOSCA certification of seeds of, 165
  nitrogen fixation and, 79
  nutrients in greens of, 281
  seed conditioning in, 344
  seeding guide for, 175
  seed longevity in, 137
Mustard, black, 113, 121
  germination requirements for, 154
  seeding rates in, 171
Mustard, brown, 345
Mustard, India, 154, 161
Mustard, spinach, 161
Mustard, white, 113, 121
  frost resistance in, 176
  germination requirements for, 154
  seed conditioning in, 346
  seeding rates in, 171
  seed longevity in, 137
Mutton, 435, see also Sheep
Mycorrhizae, 226
Myristica fragrans, see Nutmeg

N

Nandina domestica, see Bamboo, heavenly
Napiergrass
  annual productivity of, 207
  germination requirements for, 154
  maximum yields of, 198
  nutrient uptake in, 91
  photosynthesis in, 198
Nasturtium officinale, see Cress, water
NDF, see Neutral detergent fiber
Nectarine
  nutrients in, 280
  storage requirements of, 273
Needlegrass, green, 137
Nematodes
  biomass of, 77
  phytoparasitic, 258

in soybean, 258, 259
Nerium oleander, see Oleander
Neutral detergent fiber (NDF), 355, 359
Newsprint, 397
Niacin
  in alfalfa, 367
  in animal feeds, 308, 329
  in beef, 440
  in cereal grains, 302, 317
  in chickens, 451
  in corn, 328, 374
  in dairy products, 443
  in fruits, 280
  in hominy feed, 330
  in pork, 441
  in poultry feeds, 426
  in soybean meal, 290
  in sunflower meal, 290
  in turkey, 450
  in vegetables, 281
  in wheat, 310
Nickel
  in milk, 445
  in secondary effluents, 86
  in sludge, 85, 87
Nicotiana tabacum, see Tobacco
Night flowering catchfly, 344
Nightshade, black, 263
Nitrate, 106
Nitrogen, 87, see also Carbon-nitrogen ratio
  in above-ground portion of crops, 73—74
  amounts needed for plant production, 88
  in beans, 51
  in corn, 93, 326
  deficiency of, in corn, 94
  in fertilizers, 98—100, 106, 107
  handling and, 83
  liquid, 104—105
  losses of, 83, 84
  in manure, 82
  manure and, 83, 84
  mineralization and release of, 76
  in organic materials, 83
  in root material, 75
  in secondary effluents, 86
  in sewage sludge, 85
  in soils, 45, 68
  storage and, 83
  in tobacco, 386, 387
  uptake of, 89, 90—92, 96, 97
Nitrogen fixation, 78, 79, 80
Nitrogen sources for fertilizers, 48
Nodulated soybean, 80
Nonerosive velocities in canals, 193
Nutall alkali grass, 64
Nutmeg, 113, 121
Nutrients, 47, see also Elements; Minerals; Vitamins; specific types
  in above-ground portions of crops, 73—74, 93
  in alfalfa, 356, 357
  in animal feeds, 307, 308, 329, 393—395

in beans, 51
in beef, 440
in chickens, 451
conversion factors for, 460
in corn, 49
in dairy cattle feed, 412—413
in dairy products, 443
digestible, 356, 357
in ewe feeds, 423
in fruit, 280
in hominy feed, 330
in horse feed, 429
horse requirements of, 428
in lamb feeds, 422
losses of, 361
major, 87, see also specific types
in manures, 82
in milk, 443
in pork, 441
in poultry feeds, 427
secondary, 87
in sewage sludge, 85
in soils, 41—42, 45
in sweet potato, 279
total digestible, see Total digestible nutrients (TDN)
trace, 87, see also specific types
in turkey, 450
uptake of (plants), 89, 90, 90—92, 91, 92, 96, 97
in vegetables, 281
Nutritional disorders in livestock, 269
Nutritional values, see also Nutrients
  of corn, 320
  of corn silage, 374
  of sorghum, 320
  of soybean meal, 288, 290
  of sunflower meal, 290
Nuts, see also specific types
  amino acids in, 282
  conversion factors for shells of, 73
  diseases in, 250
  fungicides and, 246, 247
  protein in, 282
  toxic compounds from, 283
  weed losses in, 260
Nutsedge, yellow, 263

O

Oak, 75
Oak, cork, 113, 121
Oatgrass, tall
  germination requirements for, 154
  maximum growth rate for, 214
  seed conditioning in, 345
  seed longevity in, 137
Oatmeal
  carbohydrates in, 298
  fatty acids in, 303
Oats, see also specific types
  amino acids in, 316
  AOSCA certification of seeds of, 165

average composition of, 297
carbohydrates in, 299
chromosome number for, 113
composition of, 311, 317
conversion factors for residues of, 72
crop rotation and, 228
digestion coefficients of, 305
diseases in, 226, 248, 250
distribution factors for, 121
drying rate of, 334
fatty acids in, 303
forage yields of, 311
frost resistance in, 176
germination requirements for, 154
grades of, 314
grain size of, 295
harvesting of, 375
humidity and, 335
insects in, 236
kernels
   dimensions of, 332, 333
   physical characteristics of, 315
lime response in, 45
maximum growth rate for, 214
minerals in, 301, 317
moisture content of, 331, 332
nematodes in, 259
nutrients in, 73, 394
nutrient uptake in, 90
optimum pH for, 44
physical properties of, 295
protein in, 300, 316, 369
proximate analysis of, 296, 371
radiation use efficiency in, 214
relative feeding values of, 305
roots of, 219
salt and productivity of, 62
salt tolerance of, 55
seed conditioning in, 345
seeding rates in, 171
seed longevity in, 137
silage, 91, 378, 394
straw, 75, 371
target yield for, 113
transpiration ratio constants for, 186
vitamins in, 302, 317
waterlogging tolerance in, 197
water requirements of, 185
Oats, bristle, 138
Oats, red, 138
Oats, wild, 138, 222
Oat silage
   nutrients in, 394
   nutrient uptake in, 91
   TDN for, 378
Oat straw
   carbon-nitrogen ratio in, 75
   proximate analyses of, 371
Ocimum basilicum, see Basil
Oenothera biennis, see Primrose, evening
Oil crops, 90, see also specific types

Oils, see also specific types
   in secondary effluents, 86
   seed rates for, 218
   spacing of, 218
Oilseeds, see also specific crops
   carbon dioxide uptake in, 204
   crop factors for, 190
   protein in, 285
   seed rates for, 218
   spacing of, 218
Okra, 113, 121
   AOSCA certification of seeds of, 165
   chromosome number for, 113
   chilling injury to, 278
   freezing sensitivity of, 277
   germination requirements for, 161
   salt and productivity of, 56
   seed conditioning in, 345
   seeding guide for, 175
   seed longevity in, 138
   storage requirements of, 275
Olea europaea, see Olive
Oleander, 59
Olium perenne, see Ryegrass, perennial
Olive, 113, 121
   chilling injury to, 278
   salt and productivity of, 62
   storage requirements of, 273
   waterlogging tolerance in, 197
One-year cropping systems, 225
Onion, 113, 121, see also specific types
   AOSCA certification of seeds of, 165
   crop factors for, 190
   crop rotation and disease control in, 252
   diseases in, 226, 250
   freezing sensitivity of, 277
   fungicides and, 246
   germination requirements for, 161
   humidity effects on, 177
   insects in, 236
   nutrients in, 74, 281
   nutrient uptake in, 91
   optimum pH for, 44
   roots of, 219, 222
   salt and productivity of, 56
   salt tolerance of, 55
   seed conditioning in, 345
   seeding guide for, 175
   seed longevity in, 138
   seed rates for, 217
   storage requirements of, 275
   toxic compounds from, 283
   waterlogging tolerance in, 197
Onion, Beltsville bunching, 138
Onion, green, 275
Onion, Welsh
   germination requirements for, 162
   seed longevity in, 138
Onobrychis
   sativa, see Sainfoin
   viciifolia, see Sainfoin

Orange, see also specific types
in Arizona
chilling injury to, 278
storage requirements of, 273
in California
chilling injury to, 278
storage requirements of, 273
crop factors for, 190
diseases in, 249, 250
in Florida, 273
freezing sensitivity of, 277
fungicides and, 246
nutrients in, 74, 280
nutrient uptake in, 92
salt and productivity of, 56
in Texas, 273
Orange, sour, 113, 121, 138
Orange, sweet, 113, 121, 138
Orchardgrass, 113, 121
germination requirements for, 154
harvesting of, 375
nutrient uptake in, 91
proximate analyses of, 365, 366
roots of, 220
salt and productivity of, 62
salt tolerance of, 55
seed conditioning in, 346
seeding rates in, 173
seed longevity in, 138
Orders of soils, 29—30, 31, see also specific types
Organic matter, 83, see also specific types
carbon-nitrogen ratio of, 75
distribution of, 68
formation of, 67
Ornamental crops, see also specific types
diseases in, 250
Ornithopus sativus, see Serradella
Oryza sativa, see Rice
Oryzopsis
hymenoides, see Ricegrass, Indian
miliacea, see Smilograss
Osmotic pressure, 54
Osteospermum ecklonis, (Osteospermum), 139
Oxisols, 31—33
lime response in, 46
phosphorus in, 95
Oxygen, 87
Oyster shell, 394

P

Pacific Islands, 9
Paeonia suffruticosa, see Peony
Pak-choi, 162
Palaquium gutta, see Gutta percha
Palm, African oil, 113, 121, see also Palm, oil
Palm, date, 113, 121, 197
Palm, oil
annual productivity of, 211
maximum growth rate for, 215
mean growth rate for, 211

nutrient uptake in, 90, 92
radiation use efficiency in, 215
seed rates for, 218
spacing of, 218
Palm, sugar, 113, 121
Pan evaporation, 187—188, 189
Pangolagrass, 113, 121
annual productivity of, 207
carbon dioxide uptake in, 203
maximum growth rate for, 214
growth rate for, 207, 214
nutrient uptake in, 91
roots of, 220
Panicgrass, blue, 113, 121
germination requirements for, 154
roots of, 220
Panicgrass, green, 154, 203
Panicum
antidotale, see Panicgrass, blue
coloratum, see Kleingrass
maximum, see Guineagrass; Panicgrass, green
miliaceum, see Millet, proso
obtusum, see Mesquitegrass, vine
ramosum, see Millet, browntop
virgatum, see Switchgrass
Panothenic acid, 290
Pansy, 139
Pantothenic acid
in alfalfa, 367
in animal feeds, 308, 329—330
in beef, 440
in cereal grains, 302
in chickens, 451
in corn, 328
in dairy products, 443
in pork, 441
in poultry feeds, 426
in triticales, 310
in turkey, 450
in wheat, 310
Papaver somniferum, see Poppy, opium
Papaya, 113, 121
chilling injury to, 278
seed longevity in, 139
storage requirements of, 273
Paper, cellulose digestibility, 397
Paragrass, 112, 119
annual productivity of, 208
mean growth rate for, 208
nutrient uptake in, 91
Parasites, 269, see also specific organism
Parkia filicoidea, see Bean, African locust
Parsley, 114, 121
freezing sensitivity of, 277
germination requirements for, 162
seeding guide for, 175
seed longevity in, 139
storage requirements of, 275
Parsnip, 114, 121
crop rotation and disease control in, 252
freezing sensitivity of, 277

germination requirements for, 162
seeding guide for, 175
seed longevity in, 139
storage requirements of, 275
yields of, 216
Parthenium argentatum, see Guayule
Paspalum
  dilatatum, see Dallisgrass
  notatum, see Bahiagrass
  plicatulum, see Paragrass
  urvillei, see Vasseygrass
Passionfruit, 273
Pastinaca sativa, see Parsnip
Pasture crops, see also specific crops
  diseases in, 250
  fungicides and, 246
  weed losses in, 260
Pasture soils, 76
Pea, see also specific types
  crop rotation and disease control in, 252
  diseases in, 226, 250
  drying rate of, 334
  freezing sensitivity of, 277
  frost resistance in, 176
  fungicides and, 246
  germination requirements for, 162
  insects in, 237
  nitrogen fixation and, 79
  nutrient uptake in, 91
  salt and productivity of, 59
  salt tolerance of, 55
  seed conditioning in, 344
  seeding guide for, 175
  seed rates for, 217
  toxic compounds from, 283
  waterlogging tolerance in, 197
  yields of, 216, 453
Pea, Austrian winter, 345
Pea, chick, 110, 118
  germination requirements for, 150
  seed conditioning in, 346
  seeding rates in, 171
  transpiration ratio constants for, 186
Pea, cow, 111, 118
  AOSCA certification of seeds of, 164
  germination requirements for, 151, 160
  insects in, 237
  nematodes in, 259
  nitrogen fixation and, 79
  nutrients in above-ground portion of, 74
  protein in, 369
  salt and productivity of, 58
  seed conditioning in, 346
  seeding rates in, 171
  seed longevity in, 130
  storage requirements of, 275
  transpiration ratio constants for, 186
Pea, field
  AOSCA certification of seeds of, 164
  germination requirements for, 154
  optimum pH for, 44

roots of, 220
seed conditioning in, 345
seeding rates in, 171
transpiration ratio constants for, 186
Pea, flat, 164
Pea, garden, 114, 121
  nutrients in, 281
  seed longevity in, 139
  storage requirements of, 275
  target yield for, 114
Pea, green, see Pea, garden
Pea, pigeon, 114, 121
  seeding rates in, 174
  seed longevity in, 139
Pea, rough, 114, 121
  germination requirements for, 155
  seeding rates in, 174
Pea, southern, see Pea, cow
Pea, sweet, 144
Peach, 114, 121
  conversion factors for pits of, 73
  diseases in, 250
  freezing sensitivity of, 277
  fungicides and, 246
  nutrients in, 74, 280
  nutrient uptake in, 92
  optimum pH for, 44
  root length of, 222
  salt and productivity of, 56
  storage requirements of, 273
  waterlogging tolerance in, 197
Peanut, 114, 121, 139
  amino acids in, 282
  AOSCA certification of seeds of, 165
  carbon dioxide uptake in, 205
  conversion factors for pits of, 73
  crop residue management and disease control in, 256
  crop rotation and, 228
  diseases in, 250
  frost resistance in, 176
  fungicides and, 246
  germination requirements for, 154
  gross composition of, 288
  insects in, 236
  lime response in, 45
  nematodes in, 259
  nitrogen fixation and, 79
  nutrients in above-ground portion of, 74
  nutrient uptake in, 90
  optimum pH for, 44
  protein in, 282, 285
  proximate analysis of hulls of, 371
  salt and productivity of, 59
  seeding rates in, 172
  vitamins in, 289
Peanut, Spanish, 139
Peanut, Valencia, 139
Peanut oil, 293
Pear, 114, 121
  diseases in, 250

freezing sensitivity of, 277
fungicides and, 246
nutrients in, 280
root length of, 222
salt and productivity of, 56
seed longevity in, 139
storage requirements of, 273
waterlogging tolerance in, 197
Pecan, 114, 121
    amino acids in, 282
    conversion factors for shells of, 73
    fungicides and, 246
    protein in, 282
    proximate analysis of hulls of, 371
    weed losses in, 260
Pennisetum
    americanum, sec Millet, pearl
    clandestinum, see Kikuyugrass
    glaucum, see Millet, pearl
    purpureum, see Elephant grass; Napiergrass
    typhoides, see Millet, bulrush
Peony, 139
Pepper, see also specific types
    AOSCA certification of seeds of, 165
    carbon dioxide uptake in, 205
    crop rotation and disease control in, 252
    diseases in, 226
    fungicides and, 246
    germination requirements for, 162
    humidity effects on, 177
    insects in, 237
    salt and productivity of, 59
    seeding guide for, 175
    seed longevity in, 139
    storage requirements of, 275
Pepper, bell
    nutrient uptake in, 91
    salt tolerance of, 55
Pepper, black, 114, 121, 345
Pepper, Indian long, 114, 121
Pepper, red, 345
Pepper, sweet
    chilling injury to, 278
    freezing sensitivity of, 277
    nutrients in, 281
    storage requirements of, 275
Pepper, tabasco, 114, 121, 172
Peppergrass, 345
Peppermint, 114
    crop rotation and disease control in, 252
    distribution factors for, 121
Persea americana, see Avocado
Persimmon, common, 114, 121
Persimmon, Japanese, 114, 121, 273
Pests, losses due to, 231—265, see also specific pests
    losses in animals, 266—272
    losses in plants, 231—265
PET, see Potential evapotranspiration
Petroselinum spp., see Parsley
Petunia (*Petunia* spp.), 139
pH, 117—123

availability of plant nutrients and, 45
lime response and, 45
limestone and raising of, 47
soil, see Soil pH
Phalaris
    arundinacea, see Canarygrass, reed
    canariensis, see Canarygrass
    stenoptera, see Harding grass
    tuberosa, see Harding grass
Phaseolus
    acutifolius, see Bean, tepary
    angularis, see Bean, adzuki
    atropurpureus, see Siratro
    aureus, see Bean, mung
    coccineus, see Bean, scarlet runner
    lunatus, see Bean, lima
    vulgaris, see Bean; Bean, field; Bean, French; Bean, garden; Bean, green; Bean, lima; Bean, Mexican; Bean, navy
Phenol, 86
Phleum spp., see Timothy
Phoenix dactylifera, see Palm, date
Phosphate rock, 394
Phosphoric acid
    in organic materials, 83
    swine manure and, 84
    uptake of, 89, 90—92
Phosphorus, 87
    in above-ground portion of crops, 73—74
    amounts needed for plant production, 88
    in animal feeds, 329
    in beans, 51
    in cereal grains, 317
    in chickens, 451
    in corn, 93, 95, 328
        leaves, 49—50
        silage, 374, 375
    in crop residues, 71
    in dairy cattle feed, 412, 413
    in dairy products, 443
    dietary requirements
        heifers, 420
        horses, 428
        steers, 409
    in fertilizers, 98—100, see also Fertilizers
    fertilizers and, 48
    in fruits, 280
    in hominy feed, 330
    in horse feeds, 429
    in limestone, 47
    in manure, 82
    in milk, 444
    in pork, 441
    potassium and, 48
    in poultry feeds, 426
    in red clover, 95
    in root material, 75
    in secondary effluents, 86
    in sewage sludge, 85
    in slags, 48
    in soils, 45, 68, 95

sources for, 48
in soybean products, 287
in sunflower meal, 290
in turkey, 450
uptake of, 96, 97
in vegetables, 281
in wheat, 95
Photosynthesis
conversion factors, 459
productivity, 198
rate, 198
Phragmites
australis, see Reed, common
communis, see Fescue, tall
Physalis pubescens, see Tomato, husk
Phytoparasitic nematodes, 258
Picea
abies, see Spruce, Norway
engelmannii, see Spruce, Engelmann
excelsa, see Spruce, Norway
glauca, see Spruce, white
sitchensis, see Spruce, Sitka
Pigeons, 447
Pigs, see Swine
Pigweed, 263
roots of, 219
transpiration ratio constants for, 185
Pimenta racemosa, see Bay rum tree
Pine, see also specific types
carbon-nitrogen ratio in needles of, 75
diseases in, 226
heat killing temperature for, 177
Pine, Austrian, 139
Pine, loblolly, 140
Pine, lodgepole, 140
Pine, mugo, 140
Pine, Norway, 140
Pine, ponderosa, 140
Pine, Scotch
annual productivity of, 211
mean growth rate for, 211
root length of, 222
seed longevity in, 140
Pine, shortleaf, 140
Pine, slash, 90
Pineapple, 114, 121
chilling injury to, 278
lime response in, 45
nutrients in, 394
nutrient uptake in, 92
optimum pH for, 44
seed rates for, 218
spacing of, 218
storage requirements of, 273
Pinion nuts, 345
Pinus
contorta, see Pine, lodgepole
echinata, see Pine, shortleaf
montana, see Pine, mugo
mugo, see Pine, mugo
nigra, see Pine, Austrian

ponderosa, see Pine, ponderosa
radiata, see Spruce, sitka
resinosa, see Pine, Norway
sylvestris, see Pine, Scotch
taeda, see Pine, loblolly
Piper
longum, see Pepper, Indian long
nigrum, see Pepper, black
Pistachio, 113, 121
amino acids in, 282
protein in, 282
Pistacia vera, see Pistachio
Pisum sativum, see Pea; Pea, field; Pea, garden
Pittosporum tobira (Pittosporum), 56
Plantago lanceolata, see Plantain, buckhorn
Plantain, black, 346
Plantain, buckhorn
seed conditioning in, 346
seed longevity in, 140
Plant hardiness zones, 20
Planting time, 241
Plant pathogens, 246—247
field crops, 248—250, 257
forage/pasture crops, 250
fruit crops, 249—250
ornamentals, 250
vegetable crops, 249—250
Plant populations, 217—218
response curves of, 216
Plant toxins, 270, see also specific types
Plum, see also specific types
conversion factors for pits of, 73
freezing sensitivity of, 277
fungicides and, 246
nutrients in, 280
salt and productivity of, 56
storage requirements of, 273
waterlogging tolerance in, 197
Plum, American, 140
Plum, common, 114, 121
Plum, damson, 114, 121
Plum, Japanese, 114, 121
Plum, Klamath, 114, 121
Plum, natal, 64
Poa
arachnifera, see Bluegrass, Texas
bulbosa, see Bluegrass, bulbous
compressa, see Bluegrass, Canada
glaucantha, see Bluegrass, glaucantha
nemoralis, see Bluegrass, wood
nevadensis, see Bluegrass, Nevada
pratensis, see Bluegrass, Kentucky
trivialis, see Bluegrass, roughstalk
Poisonous plants, 271, see also specific types
Polygonum persicaria, see Ladysthumb
Pomegranate
distribution factors for, 121
salt and productivity of, 62
storage requirements of, 273
Poppy, 176
Poppy, opium, 114, 121

Pork, see also Swine
  amino acids in, 442
  composition of, 441
  grades of, 434
  protein in, 441, 453
  retail yields of, 438
Potash, 106, see also Potassium
  in organic materials, 83
  swine manure and, 84
  uptake of, 89, 90—92
Potassium, 87
  in above-ground portion of crops, 73—74
  amounts needed for plant production, 88
  in animal feeds, 307, 308, 329, 330, 412, 413
  in beef, 440
  in cereal grains, 317
  in chickens, 451
  in corn, 93, 328
    leaves, 49—50
    silage, 374, 375
  in crop residues, 71
  in dairy products, 443
  in fertilizers, 98—100, 106, see also Fertilizers
  in limestone, 47
  in manure, 82
  in milk, 444
  phosphorous sources for, 48
  in pork, 441
  in root material, 75
  in secondary effluents, 86
  in sewage sludge, 85
  in soils, 41—42, 45, 96
  in sorghum, 96
  in soybean products, 287
  in turkey, 450
  uptake of, 96, 97
Potato, 114, 121
  annual productivity of, 210
  chilling injury to, 278
  crop factors for, 190
  crop rotation and disease control in, 252
  diseases in, 226, 249, 250
  early, 275
  freezing sensitivity of, 277
  frost resistance in, 176
  fungicides and, 246
  growth rate for, 210, 214
  heat killing temperature for, 177
  insects in, 237
  irrigation of, 182
  land requirements for, 454
  late, 275
  lime response in, 45
  nutrients in, 74, 281
  nutrient uptake in, 89, 91
  optimum pH for, 44
  radiation use, efficiency in, 214
  roots of, 221
  salt and productivity of, 59
  salt tolerance of, 55
  seeding rates in, 172

seed longevity in, 140
seed rates for, 217
storage requirements of, 275
toxic compounds from, 283
transpiration ratio constants for, 186
waterlogging tolerance in, 197
water requirements of, 185
weed losses in, 265
yields of, 216, 453
Potato, sweet, see Sweet potato
Potential evapotranspiration (PET), 190
Poultry, see also Chickens; specific types
  arthropods in, 269
  classes of, 446—447
  corn nutritive value in, 320
  diseases in, 269
  environmental conditions required by, 402
  feeds for, 426
  metabolic energy, 398—400
  nutrients in manure from, 82
  nutritional disorders in, 269
  parasites in, 269
  quality of, 448—449
  sorghum nutritive value in, 320
  soybean products in feeds for, 286
  space requirements for, 402
  spermatogenesis in, 392
  temperature requirements of, 403
  total energy requirements of, 400
  toxicity of cottonseed meals in, 292
  vitamin A in feeds for, 368
Prairies, 68
Precipitation
  average inches of, 2—12, 13—15
  comparisons of different years, 13—15
  plant requirements, 117—123
Predators, 272
Preservatives, 380
Pressing, of alfalfa, 370
Primrose, evening, 131
Privet, Texas, 59
Productivity
  comparative, 453-454
  livestock
    cattle
      beef, 409—410
      dairy, 410—418
    horses, 427—429
    poultry, 423—426
    sheep, 422—423
    swine, 418—421
  plants, 198—218, see also specific crop
    annual, 207—211
    photosynthetic, 198
    radiation utilization, efficiency, 212—215
    salt concentration and, 56—65
Proso, see Millet, proso
Protein, 483, see also specific types
  in alfalfa, 357, 365, 367, 368, 370
  in animal feeds, 329, 393—395, 396
  in beef, 440, 453

in cereal grains, 300
in chickens, 451
in corn silage, 377
in dairy cattle feed, 412
in dairy products, 443
degradable, 396
digestibility of, 357
in eggs, 453
extraction of, 369
farm animals, production efficiency of, 453
farm animals, requirements for, 401, 409, 420, 428
in fruits, 280
in grasses, 354, 359, 365
in hay, 354
in horse feeds, 429
leaf, 370
in legumes, 354, 365
in milk, 453
in nuts, 282
in oats, 316
in oilseeds, 285
in peanut, 288
in pork, 441, 453
in roughages, 371
in sheep, 357
in silages, 378
soluble, 396
in sorghum, 385
in soybean, 287
in tobacco, 386
in triticales, 309
in turkey, 450
in vegetables, 281
in wheat, 336
Protozoa, 77
Proximate analysis, 483
of alfalfa, 365—366, 367
of alfalfa meals, 365—366
of animal feeds, 329
of corn silage, 374, 378
of grasses, 365—366
of hominy feed, 330
of legumes, 365—366
of millet, 319
of roughages, 371
of sorghum, 319
of sorghum silage, 378
Prune
fungicides and, 246
salt and productivity of, 56
storage requirements of, 273
Prunus
americana, see Plum, American
armeniaca, see Apricot
avium, see Cherry, sweet
cerasus, see Cherry, sour
domestica, see Plum; Plum, common; Prune
dulcis, see Almond
insititia, see Plum, damson
persica, see Peach
salicina, see Plum, Japanese

subcordata, see Plum, Klamath
Psathyrostachys juncea, see Wildrye, Russian
Pseudotsuga
menziesii, see Fir, Douglas
taxifolia, 222
Psophocarpus tetragonolobus, see Bean, winged
Psyllium, 346
Puccinellia nuttaluana, 64
Pueraria
lobata, see Kudzu
phaseoloides, see Kudzu, tropical
thunbergiana, see Kudzu
Puerto Rico, 38
Pumpkin, 114, 121
chilling injury to, 278
germination requirements for, 162
roots of, 220
seed conditioning in, 346
seeding rates in, 172
storage requirements of, 275
transpiration ratio constants for, 186
Punica granatum, see Pomegranate
Pyracantha draperi (Pyracantha), 59
Pyrethrum, 141
Pyridoxine, see also Vitamin B$_6$
in alfalfa, 367
in animal feeds, 329
in corn, 328
Pyrus spp., see Pear

Q

Quercus suber, see Oak, cork
Quince, 114, 121, 273
Quinine, 114, 121

R

Rabbits
food capacity of, 401
toxicity of cottonseed meals in, 292
Radiation
efficiency of use of, 212—215
extraterrestrial, 183
milk production and, 416
Radioactivity, 460
Radish, 114, 121
freezing sensitivity of, 277
germination requirements for, 162
humidity effects on, 177
nutrients in, 281
roots of, 221
salt and productivity of, 59
salt tolerance of, 55
seed conditioning in, 345
seed longevity in, 141
storage requirements, 275
Ragi, see Millet, finger
Ragweed, common, 263
Ragweed, giant, 263

Rainfall, see also Precipitation
  average annual, 227, 228
  average inches of, 2—12, 13—15
  comparisons of different years, 13—15
Ramie, 114, 121
Rams, see Sheep
Random numbers, 471
Rangeland crops, 260, see also specific crop
Ranunculus repens, see Buttercup, creeping
Rape, 114, 121, see also specific types
  AOSCA certification of seeds of, 165
  germination requirements, 155
  insects in, 237
  moisture content of, 332
  salt tolerance of, 55
  seed, see Rapeseed
  seeding rates in, 172
  transpiration ratio constants for, 186
Rape, annual, 155
Rape, bird, 155
Rape, dwarf essex, 346
Rape, German, 346
Rape, turnip, 155
Rape, winter, 155
Rapeseed
  drying rate of, 334
  nutrient uptake in, 90
  oil, tocopherols in, 293
Raphanus sativus, see Radish
Raspberry, see also specific types
  nutrients in, 280
  salt and productivity of, 56
  storage requirements of, 273
  waterlogging tolerance in, 197
Raspberry, black, 114, 122
Raspberry, red, 44, 114, 121
Rats
  toxicity of cottonseed meals in, 292
  vitamin A in feeds for, 368
Redtop
  germination requirements for, 155
  seed conditioning in, 345
Reed, common, 114, 122
Reed canarygrass, see Canarygrass, reed
Relative feeding values of cereal grains, 305
Relative humidity, see Humidity
Reproductive responses in ewes, 391
Reproductive yields, 216
Rescuegrass
  germination requirements for, 155
  salt and productivity of, 64
Reseda odorata, see Mignonette, common
Reservoirs, 266, 267
Residues, see also specific types
  animal, see Animal residues
  of field crops, 72
  plant, see Crop residues
Rheum
  palmatum, see Rhubarb, Chinese
  rhabarbarum, see Rhubarb
  rhaponticum, see Rhubarb

Rhizotron conditions, 222
Rhodesgrass, 114, 122
  carbon dioxide uptake in, 203
  germination requirements for, 155
  salt and productivity of, 62
  salt tolerance of, 55
  seeding rates in, 173
Rhubarb, 114, 122
  germination requirements for, 162
  optimum pH for, 44
  storage requirements of, 275
  toxic compounds from, 283
Rhubarb, Chinese, 114, 122
Ribes
  nigrum, see Currant, black
  rubrum, see Currant, red
  uva-crispa, see Gooseberry, European
Riboflavin, see Vitamin B_2
Rice, 114, 122, see also specific types
  annual productivity of, 210
  AOSCA certification of seeds of, 165
  bran, 298, 394
  bran oil, 304
  brown, 300, 303
    carbohydrates in, 298, 299
    composition of, 317
    vitamins, 302, 317
  carbon dioxide uptake in, 200, 206
  cellulose in hulls of, 397
  conversion factors for residues of, 72
  crop factors for, 190
  diseases in, 226, 250
  frost resistance in, 176
  fungicides and, 246
  germination requirements for, 155
  grain size of, 295
  growth rate, 210, 213
  insects in, 238
  minerals in, 301
  nitrogen fixation and, 78, 79
  nutrients in above-ground portion of, 73
  nutrient uptake in, 90
  optimum pH for, 44
  photosynthesis in, 198
  polished, 298
    composition, 317
    protein in, 300
    vitamins, 302, 317
  physical dimensions of kernels of, 333, 334
  proximate analysis of, 296, 371
  radiation use, efficiency in, 213
  residue production of, 70
  salt tolerance of, 55
  seed conditioning in, 346
  seeding rates in, 172, 217
  seed longevity in, 141
  straws, 371
  transpiration ratio constants for, 186
  waterlogging tolerance in, 197
  weed losses in, 260, 265
  white, 298, 301, 303

Rice, paddy
  physical properties of, 295
  salt and productivity of, 59
  seed conditioning in, 346
Rice, wild, 114, 122
  composition of, 317
  minerals in, 317
  physical properties of, 295
  proximate analyses of, 296
  vitamins in, 317
Ricegrass, Indian, 114, 122
  germination requirements for, 155
  · seeding rates in, 173
Ricinus communis, see Castorbean
Roots
  depth of, 219—221
  length of, 222
  spread of, 219—221
  suberization of harvested, 226
Rorippa nasturtium-aquaticum, see Cress, water
Rose (*Rosa* spp.), 57
Roselle, 141
Rosmarinus lockwoodii (Rosemary), 64
Roughages, 371, see also specific types
Rubber, 114, 122
  diseases in, 226
  nutrient uptake in, 92
  seed rates for, 218
  spacing of, 218
Rubidium, 445
Rubus
  idaeus, see Raspberry; Raspberry, red
  occidentalis, see Raspberry, black
  spp., see Blackberry
  ursinus, see Boysenberry
Rumex
  acetosa, see Sorrel
  crispus, see Dock, curly
Runoff, 227, 228
Rutabaga, 114, 122
  freezing sensitivity of, 277
  germination requirements for, 162
  seeding rates in, 172
  storage requirements of, 275
Rye, 114, 122, see also specific types
  AOSCA certification of seeds of, 165
  carbohydrates in, 298, 299
  carbon-nitrogen ratio, 75
  composition of, 311
  conversion factors for residues of, 72
  diseases in, 226
  drying rate of, 334
  fatty acids in, 304
  flour, 298
  forage yields of, 311
  germination requirements for, 155
  grain, size of, 295, 332
  humidity and, 335
  kernel dimensions, 332
  minerals in, 301
  moisture content of, 331

nematodes in, 259
nutrients in above-ground portion of, 73
optimum pH for, 44
physical properties of, 295
protein in, 300
proximate analysis of, 296, 365
residue production of, 70
roots of, 221, 222
salt and productivity of, 62
salt tolerance of, 55
seed conditioning in, 346
seeding rates in, 172
seed longevity in, 141
transpiration ratio constants for, 186
vitamins in, 302
water requirements of, 185
Rye, wild, see Wildrye
Ryegrass, see also specific types
  carbon dioxide uptake in, 201
  maximum growth rate for, 213
  nutrient uptake in, 91
  radiation use, efficiency in, 213
  seed conditioning in, 345
  stage of maturity of, 377
Ryegrass, annual, 114, 122
  germination requirements for, 155
  seeding rates in, 173
  seed longevity in, 141
Ryegrass, perennial, 114, 122
  annual productivity of, 209
  germination requirements for, 155
  mean growth rate for, 209
  salt and productivity of, 62
  salt tolerance of, 55
  seeding rates in, 173
  seed longevity in, 141
  yields of, 216
Ryegrass, wimmera, 155
Rye straw
  carbon-nitrogen ratio in, 75
  proximate analysis of, 371

S

Saccharum spp., see Sugarcane
Safe-seeding dates, 242
Safflower, 115, 122
  AOSCA certification of seeds of, 165
  frost resistance in, 176
  germination requirements for, 155
  insects in, 238
  meal, 394
  oil, 293
  salt and productivity of, 62
  seeding rates in, 172
  seed longevity in, 141
Sage, 115, 122, 141
Sainfoin, 115, 122
  AOSCA certification of seeds of, 165
  germination requirements for, 156
  seeding rates in, 174

seed longevity in, 142
Salsify
  freezing sensitivity of, 277
  germination requirements for, 162
  seed longevity in, 142
  storage requirements of, 275
Salsola
  kali, 222
  pestifer, see Thistle, Russian
Salt
  concentration of, 54
  in dairy cattle feed, 412, 413
  in fertilizers, 101
  field crop tolerance for, 55
  productivity and, 56—65
  in soils, 54
  tolerance for, 55
Saltgrass, 64
Salvia officinalis, see Sage
Sambucus canadensis, see Elder, American
Sand, 37, 193, 194, 196
  corn and, 50
  design velocities in drain tile and, 197
  limestone and pH of, 47
  mean velocity values and, 192
  nitrogen in, 68
  nonerosive velocities and, 193
  phosphorus in, 68
  salt percentage in, 54
Sandbur, 263
Sanguisorba minor, see Burnet
Sanitation, 226
Satureja
  hortensis, see Savory, summer
  montana, see Savory, winter
Savory, summer, 115, 122, 142
Savory, winter, 115, 122
Sawdust, 75
Scabiosa (*Scabiosa* spp.), 142
Schizachyrium scoparium, see Bluestem, little
Secale cereale, see Rye
Secondary effluents, 86, see also specific types
Secondary nutrients, 87, see also Nutrients; specific
    types
Seeding guide, 175
Seeding rates, 171—174
Seeds, see also specific types
  AOSCA certification of, 164—165, 169, 170
  conditioning of, 341—345
  defined, 484
  insect pests, 351—352
  longevity of, 124—146
  ranges of rates of, 217—218
  screens, 343—346
  sieves, 342
  separation of grasses and grains from, 349, 350
  weight relationships, 263
Seepage rates in canals, 192
Selenium
  in animal feeds, 307, 308, 397
  in corn, 328

  in dairy cattle feed, 412, 413
  in milk, 445
  in soybean products, 287
  toxicity, 283
Semen quality in rams, 391
Separation of grasses, grains and seeds, 349, 350
Separators, 347, 349
Serradella, 115, 122
  seeding rates in, 174
  seed longevity in, 142
Sesame, 115, 122
  frost resistance in, 176
  germination requirements for, 156
  roots of, 221
  seed conditioning in, 345
  seeding rates in, 172
  seed longevity in, 142
Sesame oil, 293
Sesamum indicum, see Sesame
Sesbania (*Sesbania exaltata*)
  germination requirements for, 156
  salt and productivity of, 59
  seed conditioning in, 346
Setaria
  italica, see Millet, common; Millet, foxtail; Millet,
      German; Millet, golden; Millet, Hungarian;
      Millet, Italian; Millet, Siberian; Millet, white
      wonder
  sphacelata (Setaria), 208
Sewage sludge nutrients, 85
Shade trees, 250, see also specific trees
Sheep, see also Lambs
  arthropods in, 269
  composition of manure from, 82
  digestibility of dry matter in, 405
  digestion coefficients for, 356
  diseases in, 269
  estrus in, 389
  feed for, 368, 422
  food capacity of, 401
  gestation period in, 389
  infectious diseases in, 266
  metabolic energy use by, 400
  metabolizable energy requirements of, 398, 399
  nutrients in feed for, 423
  nutrients in manure from, 82
  nutritional disorders in, 269
  parasites in, 269
  predator losses in, 272
  protein digestibility in, 357
  protein requirements of, 401
  reproductive responses in, 391
  residues of, 81
  semen quality in, 391
  sorghum digestibility in, 313
  temperature requirements of, 403, 404
  thermal stress in, 391
  triticale digestibility in, 313
  yields of, 453
SI units, conversion factors, 458—460
Sieves, 341, 342

Silage, 311, see also specific plant sources
  additives for, 380
  characteristics of, 373
  dry matter in, 378
  hay, 382
  nutrient uptake in, 91
  preservatives for, 380
  protein in, 378
  stage of maturity for harvesting of, 375
  TDN for, 378
  triticale, 311
Silene alba, see Campion, white
Silicon
  in limestone, 47
  in milk, 445
  in slags, 48
Silos, 382, 383
Silt, 37, 193
  crop rotation tillage and, 224
  design velocities in drain tile and, 197
  limestone and pH of, 47
  mean velocity values and, 192
  nutrients in, 68
Silver, 445
Silver bells, 142
Silverberry, 59
Simmondsia chinensis, see Jojoba
Sinapis alba, see Mustard, white
Siratro, 203
Sisal, 115, 122, 190
Slags, 48
Sludge metals, 87
Smartweed, Pennsylvania, 263
Smilograss, 115, 122
  germination requirements for, 156
  seeding rates in, 173
Snapdragon, 142
Sodium
  in animal feeds, 307, 308, 412, 413
  in beef, 440
  in chickens, 451
  in corn, 328
  in dairy products, 443
  in limestone, 47
  in milk, 445
  in pork, 441
  in secondary effluents, 86
  in sewage sludge, 85
  in soils, 41—42
  soils affected by, 53
  in soybean products, 287
  in turkey, 450
Sodium tripolyphosphate, 394
Softwood trees, 206, see also specific types
Soil amendments, 226
Soil organisms, 77, see also specific types
Soil pH, 41—42, 43
  for alfalfa, 49
  for beans, 51
  for corn, 49—50
  defined, 482

  optimum, for plants, 44
  sulfur and, 66
Soil regions of world, 34
Soils, 29, see also specific types
  acid, see Acid soils
  acidification of, 226
  aggregation of, 224, 225
  arable, 76
  azonal, 30
  calcimorphic, 29, 30
  carbon-nitrogen ratios in horizon samples of, 76
  cation-exchange capacities of, 41—42, 87
  chloride in, 191
  composition of, 35, 38
  criteria for subdivision of, 32
  crop rotation and, 224
  disturbed, 224
  exchangeable cations in, 41—42, 53
  halomorphic, 29, 30
  hydromorphic, 29, 30
  intrazonal, 29, 30
  land surface percentages for, 33, 34
  lateritic, 29, 30
  lime, 226
  loss of, 227, 228
  Marbut classification of, 29
  nitrogen in, 68
  nutrients in, 41—42, 45
  orders of, 29—30, 31, see also specific types
  pasture, 76
  pH of, see Soil pH
  phosphorus in, 68, 95
  potassium in, 96
  salt percentage of, 54
  shapes of structures of, 36
  size classes of structures of, 36
  sodium-affected, 53
  structure of, 35, 36
  suborders of, 29—30, 31
  temperate, 77
  texture classification of particles of, 37
  three-phase physical model of, 35
  types of structures of, 36
  zonal, 29, 30
Soil-water characteristics, 196
Soil water stress, 179
Solanum
  melongena, see Eggplant
  tuberosum, see Potato
Sorghastrum nutans, see Indiangrass
Sorghum, see also Sudangrass; specific types
  amino acids in, 320
  in animal feeds, 322
  annual productivity of, 207, 210
  AOSCA certification of seeds of, 165
  average composition of, 297
  carbohydrates in, 299
  carbon dioxide uptake in, 201
  chromosome number for, 115
  conversion factors for residues of, 72
  crop factors for, 190

crop rotation and, 229, 254
digestibility of, 313
disease control, 254, 256
diseases in, 250
distribution factors for, 122
dry milling process for, 321
feedlot performance and, 311
fungicides and, 246
germination requirements for, 156
grades of, 318
grain designations for, 318
harvesting of, 375
insects in, 238
irrigation of, 182
kernel dimensions, 295, 332
lime response in, 45
market classes for, 318
maximum yields of, 198
mean growth rate for, 207, 210
minerals in, 301
moisture content of, 331, 381
nematodes in, 259
nutrients in above-ground portion of, 73
nutrient uptake in, 89, 90, 91
nutritive value of, 320
optimum pH for, 44
pests of, 244—245
photosynthesis in, 198
physical properties of, 295, 324
potassium in, 96
processing methods for, 322
protein in, 300, 385
proximate analysis of, 296, 319, 371
residue production of, 70
salt and productivity of, 62
salt tolerance of, 55
seed conditioning in, 346
seeding rates in, 172, 217
seed longevity in, 142
silage, 378, 394
spacing of, 217
steer intake of, 312
stover, 371, 378
target yield for, 115
transpiration ratio constants for, 185
vitamins in, 302
water requirements of, 185
weed losses in, 264, 265
Sorghum, almum, 115, 122, 156
Sorghum, sweet
   composition of, 385
   germination requirements for, 156
   seed conditioning in, 346
Sorghum silage
   nutrients in, 394
   proximate analyses of, 378
   TDN for, 378
Sorghum stover, 371, 378
Sorgo, 172, 185
Sorghum
   almum, see Columbus grass; Sorghum, almum

   bicolor, see Sorghum
   halepense, see Johnsongrass
   sudanense, see Sudangrass
   vulgare, see Broomcorn
Sorrel
   germination requirements for, 162
   seed conditioning in, 344, 346
Soybean, see also specific types
   amino acids in, 287
   annual productivity of, 211
   AOSCA certification of seeds of, 165, 170
   carbon dioxide uptake in, 204
   cellulose in hulls of, 397
   chromosome number for, 115
   composition of, 286
   conversion factors for residues of, 72
   crop factors for, 190
   crop rotation and, 229
   disease control, 254, 256
   diseases in, 248, 250
   distribution factors for, 122
   earthworms and, 225
   frost resistance in, 176
   fungicides and, 246
   germination requirements for, 156, 162
   grading of, 285
   growth rate, 211, 213
   insects in, 238, 351—352
   leaf area destruction of, 231
   lime response in, 45
   minerals in, 287
   mites in, 351—352
   moisture content of, 331
   nematodes in, 258, 259
   nitrogen fixation and, 79, 80
   nodulated, 80
   nutient amounts needed for production of, 88
   nutrients in above-ground portion of, 74
   nutrient uptake in, 89, 90, 97
   optimum pH for, 44
   per-acre number of plants of, 232
   photosynthesis in, 198
   products from, 286, 287
   protein in, 285, 287
   proximate analysis of hulls of, 371
   radiation use efficiency in, 213
   residue production of, 70
   roots of, 220, 222
   salt and productivity of, 62
   seed longevity in, 143
   seed rates for, 217
   seed size, 172, 333, 346
   spacing of, 217
   target yield for, 115
   transpiration ratio constants for, 186
   waterlogging tolerance in, 197
   water requirements of, 185
   water stress effects on, 179
   weed losses in, 260—265
   yield loss from dead plants of, 231
   yields of, 97, 216

Soybean, perennial, 115, 122
Soybean meal
  composition of, 290
  nutrients in, 394, 395
  nutritional value of, 288, 290
Soybean oil, 293
Soybean straw, 371
Space requirements for farm animals, 402
Spacing of seeds and plants, 217—218
Spearmint, 115, 122
Specific gravity, of grain, 342
Spelt, 115, 122, 156
Spergula maxima, see Spurrey
Spermatogenesis in fowl, 392
Spices, toxic compounds from, 283
Spinach, 115, 122, see also specific types
  crop rotation and disease control in, 254
  freezing sensitivity of, 277
  germination requirements for, 162
  nutrients in, 74, 281
  optimum pH for, 44
  salt and productivity of, 59
  salt tolerance of, 55
  seed conditioning in, 346
  seed longevity in, 143
  storage requirements of, 275
  toxic compounds from, 283
Spinach, New Zealand
  germination requirements for, 162
  seed longevity in, 143
Spinacia oleracea, see Spinach
Spodosols, 31—33
Sporobolus
  airoides, see Alkali sacaton
  cryptandrus, see Dropseed, sand
Sprinkler systems in irrigation, 194
Spruce, 177
Spruce, Engelmann, 143
Spruce, Norway, 143
Spruce, Sitka, 143, 206
Spruce, white, 143
Spurrey, 143
Squash, see also specific types
  crop rotation and disease control in, 254
  germination requirements for, 162
  insects in, 239
  roots of, 220
  salt and productivity of, 59
  salt tolerance of, 55
  seed conditioning in, 346
Squash, hard-shell, 278
Squash, Hubbard, 186
Squash, mixta, 115, 122
Squash, summer
  freezing sensitivity of, 277
  storage requirements of, 275
Squash, winter
  freezing sensitivity of, 277
  storage requirements of, 275
St. Augustine grass, 112, 119
Stargrass, 115, 122

Statistical tables, 465—471
Statistical terms, 461—462
Steers, see also Bulls; Cattle; Cattle, beef
  cold temperature effects on, 408
  composition of manure from, 82
  dietary requirements, 409
  disassembly process for, 437
  feed efficiency in, 407
  feed intake of, 312
  feedlot performance of, 311
  food capacity of, 401
  growth rate of, 312
  land requirements for, 454
  performance of, 406, 408
  TDN requirements of, 409
  residues of, 81
  total energy requirements of, 400
  weight gain in, 407
Stenotaphrum secundatum, see St. Augustine grass
Stipa
  tenacissima, see Esparto
  viridula, see Needlegrass, green
Stizolobium deeringianum, see Bean, velvet
Stones, 192
Storage requirements
  of flowers, 276
  of fruits, 273
  of hay, 364
  of vegetables, 274—275
Stover
  corn, 371, 393
  sorghum, 371, 378
Strawberry, 115, 122, see also specific types
  chloride in soil of, 191
  diseases in, 250
  fungicides and, 246
  insects in, 239
  nutrients in, 280
  optimum pH for, 44
  salt and productivity of, 57
  seed longevity in, 143
  storage requirements of, 273
  toxic compounds from, 283
  waterlogging tolerance in, 197
  weed losses in, 260
Strawberry, Chilean, 115, 122, 220
Strawberry, Virginia, 115, 122, 220
Stress, see also specific types
  evaporative demand, 184
  soil water, 179
  thermal, 391
  water, 178, 179, 184
Strontium, 445
Studentized ranges, tables of, 469, 470
Stylo, Townsville, 115, 122
Stylosanthes
  guyanesis, see Bermudagrass
  humilis, see Lucerne, Townsville; Stylo, Townsville
Suberization of harvested roots, 226
Suborders of soils, 29—30, 31, see also specific types
Sucrose solubility, 386

Sudangrass, 115, 122, see also Sorghum
  annual productivity of, 208
  frost resistance in, 176
  germination requirements for, 156
  growth rate, 208, 212
  harvesting of, 375
  maximum yields of, 198
  nutrients in, 395
  nutrient uptake in, 91
  photosynthesis in, 198
  radiation use, efficiency in, 212
  salt and productivity of, 63
  salt tolerance of, 55
  seed conditioning in, 346
  seeding rates in, 173
  seed longevity in, 143
  target yield for, 115
  TDN for, 378
  transpiration ratio constants for, 185
Sugarbeet, see Beet, sugar
Sugarcane, 115, 122
  annual productivity of, 207
  carbon dioxide uptake in, 204
  crop factors for, 190
  diseases in, 226, 250
  growth rate, 207, 212
  insects in, 239
  lime response in, 45
  maximum yields of, 198
  nematodes in, 259
  nutrients in above-ground portion of, 74
  nutrient uptake in, 90, 92
  optimum pH for, 44
  photosynthesis in, 198
  radiation use, efficiency in, 212
  salt and productivity of, 59
  seed rates for, 218
  target yield for, 115
  waterlogging tolerance in, 197
Sugar crops, see also specific crop
  carbon dioxide uptake in, 204
  land requirements for, 454
Sugars, see also specific types
  relative sweetness of, 386
  in tobacco, 387
Sulfate, 86
Sulfur, 87
  in above-ground portion of crops, 73—74
  amounts needed for plant production, 88
  in animal feeds, 307, 308, 330
  in corn, 93, 328, 375
  in dairy cattle feed, 412, 413
  in fertilizers, 98—100, 106
  in limestone, 47
  in manure, 82
  in milk, 444
  mineralization and release of, 76
  in root material, 75
  in sewage sludge, 85
  in slags, 48
  soil pH and, 66

  in soils, 45
  in soybean products, 287
  uptake of, 89, 90—92, 97
Sulfuric acid, 66
Sunflower, 115, 122, 263, see also specific types
  AOSCA certification of seeds of, 165
  carbon dioxide uptake in, 205
  frost resistance in, 176
  germination requirements for, 156
  insects in, 239
  maximum growth rate for, 212
  nutrient uptake in, 90
  pests in, 243
  photosynthesis in, 198
  protein in, 285
  radiation use, efficiency in, 212
  roots of, 220
  salt tolerance of, 55
  seed conditioning in, 346
  seed longevity in, 143
  seed rates for, 218
  silage, 172
  target yield for, 115
  waterlogging tolerance in, 197
Sunflower meal, 290
Sunflower oil, 293
Sweet pea, 144
Sweet potato, 114, 121, see also Yam
  annual productivity of, 210
  chilling injury to, 278
  crop factors for, 190
  crop rotation and disease control in, 254
  diseases in, 226
  freezing sensitivity of, 277
  fungicides and, 246
  growth rate, 210, 214
  insects in, 239
  lime response in, 45
  nutrients in, 74, 279, 281
  nutrient uptake in, 92
  radiation use, efficiency in, 214
  salt and productivity of, 59
  seed rates for, 217
  storage requirements of, 275
Swine, 320, see also Pork
  arthropods in, 269
  composition of manure from, 82
  digestion coefficients of cereal grains for, 305
  diseases in, 269
  environmental conditions required by, 402
  estrus in, 389
  excreta production in, 81
  feed consumption of, 420
  food capacity of, 401
  gestation period in, 389
  heat production of, 419
  infectious diseases in, 266
  land requirements for, 454
  manure of, 82—84
  metabolism of, 400, 419
  metabolizable energy requirements of, 398, 399

nutritional disorders in, 269
parasites in, 269
protein requirements of, 401
relative feeding values of cereal grains for, 305
residues of, 81
simplified balanced rations for, 421
soybean products in feeds for, 286
space requirements for, 402
temperature requirements of, 403, 404, 418
total energy requirements of, 400
toxicity of cottonseed meals in, 292
ventilation requirements for, 403
vitamin A in feeds for, 368
water requirements of, 421
weight gain in, 420
yields of, 453
Switchgrass, 115, 122
germination requirements for, 157
nutrient uptake in, 91
seeding rates in, 173
seed longevity in, 144
target yield for, 115
Symbiotic nitrogen fixation, 80
Symphytum spp., see Comfrey
Syzygium aromaticum, see Clove

T

Tagetes
erecta, see Marigold
patula, see Marigold
Tampala, 144
Tangerine, 115, 122
nutrients in, 280
storage requirements of, 273
Tapioca
annual productivity of, 209
growth rate for, 209, 214
radiation use efficiency in, 214
Taraxacum officinale, see Dandelion
Target yields, 109—116
Taro, 115, 122
TDN, see Total digestible nutrients
Tea, 115, 122
root length of, 222
seed longevity in, 144
seed rates for, 217
spacing of, 217
Teff, 115, 122
Temperate grasses, 359, see also specific types
cellulose in, 397
seed rates for, 217
Temperate soils, 77
Temperature, 117—123, see also specific effects
chickens and, 427
freezing dates, in plants, 16—19
frost resistance, 176
heat killing, 177
Teosinte, see also Corn
chromosome number for, 115
distribution factors for, 122

target yield for, 115
transpiration ratio constants for, 185
Testosterone in bulls, 390
Tetragonia tetragonioides, see Spinach, New Zealand
Texture classification of particles of soils, 37
Thea spp., see Tea
Theobroma cacao, see Cacao; Cocoa
Thermal stress, 391
Thermotherapy, 226
Thiamine, see Vitamin B_1
Thistle, Canada, 263
Thistle, Russian, 185
Thuja
orientalis, see Arborvitae
plicata, see Cedar, western red
Thyme, common, 115, 122, 144
Thymus vulgaris, see Thyme, common
Ticks, 268
Tillage, 223, 224
Timothy, 115, 122
carbon-nitrogen ratio in, 75
dry matter losses in, 361
germination requirements for, 157
harvesting of, 375
maximum growth rate for, 213
nutrient losses in, 361
nutrients in, 74, 395
nutrient uptake in, 91
radiation use, efficiency in, 213
roots of, 220
salt and productivity of, 59
seed conditioning in, 346
seeding rates in, 173
seed longevity in, 144
Tin, 445
Titanium, 445
Tobacco, 115, 122, see also specific types
AOSCA certification of seeds of, 165
carbon dioxide uptake in, 206
chemical analyses of, 386, 387
crop factors for, 190
crop rotation and disease control in, 254
diseases in, 226, 248, 250
frost resistance in, 176
fungicides and, 246
germination requirements for, 157
insects in, 239
nematodes in, 259
nutrients in above-ground portion of, 74
optimum pH for, 44
seed conditioning in, 346
seeding rates in, 172
seed longevity in, 144
spacing of, 218
waterlogging tolerance in, 197
Tobacco, burley, 92
Tobacco, flue cured, 92
Tobacco, rustic, 176
Tocopherols
in alfalfa, 367
in corn, 327

in vegetable oils, 293
Tomato, 115, 122
  AOSCA certification of seeds of, 165
  carbon dioxide uptake in, 205
  carbon-nitrogen ratio in leaves of, 75
  chilling injury to, 278
  crop factors for, 190
  crop rotation and disease control in, 254, 255
  diseases in, 249, 250
  freezing sensitivity of, 277
  frost resistance in, 176
  fungicides and, 246
  germination requirements for, 162, 163
  heat killing temperature for, 177
  insects in, 239
  nutrients in, 74, 281
  nutrient uptake in, 92
  optimum pH for, 44
  roots of, 220, 222
  salt and productivity of, 59
  salt tolerance of, 55
  seed conditioning in, 346
  seed longevity in, 144
  storage requirements of, 275
  toxic compounds from, 283
  waterlogging tolerance in, 197
  weed losses in, 264, 265
Total digestible nutrients (TDN), 379, 485
  in corn silage, 376
  in dairy cattle, 414, 415
  heifer requirements for, 420
  in horse feeds, 429
  horse requirements of, 428
  for silages, 378
  steer requirements for, 409
Tower silos, 382
Toxic compounds, 283, see also specific types
Toxic conditions, 270
Toxins, plant, 270
Trace nutrients, 87, see also Nutrients; specific
    nutrients
Trachelospermum jasminoides, see Jasmine, star
Tragopogon porrifolius, see Salsify
Transpiration ratio constants, 185—186
Tree crops, 211, see also specific plants
Trefoil, see also specific types
  AOSCA certification of seeds of, 164
  germination requirements for, 157
  seed conditioning in, 344
  seeding rates in, 174
Trefoil, big, 115, 122
  salt and productivity of, 59
  salt tolerance of, 55
Trefoil, birdsfoot, 115, 122
  harvesting of, 375
  nutrient uptake in, 91
  salt and productivity of, 63
  salt tolerance of, 55
  seed conditioning in, 346
  seed longevity in, 135
Trefoil, narrowleaf, 115, 122

Trifolium
  alexandrium, see Clover, berseem
  ambiguum, see Clover, kura
  campestre, see Clover, large hop
  dubium, see Clover, small hop
  fragiferum, see Clover, strawberry
  glomeratum, see Clover, cluster
  hirtum, see Clover, rose
  hybridum, see Clover, alsike
  incarnatum, see Clover, crimson
  lappaceum, see Clover, lappa
  nigrescens, see Clover, ball
  pratense, see Clover, red
  procumbens, see Clover, large hop
  repens, see Clover, ladino; Clover, white
  resupinatum, see Clover, Persian
  semipilosum, see Clover, Kenya
  spp., see Clover
  subterraneum, see Clover, subterranean
  vesiculosum, see Clover, arrowleaf
Triticale, 115, 123
  AOSCA certification of seeds of, 165
  carbohydrates in, 299
  composition of, 311
  digestibility of, 313
  fatty acids in, 304
  germination requirements for, 157
  grain size of, 295
  minerals in, 301
  nutritional value, 309—311
  protein in, 300, 309
  proximate analysis of, 296
  steer intake of, 311, 312
  vitamins in, 310
Triticosecale, see Triticale
Triticum
  aestivum, see Wheat; Wheat, bread; Wheat,
    common
  compactum, see Wheat, club
  dicoccum, see Wheat, emmer
  durum, see Wheat, durum
  monococcum, see Wheat, einkorn
  polonicum, see Wheat, Polish
  spelta, see Spelt
  turgidum, see Wheat, poulard
  vulgare, 222
Tropical crops, 92, see also specific crop
Tropical grasses, see also specific types
  cellulose in, 397
  lime response in, 45
Tsuga heterophylla, see Hemlock, western
Tumbleweed, 185
Tung-oil tree, 115, 123
Turf grass, 91
Turkeys
  amino acids in, 442
  classes of, 446
  composition of, 450
  composition of manure from, 82
  egg production in, 426
  metabolizable energy requirements of, 398, 399

nutrients in, 450
protein requirements of, 401
residues of, 81
temperature effects on, 426
ventilation rates for, 431
Turmeric, 115, 123
Turnip, 115, 123
freezing sensitivity of, 277
frost resistance in, 176
germination requirements for, 163
nutrients in, 74, 281
roots of, 219
seed conditioning in, 345
seeding rates in, 172
seed longevity in, 145
storage requirements of, 275
transpiration ratio constants for, 186
Typha spp., see Rice

U

Ulmus americana, see Elm, American
Ultisols, 31—33
lime response in, 46
phosphorus in, 95
potassium in, 96
United States
average inches of precipitation in, 2—12, 13—15
climatic zones in, 1
conversions, of measurements, 455—460
crop residue production in, 70
dairy cattle breeds in, 411
freezing temperature in, 16—19
growing degree days in, 21—27
pan evaporation in, 187—188
soil composition in, 38
Urea, 395

V

Vaccinium
ashei, see Blueberry, rabbiteye
augustifolium, see Blueberry, lowbush
corymbosum, see Blueberry, highbush
macrocarpon, see Cranberry
Valerianella locusta, see Cornsalad
Vanadium
in limestone, 47
in milk, 445
Vanilla, 115, 123
Vanilla planifolia, see Vanilla
Vapor pressure deficit (VPD), 177
Vaseygrass, 157
Veal, 437, see also Cattle
Vectors, in disease transmission, 266—267
Vegetable oils, 293, see also specific types
Vegetables, see also specific crop
cellulose in, 397
chilling injury to, 278
crop factors for, 190
crop rotation and disease control in, 255

diseases in, 226, 250
freezing sensitivity of, 277
fungicides and, 246
nutrients in, 281
nutrients in above-ground portion of, 74
nutrient uptake in, 91—92
optimum pH for, 44
salt tolerance of, 55
seed rates for, 217
storage of, 274—275
toxic compounds from, 283
water content of, 274—275
weed losses in, 260
yields of, 216
Veldtgrass, 157
Velvetgrass
germination requirements for, 157
seed longevity in, 145
Velvetleaf, 263
Ventilation, 402, 403, 431
Verbena spp., (Verbena), 145
Vernalgrass, sweet, 157
Vertisols, 31, 32
land surface percentage for, 33
potassium in, 96
Vetch, see also specific variety
AOSCA certification of seeds of, 165
crown-, see Crownvetch
frost resistance in, 176
lime response in, 45
milk-, see Milkvetch
nematodes in, 259
nitrogen fixation and, 79
nutrients in, 395
seed conditioning in, 346
transpiration ratio constants for, 186
Vetch, common, 115, 123
germination requirements for, 157
salt and productivity of, 59
seeding rates in, 174
Vetch, hairy, see Vetch, winter
Vetch, Hungarian, 157
Vetch, kidney, 145
Vetch, monantha, 157
Vetch, narrowleaf, 157
Vetch, purple
distribution factors for, 123
germination requirements for, 157
seeding rates in, 174
Vetch, spring, 176
Vetch, winter, 115, 123
germination requirements for, 157
seeding rates in, 174
Vetch, woolypod, 157
Viburnum (*Viburnum* spp.), 60
Vicia
angustifolia, see Vetch, narrowleaf
articulata, see Vetch, monantha
benghalensis, see Vetch, purple
dasycarpa, see Vetch, woolypod
faba, see Bean, broad; Bean, horse

monantha, see Vetch, monantha
pannonica, see Vetch, Hungarian
sativa, see Vetch, common
villosa, see Vetch, winter
Vigna
  radiata, see Bean, mung
  sesquipedalis, see Bean, asparagus
  unguiculata, see Bean, yard-long; Pea, cow
Vineyards, 190, see also Grape
Viola tricolor, see Pansy
Vitamins
  in alfalfa, 367
  in animal feeds, 307, 308, 329, 393—395
  in beef, 440
  in cereal grains, 302, 317
  in chickens, 451
  in corn, 328
  in corn silage, 374
  in dairy cattle feed, 412, 413
  in dairy products, 443
  in hominy feed, 330
  horse requirements of, 428
  in peanut, 289
  in pork, 441
  in poultry feeds, 426
  in triticales, 310
  in turkey, 450
  vitamin A
    in alfalfa, 368
    in cereal grains, 317
    in chickens, 451
    in corn, 328
    in dairy products, 443
    in fruits, 280
    in livestock rations, 330, 412, 413, 427, 430
    livestock requirements, 409, 420, 429
    in peanut, 289
    in turkey, 450
    in vegetables, 281
  vitamin B, see also specific types
    in alfalfa, 367
    in animal feeds, 308
    in cereal grains, 302
    in corn, 328
    in peanut, 289
  vitamin B_1
    in alfalfa, 367
    in beef, 440
    in cereal grains, 302, 317
    in chickens, 451
    in corn, 328
    in dairy products, 443
    in fruits, 280
    in livestock feeds, 308, 329, 330, 426
    in pork, 441
    in soybean meal, 290
    in sunflower meal, 290
    in triticales, 310
    in turkey, 450
    in vegetables, 281
    in wheat, 310

vitamin B_2
  in alfalfa, 367
  in beef, 440
  in cereal grains, 302, 317
  in chickens, 451
  in corn, 328
  in dairy products, 443
  in fruits, 280
  in livestock feeds, 308, 329, 330, 426
    in pork, 441
  in soybean meal, 290
  in sunflower meal, 290
  in triticales, 310
  in turkey, 450
  in vegetables, 281
  in wheat, 310
vitamin B_6, see also Pyridoxine
  in beef, 440
  in cereal grains, 302
  in chickens, 451
  in dairy products, 443
  in pork, 441
  in triticales, 310
  in turkey, 450
  in wheat, 310
vitamin B_{12}
  in animal feeds, 308
  in beef, 440
  in chickens, 451
  in dairy products, 443
  in pork, 441
  in poultry feeds, 426
  in turkey, 450
vitamin C
  in cereal grains, 317
  in dairy products, 443
  in fruits, 280
  in peanut, 289
  in vegetables, 281
vitamin D
  in corn silage, 374
  in dairy cattle feed, 412, 413
  in peanut, 289
  in poultry feeds, 426
vitamin E
  in cereal grains, 302
  in corn, 328
  in dairy cattle feed, 413
  in peanut, 289
  in poultry feeds, 426
vitamin K
  in alfalfa, 367
  in peanut, 289
  in poultry feeds, 426
in wheat, 310
in wild rice, 317
Vitis
  rotundifolia, see Grape, muscadine
  spp., see Grape
  vinifera, see Grape, wine
Voandzeia subterranea, see Groundnut, Bambarra

VPD, see Vapor pressure deficit
Volume, conversion factors, 455

W

Walnut, see also Walnut, black
   amino acids in, 282
   conversion factors for shells of, 73
   crop factors for, 190
   protein in, 282
   weed losses in, 260
Walnut, black, 115, 123
Warm soil planting, 226
Water content, see also Moisture content
   of fruits, 273
   of vegetables, 274—275
Waterlogging tolerance, 197
Watermelon, 116, 123
   AOSCA certification of seeds of, 165
   chilling injury to, 278
   fungicides and, 246
   germination requirements for, 163
   nutrients in, 280
   seed conditioning in, 346
   seed longevity in, 145
   storage requirements of, 275
   transpiration ratio constants for, 186
Water requirements, of crops and weeds, 185
Water stress, 178, 179, 184
Water use, 180, 181
Weeds, see also specific plant
   control of, 226
   defined, 486
   density of, 261
   losses from, 260—262
   optimum pH for, 44
   seeds, 263
   soybean losses and, 261, 262
   spacing and yields of, 264
   water requirements of, 185
   yields of, 264, 265
Weevils, 233, 238, 351
Wet milling of corn, 329
Wheat, see also specific types
   aeration of, 338
   annual productivity of, 211
   AOSCA certification of seeds of, 165
   carbon dioxide uptake in, 199
   composition of, 306, 311
   crop rotation and, 230
   diseases in, 226, 248, 250
   drying of, 335, 336
   forage yields of, 311
   fungicides and, 246
   grades of, 306
   growth rate, 211, 214
   harvesting of, 375
   heat killing temperature for, 177
   humidity and, 177, 336
   insects in, 240, 242
   irrigation of, 182

land requirements for, 454
minerals in, 301
nematodes in, 259
nitrogen fixation and, 79
nutient amounts required of, 88
nutrients in above-ground portion of, 73
nutrient uptake in, 89, 90
optimum pH for, 44
phosphorus in, 95
physical properties of, 295
products, see specific product
protein in, 300, 336
proximate analysis of straw from, 371
radiation use, efficiency in, 214
residues of, 70, 72, 256
roots of, 221
salt and productivity of, 63
salt tolerance of, 55
seed conditioning in, 346
seed longevity in, 145
seed rates for, 217
stage of maturity of, 377
TDN for, 378
transpiration ratio constants for, 185
vitamins in, 302, 310
waterlogging tolerance in, 197
water requirements of, 185
weed losses in, 260
yields of, 95, 216, 453
Wheat, bread, 116, 123
   grain size of, 295
   proximate analysis of, 296
   seeding rates in, 172
   target yield for, 116
Wheat, bulgur, 298
Wheat, club, 116, 123
   germination requirements for, 158
   seeding rates in, 172
Wheat, common
   germination requirements for, 158
   transpiration ratio constants for, 185
Wheat, durum, 116, 123
   carbohydrates in, 298
   fatty acids in, 304
   germination requirements for, 158
   grain size of, 295
   moisture content of, 331
   proximate analysis of, 296
   roots of, 221
   seeding rates in, 172
   transpiration ratio constants for, 185
Wheat, einkorn, 111, 119
Wheat, emmer, 111, 119, 185
   germination requirements for, 151
Wheat, hard, see also specific varieties
   kernel dimensions, 332, 333
   physical properties of, 324
Wheat, hard red spring
   carbohydrates in, 298, 299
   fatty acids in, 304
   moisture content of, 331

Wheat, hard red winter
  carbohydrates in, 298
  composition of, 317
  fatty acids in, 304
  moisture content of, 331
  relative feeding values of, 305
Wheat, hard winter, 297, see also specific varieties
Wheat, Polish, 158
Wheat, Poulard, 158
Wheat, soft, see also specific varieties
  average composition of, 297
  nutrients in, 395
  kernel dimensions, 332, 333
Wheat, soft red winter
  carbohydrates in, 298
  fatty acids in, 304
  moisture content of, 331
Wheat, spring, see also specific types
  crop rotation and, 230, 255
  frost resistance in, 176
  seed longevity in, 145
Wheat, white
  average composition of, 297
  carbohydrates in, 298
  fatty acids in, 304
  moisture content of, 331
Wheat, winter, see also specific types
  crop residue management and disease control in, 256
  crop rotation and, 230, 255
  residue production of, 70
  seed longevity in, 145
Wheat bran
  nutrients in, 395
  vitamins in, 302
Wheat flour
  carbohydrates in, 298
  fatty acids in, 304
  protein in, 300
  vitamins in, 302
Wheat germ
  protein in, 300
  vitamins in, 302
Wheat gluten, 300
Wheatgrass, beardless, 158
Wheatgrass, bluebunch, 116, 123
Wheatgrass, crested, 116, 123, 158
  salt and productivity of, 65
  seed conditioning in, 346
  seeding rates in, 173
Wheatgrass, fairway, 116, 123
  germination requirements for, 158
  salt and productivity of, 65
  seeding rates in, 173
  seed longevity in, 145
Wheatgrass, intermediate, 116, 123
  germination requirements for, 158
  seed longevity in, 145
Wheatgrass, pubescent, 158
Wheatgrass, Siberian, 158

Wheatgrass, slender
  germination requirements for, 158
  salt and productivity of, 63
  seed longevity in, 145
Wheatgrass, standard crested, 158
Wheatgrass, streambank, 158
Wheatgrass, tall, 116, 123
  germination requirements for, 158
  salt and productivity of, 65
Wheatgrass, western, 116, 123
  germination requirements for, 159
  roots of, 219
  salt and productivity of, 63
  salt tolerance of, 55
  seed conditioning in, 346
  seeding rates in, 173
  seed longevity in, 145
Wheat mill feed values, 308
Wheat millrun, 395
Whole wheat values in animal feeds, 307
Wildrye, altai, 65
Wildrye, beardless, 63
Wildrye, blue, 146
Wildrye, Canada, 116, 123
  germination requirements for, 159
  roots of, 220
  salt and productivity of, 63
  seeding rates in, 173
  seed longevity in, 146
Wildrye, giant, 116, 123
Wildrye, Russian, 116, 123
  germination requirements for, 159
  salt and productivity of, 65
  seed longevity in, 146
Willow, 197
Wind, 188, 189
  effects on milk, 416
Wintergreen, 116, 123
Wood, see also specific plants
  cellulose in, 397
Woodlands, 78
Woven-wire cloth sieves, 341

X

Xanthophylls, 367
Xylosma senticosa, (Xylosma), 60

Y

Yam, 275, see also specific types; Sweet potato
Yam, composite, 116, 123
Yam, eboe, 116, 123
Yam, winged, 116, 123
Yields
  of alfalfa, 49, 97, 358, 361
  of beans, 51
  of corn, 49, 95, 96, 180
  of farm animals, 453
  forage, 311

graded, 216
maximum, 198
plant population response curves for, 216
of red clover, 95
reproductive, 216
safe-seeding dates and, 242
of soybean, 97
target, of selected economic plants, 109—
116
of vegetables, 216
of weeds, 264, 265
of wheat, 95

Z

Zea mays, see Corn
Zinc, 87
in above-ground portion of crops, 73—74
amounts needed for plant production, 88
in animal feeds, 307, 308
in cereal grains, 317
in chickens, 451

in corn, 93, 328
in corn silage, 375
in dairy cattle feed, 412, 413
in dairy products, 443
in limestone, 47
in manure, 82
in milk, 445
in pork, 441
in secondary effluents, 86
in sewage sludge, 85
in sludge, 87
in soils, 45
in soybean products, 287
in turkey, 450
uptake of, 97
Zingiber officinale, see Ginger
Zinnia elegans, (Zinnia), 146
Zizania palustris, see Rice, northern wild
Zonal soils, 29, 30
Zoysia
   japonica, see Lawngrass, Japanese
Zoysia matrella, see Manilagrass